Measurement and Instrumentation
Trends and Applications

Measurement and Instrumentation
Trends and Applications

Editors

M. K. Ghosh
S. Sen
S. Mukhopadhyay

Department of Electrical Engineering
Indian Institute of Technology
Kharagpur - 721302, India

Taylor & Francis
Taylor & Francis Group
Boca Raton London New York

CRC is an imprint of the Taylor & Francis Group,
an informa business

Ane Books India

Measurement and Instrumentation: Trends and Applications

© Ane Books India

First Published in 2008 by

Ane Books India

4821 Parwana Bhawan, 1st Floor
24 Ansari Road, Darya Ganj, New Delhi -110 002, India
Tel: +91 (011) 2327 6843-44, 2324 6385
Fax: +91 (011) 2327 6863
e-mail: anebooks@vsnl.com
Website: www.anebooks.com

For

CRC Press
Taylor & Francis Group
6000 Broken Sound Parkway, NW, Suite 300
Boca Raton, FL 33487 U.S.A.
Tel : 561 998 2541
Fax : 561 997 7249 or 561 998 2559
Web : www.taylorandfrancis.com

For distribution in rest of the world other than the Indian sub-continent

ISBN-10 : 1 42007 432 6
ISBN-13 : 978 1 42007 432 1

British Library Cataloguing in Publication Data
A catalogue record for this book is available from the British Library

Printed at Gopsons Papers Ltd., Noida

Preface

The importance of Measurement and Instrumentation (M&I) in modern science and technology cannot be overemphasized. On the other hand, principles and methods of measurement have been profoundly influenced by developments in science and technology in areas ranging from materials to computing. This has established M&I as an independent and interdisciplinary technical specialization and in turn, created demand for resources for training and research. This book was created out of such experiences of training and research during the last three decades in frontier areas of M&I at IIT Kharagpur. It attempts to discuss issues related to design and implementation of the building blocks for measurement systems starting from basic principles, materials and structure of sensors, electronics as well as computing. It also illustrates their applications in several engineering contexts.

The book provides an introductory exposition to conventional sensing and signal processing techniques in chapters 1, 2, 3, 4, 7 and 11. Several other modern sensing methods using chemical and biochemical principles, microwave as well as the micro electromechanical system (MEMS) sensors, have been discussed in chapters 5, 6, 8, 9 and 10. Chapters 12 and 13 present an exposition of techniques of data communication and filter design. Chapters 14 and 15 discuss two applications of image-based instrumentation, one in the area of biomedical engineering and the other in remote sensing. Signal estimation techniques are nowadays increasingly being applied to instrumentation. Chapters 16, 17 and 18 describe such applications of sensor fusion and signal estimation in areas of chemical process control, manufacturing and electric power systems. Finally in part F on applications, Chapter 19 describes application of distributed sensing in automation for irrigation and Chapter 20 discusses VLSI applications on signal conditioning and processing for medical instrumentation. Instrumentation systems used for robotics are described in Chapter 21, while those used in bioprocess instrumentation are described in Chapter 22. Finally the measurement of two phase flow parameters have been discussed in Chapter 23.

Instrumentation systems are available in bewildering diversity in form of the measurand, measurement principles, technologies, range as well as quality. It is therefore, perhaps impossible to bring every type within the scope of any one volume. This book is no exception as some of the other important topics such as ultrasonic instrumentation, analysis instrumentation, to mention just a few, have not been dealt with here.

An edited volume must be credited first to the authors. We wish to record our appreciation and thanks for the efforts put in by the authors. We would also like to thank Continuing Education Cell and Department of Electrical Engineering, IIT Kharagpur, and Ane Books India for their support in publishing this volume.

Finally we thank Mr. Rohan Singh for his assistance related to word processing during the course of compilation of the volume.

M.K. Ghosh
S. Sen
S. Mukhopadhyay

Contents

Part – B
Sensors

Part – C
Signal Conditioning and Processing

Part – D
Image Based Instrumentation

Part – E
Intelligent and Virtual Instrumentation

Part – F
Applications

Contributors

Adhikari, B.
Professor, Materials Science Centre
Indian Institute of Technology
Kharagpur, 721302, India

Banerjee, S. (Ms.)
Professor, Department of Electronics
and Electrical Communication
Engineering
Indian Institute of Technology
Kharagpur, 721302, India

Barua, A.
Professor, Department of Electrical
Engineering
Indian Institute of Technology
Kharagpur, 721302, India

Basu, S.
Former Professor, Materials Science
Centre.
Indian Institute of Technology
Kharagpur, 721302, India

Bhattacharya, P.
Research Fellow, Applied Statistics Unit
Indian Statistical Institute
Kolkata 700108, India

Bhattacharya, S. (Ms.)
Scientist D, Range Safety Division,
Integrated Test Range,
Chandipore 756025, *India*

Biswas, K. (Ms.)
Senior Lecturer, Department of
Instrumentation and Electronics
Engineering
Jadavpur University
Kolkata-700098, India

Chattopadhyay, A.B.
Professor, Department of Mechanical
Engineering
Indian Institute of Technology
Kharagpur, 721302, India

Das, B.
Former Research Fellow,
Department of Electronics and Electri-
cal Communication Engineering
Indian Institute of Technology
Kharagpur, 721302, India

Das, P.K.
Professor, Department of Mechanical
Engineering
Indian Institute of Technology
Kharagpur, 721302, India

Dutta, P.K.
Professor, Department of Electrical
Engineering
Indian Institute of Technology
Kharagpur, 721302, India

Ghosh, M.K.
Former Professor, Department of
Electrical Engineering
Indian Institute of Technology
Kharagpur, 721302, India

Joshi, A.
Former Research Fellow,
Department of Agricultural and Food
Engineering
Indian Institute of Technology
Kharagpur, 721302, India

Kal, S.
*Professor, Department of Electronics
and Electrical Communication Engi-
neering
Indian Institute of Technology
Kharagpur, 721302, India*

Mallick, A.K.
*Former Professor, Department of
Electronics and Electrical Communica-
tion Engineering
Indian Institute of Technology
Kharagpur, 721302, India*

Mishra, A.
*Former, Research Fellow, Department
of Electrical Engineering
Indian Institute of Technology
Kharagpur, 721302, India*

Mukhopadhyay, S.
*Professor, Department of Electrical
Engineering
Indian Institute of Technology
Kharagpur, 721302, India*

Pal, P.S.
*Former Research Fellow, Department
of Electrical Engineering
Indian Institute of Technology
Kharagpur, 721302, India*

Patra, A.
*Professor, Department of Electrical
Engineering
Indian Institute of Technology
Kharagpur, 721302, India*

Perla, R.
*Former Graduate Student, Department
of Electrical Engineering
Indian Institute of Technology
Kharagpur, 721302, India*

Pradhan, A.K.
*Associate Professor, Department of
Electrical Engineering
Indian Institute of Technology
Kharagpur, 721302, India*

Routray, A.
*Associate Professor, Department of
Electrical Engineering
Indian Institute of Technology
Kharagpur, 721302, India*

Roy, S.
*Former Research Fellow, Materials
Science Center
Indian Institute of Technology
Kharagpur 721302, India*

Sen, S.
*Professor, Department of Electrical
Engineering
Indian Institute of Technology
Kharagpur, 721302, India*

Sengupta, D.
*Professor, Applied Statistics Unit,
Indian Statistical Institute
Kolkata 700108, India*

Sengupta, S.
*Former Professor, Department of
Geology and Geophysics
Indian Institute of Technology
Kharagpur, 721302, India*

Sinha, S.
*Former Professor, Department of
Electrical Engineering
Indian Institute of Technology
Kharagpur, 721302, India*

Tiwari, K.N.
*Professor, Department of Agricultural
and Food Engineering.
Indian Institute of Technology
Kharagpur, 721302, India*

PART - A

Introduction

Chapter 1

Evolution and Trends in Measurement and Instrumentation

M.K. Ghosh

Abstract

A brief outline of the evolution of measurement and instrumentation – its concept, technology and scope along with present trends have been presented in this chapter. The impact of development of new concepts of science and technology on measurement and instrumentation is two fold– demand for new measurement methods and devices to support the development, as well as design of new sensors, instrument peripherals and even techniques based on the new ideas coming up in other disciplines. Thus, today's measurement technology has evolved from mere quantification of a parameter in a deterministic environment to estimation or precise quantification of single or multiple parameters in an uncertain or even an unknown environment enabling the scientists and engineers to deal with new technical challenges. The chapter further elaborates the scope of this monograph which presents the major trends in measurement and instrumentation with occasional reference to research work carried out at IIT Kharagpur.

1.1 Evolution: Kelvinian Concept of Measurement

In a modern society, measurement, which means quantification of a parameter, is the way of life. In the words of Lord Kelvin:

'...........when you can measure what you are speaking about, and express it in numbers, you know something about it; but when you can not measure it, when you can not express it in numbers, your knowledge is of a meager and unsatisfactory kind'.

This is the concept on which human civilization in general, and science and technology in particular, are based. All physical parameters are measurable. The concept of measurement vis-à-vis quantification is so deep rooted that today there are attempts to measure i.e. quantify abstract quantities like intelligence.

Measurement involves two major aspects:
- Measurement environment, and
- Data processing including standardization.

Depending on the type of measurement environment, a measurement can be either of direct or indirect type.

In direct measurement the parameter to be measured (the measurand) is compared directly with a precalibrated standard. Measurement of length of a cloth piece using a yardstick or a tape (precalibrated standard) is the most common example.

In an indirect type of measurement the measurand modulates the environment so as to generate another signal that is more convenient to handle. The element on which modulation is effected is known as a sensor or probe. The measurand (modulating parameter) is converted into another physical parameter pertaining to the properties of the sensor. The readout or signal processing device requires knowledge of the explicit relationship between these two. For example, a bimetallic thermometer modulates the bimetallic strip and produces mechanical displacement of the tip or the element, as a function of temperature rise.

In the second stage measurements are associated with processing and standardization of the measured data. In manual measurement, the human sense organs and the brain perform this function. The standardization is done by comparing the measured data with a precalibrated standard using various techniques. Since electrical signal processing is very well studied and understood, and the end users (devices) of measurement are usually electrical meters or apparatus, the modern practice is to process the measured data finally into an electrical signal whose form and level are universal and flexible for further use. The measurement principles as described above are illustrated with the help of functional block diagrams in Fig.1.1.

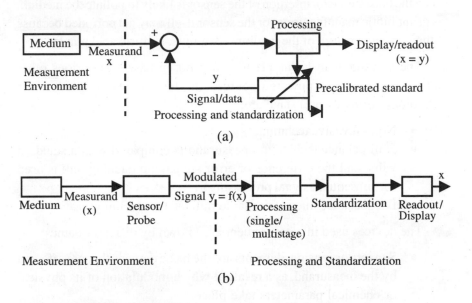

Fig. 1.1 Techniques of Measurement: (a) Direct, (b) Indirect

Basically any measurement involves the following aspects:

(a) Measurement principle or technique (science)
(b) Measuring device (technology)
(c) Quality of measurement or performance analysis of the measuring device (evaluation)

The measurement environment consists of the medium, the measurand, and the conditions under which measurement is to be performed. Together with these, the degree of perfection desired in the measurement process, decides the methodology of measurement and finally the technology of measurement to be adopted to implement the technique. On the other hand if the technology of measurement is chosen a priori, the measurement principle might have to be also chosen accordingly.

Measurement techniques are generalized as

• Invasive or contact type
• Non-invasive or contact-less type

The common scheme for measurement adopts invasive techniques. However, non-invasive techniques are unique in cases where the measurand or its source is very low powered (example, speed of a low wattage rotational body), the measurement environment is inaccessible (such as hot rolling mill) or has an extreme characteristics for insertion of a sensor (example, inside the blast furnace), insertion of the sensor is likely to pollute the medium (as in antibiotic manufacturing) or the sensor itself may get corroded because of the corrosive nature of the medium (example, sulfuric acid plant).

The measurement technique may be chosen based on various other factors, e.g., for low powered sources the techniques adopted at the technology (or hardware) stage are:

- Non - invasive technique
- Null balance technique (measurand is compared with a standard value and the difference or error signal approaching null is used for subsequent signal processing),
- Feedback technique (similar to null balance technique).

The devices used in measurement are known by different names:

- Sensors or sensing elements are the basic devices that are affected by the measurand, as a result of which modulation of its physical or chemical parameters take place.
- Probes are sensors inserted into the measurement environment to derive the signal.
- Transducers are in a general terminology used for describing any device that converts energy from one form to another. However, in measurement transducers may mean sensors and subsequent stages where measurand related signal may be transformed into a suitable (usually electrical) form.
- Instruments signify that the measuring device is complete with processing and display units, and usually ready for interfacing with external environment.

Instrumentation is the technology of measurement. The phenomenal advances in science and technology over the last few decades have its impact on Instrumentation. The growth of technology and that of instrumentation are complementary. Advances in one have aided the advances in the other. It is no wonder that a modern industrial plant is so much dependent on instrumentation that over 30 percent of its capital outlay is on instrumentation alone. Again, many of the epoch making scientific truths are nowadays possible to be verified as the right kind of instruments could be develope developed using modern technology. A multidisciplinary field like instrumentation is heavily dependent on development of technology and materials. Interestingly, a new technology in an unrelated field sometimes indirectly helps in solving many problems in instrumentation. The demand created by new processes and production techniques is very often responsible for evolution of new sophisticated instruments. Thus, use of robots in automated manufacturing systems call for development of new sensors, transducers and even techniques of measurements, which are typically compatible with the robotic environment.

Though there has been development in both scientific and industrial – or process – instrumentation, scientific instruments have a growth rate, which is much slower than that of the latter. The main reason is that development of scientific instruments is cost intensive and they have a limited clientele.

1.1.1 Instrumentation: Many Facets

The criteria to assess advancement in instrumentation are based on:

(1) Performance characteristics – static and dynamic such as accuracy, precision, dynamic (faster) response,

(2) Ability to measure parameters which could not be otherwise measured effectively (example, distributed temperature sensing),

(3) Capability to handle a volume of measurands (as in space control),

(4) Compatibility with other systems (example, multistage process control),

(5) Smartness and intelligence in the instrument performance,

(6) User friendly features (enabling ease of interface and maintenance),

(7) Complexity in retrieval of information of data from noisy background etc

To achieve a specific goal, scientists and technologists often have to wait for the development in the areas of general technology, materials, and even for soft tools like mathematical and computational methods. Looking at the functional features of modern instrumentation, advancement in all of the following aspects has been observed. These are.

(1) Sensing,

(2) Signal processing and transmission,

(3) End devices and interfacing,

(4) Techniques and concepts in information retrieval,

(5) General features,

(6) Application areas.

1.1.2 Sensors: Techniques and Technology

The sensor is the most important link in the chain of any instrumentation system as the performance of the system cannot be any better than that of the sensor itself. Many classical sensors like thermistor, thermocouple or pirani gauge still occupy a position in industrial instrumentation. However, with the introduction of new techniques and technology in signal processing, improved versions of instruments have been developed around the classical technologies based on radioisotopes, ultrasound, magnetostriction, laser and Hall effect. Phenomenal advances observed in sensor technology have been due to advances in material science, integrated circuits (IC) and photonics. The simplest way of classifying sensors is to group them according to the parameters or variables they respond to i.e., based on application as it is convenient to users. Thus, the sensors are classified as temperature sensors, pressure sensors, acceleration sensors, gas sensors etc.

For designers the classification based on material and / or principle of operation is more useful. Even then, for the same group of sensors, there can be areas of overlap with other groups. Thus, a spectroscope, a widely used system for chemical analysis can be classified as an optical device or as a chemical device. A broad classification of sensors is based on technological development presented below:

(a). Early sensors

- Mechanical
- Pneumatic
- Hydraulic
- Fluidic

(b). Classical sensors:

- Electrical and Electronic
- Magnetic
- Ultrasonic
- Nucleonic
- Microwave

(c) Modern sensors:

- Solid state
- Polymer

- Chemical
- Biosensors
- Optical
- Microsensors

(a) Early sensors

Most of these sensors are either obsolete or have limited use in spite of their low cost and/or ruggedness. Bimetallic strips are temperature sensors which can be found as inexpensive thermal relays for overload protection of motors. Governors have limited use as speed sensors. Pneumatic proximity sensors find applications even in robotics. Fluidic sensors like wall-attachment amplifiers [1, 2], once visioned as competitors of electronic sensors, could not fare upto the expectation.

(b) Classical sensors

With the development of technology, more and more physical and chemical properties of materials have been exploited to develop innovative sensors. The properties of interaction between radiation waves and matter are intelligently applied to solve problems in sensing. Many of these sensors have been studied and developed to perfection and most of them have electrical signals at the output stage. They not only continue to be used even with phenomenal advances in sensor technology, constant research and development work, based on these principles, is still being carried out to develop new sensors and their applications. The common textbooks on sensors extensively deal with this group of sensors, which are discussed in brief in the following paragraphs.

(1) Electrical and Electronic sensors

Traditionally the earliest sensors successfully and commercially exploited are electrical sensors that undergo modulation of electrical parameters like resistance, inductance, conductance, capacitance, etc. with change in measurand. Apart from this there are other types of electrical sensors, e.g.,

- Electromagnetic
- Electromechanical
- Electrostatic
- Piezo-electric
- Thermoelectric
- Electro chemical, etc.

where modulation of electrical characteristics is the basis for operation of the sensor.

The main advantage of electrical sensors is that they have an output (current, voltage, resistance, etc.), which offers flexibility in interfacing, signal processing and handling. However, noise in conventional electrical transducers is its biggest drawback. The other disadvantage of electrical sensors is its size, which has been solved to a great extent by introducing new materials and technology, as in thin film sensors.

General electrical sensors are of passive type. But there are active electrical sensors like piezo-electric transducers, thick film sensors (example, pH electrodes – electrochemical), thermocouples (thermoelectric) that are conventional but reliable and inexpensive. Novel acoustic sensors based on electrostatic charge modulation have opened up a new vista in audio instruments.

The chapter on electrical sensors deals with various electrical sensors with special emphasis on capacitive sensors which find wide range of applications in present day instrumentation. Electronic sensors are mainly semiconductor sensors. Though vacuum tube technology is considered to be obsolete, in certain areas (e.g. Photo multiplier tubes and CRT), they are still popular. Electronic sensors like IC sensors using silicon technology are increasingly being improved upon and used in very innovative ways. Smart and intelligent sensors have been presented in subsequent sections.

(2) Magnetic sensors

Magnetic and electromagnetic sensors are also treated as a subset of electrical sensors. There is a variety of magnetic sensors which finds applications in industries and chemical processes, NDT (non – destructive testing), in many scientific investigations as in submarine, geophysical prospecting and bio applications. They are used in process industries to measure pressure, strain, temperature, flow, gas, speed, rotation and also various electrical quantities.

These sensors can be classified into three groups depending on how measurand is related to the magnetic signal present in the measurement process.

(a) Measurand modulating the external magnetic field.
      These sensors are basically non-invasive (non-contact) type, such as, inductive proximity sensors and reluctance modulation sensors (the toothed wheel speed sensors, one of the few intrinsic digital sensors).

(b) Measurand modulating magnetic properties of a secondary transducer.

There are a number of sensors in this category, such as, magnetostrictive, magneto-resistive, magneto-optical sensors, which are used in sensing force, stress displacement, speed, fluid level and various other parameters. Hall effect transducers used for measurement of angular displacement and current.

(c) Based on magnetic characteristics of measurand.

The crux of the problem in magnetic sensor technology is to measure the magnetic field modulated, induced in secondary elements or generated by the measurand. In this respect magnetic sensors are quite different from other categories of sensors with identical measurement configuration. The properties of magnetic field which are measured are: presence (magnitude), direction and rotation. The sensing devices (of magnetic field) can be classified depending on strength of the magnetic field, such as, weak ($< 10^{-9}$ T), medium (10^{-4} to 10^{-3} T) and strong ($> 10^{-3}$ T). The devices which are used for sensing of magnetic field are: search coil, Flux gate, nuclear precession, Fibre Optic (interferometric), SQUID (superconducting interference devices), etc. SQUID ($<< 10^{-11}$ T) and Fibre Optic devices can be used for sensing weak magnetic fields (as in bio-magnetism, localized flux in mechatronics). Hall effect and Magneto-Optical devices are used for strong magnetic fields, and most of others for weak to medium fields and search coil as a true universal magnetic field sensor.

Advancement in magnetic sensors is very prominent in the area of magnetic field sensing. In addition to magnetic sensors, magnetic imaging is used for Nondestructive Testing (NDT) and biomedical applications.

(3) Ultrasonic sensors

Ultrasonic or Sonar technique of measurement is a very powerful tool in biomedical, NDT, marine and industrial applications. The heart of these sensors is the piezoelectric crystal which, when subjected to electrical oscillation (charge or voltage), generates ultrasound (mechanical vibration of medium) and vice-versa. The former is the principle of an ultrasonic signal transmitter and the latter of a receiver. The medium is placed in between the transmitter (T_x) and receiver (R_x). In many applications using pulse mode signal, same unit is used as a transmitter-receiver. Non-invasive measurement is possible. Two distinct groups of instruments are developed using ultrasonic technique, e.g., (1) ultrasonic transducers and (2) ultrasonic imaging. Ultrasonic transducers are different from piezoelectric crystal based sensors studied under electrical sensors, where electrical voltage (or charge) proportional to the measurand in the form of pressure, motion or vibration is obtained across the faces of the piezo-crystal.

The common crystals used are quartz (SiO_2), Barium-titanate (BaT_iO_3), Lead Zirconate, ADP, etc. for measurement of vibration, dynamic pressure, acceleration. Rochelle salt is used for inexpensive signal retrieval and recording. In the ultrasonic transducers use of $T_x - R_x$ pair is mandatory. Hence the measurement of the parameter is based on modulation of ultrasound (MHz frequency) generated by T_x. The ultrasonic signal is modulated by the measurand and received by R_x (Fig. 2). Two configurations are used:

(a) Through transmission technique (continuous mode using separate $T_x - R_x$ pair as in Fig. 1.2), and.

(b) Reflection technique (pulse mode of operation using the same piezoelectric crystal as T_x and R_x).

The instruments based on ultrasound are thickness gauge, instrument for material testing in NDT (presence of crack or void inside and examination of surface finish), detector for liquid or slurry in a closed pipeline, flow and level measurement, measurement of depth of sea bed (sonar technique), detection of underwater object, proximity sensing, etc. ultrasonic imaging (2D and 3D) is now widely used in medical diagnostics, (echocardiogram, ultrasonography) and material testing (NDT).

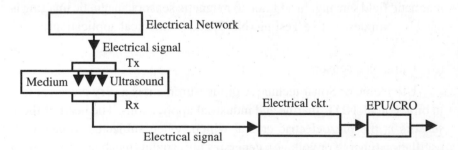

Fig. 1.2 Schematic of ultrasonic measurement (direct transmission)

(4) Nucleonic Gauges

Nucleonic sensors (or commonly known as Nucleonic gauges) are based on modulation of nucleonic radiation by the measurand. The nucleonic sources used are α, β, γ or neutron sources. The choice of sources depends on their half-life, half-distance and activity. Two basic principles of modulation are (1) ionization of medium (using α sources), and (2) modulation of radiation intensity. Ionization technique has limited use for identification of gas, vacuum pressure measurement and few other variables based on the configuration of the ionization gauge and its output circuitry [3]. The detector used is ionization chamber (α source inside). The intensity modulation technique is schematically shown in Fig. 1.3.

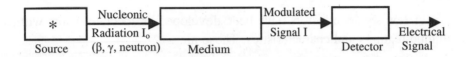

Fig. 1.3 Schematic diagram of a nucleonic method of measurement

The sources generally used are β, γ, neutron (for moisture measurement in grains, sintering plant). The detectors used are proportional counter, Geiger Muller Counter and Scintillation counter. Photographic films are used for recording and safety of operators. Nucleonic gauges are basically non-invasive devices widely used in industrial environments for measurement of thickness, density, flow, speed and many other variables. Innovative modulation technique makes them suitable for measurement of any physical parameter [4].

(5) Microwave sensors

Microwave technique of measurement, in its limited use, is a powerful non-invasive technique. The measurement principle based on modulation of microwave signal is similar to that of ultrasonic technique of measurement. A few important applications are range finding (long distance measurement), location of an object from another object with relative motion (used in robotics, navigation and railway-safety), detection of speed of a moving object (using Doppler shift), and geological prospecting.

(c) Modern sensors.

This group of sensors is not necessarily the youngest in the family of sensors but they, by virtue of their potential and versatility, have drawn the attention of the researchers. They are deemed to be sensors of the future.

(1) Solid state sensors

These sensors exploit the electronic properties of solids for measurement. They are mostly made of semiconductor materials.

The basic features of the semiconductor materials are occurrence of a forbidden energy gap for charge carriers and a filled valence band. They widely differ in compositions and are used extensively for developing semiconductor sensors for measurement of temperature, force, pressure, strain, humidity etc. The semiconductor materials belonging to Groups III-IV and II-VI, such as *GaAs, AlS, InSb, InAs, CdS, CdSe, ZnO, ZnS* are used to make interesting sensors.

A few semiconductors have been developed as special sensors with unique features and characteristics. These are:

(a) Metal oxide sensors

(b) Silicon sensors

(c) Hall effect sensors

• Metal Oxide Semiconductor (MOS) sensors

Metal oxide semiconductor sensors appear to be promising as low cost gas sensors as they are sensitive to gas with or without catalysts. Development of gas sensors for H_2, H_2S, O_2, CO, CH_4, using SnO, ZrO_2, WO_3, and ZnO have been successful.

• Silicon sensors

Though silicon is also a semiconductor material, the exploitation of the silicon planar processing technology and excellent mechanical properties of silicon (tensile strength, young's module, hardness, strength to weight ratio), has enabled the researchers to develop a new class of sensors in the form of chips that are sensitive to radiation signals (electro magnetic, nuclear, optical), mechanical signals (pressure, stress, acceleration, flow, displacement), thermal signal, chemical signal, etc. IC sensors like AD590 (temperature sensor) are used as calibration standards. Advances in silicon planar technology have led to design of integrated sensors, combining sensors and processing circuits on a single chip for commercial exploitation. The miniaturization in its turn has given birth to new generation of sensors; 'smart sensors' and 'intelligent sensors'.

The primary activities in development of thin film sensors can be observed in the areas of semiconductor or IC sensors. However, thin film sensors encompass quite a few different materials. Techniques have been developed to deposit thin layers of resistive, piezoelectric, semiconducting or magnetic materials on suitable substrates like silicon. Thin film *NiCr* strain gauges, platinum temperature sensors, capacitive aluminum displacement sensors, thin film thermocouples, *Ni-Fe* film magnetic sensors, etc have been developed as inexpensive good quality sensors. Thick film thermistors are basically semiconductors. However, thick film *p*H sensors (electrodes) belong to other group of sensors.

- Hall effect sensors

There are certain characteristics, which are exhibited by both semiconductors and metals. These are Hall effect sensors. When a transverse magnetic field is applied to a current carrying conductor, an electric field (Hall field) perpendicular to the direction of both the magnetic field and current appears to counterbalance the Lorenz force acting on the bulk of the current carriers. The magnitude of the Hall field is proportional to the product of the magnetic field strength and current. Many materials possess Hall effect characteristics. However, Hall effect is much more pronounced in semiconductors (compounds of Groups III-IV, like Ge, and As, Si, Bi, etc) than in metals (like Fe, Sn). Hall effect sensors are small in size and can be used for sensing magnetic field, angular rotation, and measurement of power.

(2) Polymer sensors

The variation of conductivity, permittivity and mass of certain polymers in presence of selective gases makes them candidates for futuristic gas sensors. Polymers sensitive to H_2S, CO, CH_4 and moisture will be useful as sensors in pollution monitoring, humidity measurement and industrial environments. Novel gas sensors or 'electronic noses' have been developed using arrays of poly pyrole (conducting polymers) or chemiresistors [5, 6].

(3) Chemical sensors

A chemical sensor may be defined as a simple-to-use, robust device that is capable of reliable quantitative or qualitative recognition of atomic, molecular or ionic species. Its functional parts constitute: (1) a zone of selective chemistry, and (2) a virtually non-specific transducer. A chemical sensor selectively interacts with the measurand and transforms a chemical parameter (say, *p*H) into a chemical or physical (electrical) signal [7].

The chemical sensor technology is conceptually an interdisciplinary subject and hence it has so much of commonness with other types of sensor technology that it often becomes difficult to identify it based on sensitivity chemistry alone. Table.1.1 gives a few examples.

Table. 1.1 Intersection of Chemical sensors with other types of sensors

Sensor	Selective chemistry	Also classified as
• pH electrode	Variation of electric potential across electrode membrane with pH change	Electrical/thick film sensor.
• Liquid crystal thermometer	Colour change due to reflectance variation with structure change	Optical sensor.
• Metal Oxide semiconductor	Variation of electrical resistance in presence of gas	Solid state sensor.
• Polypyrrole	Conductivity change due to recognition of molecular species. (mass sensor)	Polymer sensor.
• Spectroscopy	Change in optical properties with chemical reaction	Optical sensor.

(4) Biosensors

There exists fuzziness in the definition of biosensors, the sensors of the twenty first century. Dictionary meaning of bio or biological is 'of living being'. Hence, in conformity with the present classification of sensors, this group of sensors should include either those using biomaterials or those, which exploit modulation of biological characteristics and phenomena corresponding to various measurands. However, the terminology is used in literature to designate the sensors based on the following:

(a) Sensors using biotic or living organisms

An automated and continuous water quality monitoring system or biological Early Warning System (BEWS) developed by Biological Monitoring. Inc., USA utilizes fish, electronics and a computer. Individual small fish are housed within 500 ml monitoring chambers (eight or twelve in number) through which water under investigation is pumped. The technology relies upon the fact that all fish generate a microvolt level bioelectric field resulting from their neuromuscular activities. Non-contact submerged electrodes located within each monitoring chamber receive these signals. The data represent the fish's ventilatory and certain locomotor behaviour. Abnormal behaviour of sufficient number of fish simultaneously is an indicator of existence of toxic condition.

In other instances piezoelectric characteristics of banana leaves, proximity sensitivity of touch-me-not leaves have the promise and potential of becoming innovative sensors. Living cells, organs, tissues of plants and animals are expected to contribute to a galaxy of miniaturized and very powerful sensors and also to future computer technology (bio-computing).

(b)   Sensors using dead organisms or biochemical materials.

In a broad sense all sensors (like polymer sensors), which use organic materials for generation of signals proportional to the measurands belong to this class of biosensors. The mechanism of sensing involves certain chemical reaction (organic chemistry). Hence they are also termed as chemical or bio-chemical sensors. Quite a few gas sensors belong to this category.

(c)   Sensors whose end use is for assessment of biological phenomena.

Many sensors used in biomedical applications (diagnostics, analysis and monitoring) e.g. immunoassay system for *HIV* or antibody detection, test strips for diabetic detection, disposable sensing electrodes using *Ag-AgCl* conductors on polymer substrates for *EGG* and *EEG*, etc. are classified by many as biosensors. On the other hand, this class also includes a few sensors used in drug detection like rapid detection of antibiotics in milk.

(d)   Sensors using the concept of biological reactions or biological functional mechanism.

This group has probably the weakest link with biotic or living organisms. The sensors utilize neurological reactions; imitate functions of living organisms like vision, olfaction, but not necessarily using biomaterials.

Most of the biosensors of today as reported in literature are those sensors, which use biomaterials i.e. organic materials for sensing a measurand. The main focus in the present volume is on this group of biosensors or biochemical sensors.

(5) Optical sensors

Modulation of optical properties by different measurands has been investigated in classical optics using bulk components. However, with the introduction of optoelectronics (lasers, *LEDs*, *PIN* photo diode, *APD*, optical fibre, etc.), very innovative and intricate optical sensors have been developed. On the other hand, the development of various optical imaging techniques and apparatus has led to a very special class of accurate and novel instrumentation system. Spectroscopy has been perfected to a great level and is now an inseparable part of analytical instrumentation. Today opto-electronics provide solutions to miniaturization in sensing, data transmission and instrumentation leading to new concepts and products like Integrated Optics. Optical fibre based sensors (OFS) can measure a wide variety of process variables when any one of the optical properties, e.g., intensity, phase, frequency, wavelength, polarization, evanescent electic field or scattering loss is modulated. Interferometric optical fibre sensors are highly sensitive (for example, can measure temperature variation $\sim 10^{-8}$ °C, gyroscopic rotation $\sim 0.1°/24$ hrs). Optical fibre based scattering modulated distributed temperature sensors are capable of measuring temperature variation of the optical fibre every few metre of its length. OFS is versatile and can be used for sensing almost all types of physical variables. They facilitate transmission of measured data simultaneously without much of attenuation and electromagnetic interference.

In addition to OFS, optical sensors are being developed in other directions as well, such as, laser based instrumentation. Image based instruments like 2D photography, 3D videography, 3D holograms, optical fibre based interferometric images, night vision system, etc. They are being used in industrial, biomedical, environmental and defence applications.

(6) Microsensors or Micro Electro Mechanical Sensors (MEMS) [8, 9]

The technological need for varied applications like robotics, space vehicles and compactness has given rise to a separate class of .sensors known as MEMS or Microsensors. The thin film sensors belong to this group. The basic difference is that MEMS are mostly electromechanical sensors and they embrace all kinds of sensor principles. The objective of this group of sensors is: miniaturization, integration (of sensor, signal processing and actuator together along with other features like self-calibration, fault diagnostics, etc) and develop sensor arrays (to improve reliability or incorporate multiple function). The MEMS industry has an estimated market of US$ 10 billion. The annual increase in the market volume is 20%.

Microsensors have lower manufacturing costs (mass production and less materials). They are based on capacitance variation, optical signal variation, magnetic signal variation, piezo-resistance variation, chemical structure variation, and even on bioproperty variation. The principles and the applications appear to be vast. At present most of them are developed as pressure, acceleration, displacement, temperature, sensors etc. In future, biochemical and other forms of sensing can also be foreseen.

1.1.3 Signal Conditioning and Signal Processing

The role of signal conditioning and signal processing in instrumentation cannot be overemphasized. What began with the simple conversion of the primary sensor output into a usable form or for elimination of noise using a filter, has now grown into a very complex and sophisticated technology. Level and nature of processing circuits employed in instrumentation depends on factors related to

- The sensor, its type, its signal level etc.
- Mode of signal handling – analog or digital,
- Ambient noise and interference,
- Signal transmission – distance, transmission environment and interference,
- End devices, interfacing mode and application areas,
- Enhanced characteristics, capabilities and features of Instrumentation.

A typical data acquisition system (DAS) (Fig. 1.4) consists of signal conditioner, multiplexer, data converter, data processor, data transmitter, storage and display system. The subsystems vary. In many commercial transducers, the primary sensor output is conditioned and processed so as to get a standard output of 4-20 mA. Digital data communication in form of serial or parallel data transfer is nowadays the preferred mode of data transmission.

Signal conditioner is a generalized terminology to designate many functional components required in a DAS. A few of these are secondary sensors (for conversion of primary sensor output into a readable output, usually electrical), amplifier for amplification of low power or low level output of a sensor), attenuator (for high sensor output), rectifier (for ac to dc conversion), modulator (for dc to ac conversion), filter (for filtering the noise at the measurand environment or at the transmission stage), linearizer (for linearization of sensor output), etc.

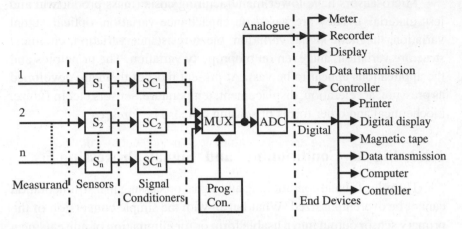

Fig. 1.4 Generalised Data Acquisition System (DAS)

For acquisition of a large volume of data simultaneously (in a short interval of time) and to have economy in number of acquisition paths, a single channel is used employing a multiplexer (MUX). If analog signal is desired, it is obtained from the output of the MUX and then fed to the end devices depending on desired use of the instrumented data.

On the assumption that the sensor outputs are analog signals, the multiplexer output is digitized using an Analog to Digital converter (ADC). The programmed controller synchronizes the MUX with the ADC. The digital output is fed to the end devices as per objective of instrumentation.

Present day signal processing can be (1) hardware based, (2) software based, or (3) a hybrid one – combination of both. However, even for the software based processing a minimum of hardware subsystems are required. To elaborate it we consider a simple flow transmitter. Very common practice is to use a venturi tube followed by a differential pressure transducer. The differential pressure output is fed to a hardware square rooting circuit and its output is proportional to the flow rate. Hence quite often the differential pressure signal is first converted into an equivalent electrical signal and then square rooting becomes easier. The square rooting circuit can be hardware based using a nonlinear circuit, which has a characteristic of $y = (x)^{0.5}$. However that involves approximation. It is now done in a much simpler and elegant way by inputting the differential pressure (after conversion into electrical signal in digital form) to a microprocessor. It serves as the square root extractor as well as the flow indicator. This increases accuracy, linearity and flexibility without adding to extra cost.

In majority of the modern instrumentation systems software is indispensable. Intelligent sensors, image based sensing are largely dependent on the software based signal processing. Software based filters are found to be more effective than their hardware counterparts.

Hardware based signal processing can be (1) analog or (2) digital. However, software based signal processing has to be digital. Subsystems of the signal processing and transmission are chosen based on the method adopted for processing and its complexity. The present day developments are system-on-chip circuits for the necessary signal conditioning followed by a processor for signal processing.

1.1.4 Signal Transmission

Telemetry or remote metering has reached a new height of advancement. Classical telemetry i.e., transmission of signal through a pair of conductors required adoption of means to overcome Electromagnetic Interference / Radio Frequency Interference (EMI / RFI). In its limited use it is still practiced locally either by current conversion (4-20 mA) and transmission or by voltage transmission through axial core and shielded cables. In the current transmission technique the signal in the form of a voltage is obtained across a 250 ohm resistor (1-5 volts). In its advanced form, wired telemetry is implemented through PSTN (Public Switching Telephone Networks) and LAN (Local Area Network). PSTN is convenient as it utilizes the existing telephone network for data communication. Fibre Optic LAN has a large bandwith and low attenuation that makes signal transmission more efficient. Both analog and digital signals can be transmitted through wire telemetry. The modern practice is to use digital signal transmission as corruption of data with noise is minimized, security and confidentiality of data (coders are available for enhancement of security) are increased and compatibility of data with computers is guaranteed.

Wireless telemetry started with microwave based signal transmission. Its range has been expanded to much larger distances by using satellite communication. The latter has made it possible to transmit and receive signals from outer space, for remote sensing, tele-surgery, tele-imaging, environmental studies, weather forecasting, defence, etc.

1.1.5 End Devices and Interfacing

Development in this area of broader field of signal processing is comparatively less. Meters (mostly voltmeters graduated in terms of measurand) are still being used because of their low cost and analog indication (a concept about which a human has psychological obsession!). However, the potentiometric strip chart recorders, oscillographs have faded out. Today PC monitor, Visual Display Screen, LCD (Liquid Crystal Display) are widely used for display. These software based display devices, though digital in nature are used to give digital or analog (graphical form) indication. Virtual instrumentation principle is used for display in any form.

For recording and storage of data there are data loggers, printers compatible with the display devices, Video and CD cassettes. Magnetic recorders have now limited use.

1.2 The Trends: Taking up New Challenges

Advances in technology have a direct influence on modern instrumentation, proposing new challenging tasks in addition to the more traditional ones. The classical concept of instrumentation system that starts with a probe and ends in a display or recorder has undergone a vast change in techniques of measurement and data retrieval. Incomplete and incoherent information are collated to construct the true information about the process and its measurands. Interfaced discrete components have been replaced by compact modular units that have in built multifunctional blocks including signal processing, logic, computational and decision making capability. Quite obviously, this advancement in modern instrumentation has expanded its scope of application as given in following sections.

1.2.1 Smart and Intelligent Sensors

When signal available from the sensor is small, signal processing, to keep the signal-to-noise (S/N) ratio large, in presence of a noisy environment becomes formidable. Integration of processing circuits with the sensor circuits gives an easier solution to this problem. Such sensors are also faster, more accurate and linear and smaller in size. They are more compatible with data highways and computers. These sensors are referred to as "smart sensors". Smart sensors with decision making features are known as "intelligent sensors". They incorporate the features of (1) auto compensation or adaptability to ambient and process conditions, (2) simultaneous measurement of two or more variables, and (3) implementation of self-diagnostic routines.

The majority of the smart and intelligent sensors are obtained from micromachined silicon sensors. Pressure sensors, temperature sensors, humidity sensors have been fabricated, which find applications in industrial processes, automobiles, and biomedical instrumentation. Reference [10] describes a sensor with intelligent interface which enables the range and damping time constant to be set via a communication port.

Reference [11] describes a smart temperature sensor fabricated by a standard CMOS process. The sensor is designed for low power operation.

For assessment of indoor thermal environment, a smart anemometer (air flow sensor) has been designed [12] which consists of four heating resistors, four bridge branches, a voltage-to frequency converter based on a relaxation oscillator, all integrated onto the same chip using a 3 GHz biopolar process.

Smart sensors have been successfully fabricated using integrated optical technology. Reference [13] describes a smart photo sensor consisting of two photodiodes implanted into shallow and substrate regions. The sensor successfully encodes the intensity and colour of the incident light into a frequency form.

Reference [14] describes a smart 3-D imaging system fabricated by CMOS process. It uses position sensitive device (PSD) arrays and is capable of real time signal processing.

1.2.2 Image Based Instrumentation [15]

2-D image or photography invented in early nineteenth century has come out of its age in forms, quality and application. Modern imaging is no more mere photography and is available in different forms e.g.

(1) 2D or 3D (stereo)

(2) Still (off-line) and continuous (on-line)

(3) (i) Optical:

Visible – black and white and colour photography

Infrared – black body or thermal, night-vision

Laser based – holography and interferometry

(ii) X-ray

(iii) Ultrasonic

(iv) Magnetic

(v) Multi Sensor based – using discrete sensors in some non-expensive and limited applications.

Modern imaging techniques differ in the image acquisition stage but the image processing i.e. the analysis and manipulation of a (usually digitized) image, especially to improve its quality, is almost identical. It is, by and large, software based. It has improved perceptibility when the image suffers from low modulation depth, lack of resolution or excessive noise.

Instrumentation, based on imaging techniques, is found to be convenient to use in the following areas:

1. For simultaneous mapping or assessing varying parameters at different locations of a large object, system or area,

2. non-invasive measurement,

3. measurement in inaccessible areas,

4. data storage in compact and miniaturized form.

A few specific areas of application are

(1) Remote sensing – Satellite or aerial images are used for geographical, environmental, pollution, and natural resource-survey, and defence applications.

(2) Simultaneous assessment or mapping of a distributed process parameter as in a blast furnace temperature mapping using thermal imaging technique.

(3) Data storage in compact form like holographic memory for computers and similar devices

(4) Biomedical instrumentation. A large number of diagonstic apparatus are based on imaging techniques. X-ray, magnetic and ultrasonic images are widely used as non-invasive techniques of diagnosis of different diseases or abnormal conditions of human organs not visible from outside. Endoscopy, laparoscopy, bronchoscopy are imaging using fibre optic for inspection of inaccessible human organs. Imaging technique is also used for measurement of muscular motion or cardiac motion.

Image processing consists of the following functional stages:

(1) Imaging system
    (a) Acquiring and storing the images,
    (b) representation schemes and
    (c) encoding and decoding of images for different transmission and less storage space.

(2) Image enhancement

From bad quality, poor contrast and noisy images to derive good quality, high contrast images.
(3)   Segmentation of images
      Used extensively in computer vision and robotics
(4)   Image classification
(5)   Image restoration
      Using mathematical model and digital signal processing the true image can be restored from the degraded one.
(6)   Image interpretation
      This includes identification of objects and assessment of parameters of interest.

1.2.3 Multisensor Probe [16, 17]

Application of multisensors in instrumentation is needed in different fields e.g.

(1) Multiple sensors for measurement of different parameters like pressure, temperature, etc. This happens to be the usual measurement environment.
(2) Multiple identical sensors for mapping or imaging. This concept is still in practice where acquisition of image is difficult or expensive. For example, detection of methane or CO concentration at different points in an underground mine requires multi-identical sensors.

1.2.4 Kelvinian Concept Revisited

Technological advancement has a positive feedback effect on various disciplines of science and technology and in particular on advancement of Measurement and Instrumentation. Advancement of technology creates demand on development of instrumentation science and engineering, and the developments in Measurement and Instrumentation further augment technological advancement creating more demands on instrumentation development, and so on till a balance is achieved. This is supported by new materials and even new scientific theories. The effect is so much irresistible that new concepts of measurement have to be evolved enabling inclusion of very complex or unknown measurement environment. Thus, measurements under extreme uncertainties, avoided till suitable sensors are available, have been ventured by redefining the Kelvinian concept of measurement, which traditionally aims at quantification of the measurand with emphasis on accuracy. In modern instrumentation, identification of the medium and/or its parameter, and estimation of measurands are gaining importance. The development in this direction is different from the ones discussed in the foregoing sections in the sense that they (smart and intelligent sensors, image based sensors and multisensory probes) are still based on Kelvinian definition of measurement. It is only the technology of quantification and achievement of accuracy that is new in them. But in the second group, measurement is synonymous with estimation that leads to quantification within a tolerance band. Surprisingly the measurements based on these concepts often lead to highly accurate quantification (Kelvinian measurement!). These new concepts of estimating the measurands are essentially methods of intelligent computation based on raw data received from noisy environment or through cheap and not so accurate sensors. Strangely enough these methods of data processing are so akin to natural human approach! In technological environment, they are based on soft computing technologies suitable for enhancing characteristics, features and capabilities of the instrumentation. There are two basic groups of this class of instrumentation.

I. Concepts based on natural human reasoning

They are able to identify the solution by means of typical human approaches such as similarity, analogy, interpolation, extrapolation, generalization, inference and multiple goal optimizations. The common methods are:

- Fuzzy logic (FL)
- Artificial neural network (ANN) or Artificial Intelligence(AI)
- Expert system

II. Concepts based on statistical analysis

Kelvinian concept of statistical measurement has been practiced for quantification of raw data extracted from a randomly varying noisy environment. The accuracy of measurement is determined from statistical treatment of apparently accurate measured data. However, the accuracy of measured data (statistical error) only indicates the probability of the true reading lying within an error band. The statistical estimation techniques are, on the other hand, applied during the measurement process so as to estimate the parameter. Its objective is more to arrive at the solution closely rather than analyze the error in measurement from a volume of measured data. These techniques are also highly computer oriented. Different methods are:

- Estimation techniques
- Correlation techniques

1.2.4.1 Fuzzy Logic Based Measurement [16,18,19]

The natural instinct of a human is to measure a parameter in a fuzzy manner. In all natural process of measurements be it a distance, or a temperature, human sensory organs estimate the measurand and update the value based on the error (also estimated) as a result of first estimation. The first estimation or reference is subjective and so is subsequent updating. Once the reference is fixed, subsequent measurements and updating are qualitative in nature. The logic on which it is based is known as fuzzy logic (FL) and the measurement fuzzy measurement. Fuzzy systems are particularly well-suited to manage uncertainty.

Fuzzy logic is a non-deterministic data treatment but it is different from the stochastic method in the sense that the deduction about the system behaviour is done (1) in stochastic systems by evaluating probability density function associated with each possible variable, and (2) in fuzzy logic systems by application of suitable combinatorial rules (membership function). Framing the rules and identification of the variables require a prior knowledge of some structural and physical description of the system.

Fuzzy reasoning is a powerful data processing. Each measured data is assigned a degree of membership function, a real number in the range [0,1], where 1 denotes full membership and 0 denotes no membership. In overlapping domains, it is possible that an input value may have nonzero membership degree in more than one set. Fuzzy rules are framed in shorter forms with antecedent clauses (input conditions) and consequent clauses (output actions). The deduction about the measurand is finally processed using 'possibility function', 'admissibility function', etc.

1.2.4.2 Artificial Neural Network [19, 20, 21, 22]

Artificial Neural Networks (ANN) provides a completely new and unique way to look at information processing. They are adaptive which take data and learn from it. They infer solutions from the data presented to them, after capturing quite subtle relationships. Two different groups of people are exploring neural networks. The first group is composed of biologists, physicists and scientific psychologists who work toward developing a neural model that accurately mimics the behaviour of the brain. The second group consists of engineers who are concerned with development of "artificial" neurons that can be interconnected to form networks with interesting and powerful computational capabilities. The second group uses both biological and nonbiological concepts. Thus, today's ANN is fundamentally different from biological science, traditional computer or standard software techniques (as it does not depend on programmer's prior knowledge of rules).

Artificial Neural networks have several distinct special features, e.g. (1) they are adaptive. They take data and learn from it. They infer solutions to any new set of data often capturing quite subtle relationships. (2) They can handle imperfect data providing a measure of fault tolerance. (3) They can generalize by processing data that broadly resembles the data they were trained. In practical measurements it automatically eliminates noise part of the data. (4) The networks are nonlinear. Hence they can capture complex interactions among the input variables in a system. Processes requiring multi-sensor fusion are immensely benefited. (5) Neural networks do parallel processing and hence numerous identical independent operations can be executed simultaneously resulting in higher speed and economy.

Artificial Neural networks are useful for pattern recognition – for classification or function estimation. A neural network classifier combined with a camera has been developed for grading potatoes, read zip codes, match finger prints to a data base; combined with a microphone it can diagnose engine trouble, identify underwater sounds, or generate signals to counteract vibration noise. Calibration of sensors or systems is another area in which neural networks can be useful. Low cost sensors and computers aided by neural networks made process control faster and more accurate. However, effectiveness of the neural networks depends on the supervised learning method or training. Back propagation method is that in which an output error signal is fedback through the network, altering connection weights between nodes in the network so as to minimize that error. Neural networks can consume huge amounts of computer time – two months, for example – particularly during training [24]. For systems the intelligence imparted through adoption of NN is also known as artificial intelligence (AI).

1.2.4.3 Expert Systems

Many of us have come across experienced people who can predict a phenomenon or estimate a parameter based on the experience or expertise developed by them. Some work on intuitive reasoning and some on logic based on their experience. Such predictions and estimates are mainly based on personal databases and natural with human. Expert systems are auxiliary tools in measurement and instrumentation for intelligent data processing, especially with the objective of optimization and analysis of the behaviour of instruments and systems. The system is based on building a database comprising input variables and the output variables deduced from the expertise of the designer of the system. The expertise available can be based on complex mathematical relationship or empirical and statistical rules. The goal is to capture the behaviour and the choices of experts so that they can be autonomously reproduced on new cases.

Expert systems in measurement and instrumentation [18,19] when applied to a specific domain (interpretation system) may be considered as high-level sensors. Such systems are being used in speech recognition, image analysis, bio-medical and process (manufacturing) diagnostics, and in process monitoring. Expert systems are being used in design and planning of a system, calibration of instruments and optimization of measurement strategies. However, presently available form of expert systems suffer from a few deficiencies e.g. (1) they are not able to operate in the real-time environments and hence can not be used for on-line operation of dynamic systems, (2) the technique is highly subjective (quite natural!), and (3) complexity of database and inference (or interpretation) may demand large memory.

1.2.4.4 Concepts Based on Statistical Analysis

Statistical signal processing methods are being widely used in estimation from noisy data and are known as Estimation theory. They are being applied to control systems for parameter estimation (system identification based on observation or measured data) and state estimation (estimation of signals of interest from measurements that are corrupted with noise – also known as signal estimation or tracking).

(a) Estimation techniques in measurement and instrumentation are already being used in image processing, data transmission, sensor validation, etc. It can be classified as

(1) Static estimation (as in image processing), and

(2) Dynamic estimation (as in process control).

In instrumentation statice stimation is often appropriate. The crux of the estimation technique is to select a performance index (a scalar quantity) based on the difference between 'estimate' and the 'measured data' and optimize it (usually minimize). The popular estimate is the 'Least Square" (LS) estimate. The more challenging are the dynamic estimation techniques.

Dynamic estimation can be classified as

(1) Conventional filtering technique -

     (a) hardware based – analogue and digital filters

     (b) software based, and

(2) Estimators

The gains in the estimators are manipulated depending on the data in suitable ways based on stochastic properties of the signals and noise components involve to optimize some performance index which is also a specification of statistical characteristics of filtered output. Estimators are essentially computer based techniques.

Some of the different forms in use in dynamic model based estimation are:

- Adaptive filters
- Kalman filters
- Weiner filters
- Levinson filters
- Levinson – Durbin filters
- Stochastic gradient algorithm
- Least Mean Square (LMS) algorithm

The Kalman filter is a very popular technique. It is a recursive algorithm for estimation of the state given the values of the observed outputs and based on state space model. Application of estimators in measurement of instrumentation is an emerging area of multi-sensor systems of measurement and control.

(b) Correlation technique:

Cross correlation technique is already an established method of signal processing in instrumentation e.g. ultrasonic method of liquid flow measurement, optical method of width measurement in the manufacturing of plates. The technique is suitable for static and dynamic measurements. Cross correlation technique as an estimation technique is faster than recursive methods of estimation. The method is suitable for image processing or mapping.

1.2.5 Virtual Instruments [19]

Virtual instruments are basically computer based "soft instruments" that can realize the conventional measuring instruments as well as merge several instruments into only one operator interface. These simulation driven instruments cost nothing and require no maintenance as it does not have any hardware structure. Virtual Instruments (VI) create user-friendly environment for training, design and situations where man-machine operations are to be integrated e.g. in telemedicine.

Virtual instruments use powerful graphic libraries and compilers to configure a 'virtual front panel' on the computer screen. All the traditional controls like knobs, switches, push buttons, etc. are represented graphically and operated by a pointing device e.g. cursor, mouse. Indicators like scope display, strip charts, LED's are also mimicked. Operation of the instruments like inputting external signals, establishment of connections between different terminals, interfacing with other devices can be done with the help of the pointing device and complete system, circuit or operation diagram is possible to be drawn. The graphic compiler translates the specific operation related algorithms into executable computer code.

Use of VI does not require any knowledge of programming. However, building a virtual instrument does. Techniques based on Lab Windows or CVI (C programming for Virtual Instruments) are available in literature. Prentice Hall PTR and National Instruments have published books in the National Instruments Virtual Instrumentation Series on sensors, signal processing, graphical programming, technique, DAS, VI programming technique and applications [25].

1.3 The Monograph: Its Scope

IIT Kharagpur, with its multidisciplinary programmes, has been successfully conducting programmes, on M & I at all levels over a few decades along with its researches in M & I leading to innovative ideas and products. All the areas of M & I discussed under the present chapter have been either well studied or being researched upon in various departments of IIT Kharagpur. The scope and principles of many of these M & I areas are presented in subsequent chapters with occasional reference to the work done at IIT Kharagpur. An attempt has been made to cover the conventional areas which are well developed along with new concepts, technological developments and applications of M&I.

The monograph presents a detailed account of different classes of sensors starting classical to the state-of-the-art sensors. However, the absence of separate chapters on ultrasonic sensors, magnetic sensors, smart and intelligent sensors is regretted.

Similarly, there is a need for a separate chapter on signal processing, more specifically digital signal processing, to make the presentation in the area of signal conditioning and processing techniques and trends complete. The relevant part (C) remains more as a class of case studies. Chapters on Virtual Instrumentation, Fuzzy logic based sensors and simulation (part E) are also lacking in this monograph.

In the part on Applications (F) a wide coverage has been given starting from instrumentation in unconventional areas (irrigation) to an interesting new area like Bioprocess (modeling and instrumentation).

1.4 References

1. Anderson, B.W. The Analysis and Design of Pneumatic systems, John Wiley

2. Blestering, C.A. Fluidic System Design, Wiley Interscience.

3. Schumilovskii, N.N. and Mell'ttser, L.V. Radioactive Isotopes in Instrumentation and Control, Pergamon Press, 1964.

4. Piraux, H. Radioisotopes and their Industrial Applications. Philips Technical Library, 1964,

5. Pelose, P. and Persaud, K. Gas sensors: Towards an Artificial Nose, NATO ASI Series, Vol.43, Sensors and Sensory Systems for Advanced Robots, Penkin, P (Ed).

6. Zee, F and Judy, J.W. – Sensors and Actuators: Chemical, Vol 72(2), 2001.

7. Edmonds, T.E. (Ed) – Chemical Sensors, Blackie, London, 1988.

8. Muller, R.S. et al (Ed). Microsensors, IEEE Press, NY, 1990.

9. Gardner, J.W. – Microsensors – Principles and Applications, John Wiley and Sons, NY, 1994,

10. Petriu, E.M. (Ed). Instrumentation and Measurement Technology and Applications. Chapter 8 on Smart Sensor Integration Techniques, IEEE, NY, pp.229-232.

11. Ibid, pp. 224-228.

12. Ibid, pp. 218-223.

13. Ibid, pp. 241-244.

14. Ibid, pp. 245-248.

15. Ibid, Chapter 11 on Image Processing, pp.319-356, and Chapter 12 on Miscellaneous Applications, pp.357-410.

16. Ibid, Chapter 14 on Soft Computing and Virtual Environment – Based Applications, pp. 464-470.

17. Murthy, AVSN. Optical Proximity Sensor for Robot Safety. M.Tech. Dissertation, E.E. Dept., IIT Kharagpur, December 1992.

18. Petriu, E.M. (Ed). Instrumentation and Measurement Technology and Applications. Chapter 14 on Soft Computing and Virtual Envoronment – Based Applications, pp.499-504.

19. Ibid, pp. 453-457.

20. Ibid, pp. 477-485.

21. Hush, D.R. and Horne, B.G. – Progress in Supervised Neural Networks. IEEE Signal Processing Magazine, January 1993, pp.7-39.

22. ——— Special Issue on Neural networks in Control Systems. IEEE Control System Society Magazine, Vol 12, No. 2, April 1992.

23. ——— Special issue on Neural Network Applications. IEEE Transactions on Industrial Electronics, Vol 39, No. 6, December 1992.

24. Hammerstrom, D. – Working with Neural Networks. IEEE Spectrum, July 1993, pp. 46-53.

25. ——— National Instruments Virtual Instrumentation Series, Prentice Hall PTR and National Instruments, www.phptr.com/bookscri/virtual.html.

| Chapter 2 |

Instrument Performance Evaluation

S. Sen

Abstract

Any measurement system has a set of specifications from which one can get an idea about the performance of the system. Broadly speaking, the specifications define the characteristics of the instrument under different input conditions, e.g. static, time-varying and random. One has to be familiar with these terminologies, should be thorough with the operation of the instrument, the methods for eliminating the undesirable errors and also how to calibrate the instrument at regular intervals. This chapter attempts at providing a broad overview of the techniques for evaluating the instrument performance. It also describes the statistical methods for analysis of instrument performance.

2.1 Static and Dynamic Characteristics

The performance of an instrument or a measuring instrument is evaluated based on its *static* and *dynamic* characteristics. The *static* characteristics refer to the case where the different inputs to the system are either held constant or varying very slowly with respect to time and we are interested with the steady state value of the output. On the other hand, *dynamic* characteristics refer to the performance of the system when the inputs are varying with time.

2.1.1 Static Characteristics

The catalogue of an instrument would describe the performance of the instrument in terms of a number of parameters. The items that can be classified under the heading static characteristics are:

Range (or span): It defines the maximum and minimum values of the inputs or the outputs for which the instrument is recommended to be used. For example, for a temperature measuring instrument the input range may be 100-500 °C and the output range may be 4-20 mA.

Sensitivity: It can be defined as the ratio of the *incremental output* and the *incremental input*. While defining the sensitivity, we assume that the input-output characteristic of the instrument is approximately linear in that range.

Linearity: Linearity is actually a measure of nonlinearity of the instrument. When we talk about sensitivity, we assume that the input/output characteristic of the instrument to be approximately linear. But in practice, it is normally nonlinear, as shown in Fig.2.1. The *linearity* is defined as the maximum deviation from the linear characteristics as a percentage of the full scale output. Thus,

$$\text{Linearity} = \frac{\Delta O}{O_{max} - O_{min}} \qquad (1)$$

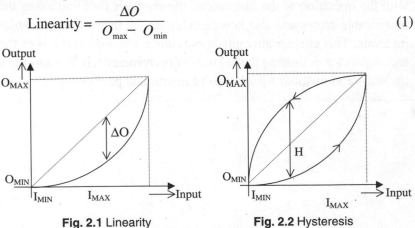

Fig. 2.1 Linearity   **Fig. 2.2** Hysteresis

Hysteresis: Hysteresis exists not only in magnetic circuits, but in instruments also. For example, the deflection of a diaphragm type pressure gage may be different for the same pressure, but one for increasing and other for decreasing, as shown in Fig.2.2. The *hysteresis* is expressed as the maximum hysteresis as a full scale reading, i.e., referring Fig.2.2.

$$\text{Hysteresis} = \frac{H}{O_{max} - O_{min}} X100 \tag{2}$$

Resolution: In some instruments, the output increases in discrete steps, for continuous increase in the input, as shown in Fig.2.3. *Resolution* indicates the minimum detectable input variable. For example, an eight-bit A/D converter with +5V input can measure the minimum voltage of $\frac{5}{2^8 - 1}$ or 19.6 *mv*. Referring to Fig.2.3, *resolution* is defined as:

$$Resolution = \frac{\Delta I}{I_{max} - I_{min}} X100 \tag{3}$$

The quotient between the measuring range and resolution is often expressed as *dynamic range* and is defined as:

$$\text{Dynamic range} = \frac{measurement\ range}{resolution} \tag{4}$$

and is expressed in terms of dB. The dynamic range of an *n*-bit ADC, comes out to be approximately *6n* dB.

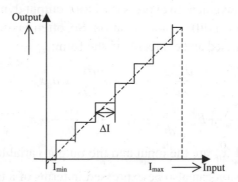

Fig. 2.3 Resolution

Accuracy: Accuracy indicates the closeness of the measured value with the actual or true value, and is expressed in the form of the *maximum error* (= *measured value - true value*) as a percentage of full scale reading. Thus, if the accuracy of a temperature indicator, with a full scale range of 0-500 °C is specified as $\pm$ 0.5%, it indicates that the measured value will always be within $\pm 2.5\,^\circ C$ of the true value, measured through a standard instrument during the process of calibration. Thus if it indicates a reading of $250\,^\circ C$, the error will also be $\pm 2.5\,^\circ C$, i.e. $\pm 1\%$ of the reading.

Precision: Precision indicates the repeatability or reproducibility of an instrument (but does not indicate accuracy). If an instrument is used to measure the same input, but at different instants, spread over the whole day, successive measurements may vary randomly. The random fluctuations of readings, (mostly with a Gaussian distribution) is often due to random variations of several other factors which have not been taken into account, while measuring the variable. A precision instrument indicates that the successive reading would be very close, or in other words, the standard deviation σ_e of the set of measurements would be very small. Quantitatively, the precision can be expressed as:

$$Precision = \frac{measured\ range}{\sigma_e} \tag{5}$$

2.1.2 Dynamic Characteristics

Dynamic characteristics refer to the performance of the instrument when the input variable is changing rapidly with time. For example, the human eye cannot detect any event whose duration is more than one-tenth of a second; thus dynamic performance of the human eye cannot be said to be very satisfactory. The dynamic performance of an instrument is normally expressed by a differential equation relating the input and output quantities. It is always convenient to express the input-output dynamic characteristics in form of a linear differential equation. So, often a nonlinear mathematical model is linearised and expressed in the form:

$$a_n \frac{d^n x_0}{dt^n} + a_{n-1} \frac{d^{n-1} x_0}{dt^{n-1}} + \cdots + a_1 \frac{dx_0}{dt} + a_0 x_0$$
$$= b_m \frac{d^m x_i}{dt^m} + b_{m-1} \frac{d^{m-1} x_i}{dt^{m-1}} + \cdots + b_1 \frac{dx_i}{dt} + b_0 x_i \tag{6}$$

where x_i and x_0 are the input and the output variables respectively. The above expression can also be expressed in terms of a transfer function, as:

$$G(s) = \frac{x_0(s)}{x_i(s)} = \frac{b_m s^m + b_{m-1} s^{m-1} \cdots + b_1 s + b_0}{a_n s^n + b_{n-1} s^{n-1} \cdots + a_1 s + a_0} \tag{7}$$

Normally $m<n$ and n is called the order of the system. Commonly available sensors normally follow either *zero-th order, first order* or *second order* dynamics. Here are a few such examples:

Potentiometer: Displacement sensors using potentiometric principle (Fig.2.4.) have no energy storing elements. The output voltage e_o can be related with the input displacement x_i by an algebraic equation:

$$e_o(t)x_t = Ex_i(t); \quad or, \quad \frac{e_o(s)}{x_i(s)} = \frac{E}{x_t} = constant \tag{8}$$

where x_t is the total length of the potentiometer and E is the excitation voltage. So, it can be termed as a *zeroth order system.*

Thermocouple: A bare thermocouple (Fig.2.5) has a mass (m) of the junction. If it is immersed in a fluid at a temperature T_f, then its dynamic performance relating the output voltage e_o and the input temperature T_f, can be expressed by the transfer function:

$$\frac{e_o(s)}{T_f(s)} = \frac{K_v}{1 + s\tau} \tag{9}$$

where, K_v = steady state voltage sensitivity of the thermocouple in V/ °C.

τ = time constant of the thermocouple = $\dfrac{mC}{hA}$

m = mass of the junction
C = specific heat
h = heat transfer co-efficient
A = surface area of the hot junction.

Hence, the bare thermocouple is a first order sensor. But if the bare thermocouple is put inside a metallic protective well (as it is normally done for industrial thermocouples) the order of the system increases due to the additional energy storing element (thermal mass of the well) and it becomes a second order system.

Seismic Sensor: Seismic sensors (Fig.2.6.) are commonly used for vibration or acceleration measurement of foundations. The transfer function between the input displacement x_i and output displacement x_o can be expressed as:

$$\frac{x_o(s)}{x_i(s)} = \frac{Ms^2}{Ms^2 + Bs + K} \tag{10}$$

where: M = mass of the seismic body
B = damping constant
K = spring constant

From the above transfer function, it can be easily concluded that the seismic sensor is a *second order system.*

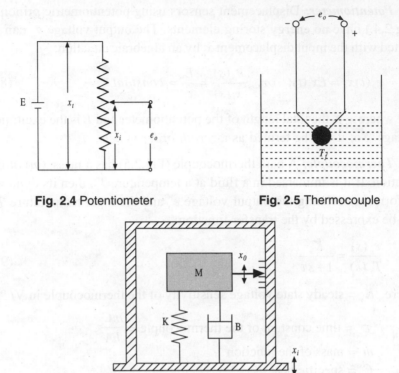

Fig. 2.4 Potentiometer   **Fig. 2.5** Thermocouple

Fig. 2.6 Seismic sensor

2.1.3 Evaluation of Dynamic Performance

The output of the sensor is connected to the signal conditioning circuit. The transfer function of the signal conditioning circuit can also be brought out similarly and the overall transfer functions of the measuring system can be computed. The performance of the measuring system under time varying input signals is a major concern to the designer as well as the user. The sensor is normally subjected to different types of time varying inputs. The input could be either *sinusoidal*, *step*, *ramp* or *impulse* type. The performance of the measuring system under dynamic conditions is normally expressed in terms of different parameters, e.g., *bandwidth*, *cut off frequency*, *settling time*, *peak overshoot*, *steady state error* etc. These parameters define the range of the time varying signals that the instrument is capable of measuring within a tolerable range of accuracy.

2.2 Deterministic and Stochastic Measurement

In most of the cases of measurement, it is assumed, that the variable we are measuring is a well-defined quantity, either remaining constant during the period of measurement, or varying slowly in a deterministic fashion. For example, the temperature inside a furnace may be varying slowly in a particular manner and we are interested in measuring the profile of the temperature variation. In this case, the effects of other interfering inputs are small and our main concern is to reduce the error in measurement introduced due to the interfering inputs. The type of measurements we mostly deal with is the *deterministic measurement.*

On the other hand, there are few cases where the measuring input itself is a randomly varying quantity, or the input signal is corrupted with noise. *Stochastic measurement* refers to the cases where, we want to use the stochastic nature of the signal for the measurement. For this, often one has to perform the *cross-correlation* or *autocorrelation* of the signal. The autocorrelation operation can be used to determine the time period of a periodic signal that is corrupted with noise. The cross-correlation technique is often used to measure the velocity of a fluid under turbulent condition. Here two sensors are placed at two different locations along the stream. The turbulence of the stream here is the tagging signal and the time to travel for the tagging signal from one sensor to the second one is estimated by performing the cross-correlation operation with a variable time delay. The time delay for which the cross-correlation function is maximum, will be the time taken for the tagging signal to travel from one sensor to the other. Knowing the distance between the sensors, the velocity of the fluid can easily be calculated.

2.3 Error Analysis

The term *error* in a measurement is defined as:

Error = Instrument reading - true value.

The errors in an instrument reading, may be classified in to two categories as:

1. Systematic errors

2. Random Errors.

Systematic errors are those that affect all the readings in a particular fashion. Zero error, and bias of an instrument are examples of systematic errors. On the other hand, there are few errors, the cause of which is not clearly known, and they affect the readings in a random way. This type of errors is known as *random errors*. There is an important difference between systematic errors and random errors. In most of the case, systematic errors can be corrected by calibration, whereas random errors can never be corrected, the can only be reduced by averaging, or error limits can be estimated.

2.3.1 Systematic Errors

Systematic errors may arise due to different reasons. It may be due to the shortcomings of the instrument or the sensor. An instrument may have a zero error, or its output may be varying in a nonlinear fashion with the input. The amplifier inside the instrument may have input offset voltage and current which will contribute to zero error. Different nonlinearities in the amplifier circuit will also cause error due to nonlinearity. Besides, systematic error can also occur due to improper design of the measuring scheme. It may arise due to the loading effect, improper selection of the sensor or the filter cut off frequency. Systematic errors can be due to environmental effect also. The sensor characteristics may change with temperature or other environmental conditions.

The major feature of systematic errors is that the sources of errors is recognisable and be reduced to a great extent by carefully designing the measuring system and selecting its components. Placing the instrument in a controlled environment may also help in reduction of systematic errors. They can be further reduced by proper calibration of the instrument.

2.4 Random Errors

It has been already mentioned that the causes of random errors are not exactly known, so they cannot be eliminated. They can only be reduced and the error ranges can be estimated by using some statistical operations. If we measure the same input variable a number of times, keeping all other factors affecting the measurement same, the same measured value would not be repeated, the consecutive reading would rather differ in a random way. But fortunately, the deviations of the readings normally follow a particular distribution (mostly normal distribution) and we may be able to reduce the error by taking a number of readings and averaging them out.

Few terms are often used to characterize the distribution of the measurement, namely,

$$Mean\ Value\ \bar{x} = \frac{1}{n}\sum_{i=1}^{n} x_i \tag{11}$$

where n is the total number of readings and x_i is the value of the individual readings. It can be shown that the mean value is the most probable value of a set of readings, and that is why it has a very important role in statistical error analysis. The *deviation* of the individual readings from the mean value can be obtained as :

$$Deviation \quad d_i = x_i - \bar{x} \tag{12}$$

We now want to have an idea about the deviation, i.e., whether the individual readings are far away from the mean value or not. Unfortunately, the *mean of deviation* will not serve the purpose, since,

$$Mean\ of\ deviation = \frac{1}{n}\sum_{i=1}^{n}(x_i - \bar{x}) = \bar{x} - \frac{1}{n}(n\bar{x}) = 0$$

So instead, *variance* or the *mean square deviation* is used as a measure of the deviation of the set of readings. It is defined as:

$$Variance \quad V = \frac{1}{n-1}\sum_{i=1}^{n}(x_i - \bar{x})^2 = \sigma^2 \tag{13}$$

The term σ is denoted as *standard deviation.* It is to be noted that in the above expression, the averaging is done over n-1 readings, instead of n readings. The above definition can be justified, if one considers the fact that if it is averaged over n, the variance would become zero when n=1 and this may lead to some misinterpretation of the observed readings. On the other hand the above definition is more consistent, since the variance is undefined if the number of reading is *one*. However, for a large number of readings ($n>30$), one can safely approximate the variance as,

$$Variance\ V = \frac{1}{n}\sum_{i=1}^{n}(x_i - \bar{x})^2 = \sigma^2 \tag{14}$$

The term *standard deviation* is often used as a measure of uncertainty in a set of measurements.

2.4.1 Propagation of Error

Quite often, a variable is estimated from the measurement of two parameters. A typical example may be the estimation of power of a d.c circuit from the measurement of voltage and current in the circuit. The question is that how to estimate the uncertainty in the estimated variable, if the uncertainties in the measured parameters are known. The problem can be stated mathematically as follows. Let,

$$y = f(x_1, x_2, \ldots, x_n) \tag{15}$$

If the uncertainty (or deviation) in x_i is known and is equal to Δx_i, $(i = 1,2,..n)$, what is the overall uncertainty in the term y?

Differentiating the above expression, and applying Taylor series expansion, we obtain,

$$\Delta y = \frac{\partial f}{\partial x_1} \Delta x_1 + \frac{\partial f}{\partial x_2} \Delta x_2 + ... + \frac{\partial f}{\partial x_n} \Delta x_n \tag{16}$$

Since Δx_i can be either +ve or -ve in sign, the maximum possible error is when all the errors are positive and occurring simultaneously. The term *absolute error* is defined as,

$$Absolute\ error : |\Delta y| = \frac{\partial f}{\partial x_1} |\Delta x_1| + \frac{\partial f}{\partial x_2} |\Delta x_2| + ... + \frac{\partial f}{\partial x_n} |\Delta x_n| \tag{17}$$

But this is a very unlikely phenomenon. In practice, $x_1, x_2, ..., x_n$ are independent and all errors do not occur simultaneously. As a result, the above error estimate is very conservative. To alleviate this problem, the cumulative error in y is defined in terms of the standard deviation. Squaring equation (17), we obtain,

$$(\Delta y)^2 = \left(\frac{\partial f}{\partial x_1}\right)^2 (\Delta x_1)^2 + \left(\frac{\partial f}{\partial x_2}\right)^2 (\Delta x_2)^2 + ... + 2 \frac{\partial f}{\partial x_1} \frac{\partial f}{\partial x_2} .(\Delta x_1 \Delta x_2) + ... \tag{18}$$

If the variations of $x_1, x_2, ...$ are independent, positive value of one increment is equally likely to be associated with the negative value of another increment, so that the sum of all the cross product terms can be taken as zero, in repeated observations. We have already defined *variance V* as the mean squared error. So, the mean of $(\Delta y)^2$ for a set of repeated observations, becomes the variance of y, or

$$V(y) = \left(\frac{\partial f}{\partial x_1}\right)^2 V(x_1) + \left(\frac{\partial f}{\partial x_2}\right)^2 V(x_2) + ... \tag{19}$$

So the standard deviation of the variable y can be expressed as:

$$\sigma(y) = \left[\left(\frac{\partial f}{\partial x_1}\right)^2 \sigma^2(x_1) + \left(\frac{\partial f}{\partial x_2}\right)^2 \sigma^2(x_2) + ...\right]^{1/2} \tag{20}$$

2.4.2 Normal (Gaussian) Distribution

The distribution curve for a number of readings of a same variable takes the nature of a histogram. But if the number of readings is increased and the interval is also decreased, then the distribution curve can be represented by a smooth curve. There are few standard distribution curves, but most common among them is *Gaussian,* or *normal* distribution curve. Here x is the measured variable and $P(x)$ is the probability that the measured value will be within a small range x and $x+dx$. For a normal distribution, $P(x)$ is expressed as:

$$P(x) = \frac{1}{\sigma\sqrt{2\pi}} e^{-\frac{(x-\mu)^2}{2\sigma^2}} \tag{21}$$

where μ and σ are the mean and standard deviation for a large number of measurements. If we plot $P(x)$ vs. x, we would obtain a bell shaped curve A typical probability distribution curve for a normal distribution is shown in Fig.2.7. Moreover,

$$\int_{-\infty}^{+\infty} P(x)\, dx = 1.0 \tag{22}$$

and

$$P(\mu) = \frac{1}{\sigma\sqrt{2\pi}} \tag{23}$$

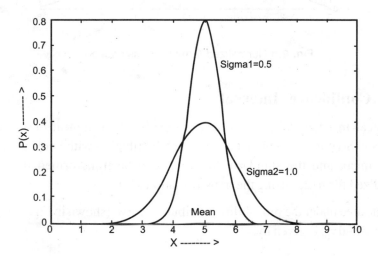

Fig. 2.7 Normal distribution

So the peak of the curve is dependent on σ. Now introducing the normalized variable,

$$z = \frac{x - \mu}{\sigma} \tag{24}$$

we can have, $P(z) = \frac{1}{\sqrt{2\pi}} e^{-\frac{z^2}{2}}$ (25)

so that, substituting (24) in (22) and recalling that $dz = dx / \sigma$,

$$\int_{-\infty}^{+\infty} P(z)\, dz = 1.0$$

A normalized Gaussian distribution curve has been shown in Fig.2.8.

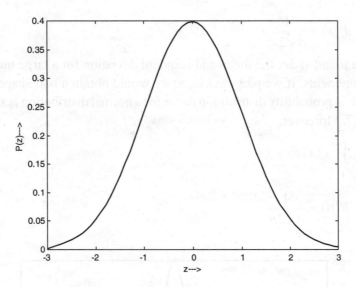

Fig. 2.8 Normalized Gaussian Distribution

2.4.3 Confidence Interval

Recall that eqn.(25) does not have any explicit term of mean or standard deviation. So it is apparent that any normal distribution with known values of the mean and the standard deviation can be transformed into the normalized form (25) using the transformation (24).

The area under the probability distribution curve (shown in Fig.2.4), for any specified z_0, is given by:

$$F(z_0) = \int_{-z_0}^{z_0} \frac{1}{\sqrt{2\pi}} e^{-\frac{z^2}{2}}\, dz \tag{26}$$

The value of $F(z_0)$ indicates the probability that the value of z shall remain within the limit $(-z_0, +z_0)$. However integration of the term in (26) is not straightforward, since it involves the use of *Gamma Functions*. So, instead, statistical tables are available, which are used for computations. Values of $F(z_0)$ corresponding to few typical values of α are given in Table 2.1.

Table 2.1

Value of z_0	Chances of falling within the limit $(-z_0, +z_0)$
0.6745	0.5
1.0	0.6828
2.0	0.9546
3.0	0.9972

To illustrate the above table, let us take an example. Suppose we have taken a large number of resistors of the same type and found that the mean value is 100Ω and the standard deviation is 2Ω. We want to find out the probability that any randomly chosen resistor will have the resistance value within the range $(96\Omega, 104\Omega)$. For this, using (24) we can easily obtain the normalized limit z_0 as $(104-100)/2$ or 2.0. Corresponding to this, from Table 2.1 we obtain the probability as 0.9546. In other words, we can say that, the probability that any randomly chosen resistor will have a resistance value within the range 96 to $104\ \Omega$ is 95.46%. The interval $(96\Omega, 104\Omega)$ for this case is known as *Confidence Interval* for probability of 95.46%.

Mathematically speaking, the *confidence interval* for a normal distribution is expressed as:

$$F(-z_0 \leq z \leq +z_0) = p \tag{27}$$

For $z_0 = 1.96$, $p = 0.95$, which indicates that there is 95% confidence that for any reading, z will lie in the interval $(-1.96 \leq z \leq +1.96)$, or

$$F(-1.96 \leq z \leq +1.96) = 0.95$$

Using (24), we can also express the confidence interval in terms of x as the interval $(\mu - 1.96\sigma \leq x \leq \mu + 1.96\sigma)$, i.e.,

$$F(\mu - 1.96\sigma \leq x \leq \mu + 1.96\sigma) = 0.95, \tag{28}$$

where μ is the mean and σ is the standard deviation.

2.4.4 Analysis with Finite Number of Samples

While discussing normal distribution in the last section, we have assumed that both the mean and standard deviation (s.d.) have been calculated based on a very large number of data. But in practice, experimenting with very large number of data is costly at the same time, time consuming. For determining the distribution of a particular type of resistors, nobody will advice of measuring resistance for few thousand samples and arrive at the mean and s.d. values. A practical solution is to pick up few samples and measure their values. But definitely this practical solution is at the cost of accuracy. Moreover, a single blunder in measurement, may cost heavily in the accuracy of measuring mean and s.d. So the questions are that, how to identify and eliminate a bad data (gross blunder in measurement), and secondly, for a limited number of samples, how to estimate the loss of accuracy for measurement of the mean. The second question can also be put forward in a different way: if we want that the mean value we obtained from our experiment with a limited number of samples to lie within a certain tolerance from the actual value, what is the minimum number of samples to be taken? We shall try to answer these questions in this section.

2.4.5 Chauvenet's Criterion

This criterion is for elimination of bad data. It states that if the deviation of a particular data from the mean value is more than a certain limit, then reject the data. In other words, if

$$\left| \frac{d_i}{\sigma} \right| \rangle R_0 \tag{29}$$

then reject that particular data. d_i and σ are defined as in eqns. (12) and (15) respectively. The value of the limit R_0 is dependent on the number of samples. The recommended values are given in Table 2.2.

So to apply Chauvenet's criterion, one should calculate the mean and s.d. for the set of readings, calculate the deviation of individual readings from the mean, and if for a particular reading is more than the maximum acceptable limit, then that reading is rejected and fresh mean and s.d. to be calculated. It is to be noted that Chauvenet's criterion can be applied only once and if several data indicate violation of the acceptable limit, then the measurement process can be suspected to be inadequate.

Table 2.2. R_0 for Chauvenet's Criterion

Number of samples (n)	Maximum acceptable deviation (R_0)
2	1.15
3	1.38
4	1.54
5	1.65
6	1.73
10	1.96
15	2.13
25	2.33
50	2.57
100	2.81
500	2.39
1000	3.38

2.4.6 Importance of the Arithmetic Mean

It has been a common practice to take a number of measurements and take the arithmetic mean to estimate the average value. But the question may be raised: *why mean?* The answer is: *The most probable value of a set of dispersed data is the arithmetic mean.* The statement can be substantiated from the following proof.

Let $x_1, x_2, x_3, ..., x_n$ be a set of n observed data. Let X be the central value (not yet specified).

So the deviations from the central value are $(x_1 - X), (x_2 - X), ...(x_n - X)$.

The sum of the square of the deviations is:

$$S_{sq} = (x_1 - X)^2 + (x_2 - X)^2 + ... + (x_n - X)^2$$
$$= x_1^2 + x_2^2 + ... + x_n^2 - 2X(x_1 + x_2 + ... + x_n) + nX^2$$

So the problem is to find X so that S_{sq} is minimum. So,

$$\frac{dS_{sq}}{dX} = -2(x_1 + x_2 + ... + x_n) + 2nX = 0$$

or,

$$X = \frac{1}{n}(x_1 + x_2 + ... + x_n) = \bar{x}$$

So the arithmetic mean is the central value in the least square sense. If we take another set of readings, we shall reach at a different mean value. But if we take a large number of readings, definitely we shall come very close to the actual value (or universal mean). So the question is, how to determine the deviations of the different set of mean values obtained from the actual value?

2.4.7 Standard Deviation of the Mean

It is apparent from the above discussions that the mean values of different sets of readings also follow a normal distribution. Here we shall try to find out the standard deviation of the mean value obtained from the universal mean or actual value.

Consider a set of n number of readings, $x_1, x_2, x_3, \ldots, x_n$. The mean value of this set expressed as:

$$\bar{x} = \frac{1}{n}(x_1 + x_2 + \ldots + x_n) = f(x_1 + x_2 + \ldots + x_n)$$

Using (4.11) for the above expression, we can write:

$$V(\bar{x}) = \left(\frac{\partial f}{\partial x_1}\right)^2 V(x_1) + \left(\frac{\partial f}{\partial x_2}\right)^2 V(x_2) + \ldots + \left(\frac{\partial f}{\partial x_n}\right)^2 V(x_n)$$

$$= \frac{1}{n^2}[V(x_1) + V(x_2) \ldots + V(x_n)]$$

Now the standard deviation for the readings $x_1, x_2, \ldots, x_n$ is defined as:

$$\sigma = \left[\frac{1}{n}[V(x_1) + V(x_2) \ldots + V(x_n)]\right]^{1/2}, \text{ where } n \text{ is large.}$$

Therefore,

$$V(\bar{x}) = \frac{1}{n^2}(n.\sigma^2) = \frac{\sigma^2}{n}$$

Hence, the standard deviation of the mean,

$$\sigma(\bar{x}) = \frac{\sigma}{\sqrt{n}} \tag{30}$$

which indicates that precision can be increased, (i.e. $\sigma(\bar{x})$ reduced) by taking large number of observations. But the improvement is slow due to the $\sqrt{n}$ factor.

2.4.8 Confidence Interval for the Universal Mean

Recall that for a normal distribution in z given in (26) the confidence interval is defined as:

$$F(-z_0 \le z \le +z_0) = p \tag{26}$$

We have already obtained in (30), the standard deviation of the mean $\bar{x}$ for finite number of samples is

$$\sigma(\bar{x}) = \frac{\sigma}{\sqrt{n}}.$$

If we denote the universal mean as m (i.e. mean for a very large number of samples), then we can write using (24),

$$z = \frac{\bar{x} - \mu}{\sigma/\sqrt{n}}$$

Note that the value of μ is still unknown to us. We are trying to estimate μ from our knowledge of $\bar{x}$ and σ. Substituting the above expression for z in (26), we obtain,

$$F(-z_0 \le \frac{\bar{x} - \mu}{\sigma/\sqrt{n}} \le +z_0) = p$$

or,
$$F(\bar{x} - \frac{z_0 \sigma}{\sqrt{n}} \le \mu \le \bar{x} + \frac{z_0 \sigma}{\sqrt{n}}) = p \tag{31}$$

From the above expression, even if the universal mean (i.e. the actual value) μ is unknown, one can estimate its value from a limited number of measurements and calculating the mean and standard deviation from the limited number of observations. The estimation of μ is expressed in terms of *confidence interval* as shown in (31). The value of p corresponding to a selected z_0 an be obtained from Table-1. For example, we can say with 95% confidence that the actual value will be in the interval:

$$(\bar{x} - \frac{1.96\sigma}{\sqrt{n}} \le \mu \le \bar{x} + \frac{1.96\sigma}{\sqrt{n}}).$$

2.4.9 Student's *t*-distribution

In the last section, we have estimated the confidence interval for the universal mean. But there is an assumption in the deduction. We have assumed that the standard deviation obtained from the limited number of samples is same as that of for a very large number of samples. As a result, the estimation becomes optimistic if the number of samples taken is small ($n < 20$). Students' *t*-distribution is used to correct the result for these cases. Here the confidence interval is written as:

$$F(\bar{x}-t(\alpha)\frac{z_0\sigma}{\sqrt{n}} \leq \mu \leq \bar{x}+t(\alpha)\frac{z_0\sigma}{\sqrt{n}}) = p \qquad (32)$$

$t(\alpha)$ is the statistic known as *Student's t* and α is the level of significance. Its value depends on the number of samples and the confidence level p. The step by step procedure to find out $t(\alpha)$ is as follows:

1. Select a confidence level p (say $p=0.95$).

2. Find $\alpha = 1 - p$ (=0.05 here)

3. Find the degrees of freedom f ($f = n$-1 for the present case), where n is the number of observations.

4. Find the value of t from the t-distribution table, corresponding to $\alpha/2$ and f. A sample t-distribution table is given in Table 2.3.

Example: For $n = 10$ and $p = 0.95$, $f=9$, and $\alpha/2 = 0.025$. Correspondingly, $t(\alpha) = 2.262$.

Therefore the confidence interval is:

$$F(\bar{x}-\frac{2.262\sigma}{\sqrt{n}} \leq \mu \leq \bar{x}+\frac{2.262\sigma}{\sqrt{n}}) = 0.95 \text{ instead of :}$$

$$F(\bar{x}-\frac{1.96\sigma}{\sqrt{n}} \leq \mu \leq \bar{x}+\frac{1.96\sigma}{\sqrt{n}}) = 0.95$$

following (31). It is obvious to mention that when n becomes very large, the value of $t(\alpha)$ tends to the limiting value (i.e. 1.96 in the present case).

Table 2.3 (*t*-distribution table)

$\alpha/2$ \\ f	0.4	0.3	0.2	0.1	0.05	0.025
1						
.						
.						
4						2.78
.						
9						2.262

2.5 Calibration and Error Reduction

It has already been mentioned that random errors cannot be eliminated. But by taking a number of readings under the same condition and taking the mean, we can considerably reduce the random errors. In fact, if the number of readings is very large, we can say that the mean value will approach the true value, and thus the error can be made almost zero. For finite number of readings, by using the statistical method of analysis, we can also estimate the range of the measurement error.

On the other hand, systematic errors are well defined, the source of error can be identified easily and once identified, it is possible to eliminate the systematic error. But even for a simple instrument, systematic errors arise due to a number of causes and it is a tedious process to identify and eliminate all the sources of errors. An attractive alternative is to calibrate the instrument for different known inputs.

Calibration is a process where a known input signal or a series of input signals are applied to the measuring system. By comparing the actual input value with the output indication of the system, the overall effect of systematic errors can be observed. The errors at those calibrating points are then made zero by *trimming* a few adjustable components or by using software corrections.

Strictly speaking, calibration involves comparing the measured value with the *standard instruments* derived from comparison with the primary standards kept at Standard Laboratories. So in an actual calibrating system for a pressure sensor (say), we not only require a standard pressure measuring device, but also a *test-bench*, where the desired pressure can be generated at different values. The calibration process of an acceleration measuring device is more difficult, since, the desired acceleration should be generated on a body, the measuring device has to be mounted on it and the actual value of the generated acceleration is measured in some indirect way.

The calibration can be done for all the points, and then for actual measurement, the true value can be obtained from a *look-up table* prepared and stored before hand. This type of calibration, is often referred as *software calibration*. Alternatively, a more popular way is to calibrate the instrument at one, two or three points of measurement and trim the instrument through independent adjustments, so that, the error at those points would be zero. It is then expected that error for the whole range of measurement would remain within a small range. These types of calibration are known as single-point, two-point and three-point calibration. Typical input-output characteristics of a measuring device under these three calibrations are shown in Fig. 2.9.

The single-point calibration is often referred as offset adjustment, where the output of the system is forced to be zero under zero input condition. For electronic instruments, often it is done automatically and is the process is known as *auto-zero* calibration. For most of the field instruments calibration is done at two points, one at zero input and the other at full scale input. Two independent adjustments, normally provided, are known as *zero* and *span* adjustments.

One important point needs to be mentioned at this juncture. The characteristics of an instrument change with time. So even it is calibrated once, the output may deviate from the calibrated points with time, temperature and other environmental conditions. So the calibration process has to be repeated at regular intervals if one wants that it should give accurate value of the measurand through out.

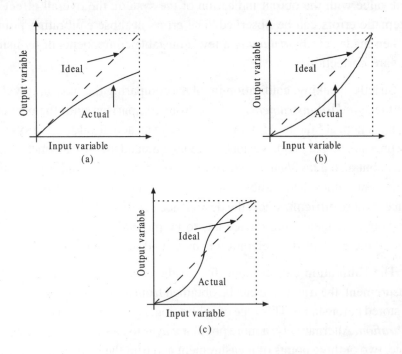

Fig. 2.9 (a) Single point calibration, (b) Two point calibration,
(c) Three point calibration

2.6 References

1. Doeblin E.O.: Measurement Systems-Application and Design, 4/e, McGraw-Hill, NY, 1995.

2. Bentley J.P.: Principles of Measurement Systems, Longman, 3/e, 1995.

3. Stout M.B.: Basic Electrical Measurements, 2/e, Prentice Hall of India, New Delhi, 1981.

4. Kreyszieg E.: Advanced Engineering Mathematics, 3/e, Wiley Eastern Ltd, New Delhi, 1980.

5. Nakra B.C. and Choudhury K.K.: Instrumentation Measurement and Analysis, Tata McGraw-Hill, New Delhi, 1985.

6. Pallas-Areny R. and Webster J.G.: Analog Signal Processing, John Wiley, NY, 1999.

7. Hollman J.P. and Gajda *jr* W.J..: Experimental Methods for Engineers, 5/e, McGraw-Hill, NY, 1989.

PART - B

Sensors

Chapter 3

Electrical Sensors

S. Sen

Abstract

This chapter addresses issues related to construction, performance and modeling of electrical sensors. A brief overview of different types of sensors has been provided. Both inductive and capacitive sensors have been discussed. Major stress is given on capacitance sensors. Different types of capacitive sensing elements and their signal conditioning circuits have been discussed. The electrostatic field equations governing the performance of the sensing element have been elaborated and methods for modeling a capacitance sensor have been discussed.

3.1 Introduction

In a way, most of the sensors can be called electrical sensors, because they convert non-electrical signals to electrical signals. But the term *electrical sensor* is normally used to describe the sensors where changes in the specific electrical parameters occur due to variations of the input signal. The change in the sensor output may be in terms of voltage, or in terms of resistance, capacitance or frequency variations. The advantage of classifying the sensors in this manner is that, the measuring circuits for a particular class of sensors are almost identical. A resistance sensor may be used for measuring temperature, strain, humidity or many other parameters, but electrically speaking, they are very similar as are their signal condition circuits.

Based on the principles of operation, the electrical sensors are normally classified into two broad categories, namely, *active sensors* and *passive sensors*. *Active sensors* are self-generating type; they convert non-electrical signals directly into electrical signals, without requiring any excitation from an additional electrical source. Examples are *thermocouple, piezoelectric sensor, pH electrode* etc. On the other hand, in a *passive sensor*, one of the electrical properties changes with the input signal. So we require an external excitation source to convert the change in the passive electrical parameter (e.g. resistance, inductance or capacitance) to a more convenient form such a voltage signal. The resistance of a strain gage changes with input strain, and we require a bridge circuit to convert it into an output voltage signal.

Apart from these two categories, one may be inclined to introduce another type of sensors, *semiconductor type sensors*. With rapidly advancing semiconductor technology, semiconductor sensors are becoming more and more popular. The major advantages of these sensors are that they are very small in size and they can be integrated with the signal processing circuits. Their measuring principles could be different, but the common thing among them is that they all use properties of semiconductors to sense the signals. A *photodiode*, a *Hall effect sensor*, or a *semiconductor temperature sensor* etc. can all be placed in this category.

In this chapter, we would mainly concentrate on the conventional passive and active electrical sensors. A broad outline of different types of active and passive sensors would be given together with their basic measuring scheme. An interested reader may go through available standard text books

[1-3], for details. The main focus of this chapter is on *capacitance sensors*, which find wide range of applications in modern day instrumentation. Different configurations of capacitance sensors applicable for measurement of a wide range of physical quantities are discussed along with typical measuring schemes.

3.2 Active Sensors

Most important types of active sensors are (a) thermocouples, (b) piezoelectric sensors and (c) pH sensors. All of them generate small voltage signals based on the energy conversion principles, and do not require any external electrical excitation.

3.2.1 Thermocouples

They are the most common among the active sensors. Different types of thermocouples are available to measure temperatures at different ranges starting from cryogenic temperatures (50K) to very high temperatures (2300K). The major advantages are small size, stability, ruggedness and moderate sensitivity over a wide range of temperature. They work on the basic principle of *thermoelectric effect* that states that if two dissimilar metals are connected and their junctions are manufactured at two different temperatures, then an *emf* would be produced depending upon the temperature difference between the *hot junction* and the *cold junction*. Theoretically any two dissimilar metals are capable of acting as a thermocouple, but from the point of view of stability, repeatability and ruggedness, only few pairs (mostly alloys) qualify to be used as commercial thermocouples. Commercial thermocouples are available in the names of different types, such as type *J, K, R, S, T* etc. These thermocouples are made of different pairs of metals or alloys and have different operating ranges and sensitivities. Several properties of thermocouples (known as *laws of thermocouple*) make the measurement of temperature easy, since a third metal can be inserted in the circuit for measurement without affecting the voltage output. For measurement of temperature, the unknown temperature forms the hot junction, while the cold junction is at the ambient temperature. To avoid error in measurement due to variation of ambient temperature, a suitable cold junction compensation scheme should be employed in the measuring circuit. The voltage generated is in millivolt range, with sensitivity of a few $\mu V/°C$, and so needs amplification for display and further processing.

3.2.2 Piezoelectric Sensors

Piezoelectric devices produce electric charge when they are subjected to mechanical deformation. The effect is reversible, that is, application of electric potential across the surfaces of a crystal also produces mechanical deformation. The applications of piezoelectric crystals are numerous, starting from gas lighters and electronic watches to ultrasonographic equipment. They are also used for measurement of displacement, force, velocity, acceleration etc. Piezoelectric properties are observed in particular type of crystals. They can be naturally available crystals (e.g. Quartz, Rochelle salt etc), or synthetically made to attain the property, such as, Lead Zirconate-Titanate (PZT), Barium Titatnate, Lithium Niobate etc. Some polymer materials, such as Polyvinylidelene Fluoride (PVDF) have also been developed, which have higher sensitivities compared to inorganic crystals, and find wide application in bio-medical areas.

Piezoelectric transducers can be operated in different modes, such as: thickness expansion (TE), length expansion (LE), thickness shear (TS) etc. A mode is dependent upon the orientation in which the crystal surface is cut, the direction of displacement and the direction in which the induced charge is measured. A particular type of crystal is operated in a mode where its charge sensitivity is maximum. The charge sensitivity is defined as the charge generated per unit force applied to the crystal. Typical values of charge sensitivities are 2.3pC/N for Quartz and 150pC/N for Barium Titanate.

Piezoelectric crystals are non-conducting. Metal films are deposited on the two opposite faces of the crystal and shielded lead wires are taken out to amplify and measure the voltage output, which, multiplied by the capacitance of the crystal will give the charge generated. The crystal is often mounted on a seismic mass and the combination works as a piezoelectric accelerometer. The measuring circuit is a particular type of amplifier, called the charge amplifier. The use of charge amplifier gives in the advantage that the amplifier output voltage is independent of the connecting cable length and its capacitance. The piezoelectric accelerometers operate over a wide range of frequencies, the upper cut-off frequency (10-20kHz) being limited by the natural frequency of the seismic mass, while the lower one (1-3Hz) is set by the charge amplifier.

3.2.3 pH Sensors

The wide use of pH electrode is well known. They measure the H^+-ion concentration of a solution. The basic principle is similar to that of a Daniel cell. When a metal is immersed in a solution of its own ion, a potential is created between the metal and its solution. This is called a half cell and the potential is governed by the Nernst's equation. Another half cell of a different metal/solution combination and a salt bridge between the two half cells constitute a full cell and the potential difference between them can be measured. Though hydrogen has the properties of a metal, since it is gaseous, it cannot work as a half cell. A special type of sodium ion-selective glass can work as a half cell, sensitive to H^+-ion concentration in an aqueous solution, and its sensitivity is very consistent in the whole range of 0-14pH (59.2mv/pH at 25°C).

The measuring cell is made of a glass bulb, which is immersed the unknown solution. The glass bulb is filled with a buffer solution and a lead wire is immersed inside the buffer solution. While, the glass electrode constitutes the measuring cell, to make it a full cell, we need another half cell, whose potential should essentially be constant. The second half cell, known as the reference cell, is made of either calomel (mercury-mercurious chloride), or silver-silver chloride, immersed in saturated KCl solution. Often these two half cells, namely the measuring and the reference half cells, are housed in the same casing, and the electrode is called a combination electrode. The schematic arrangement of a combination cell is shown in Fig.3.1.

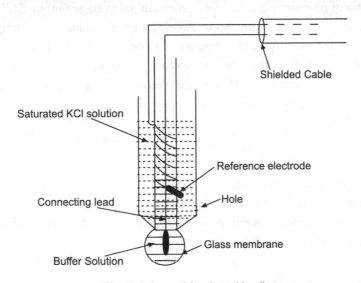

Fig. 3.1 A combination pH cell

The two lead wires, taken out of the two half cells are the output terminals of the pH probe, and should be connected to the amplifying and measuring circuits. But there is a glass membrane present in between the lead wires, which makes the internal impedance of the cell very high (few hundred megaohms). This makes the measuring circuit of the pH electrode complex and special attention is needed to overcome loading effect on this sensor output. For, the amplifier input impedance should be in order of gigaohms and special amplifier circuits are needed. A FET-input op-amp circuit can be employed, or otherwise, a potentiometric technique of measurement should be used for measuring the output voltage of the probe. The current drawn from the probe should be less than about 10^{-12}A, a special type of moisture resistant insulation is also provided to the lead wire.

3.3 Passive Sensors

As discussed earlier, we would classify the passive sensors as (i) resistive sensors, (ii) inductive sensors and (iii) capacitive sensors. The commonality among these types of sensors is that they all require external power supply to activate them. The difference is in terms of the electrical parameter varying with the input variable. The measuring circuit is thus similar for a particular type of a passive sensor. Thus, though an RTD is used for measuring temperature, and a strain gage for strain measurement, the measuring circuits, are all similar, they are all intended to measure the variation of resistance and to give a proportional voltage output. Still, there are differences, depending upon several factors, such as, the range of resistance, the magnitude of variation and nonlinearity etc. The measuring schemes thus do vary from sensor to sensor. Simple resistive transducers, like strain gage and RTD (Resistance Temperature Detector) will not be discussed here, we rather concentrate on inductive and capacitive transducers.

3.3.1 Inductance Sensor

Inductive type transducers can operate on either of the two principles: variation of self-inductance, or variation of the mutual inductance. The former type is known as the variable reluctance type, while the later type is more popularly known as LVDT.

3.3.1.1 Variable Reluctance Type

This type of sensors consists of a fixed U-shaped magnetic core, and a movable plunger or armature, a small air gap existing between the two. A coil is wound on the fixed magnetic core and the self-inductance of the coil.

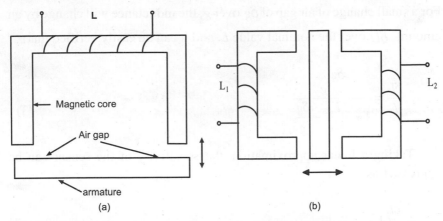

Fig. 3.2 (a) Variable reluctance type sensor, (b) Push-pull type configuration

changes with the variable air gap, thus making it a displacement sensor. Fig.3.2(a) shows the schematic arrangement of such a sensor. Here, as the air gap decreases, the self-inductance of the coil increases. However, to get an idea about the sensitivity and linearity of the sensor, we need to study the magnetic circuit of the sensor. It is well known that the self-inductance of a coil is given by the expression:

$$L = \frac{N^2}{\mathfrak{R}},\qquad(1)$$

where, N = Number of turns of the coil and

$\mathfrak{R}$ = Reluctance of the magnetic circuit.

The reluctance $\mathfrak{R}$ is actually the sum of two values: reluctance of the iron path and the reluctance of the air gap. Assuming uniform flux density throughout the air gap, one can obtain:

$$\mathfrak{R} = \frac{l}{\mu_0 \mu_s A} + \frac{g}{\mu_0 A} = \frac{1}{\mu_0 A}(g + \frac{l}{\mu_s}),\qquad(2)$$

where,

l = total length of the iron path of the flux (core and armature)

g = total air gap length

μ_s = relative permeability of the iron material (core and armature)

μ_0 = permeability in free space

Thus, if displacement signal is applied to the armature so as to reduce the air gap, the reluctance will decrease, thus increasing the self inductance.

For a small change of air gap of δg over g, the inductance will change by an amount δL over the nominal value L, and it can be easily verified that;

$$\frac{\delta L}{L} = \frac{\delta g / g}{1 + \dfrac{l}{\mu_s g} - \dfrac{\delta g}{g}} \tag{3}$$

Taking a linear approximation, the sensitivity of the sensor can be expressed as:

$$\frac{\delta L / L}{\delta g / g} = \frac{1}{1 + l / \mu_s g} \tag{4}$$

which is valid for very small $\delta g / g$ only, otherwise the expression is generally nonlinear. To increase the sensitivity, the ratio $l / \mu_s g$ should be as small as possible. Magnetic cores of very high permeability (ferrite core: $\mu_s = 1000$, mu-metal $\mu_s = 50,000$) are normally used as the core material.

3.3.1.2 Push-pull Type Variable Reluctance Sensors

These sensors are often used for displacement measurement. The basic configuration is shown in Fig. 3.2(b). Here, two identical magnetic cores share a single armature in the magnetic circuit. If the armature moves up, the inductance of the upper coil (L_1) increases, while that of the lower coil (L_2) decrease. Due to the presence of two coils, not only the overall sensitivity is doubled, but also the effect of nonlinearity can be eliminated. This is evident from the following derivation.

For an upward movement of the armature by an amount of δg, the new inductance of the upper coil can be obtained from equations (1) and (2):

$$L_1 = \frac{N^2}{\mu_0 A(g - \delta g + l / \mu_s)} = \frac{N^2}{\Re_0 (1 + K(g - \delta g))} \text{ (say)} \tag{5}$$

where,

$$\Re_0 = \mu_0 A l \Big/ \mu_s$$

and $K = \mu_s \Big/ l$

Similarly, for the lower coil, the inductance becomes,

$$L_2 = \frac{N^2}{\Re_0(1 + K(g + \delta g))} \tag{6}$$

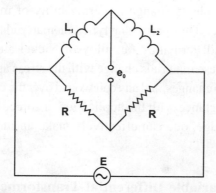

Fig.3.3 A.C. Bridge Circuit for push-pull sensor

Now, if these two coils are connected to an AC bridge as shown in Fig 3, then the expression for the unbalanced voltage comes out to be:

$$e_0 = E\left[\frac{j\omega L_2}{j\omega(L_1 + L_2)} - \frac{R}{R + R}\right] = E\left[\frac{L_2}{L_1 + L_2} - \frac{1}{2}\right]$$

Substituting for L_1 and L_2 from (6) and (7) and simplifying, yields the expression:

$$e_0 = \frac{E}{2}\left(\frac{K\delta g}{1 + Kg}\right) \tag{7}$$

The above expression clearly shows that, the effect of nonlinearity here has been mostly eliminated, and the output voltage of the bridge is directly proportional to the displacement signal (δg), without any restriction. The sensitivity also increases compared to a single element sensor. The other advantages of push-pull type sensors are that the force required to move the armature is reduced and the configuration provides immunity from external magnetic field. The nonlinearity of a push-pull type sensor can be less than 1%.

3.3.1.3 Other Types of Reluctance Sensors

The principle of varying reluctance is also used for making other types of inductance sensors. Two common examples are (i) *proximity* sensing using eddy current type sensor and (ii) *magnetostrictive* type sensor. In the first type, a U-shaped magnetic core with a coil wound over it acts as a search coil. Any magnetic material coming in the vicinity to the vicinity of the open ends of the U will change the inductance of the coil, which can be measured using the A.C. bridge circuit. The magnetostrictive type sensor works on the principle of change in permeability of magnetic core with application of force. Two particular types of materials are used for this purpose: Nickel and Permalloy (an alloy of Nickel and Iron). The B-H characteristics of these materials change with the stress applied. As a result, the permeability also changes, and any coil wound over the core will experience change in self-inductance with the application of force. By connecting the coil to an A.C. bridge, one can effectively make an inductive type force sensor.

3.3.1.4 Linear Variable Differential Transformer (LVDT)

The LVDT works on the principle of variation of mutual inductance. It is one of the most popular types of displacement sensor. It has good linearity over a wide range of displacement. Moreover the mass of the moving body is small, and the moving body does not make any contact with the static part, thus minimizing the frictional resistance. Commercial LVDTs are available with full scale displacement range of ± 0.25mm to ± 25mm. Due to the low inertia of the core, the LVDT has a good dynamic characteristics and can be used for time varying displacement measurement range.

The principle of operation of LVDT is well known, and so will not be elaborated here. It is based on the principle of variation of the mutual inductance between two coils with displacement. It consists of a primary winding and two identical secondary windings of a transformer, wound over a tubular nonmetallic (paper) former, so that a ferromagnetic core of annealed nickel-iron alloy can move through the former. The two secondary windings are connected in series opposition, so that the net output voltage is the difference between the two. The primary winding is excited by 1-10V r.m.s. A.C. voltage source, the frequency of excitation may be anywhere in the range of 50 Hz to 50 KHz. The output voltage is zero when the core is at central position (voltage induced in both the secondary windings are same, so the difference is zero), but increasing as the core moves away from the central position, in either direction. Thus, from the measurement of the output voltage only, one cannot predict, the direction of the core movement. A phase sensitive detector (PSD) is a useful circuit to make the measurement direction sensitive. It is connected at the output of the LVDT and compares the phase of the secondary output with the primary signal to judge the direction of movement. The output of the phase sensitive detector after low pass filtering becomes a d.c voltage for a steady deflection. LVDTs are widely used for position control application in areas such as aerospace where high reliability makes of a superior sensor compared to a potentiometer.

3.3.2 Capacitance Sensors

The capacitance type sensor is a versatile one; it is available in different sizes and shapes. It can also measure very small displacements in micrometer range. Often the whole sensor is fabricated on a silicon base and is integrated with the processing circuit to form a small chip. The basic principle of a capacitance sensor is well known. But to understand the various modes of operation, consider the capacitance formed by two parallel plates separated by a dielectric. The capacitance between the plates is given by:

$$C = \frac{\varepsilon_r \varepsilon_0 A}{d} \tag{8}$$

where A = Area of the plates

d = separation between the plates

ε_r = relative permittivity of the dielectric

ε_0 = absolute permittivity in free space = 8.854×10^{-12} F/m.

A capacitance sensor can be formed by varying either (i) the separation (d), or, (ii) the area (A), or (iii) the permittivity (ε_r). A displacement type sensor is normally based on the first two (variable distance and variable area) principles, while the variable permittivity principle is used for measurement of humidity, level, etc. Fig.3.4 shows the basic constructions of three types of capacitance sensors mentioned above. The variable area type and the variable permittivity type sensors give rise to linear variations of capacitance with the input variables, while a variable separation type sensor follows inverse relationship.

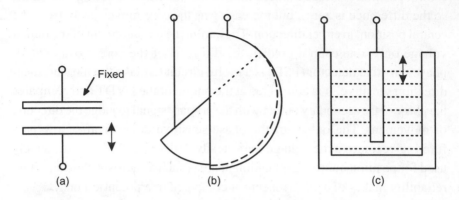

Fig. 3.4 Capacitive Sensors: (a) variable separation type, (b) variable area type, (c) variable permittivity type

For a variable separation type sensor, the change in capacitance can be estimated from eqn. (9). For a small change in displacement δd, the change in capacitance is given by:

$$C + \delta C = \frac{\varepsilon_0 \varepsilon_r A}{d + \delta d} = C\left[1 - \frac{\delta d}{d} + \left(\frac{\delta d}{d}\right)^2 - \cdots\right] \quad\quad (9)$$

which can be approximated for a small δd, by the linear approximation and the sensitivity can be expressed as:

$$\frac{\delta C}{C} = -\frac{\delta d}{d} \quad\quad (10)$$

However, for large displacement, the higher powers of $\frac{\delta d}{d}$ contribute to the nonlinearity of the characteristics.

Similar to the case of variable reluctance type sensors, push-pull type configuration is also commonly in use for capacitance sensors. The push-pull type configuration not only eliminates the effect of nonlinearity, but also increases the sensitivity and reduces the effect of stray capacitance field. Two typical push-pull type configurations, a variable distance type and a variable area type are shown in Fig. 3.5.

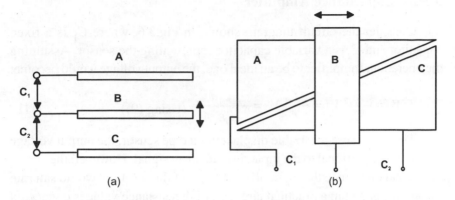

Fig. 3.5 Push-pull type capacitance sensor: (a) variable separation type, (b) variable area type

It should be noted that, the variation of capacitance in the sensor is generally very small (few *pF* only, it can be even less than a *pF* in certain cases). These small changes in capacitance, in presence of large stray capacitances existing at different parts of the circuit are difficult to sence. So the output voltage would generally be noisy, unless the sensor is designed and shielded carefully, the measuring circuit should also be capable of reducing the effects of stray fields. The push-pull type capacitance sensor can be connected to an AC bridge circuit, similar to the case of inductive sensors (Fig.3.3). In place of standard AC bridges, efficient circuits known as *capacitance amplifiers* are often used. They provide the amplification of voltage, at the same time, partial immunity from stray capacitance coupling.

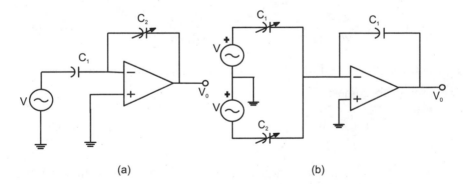

Fig. 3.6 Low input impedance type capacitance amplifier with (a) single active element, (b) push-pull sensor.

3.3.2.1 Capacitance Amplifier

Consider the circuit diagram shown in Fig.3.6, where C_1 is a fixed capacitor and C_2 is a variable capacitor representing the sensor. Assuming the operational amplifier to be an ideal one, the output voltage would become:

$$V_0 = -\frac{X_2}{X_1}V = -\frac{C_1}{C_2}V = -\frac{C_1 d}{\varepsilon_0 \varepsilon_r A}V \text{ (Using (9))} \tag{11}$$

Thus, if we use a variable displacement type sensor, the output voltage would be proportional to the input displacement signal. However, the circuit is an ideal one, since the input offset current of the op. amp. would saturate the amplifier and in a practical circuit, a high resistance value is connected in parallel with the sensor C_2.

For a push-pull type transducer, the above-mentioned circuit can be modified as shown in Fig. 3.6(b). Here we require two balanced voltage supplies V, connected in series with the center point grounded. This can be achieved by using either a center-tapped transformer, or using an inverting amplifier to obtain $-V$ from V. C_1 and C_2 are the two sensing capacitors connected in push-pull. The current through the feedback amplifier C_f would be:

$$I = V\omega(C_1 - C_2) = -V_0 \omega C_f$$

or, $$V_0 = -\frac{C_1 - C_2}{C_f}V$$

For a variable displacement type sensor, using (10), we have,

$$C_1 - C_2 = \frac{\varepsilon_0 \varepsilon_r A}{d + \delta d} - \frac{\varepsilon_0 \varepsilon_r A}{d - \delta d} = \frac{\varepsilon_0 \varepsilon_r A(2\delta d)}{d^2 - (\delta d)^2},$$

which can be approximated by neglecting $(\delta d)^2$ and the output voltage expression becomes:

$$V_0 = -\frac{\varepsilon_0 \varepsilon_r A}{d^2} \frac{V}{C_f}(2\delta d) \tag{12}$$

The above expression indicates a nearly linear relationship with the input signal δd. This type of configuration of capacitance amplifier also is known as a *Low Input Impedance Amplifier*, since the measuring circuit draws current from the source.

Two more types of capacitance amplifier circuits also are in use. They are known as *High Input Impedance Amplifier* and *Feedback type Amplifier.* Among the three, the feedback type one has been found to be the most suitable one. The typical push-pull type circuit configurations for these two types of circuits are shown in Fig. 3.7. For both the cases, the expression for the output voltage comes out to be as:

$$V_0 = \frac{C_1 - C_2}{C_1 + C_2} V \qquad (13)$$

Stray fields always cause problems in measurement for capacitance sensors. To circumvent this problem, guard electrodes are often used. The guard electrode shields the sensor from extraneous fields. A typical push-pull type variable area capacitance sensor with guard electrode arrangement is shown in Fig.3.8. Here the guard electrode moves along with the moving electrode and is at the same potential as the moving electrode, but the current through the guard electrode does not enter the measuring circuit.

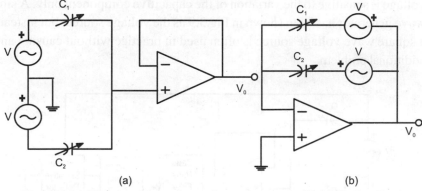

Fig. 3.7 (a) High input impedance Capacitance amplifier, (b) Feedback type capacitance amplifier

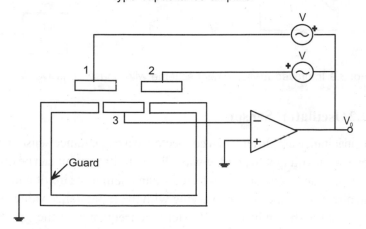

Fig. 3.8 Capacitance sensor with guard electrode arrangement

However, the measuring scheme shown in Fig.3.8 suffers from a limitation. The circuit is sensitive to variation of resistive components also. Thus the variation of leakage resistance may also cause change in output. A better circuit would be a scheme, where the output is sensitive to the variation of the capacitance component only, and not the resistive component. This can be achieved by incorporating a *Phase Sensitive Detector (PSD)* in the circuit. Such an arrangement is shown in Fig. 3.9. The schematic arrangement has been shown for a high input impedance capacitance amplifier. A band pass filter is used which allows the excitation frequency only to pass, suppressing all other noise frequencies which may cause problems in measurements. An ac amplifier amplifies the ac output voltage whose output is fed to the PSD. The PSD output is a demodulated ac voltage, which after the low pass filtering gives a dc output voltage. Assuming that the band pass filter and the ac amplifier do not contribute any phase shift for the supply frequency, it can be shown that, the average dc output voltage is sensitive to the variation of the capacitive component only. A sine wave generator has been shown in Fig.3.9 as the voltage source; but instead, a square wave voltage source is often used in practice without causing any additional problem.

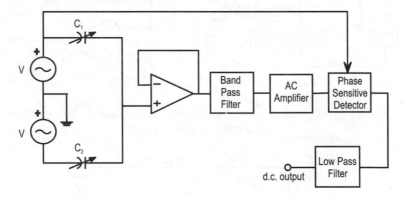

Fig. 3.9 Improved measurement scheme with phase sensitive detector.

3.3.2.2 Oscillator Sensor

Capacitance amplifier when connected to a capacitance sensor provides analog output voltage. In another way, direct digital output can be obtained, if the capacitance sensor is connected in an oscillator circuit and one can obtain direct frequency output varying with the capacitance value. Such an arrangement is shown in Fig. 3.10. Here the frequency of the square wave output is given by:

$$f = \frac{1}{2RC \ln\left(\dfrac{1+x}{1-x}\right)} \tag{14}$$

where x is a fraction of the capacitance variation in the push-pull arrangement. The advantage of the circuit is the simplicity, but the sensitivity is low, and the long-term frequency stability of the oscillator circuit is poor.

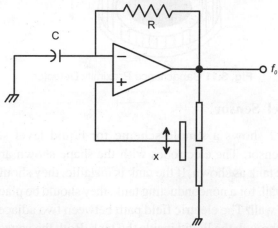

Fig. 3.10 Capacitance sensor oscillator

3.3.2.3 Application Areas for Capacitance Sensors

Capacitance sensors find wide applications for measurement of different process variables. Typical sensor constructions have been reported and elaborated in Baxter [4]. The constructions and principles of operation of few types of sensors have been elaborated below:

Proximity Detector

Capacitive proximity detectors are small in size, noncontact type and can detect presence of metallic or insulating objects in the range of approximately 0-5cm. For detection of insulating objects, the dielectric constant of the insulating object should be much larger than unity. Fig. 3.11 shows the construction of a proximity detector. Its measuring head consists of two electrodes, one circular (B) and the other an annular shaped one (A); separated by a small dielectrical spacing. When the target comes in the closed vicinity of the sensor head, the capacitance between the plates A and B would change due to the change in the electric field pattern caused by the presence of the target. This can be sensed by comparing with a fixed reference capacitor and using the measuring scheme, shown in Fig.3.9. Note that this is a simpler problem of detecting the presence of an object within a certain distance, compared to that of measuring displacement.

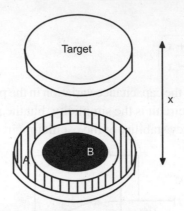

Fig. 3.11 Capacitance Proximity Detector.

Liquid Level Sensor

Fig. 3.12 shows a simple scheme for liquid level sensing using capacitance sensor. The electrodes with the shape shown are attached to the wall of the tank as shown. If the tank is metallic, they should be attached to the inside wall, for a nonconducting tank, they should be placed on outside surface of the wall. The electric field path between two adjacent electrodes is completed through the liquid inside the tank. Both the capacitors C_{DS} and C_{FS} vary with the liquid level, but the variations are not in the same way. With the measuring circuit used, the output voltage expression, is given by,

$$V_0 = -\frac{C_{DS}}{C_{FS}} V \qquad (15)$$

As the liquid level increases, the output voltage will increase from near zero to V, in a linear fashion.

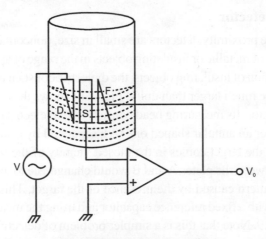

Fig. 3.12 Liquid level sensor

Linear Motion Measurement

Capacitance sensors can also be used for linear motion measurement with good linearity and sensitivity like an LVDT. Fig. 3.13 shows the schematic arrangement of a V-ramp type sensor. Here two fixed plates A and B, cut from a rectangular metallic sheet and are placed side by side as shown. The pick up, a rectangular moving plate C can slide over the fixed plates maintaining a constant air gap between the two surfaces. As the moving plate moves from left to right, the capacitance between A and C decreases, while that between B and C increases, forming a push-pull configuration. By connecting the capacitances those formed as in Fig.3.7(b), the output voltage can be obtained as a linear function of the displacement signal. The advantage of using a V-ramp configuration, over a ramp pattern as shown in Fig. 3.5(c) is that, even though the spacing between the fixed plate and the moving plate is not very uniform, the output characteristics of the sensor is unaffected.

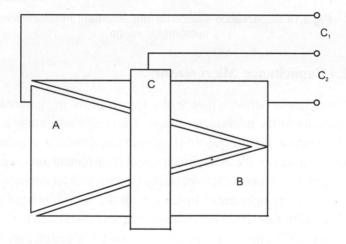

Fig. 3.13 V-ramp type linear motion sensor

Micro Displacement Measurement

The above scheme is suitable for measurement of displacement in the range of centimeters. But quiet often, the requirement of displacement measurement is in the μm range. Suitable capacitance sensors can be designed to meet this requirement. Kosel et al. [5] first proposed a comb like structure for both the static and the movable plate. The response of the output would be a periodic voltage with the period equal to the pitch of the comb. Pedrocchi et al. have proposed similar structure [6] for positioning the a micromirror for laser surgery. The schematic arrangement has been shown in Fig. 3.14. The sensor is suitable for measurement of distance with a resolution of 10 nm over 500μm range of travel.

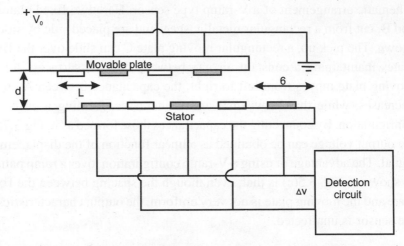

Fig.3.14 Capacitance sensor for displacement measurement in micrometer range

3.3.2.4 Capacitance Microsensor

Capacitance sensors find wide applications in the area of small displacement in the micrometer range. The major advantage is the small size of the sensor. But the major difficulty in measurement, is the mechanical fabrication to satisfy the exact dimensional requirement and connecting the sensor terminals with suitable measuring circuits with appropriate grounding and shielding arrangements. Instead, both the sensor and the processing circuit could be fabricated on a silicon base and integrated in the form of a microsensor. With the development of advanced IC technology, it has now become possible to have small size, light weight capacitance microsensors, giving a stable electrical output. The schematic diagram of a microsensor is shown in Fig.3.15. Analog Devices brought out the first commercial microsensor of this type in 1991. ADXL50 is a capacitance accelerometer, with a resolution of about 10^{-5}m/s^2 (10^{-6}g, where g is the acceleration due to gravity). The capacitance sensor consists of 42 sets of parallel plates of moving and fixed electrodes, each of width 2 μm and a gap between the plates being also 2 μm. The total sensor electrode capacitance (neglecting fringe fields) is about 0.1 pF. The measuring circuit employs a feedback principle to provide a stable d.c. output.

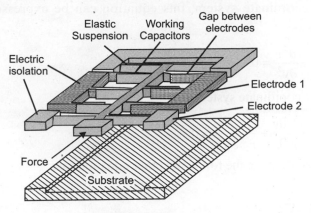

Fig. 3.15 Capacitance microsensor

3.4 Mathematical Modeling of Sensors

We have seen in last few sections that, the sensors can be of different sizes and shapes. Once the sensor is designed, it is also necessary to predict the behavior of the sensor. As a result, a proper method of mathematical modeling is an essential component of sensor design. Many of the sensors are irregular in shape, and small in size. As a result, simplified regular structure assumption is not valid in most of the case. Exhaustive analytical or numerical methods are needed for simulating the performance of the sensors. It is important to mention that, the governing mathematical equations for electrical sensors are guided by Maxwell's electromagnetic field equations. For the sake of simplicity, we shall concentrate only the electrostatic part of Maxwell's equations and restrict ourselves to the modeling of resistance and capacitance sensors only; the inductance transducers shall be left out.

In this section, the basic governing equations of electrostatic field: the Laplace equation and its associated equations for resistance and capacitance calculations are discussed. Different boundary conditions and the methods of solution are also dealt with. Lastly, an example as provided to illustrate the efficacy of the method of modeling and simulation of sensors.

3.4.1 Laplace Equation

The potential distribution inside an electrostatic field normally takes the form of Laplace equation [8].

$$\nabla^2 v = 0 \tag{16}$$

where v is the potential at any point and ∇ is the gradient operator. In

Cartesian coordinate system, this equation can be expressed in two dimensions as:

$$\frac{\partial^2 v}{\partial x^2} + \frac{\partial^2 v}{\partial y^2} = 0 \tag{17}$$

and in polar coordinate system as:

$$\frac{\partial^2 v}{\partial r^2} + \frac{1}{r}\frac{\partial v}{\partial r} + \frac{1}{r^2}\frac{\partial^2 v}{\partial \theta^2} = 0 \tag{18}$$

We also have the following other equations as,

Electric Field:

$$E = -\nabla v \tag{19}$$

Current density:

$$J = \sigma E \tag{20}$$

where σ is the conductivity of the medium.

Current:

$$I = \oiint J.dA \tag{21}$$

Charge density:

$$D = \varepsilon_0 \varepsilon_r E \tag{22}$$

Charge:

$$Q = \oiint D.dA \tag{23}$$

Now suppose, a capacitance sensor has been designed with a pair electrodes and a voltage V has been applied across them. Then to calculate the capacitance between the electrodes, finally we have to solve the Laplace equation to obtain the potential distribution in the medium then it is required to calculate the electric field on the surface of an electrode using (21), and the charge on the electrode using eqns. (23) and (24) and obtain the capacitance as:

$$C = \frac{Q}{V} \tag{24}$$

Similar is the case for resistance calculation, of the space between the electrodes is filled up by a conducting medium. Here one has to use equations (21) and (22), and obtain the resistance as:

$$R = \frac{V}{I} \tag{25}$$

3.4.2 Solution of Laplace Equation

It is seen from the above that if we can solve the Laplace equation, the capacitance (resistance) calculation is straightforward. The Laplace equation, being a partial differential equation, has to be solved with given boundary conditions appropriate for the complex geometry of sensors. The solution is unique, provided, the potential v, or its derivative in the direction of the normal to the boundary surface is defined everywhere. These two boundary conditions are known as *Dirichlet* and *Neuman* boundary conditions respectively. The general form of the solution of the equation in Cartesian coordinate system is given by:

$$v(x, y) = \sum_{k=1}^{\infty} (a e^{kx} + b e^{-kx})(c \sin ky + d \cos ky) \qquad (26)$$

The solution of Laplace equation is normally can be obtained using (i) analytical technique or (ii) numerical technique, conformal mapping in a powerful technique for analytical solution and would be explained in this section.

To illustrate the method of conformal mapping for modeling of a capacitance sensors, let us consider the example of a cylindrical capacitance sensor for active magnetic bearing spindles, as discussed in [9]. Here four segments of concentric cylindrical electrodes are placed around the metallic cylindrical rotor, to measure the rotor eccentricity, as shown in Fig. 3.16(a). In ideal condition, when both are concentric, let the radius of the sensor be b and the gap between the sensor and the rotor be δ. Now introducing the transformation:

$$t = \ln z = u + jv, \ z = x + jy = r e^{j\theta} \qquad (27)$$

the domain of interest can be transformed into parallel plates, as in Fig.3.16(b) with a gap between them as:

$g = \ln b - \ln(b - \delta) = -\ln(1 - \delta / b) \approx \delta / b$ (using first order approximation)

Then the capacitance between the rotor and a segment of the sensor, making an angle $\delta\theta$ at the center, would be (refer Fig.3.16 (b)):

$$\Delta C = \frac{\varepsilon \Delta\theta \, wb}{\delta} \qquad (28)$$

where ε is the permittivity of the medium and w is the length of the sensor. In this way, the capacitance of an annular segment of the cylindrical sensor can easily be obtained using conformal mapping.

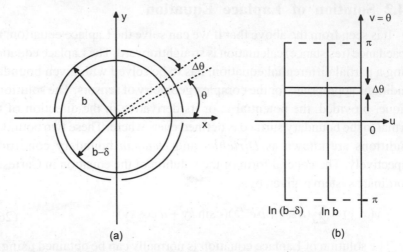

Fig. 3.16 (a) Cylindrical capacitance sensor with concentric rotor,
(b) sensor in transformed coordinate system

Now consider Fig. 3.17(a), where the rotor has a small eccentricity of α from the origin at an β. In that case, if we use the transformation (28), the straight line on the left in Fig.3.16(b) will be deformed, as shown in Fig. 3.17(b). To obtain a mathematical expression of the capacitance under the deformation, we can have:

$$t' = t_n + \Delta t = \ln[\alpha e^{j\theta} + (b-\delta)e^{j\theta}]$$

$$= \ln[(b-\delta)e^{j\theta}] + \ln\left\{1 + \frac{\alpha}{b-\delta}e^{j(\beta-\theta)}\right\} \tag{29}$$

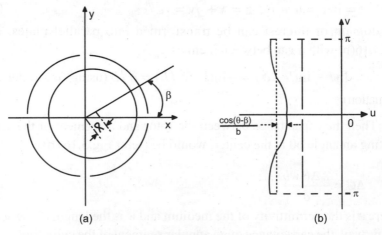

Fig. 3.17 (a) Cylindrical capacitance sensor with eccentric rotor,
(b) sensor in transformed coordinate system

The incremental term in (30), can be expressed as:

$$\Delta t = \Delta u + j\Delta v = \ln\left\{1+\frac{\alpha}{b-\delta}\,e^{j(\beta-\theta)}\right\}$$

$$= \ln\left[\left\{1+\frac{\alpha}{b-\delta}\cos(\beta-\theta)+j\frac{\alpha}{b-\delta}\sin(\beta-\theta)\right\}\right]$$

$$= \ln(R\cos\gamma + jR\sin\gamma)$$

from which, after further simplification, one can be write:

$$\Delta u = \frac{1}{2}\ln\left[\frac{\alpha^2 + (b-\delta)^2 + 2\alpha(b-\delta)\cos(\beta-\theta)}{(b-\delta)^2}\right]$$

$$\approx \frac{1}{2}\ln\left[1+\frac{2\alpha}{b-\delta}\cos(\beta-\theta)\right]$$

Therefore, $\quad \Delta u \approx \dfrac{\alpha}{b-\delta}\cos(\beta-\theta) \approx \dfrac{\alpha}{b}\cos(\beta-\theta)$ $\qquad$ (30)

(neglecting the coefficient of α^2, α- the eccentricity being small)

(as, $\ln(1+x) = x - \dfrac{x^2}{2} + \dfrac{x^4}{4} - \ldots \approx x$ (for $x\ll 1$)).

From eqn. (29), neglecting the fringing effect, the incremental capacitance ΔC would now change due the eccentricity, as:

$$\Delta C = \frac{\varepsilon w \Delta\theta}{\dfrac{\delta}{b}-\Delta u} = \frac{\varepsilon w b \Delta\theta}{\delta-\alpha\cos(\theta-\beta)} \qquad (31)$$

Now, the total capacitance between the rotor and the segment of the cylindrical sensor spreading over an angle 0 to $\pi/2$, can be obtained by integrating eqn. (32) over this range. Simplifying further, the capacitance can be expressed as:

$$C = \frac{\varepsilon b w}{\delta}\left[\frac{\pi}{2}+\frac{\alpha}{\delta}(\cos\beta + \sin\beta)\right] \qquad (32)$$

Thus, using conformal mapping, we can obtain an analytical expression of the capacitance between the segment of the cylindrical sensor and the rotor can be obtained under eccentricity, which could have been impossible to obtain otherwise.

3.5 References

1. J.P. Bentley: *Principles of Measurement Systems* (3/e), Longman, U.K., 1995.

2. E.O. Doeblin: *Measurement System Application and Design* (4/e), Mcgraw-Hill, Singapore, 1990.

3. T.R. Padmanabhan: *Industrial Instrumentation*, Springer-Verlag, London, 2000.

4. L.K.Baxter: *Capacitance Sensors Design and Applications*, IEEE Press, NJ, 1997.

5. P.B. Kosel, G.S. Munro and R. Vaughan: Capacitive transducer for accurate displacement control, *Trans. IEEE (I &M), vol. IM-30, pp.114-122,* 1981.

6. A. Pedrocchi, S.Hoen, G. Ferringo and A.Pedotti: Perspectives on MEMS in Bioengineering: a novel capacitive position microsensor, *Trans. IEEE (Biomedical Engg.), vol.47, pp.8-11,* 2000.

7. A.D. Khazan: *Transducers and Their Elements*, Prentice Hall, NJ, 1994.

8. D.J. Grifftihs: *Introduction to Electromagnetics* (2/e), Prentice Hall of India, New Delhi, 1991.

9. H. Ahn, S. Jeon and D. Han: Error analysis of the cylindrical capacitive sensing for active magnetic bearing spindles, *Trans. ASME, J. Dynamic Systems, Measurements and Control,* vol.122, pp.102-107, 2000.

Chapter 4

Optical Sensors

M.K. Ghosh

Abstract

Optical techniques have emerged as a state-of- the-art measurement technique with immense potential. Accurate and intricate measurement of parameters in industries, biomedical, defence and scientific research is heavily dependent on optical methods. Distinct classes of optical measurement techniques are (i) signal (optical) modulation environment, e.g. bulk optics, photo electronics, integrated optics based, (ii) imaging techniques like photography, thermal or IR imaging, holography and interferometry, and (iii) spectroscopy i.e. UV and IR. Brief principles and applications of the principal techniques are briefly dicussed in this chapter.

4.1 Introduction

Optical signal has many advantages, one of which is that it can be transmitted through air. Sending messages using fire and optical signalling has been used even in early days of human civilization. Optics, the science of light, has been studied and understood quite well for more than a century. However, lack of development of compatible technology made its use in sensing and instrumentation quite limited. Optical instruments like microscopes, telescopes, theodolites or even spectroscopes are in use for decades for scientific and engineering purposes. But it is in the later part of the twentieth century the revolutionary concept of Optoelectronics i.e., use of electronics in the optical context, that saw development and applications of physical devices and systems which work on the mediation of both photons and electrons in a classical and modern optical environment and opened up a new vista of horizon in the domain of sensing and instrumentation. It may be ventured to predict that the 21st century is likely to be the age of photonics i.e., the science and technology of generation, control and guidance of photons, just as the previous century had been the age of electronics i.e., the science and technology of generation, control and guidance of electrons. The vast potential of optoelectronics has even brought in different concepts of sensing and instrumentation as the technology enjoys the merits of both optics and electronics in sensing, signal processing, storage, data transmission, imaging, display and retrieval of information. The devices and systems are being increasingly used in industries, non- destructive testing, communication, biomedical and defence applications.

Study of modern optics is based on two concepts: (1) ray optics (Particle Physics) and (2) wave optics (Electromagnetic wave propagation). Both the concepts are used to explain properties and principles of optical devices and the optical phenomena. For example, the function of optical detectors is well understood through Particle Physics, but interferometry is better explained using Electromagnetic Theory (Maxwell's equations).[1,2,3,4,5,6]

In optical instrumentation it is often difficult to separate out the functional unit that serves as the sensor. The sensing, signal processing and retrieval of information are so much fused together that any such attempt to analyze it in the conventional way may be mere conceptual than a physical reality. An attempt is made to classify optical sensors and instrumentation in the following paragraphs.

4.1.1 Classification based on signal acquisition techniques

(1) Sensors based on optical signal modulation environment

    (a)   Without a waveguide

    (b)   With a waveguide

There can be a further sub-classification of these sensors depending on type of optical sources used e.g. non-coherent source, coherent source or laser, and a blackbody as the source.

(2) Sensors acquiring information that require special processing

    (a)   Interferometry (coded phase)

    (b)   2-D optical imaging (intensity distribution)

    (c)   3-D optical imaging including stereo imaging and night vision imaging

    (d)   Holography

There is also IR imaging or thermal imaging that does not require any separate optical source and depends on blackbody radiation of objects.

(3) Spectroscopy

    (a)   UV Spectroscopy

    (b)   IR Spectroscopy

For identification and analysis of elements optical spectroscopy is used.

4.1.2 Classification based on coherency of optical sources

(1) Non-coherent optical sensing

    (a)   Modulation of intensity, wavelength, scattering of optical signal

    (b)   Optical pyrometry

    (c)   Optical imaging and IR imaging

(2) Coherent or laser sensing

    (a)   Interferometry with modulation of phase, polarization, evanescent wave

    (b)   Doppler frequency shift or frequency modulation

    (c)   Optical imaging and IR imaging

    (d)   Holography

There can be further sub-classification based on types of waveguides used and the dimension (point, 2-D or 3-D) of the image patterns.

4.2 Sensors Based on Modulation of Optical Signal

Optical sensors are quite popular for their easy adoption in non-invasive environment, high degree of sensitivity to various physical and chemical changes, innovation in modulation techniques, immunity to EMI, and miniaturization using optoelectronics devices. Light is an electromagnetic phenomenon. The properties of optical signal, which can be modulated by the measurands, are:

(1) Intensity

(2) Phase

(3) Wavelength or colour

(4) Frequency

(5) Polarization

(6) Scattering, and

(7) Evanescent Electric Field

The schematic diagram of an optical sensor is shown in Fig 4.1. The optical source can be a white light source, a laser or a blackbody. The transmission medium or waveguide may be air or an optical fibre. The modulator is a device that modulates one of the properties of the optical signal as per variation of the measurand. The detector is a photoelectric detector or an optical screen with visual display. Output of the detector is a signal, usually electrical, or information that can be interpreted in terms of the measurand.

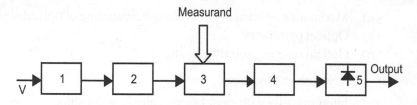

1- Optical source,   2- Optical waveguide,   3- Modulator
4- Optical waveguide,   5- Detector, V- Voltage supply

Fig. 4.1 Schematic diagram of an optical sensor

4.2.1 Sensing with air as the transmission medium (nowaveguide).

Optical sensors without any waveguide have limited use as because ambient interference like fluctuation of ambient light, temperature, humidity, refractive index, appearance of dust particles may affect the measurement. Moreover, the flexibility of the sensor is constrained by the fact that vibration and factors leading to loss of alignment between components result in erroneous results. Some of the popular sensors are described below.

(a) Optical speed sensor

This is one of the earliest optical sensors used. Light from the source is reflected from the mirror or reflector placed on the shaft on to a detector (Fig 4.2). Number of pulses per unit time is directly proportional to the speed of the shaft. For digital display a counter is used at the output of the photodetector. Alternatively, averaging the output pulses will give analogue voltage proportional to speed.

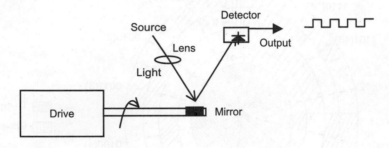

Fig. 4.2 Optical speed sensor

An accurate speed sensor having the same configuration as in Fig 4.2 uses a laser source and the doppler phase shift of the beam reflected from the rotating mirror is interpreted as a measure of the shaft speed.

(b) Optical shaft encoder

This is one of the few inherently digital sensors developed early and is still being in use quite extensively (Fig 4.3a). The perforated disc modulates optical signal. The range of the sensor for sensing shaft angle depends on the number of bits or tracks on the encoder. For a four track encoder the numbers of sources and detectors are also four each and its range of measurement is 2^4-1 or 15° (Fig. 4.3b). Encoders with Grey code, BCD codes increase versatility and accuracy of measurement.

Principle of optical encoder has been innovatively used to measure slope or angle of inclination [7].

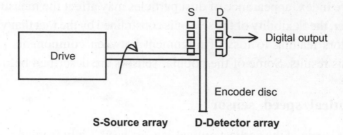

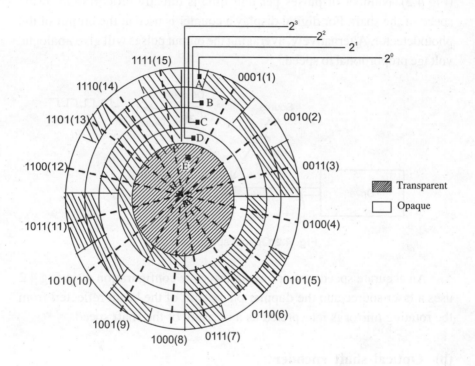

Fig. 4.3a Optical shaft encoder

Fig. 4.3b Encoder

(c) Laser based sensors

Laser has certain unique features like coherence, monochromacity, and low divergence in addition to other advantages of optical radiation from non-coherent sources.[3,4]. These features led to the development of many instruments for measurement of physical parameters, chemical analysis, range finding, etc. In addition laser plus optical waveguides (optical fibres) and laser-based holography make a series of sophisticated sensors.

(1) Laser for alignment of structures

Laser-based theodolites are used for surveying, construction of high rise structures, installation and alignment of heavy machines, and sawmill blades as because lasers have very low divergence (~ milirads) (Fig 4.4).

Fig. 4.4 Alignment of structures using laser

(2) Distance measurement

For accurate distance measurement (better than 1 mm in 1 km) in large structures, dams, bridges, geodesic survey beam modulation telemetry scheme (Fig 4.5) is being used. The instrument can be used in daylight. The phase shift of the reflected beam from the reference beam is given by

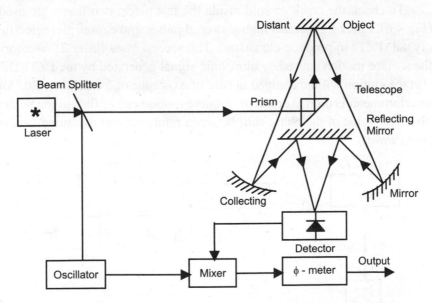

Fig. 4.5 Distance measurement using beam modulation telemetry

$$\phi = 2\pi(n_g L)/\lambda \tag{1}$$

where L = target distance, n_g = group R.I. of atmosphere. For example, a scheme uses He-Ne laser for which $\lambda = 632.8$ nm, $n_g = 1.00028$ for dry air containing 0.03% CO_2 at 150^0C and 760 torr. The average of the measurements is taken and corrections for temperature and pressure applied to achieve the required degree of accuracy. The φ-meter used for high precision measurement is a Michelson interferometer.

Pulse echo technique is used for measurement of large distances by measuring the transit time (time of flight) for round trip of a very short pulse reflected from a distant object.

Optical radar (OPDAR) or LIDAR (light detection and ranging) is used for atmospheric studies, and for motionless, slow moving or even fast moving or manoeovouring types of targets. Lunar distance (384,400 km) was measured using retroreflectors left on the surface of the moon (by Apollo 11, 14, and 15) with an accuracy of ±15 cms. In Soviet version a Ruby laser was used, for which the collecting mirror of the telescope was 2.6 m. The instrument is used in defence for measurement of distances ~ 10 km ±5 m.

(3) Laser flaw detection system

To check the crack or void inside the test piece, two lasers are used (Fig 4.6). Laser 1 generates high-powered pulses and causes piezoelectric crystal (PZT) to produce ultrasound. The second laser (laser 2) monitors the surface motion caused by ultrasonic signal generated by the PZT. The surface motion will be damped in case of existence of a crack or void. An interferometer is used to detect the relative motion rate or flaw given by the observed value of the phase shift between reference and measured beams (not shown).

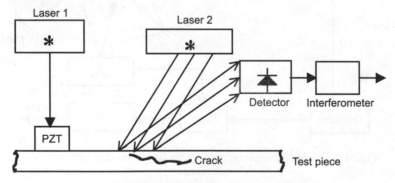

Fig. 4.6 NDT or flaw detection using laser

(4) Laser dimensional gauge

Laser dimensional gauge is a non-contact system for measurement of diameter of rods, wires and tubes in steel plants. It can be used for real time dimension monitoring as well as sample inspection of items. In the schematic diagram of the laser dimensional gauge (Fig 4.7), a low power 2-8 mw laser with well collimated and narrow (~ 1 mm dia) beam is used. The optical scanner produces a sweep of laser beam whose speed is proportional to the rotating speed of the prism. Fast scanning makes the vibration or any movement of the target negligible. A collimating lens produces parallel beams, which sweep the workspace at a linear rate proportional to prism's speed. When an object (whose diameter is to be measured) is placed in the path of the scanning beam, it obstructs the beam. With parallel scanning rays, the width of the shadow is exactly equal to the diameter of the rod. As a result of the shadow cast by the target, the detector output voltage exhibits a 'notch' whose width in time is proportional to the target width in space. If θ is the angle subtended at the prism by the collimating lens,

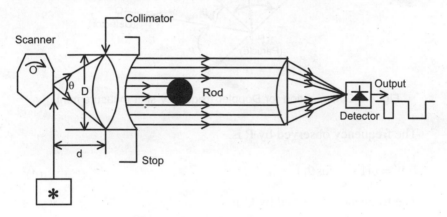

Fig. 4.7 Laser dimensional gauge

$\tan (\theta/2) = D/2d$ (2)

To produce this sweep angle θ, the rotation of the reflector surface should be $\theta/2$. For a speed of the prism = n rpm,

$t_s = (\theta/2) / (n.360^0)/60 = \theta / 12n$ sec   (3)

Example: $\theta = 8°$, D = 4cms, d = 30 cms, n = 1000 rpm for a hexagonal prism number of scans per sec is $(1000\times6)/60 = 100$. If the moving rod speed is 100 m/s, in time $t_s = 2/3$ ms, the rod will move by $100\times(2/3)\times10^{-3}$ = 0.067 m = 6.7 cms.

(5) Laser Doppler Velocimeter

In a Laser Doppler velocimeter (LDV), Doppler shifted scattered light is heterodyned (mixed) either

(1) with unshifted light obtained directly from original source, or

(2) with further scattered light having a different shift scattered through different angle and different point.

In Fig 4.8, light at a frequency υ from a laser source S falls on a particle P moving with a velocity v in the direction PR and the scattered light is received at Q, the detector or the receiver. Let both S and Q be stationary and PB be the angle bisector of <SPQ.

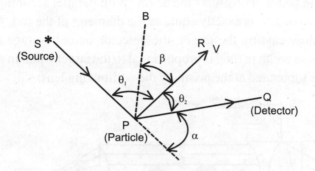

Fig. 4.8 Configuration for Doppler frequency shift measurement

The frequency observed by P is

$$\upsilon' = \upsilon(1 + \frac{v}{c} \cos\theta_1)$$ (4)

The frequency observed by Q is

$$\upsilon'' = \frac{\upsilon'}{1 - \frac{v}{c} \cos\theta_2} = \frac{\upsilon (1 + \frac{v}{c} \cos\theta_1)}{(1 - \frac{v}{c} \cos\theta_1)}$$ (5)

For normal velocities v<<c, then

$$\frac{\upsilon''}{\upsilon} \approx (1 + \frac{v}{c} \cos\theta_1)(1 + \frac{v}{c} \cos\theta_2)$$

$$\approx 1 + \frac{v}{c}(\cos\theta_1 + \cos\theta_2)$$ (6)

Hence, $\Delta\upsilon = \upsilon'' - \upsilon = \upsilon\left(\frac{\upsilon''}{\upsilon} - 1\right)$

$$= v\frac{v}{c}(\cos\theta_1 + \cos\theta_2)$$ (7)

Now $\alpha = \pi - (\theta_1 - \theta_2)$

and $\beta = (\theta_1 - \theta_2) / 2$

Equation (7) can be expressed as

$$\Delta\upsilon = \upsilon.\frac{\upsilon}{c}.2\cos\frac{\theta_1 + \theta_2}{2}\cos\frac{\theta_1 - \theta_2}{2}$$

$$= 2\upsilon\frac{1}{\lambda}\cos\beta\ \sin\frac{\alpha}{2} \qquad (8)$$

In optical beating or heterodyning, Doppler shifted scattered light and an unshifted reference light are simultaneously present on the detector Q. The detector output is proportional to intensity of the light, i.e., square of the optical electric field. If the two optical electric fields are given by

$$E' = E_1 \cos(2\pi\ \upsilon_1 t + \phi_1)$$

and $\quad E'' = E_2 \cos(2\pi\ \upsilon_2 t + \phi_2),$

the detector output I(t) is given by

$$I(t)\alpha\ (E' + E'')^2$$

$$= K[E_1 \cos(2\pi\upsilon_1 t + \phi_1) + E_2 \cos(2\pi\upsilon_2 t + \phi_2)]^2 \qquad (9)$$

Neglecting optical frequencies or more,

$$I(t) = K[\frac{1}{2}(E_1^2 + E_2^2) + E_1 E_2 \cos(2\pi(\upsilon_1 - \upsilon_2)t + (\phi_1 = \phi_2))] \qquad (10)$$

Thus, the detector output contains a DC component and an AC component. The DC component is used to determine when material is in the measurement region. The AC component is proportional to $E_1 E_2$ or $\sqrt{I_1 I_2}$ (where I_1 and I_2 are intensities of the individual beams) and it contains a beat frequency $(\upsilon_1 - \upsilon_2)$ which is directly proportional to the speed of the particle P.

Signal to noise ratio is an important factor, which determines the ability to observe the signal. Beat frequency term ($\Delta\upsilon$) is proportional to $\sqrt{I_1 I_2}$. So, in principle, if scattered light (I_1) is weak but the reference beam (I_2) is large, we can still get a strong signal (known as reference beam technique). The only constraint is the signal-to-noise ratio. With increase of intensity, photon noise dominates over the beat signal.

In an LDV, a He-Ne laser ($\upsilon = 4.7 \times 10^{14}$Hz) produces a Doppler frequency shift of $\Delta \upsilon = 32$ MHz for a particle velocity of $v = 10$ m/s. The optical beating using small detector area to maintain a coherence condition becomes a time varying interference phenomenon. The detector experiences alternating light and dark fringes and thus gives an alternating output at beat frequency.

In the differential Doppler technique two equal intensity beams are split from a single beam using a 50/50 beam splitter (Fig 4.9) and crossed to form a measurement region. When a light - scattering surface passes through the measurement region, light from each beam is scattered. The mixing, or heterodyning, of the scattered light from the two beams provides the Doppler signal that contains the surface speed information. Since the light scattered from the beams reaches the detector simultaneously, a beat is obtained of frequency equal to difference in Doppler shifts corresponding to the two angles of scattering. The beat frequency is independent of receiving direction and is given by

$$\upsilon_D = \frac{2v}{\lambda} \sin (\alpha / 2) \qquad\qquad (11)$$

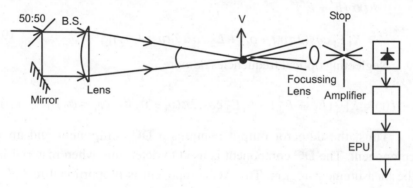

Fig. 4.9 Differential Doppler shift technique

Reference beam technique is very useful with liquid motion rate measurement. But differential Doppler technique is used when scatters are few i. e., gas flow rate measurement. Light in the latter case is collected over a wide aperture.

LDV systems have been used in cold rolling mills in both the steel and aluminium industries at the entrance and exit of mill stands to calculate differential speed of Automatic Gauge Control (AGC) utilizing massflow. They have been also tried successfully in hotstrip mills.

4.2.2 Optical Sensors using Waveguides: Optical Fibre Sensors (OFS)

The optical fibre sensors started in earnest in 1977 even though some isolated demonstrations preceded this date. Optical fibre[8,9] plays two major roles in OFS: (1) sensing or probe, and (2) transmission of optical signal. It has certain properties that make it attractive, and sometimes the most suitable, as a sensor in industrial, bio medical and scientific environments. The most important features are: (1) intrinsic safety, (2) immunity to EMI and RFI, (3) light weight and small size (1 km of uncabled fused silica fibre of 200 micron dia weighs 70g and occupies 30 cm^3), (4) low attenuation (as low as 0.1 dB/km attenuation has been achieved), (5) chemically inert, and (6) flexibility in configuration. OFS have the advantages of signal processing - optically and electronically, so that, very high resolution of measurement of parameters (for example, 10^{-8} °C) is possible.

The properties of the optical signal which are modulated by the measurand are the same as in the sensors without any waveguide (section 4.2). In OFS optical fibre plays an active role in the modulation process. There are many ways in which guided light within a waveguide may be modulated by an environmental parameter [9,10,11,12]. Based on the modulation technique the OFS can be grouped into two basic classes:

(1) Extrinsic type

(2) Intrinsic type

In the extrinsic type of OFS, the fibre serves as a light transmitter i.e. it sends the light from the source and collects the light after it is modulated by the measurand on to the detector. The modulation of optical signal takes place externally from the fibre, usually by way of an attenuation process. In the intrinsic type of sensors, the optical fibre itself takes part in sensing and modulation. The properties of light propagating along the optical fibre are modulated by interaction with the measurand, but without the light actually leaving the fibre. The modulation process may occur at a single location in the fibre, or at various locations along its length. The advantages of this class of sensors stem from the fact that no external optical interference exists at the modulator head. But it faces the major problem from 'lead sensitivity' i.e., if the optical fibre within the modulator head is capable of imposing modulation of light passing through it, so too are the parts of the fibre acting as 'feed' and 'return' paths.

(a) OFS based on intensity modulation

In any OFS, it is the intensity of one specific property of light or other that is sensed. However, in intensity modulated sensors it is the total intensity irrespective of wavelength or coherence or polarization nature of the light that is modulated and detected. The principal intensity modulation mechanisms (Fig.4.10) are

(1) moving reflectors

(2) moving mask

(3) varying refractive index, and

(4) microbending (of optical fibre)

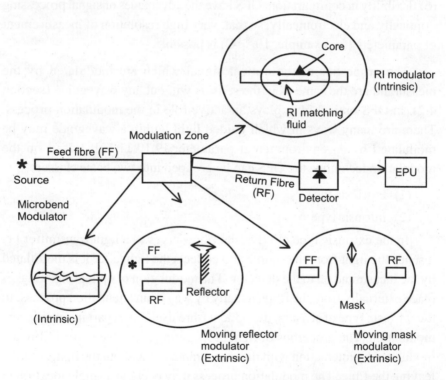

Fig. 4.10 Principal classes of intensity modulated sensors

In this type of OFS, there are, in effect, two stages in the transduction process. The measurand is caused to interact with a light intensity modulator and this transfers the information by varying the intensity of guided light. The light source is usually an LED or a laser, and both multimode and singlemode fibres are used.

(1) Moving reflector type

Proximity (very small displacement) sensors, NDT for surface finish, vibration sensors have been made based on this technique. The major requirements in such sensors are spatial orthogonality of the reflector on the fibre axes and 100% reflectivity of the reflector. Small tilting in the orientation of the reflector will have an effect on the sensitivity.

(2) Moving mask type

The moving mask technique is used to develop many sensors that are based on small motion or displacement measurement. Different configurations are:

(a) relative displacement between feed fibre (fixed) and return fibre (free) without any masking (Fig.4.11a, 4.11b),

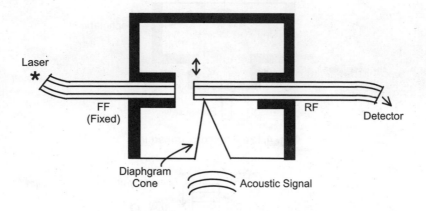

Fig. 4.11a Hydrophone

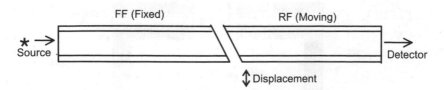

Fig. 4.11b Displacement sensor

(b) relative displacement between feed fibre (fixed) and return fibre (free) with periodic masking consisting of alternate transparent and opaque regions of equal width (Fig. 4.12), and

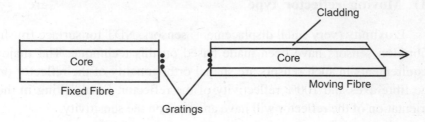

Fig. 4.12 Displacement sensor

(c) a moving (externally placed) mask between fixed feed and return fibres (Fig 4.13 a,b).

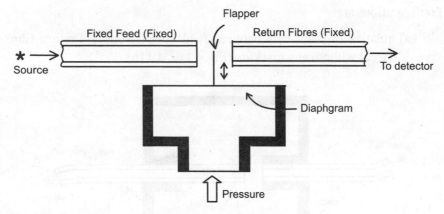

Fig. 4.13a Pressure transducer

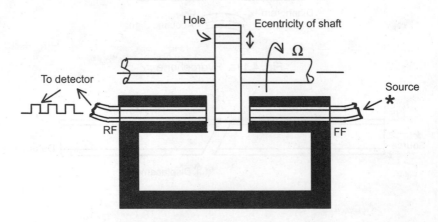

Fig. 4.13b Shaft Eccentricity meter

In Fig 4.13b the modulator is the disc with holes fitted to the shaft whose eccentricity is to be detected [13].

In all the above configurations identical optical fibres are used. Multimode fibres increase the range (because of larger diameter available) and plastic fibres can be used to this advantage. Step index fibres are usually used.

(3) Refractive index modulation type

Refractive index modulated sensors depend on modulation of the refractive index of the cladding of the fibre. Two such configurations have been exploited to design temperature sensors. In one version (Fig.4.10) the cladding from a section is removed and encapsulated with a fluid whose refractive index varies with temperature. The intensity of light variation is then a function of cladding temperature change. In another configuration, the step index fibre used has core and cladding made of glasses having different temperature coefficients. This forms the basis of an alarm system, since, at the temperature at which the core and cladding refractive indices become equal, the fibre ceases to act as a guide.

(4) Microbending type

In the microbend loss modulator (Fig.4.10), microbending causes core modes to couple to cladding and radiation modes, which decreases the core power. Clamping the fibres between a pair of jaws (of deformers) induces the microbending, which squeezes the fibre when displaced. The jaws are usually corrugated surfaces with small radii to increase the microbending effect. In a multimode fibre, the period of spatial configuration (usually of the order of an mm) is determined by whether the lower order, or higher order modes are being coupled out of the core. Sensors and transmitters based on microbending have been designed to sense pressure, flow, level, temperature, force, etc. The basic concept in the design is the innovation in generation of modulation in the deformer using the characteristics of the measurand.

(b) OFS based on phase modulation

The phase modulated OFS have a few special features:

(1) Physical configurations: a coherent light source (laser), a singlemode fibre as the waveguide, directional couplers or beam splitters for splitting or recombining the light beams, and an interferometer for optical signal processing (Fig.4.14). Basically they are intrinsic sensors.

(2) Modulation of phase is due to changes in the optical path or optical
     properties of the waveguide. Optical phase change can be measured
     as a function of time or as a function of frequency or even as a
     visual display (fringe patterns)

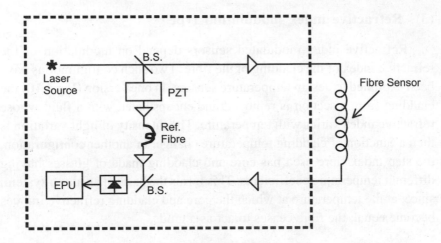

Fig 4.14 Phase modulated OFS using a Mach-Zehnder interferometer

Very high-resolution measurements are feasible using phase modulation
technique. The total phase of light path along the optical fibre is given by

$$\phi = 2\pi(L.n)/\lambda \tag{12}$$

Phase modulation is, therefore, dependent on

- Total physical length of the waveguide (L)
- The refractive index (n) and the index profile of the waveguide
- The geometrical transverse dimensions of the waveguide, and
- Sagnac effect

The total physical length of an optical fibre may be modulated by

- Application of longitudinal strain (magnetic field sensor, force sensor
- Thermal expansion (temperature sensor)
- Application of a hydrostatic pressure causing expansion via
  Poisson's ratio

The refractive index varies with

- Temperature
- Pressure and longitudinal strain via the photoelectric effect and
  guide dimensions with the changes listed in the preceding paragraph

Sagnac effect is possible even in air. However, optical fibre based gyroscope is a novel sensor based on Sagnac effect (Fig.4.15) that can measure rotation (of the order of 1° / 24 hours) induced phase difference $\Delta\phi$ is given by

$$\Delta\phi = (4\pi LR /\lambda C)\ \Omega \qquad\qquad (13)$$

where λ=optical wavelength, L=total length of the fibre coil of radius R and number of turns N $=2\pi RN$, and C=velocity of light.

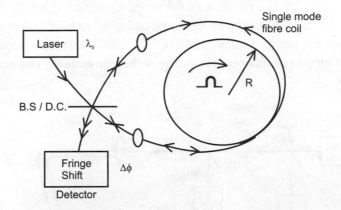

Fig. 4.15 Optical fibre gyroscope using on Sagnac interferometer

(c) OFS based on wavelength or colour modulation

The areas in which colour or wavelength modulation may be exploited are:

- Chemical analysis
- Analysis of phosphorescence and luminescence
- Analysis of absorption of colour spectra
- Analysis of blackbody radiation, and
- Optical filters where the transmission characteristics of the filter are made to be a function of external physical parameter.

These classes of OFS have a few special features common to them:

(1)   They work in either reflection mode or in transmission mode. The source of light is typically in the visible or UV range and the returned illumination is in the IR to visible regions. The critical components are the source and spectrometer (or interference filters).

(2)   The sensors based on colour modulation technique usually employ

ratio measurement technique or two-colour measurement technique. One of these wavelengths is affected by the measurand and the other is independent of the measurand. It takes care of environmental effect on the sensor. The signal is low and so a Lock-in-Amplifier is used for better signal processing.

In the temperature-sensing scheme (Fig.4.16) the fibre tip is coated with a rare-earth phosphor which is excited by ultraviolet radiation travelling along the fibre. Phosphor radiates at a number of visible wavelengths, the intensity of which is a function of the probe-tip temperature. By filtering two wavelengths (red and green, say), and taking the ratio of their intensities, probe temperature in the range of 500°C / 2500°C with an accuracy of ±10°C is measured independent of the source or ambient characteristics variation.

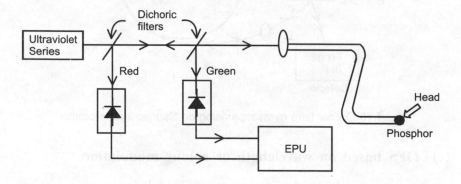

Fig. 4.16 Rare earth temperature sensor

Liquid crystal like Merck Thermochrome (Thermomagic Farbe 30/35), with its peak wavelength varying from 664 nm at 28.5°C to 461 nm at 52.5°C for a film thickness of 60 microns on fibre- tip, has been used to measure temperature of the human body. The Thermochrome is soluble in sodium silicate and when deposited on a fibre-tip it remains as a thin film.

pH sensors (in the range of pH values 3 to 13) have been developed with chemicals (Bromophenol blue or red phenol) to transmit light at 550 nm to 600 nm and at 488 nm. Similar type of sensors for methane, NO_x,CO_2 have been reported, the operation and modulation characteristics of which depend on use of compatible chemicals.

Temperature sensors using semiconductor material like Ga As (30°C to 70°C, high absorption at $\lambda_0 = 920$ nm) and Ga P (30°C to 80°C, attenuation at λ_0 = 550 nm) have been developed. Other types are Ge (1500nm), Si (1100nm) for the range of 25°C to 150°C.

A generalized and simple scheme for different types of colour modulated OFS for measurement of temperature and pH is shown in Fig. 4.17, in which the basic scheme remains unchanged and only the sensor heads are to be replaced by elements sensitive to the measurand. The dynamic characteristics of the sensors depend on the physical thickness of the element film.

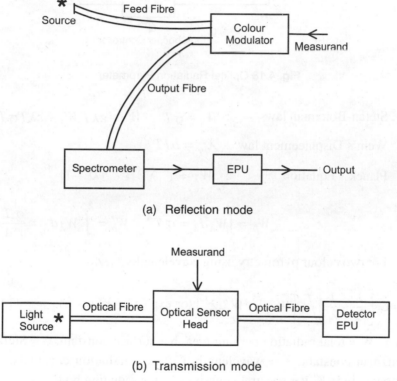

(a) Reflection mode

(b) Transmission mode

Fig. 4.17 Basic scheme of a colour modulated OFS

An optical fibre based radiation pyrometer is useful for measurement of blackbody radiation and hence measurement of temperature as high as 600°C to 3000°C. The main feature of this temperature measurement scheme (Fig. 4.18) is remote sensing where the optical fibre plays the role of an optical signal collector, the source of light being the blackbody itself. It is a self calibrated system, with its output in the ratiometric measurement mode, which can be calibrated using Stefan-Boltzman law, Planck's law and Wein's displacement law as given by the following expressions.

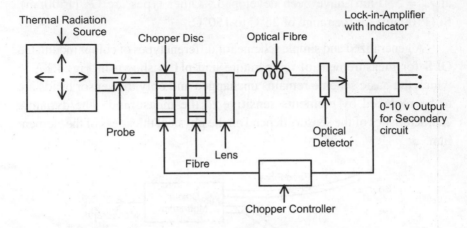

Fig. 4.18 Optical Radiation Pyrometer

Stefan-Boltzman laws   : $W_b = \sigma T^4$, $\; W = \varepsilon \lambda T . W_b = \varepsilon \lambda T \sigma T^4$

Wein's Displacement law : $\lambda_m = b / T$

Planck's Equation   : $W_\lambda = C_1 \lambda^{-5} [\varepsilon c_2 / \lambda T - 1]^{-1}$

$$W_b = \int_0^\infty W_\lambda d_\lambda = \sigma T^4, \quad W_b' = \int_0^{\lambda_m} W_\lambda d_\lambda = \frac{\sigma T^4}{4},$$

For two colour pyrometry using wavelengths λ_1, λ_2

$$T = C_2 \left[\frac{1}{\lambda_2} - \frac{1}{\lambda_1} \right] \Big/ l_n \left[W \lambda_1 \lambda_1^5 \big/ W \lambda_2 \lambda_2^5 \right] \text{ for exp } (-C_2 / \lambda T) \ll 1 \qquad (14)$$

where W = total radiation per unit area, b = surface constant, σ = Stefan - Boltzman constant, λ = wavelength, C_1, C_2 = radiation constants, T = temperature in K, ε = spectral emissivity of the radiating body.

(d)   OFS based on frequency modulation

Laser Doppler Velocimeter using optical fibre (Fig.4.19) is used for blood flow measurement. The Bragg cell is used for frequency shifting. It serves as a reference beam. The moving red cells in the blood cause a Doppler shift Δf on the incident light that is subsequently scattered back and captured by the fibre. The Doppler shifted light is made to combine with the reference light beam generating an intermediate frequency at f_1 with a side band at $f_1+\Delta f$. The magnitude and sign of Δf is a measure of the blood flow velocity and direction. Blood flow velocimeter has been developed for ranges 0.04 to 10 m/s ±5%. The relevant formulae are.

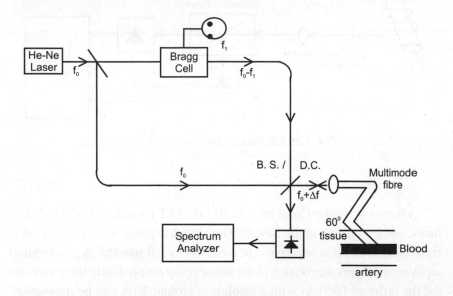

Fig. 4.19 Laser Doppler Velocimeter for blood flow

$$f_1 = f_0 /(1 - \frac{v}{c}) \approx f_0(1 + \frac{v}{c}) \tag{15}$$

$$\Delta f = f_0 - f_1 \tag{16}$$

(e) OFS based on polarization modulation

Polarization plays an important part in systems using singlemode fibres. The physical phenomena that influence the state of polarization are: Faraday rotation, electro-optic effect, photoelastic effect, electro-gyration and even by mechanical twisting or wrapping the fibre around a mendrel.

In a current sensor based on polarization modulation using Faraday effect (Fig.4.20), linearly polarized light is launched into a low birefringence optical fibre. The orientation of the light is changed in the presence of magnetic field pointing along the direction in which the light propagates (Faraday effect). The magnitude of the polarization rotation ϕ is given by

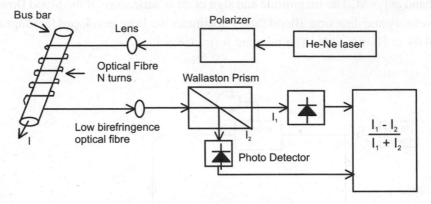

Fig. 4.20 Faraday rotation current sensor

$$\phi = v \int_l Hdl = vNI \qquad (3.17)$$

where v = Verdet Constant 3.3×10^{-4} deg/AT for silica, N = number of turns, and I = current in the busbar. The shift ϕ is positive if polarization is right handed and it is negative if polarization is left handed. A polarization analyzer measures the rotation of the linear polarization. Fairly large currents (of the order of 1000 A) with a resolution around 30 A can be measured.

(f) Voltage sensor based on Pockel's effect

Voltage sensing is possible by using a piezoelectric material (basic modulator) around which the optical fibre is wound. When a voltage is applied to the crystal, a phase difference in the light propagating through the fibre, with respect to the reference beam, takes place due to (1) fibre elongation, and (2) photo electric effect. However, the sensor exhibits temperature sensitivity which can be eliminated by ratiometric measurement. For high bandwidth operation, electrooptic effect is used as in integrated optics, microswitches, modulators. In the voltage sensor based on electro-optic effect, the refractive index of the material changes depending on polarization directions and light propagation directions through the crystal. The crystal structure and orientation give rise to (1) longitudinal electro-optic effect (also known as Pockel's effect), or (2) transverse electro-optic effect. The former induces linear birefringence when the electric field is applied along the propagation axis and the latter arises when the electric field is applied in direction transverse to propagation direction. One such crystal is lithium niobate.

(g) OFS based on scattering modulation

Scattering loss is caused by the interaction between light and particles. It takes place in various forms. Some of the scattering losses are temperature sensitive and they contribute to very sensitive and elegant distributed temperature sensors (DTS), which find applications in office buildings, naval ships, aircraft for fire protection, and in transformer condition monitoring for 'hot spot' detection. A simple configuration of a DTS based scattering in optical fibres is shown in Fig 4.21. The heart of the system is a PC controlled optical time domain reflectometer (OTDR). Two forms of scattering loss are exploited: Rayleigh and Raman. Rayleigh backscatter coefficient depends on composition and structure of the fibre core. To make the scattering (Rayleigh) loss temperature sensitive, the core is doped with Nd 3+ at regular intervals (~ 10 cms or so is achievable) along the fibre length. The special fibre is laid along the structure or building where fire hazard (or temperature rise) is intended to be monitored and controlled. The intensity of Rayleigh scattering loss is 100 to 1000 times more than Raman scattering loss. Different fibres used are liquid core (upto 160°C), solid core (with UV light curable cladding 20 to 150°C) doped with Nd 3+. Raman scattering component is highly temperature sensitive. The DTS based on Raman scattering uses low loss general purpose optical fibre.

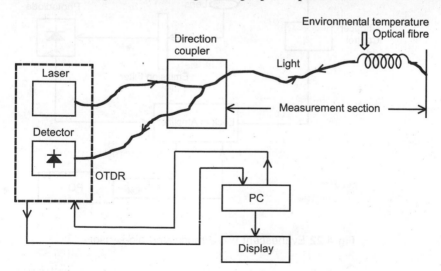

Fig. 4.21 Schematic diagram of a DTS using scattering loss modula

(h) OFS based on evanescent electric field modulation

The incident and reflected lights at the core cladding interface of an optical fibre result in a standing wave which has a finite electric field amplitude and can not go to zero instantaneously. It decays exponentially into the low index medium (cladding). The low index phase is called an evanescent wave. Evanescent fibre optic devices are used in two distinct areas: (1) Chemical and biosensors, and (2) optical couplers and sensors. In chemical industries these sensors find application for waste-water monitoring and treatment, for hazardous-waste monitoring, for sensing of hydrogen, methane and other industrial gases, in medicine and pharmacy,etc. Any chemical, for which a specific binding agent can be found out, can be measured by these techniques. Development of bio-sensors or immunoassay system is the result of fluorescence property, in presence of evanescent field, of antigen or antibody bound to the optical fibre surface (Fig. 4.22). The antibody coated probe is immersed in an aqueous (phosphate buffered saline) solution (nsoln < ncladding), increasing the waveguide mode capacity.

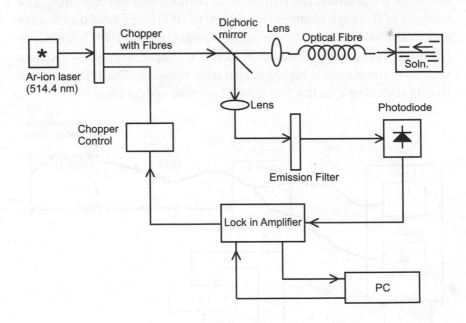

Fig 4.22 Evanescent wave modulated bio-sensor

Modulation of the laser light permits reduction of the ambient light. Using low fluorescing components and a dichroic mirror, which reflects only the wavelength of the fluorescent light, minimizes the background signal from the optical components. Other types of evanescent wave modulated devices are directional couplers.

4.3 Sensors Requiring Special Processing

In modern measurement and instrumentation the measured data are often available in forms which require special interpretation. They are very sensitive and may contain plenty of information and measured data. Truly speaking the major parts of these instruments belong to sophisticated read out devices. However, unlike general-purpose analog and digital read out devices, they are very application specific. In a sense, conventional 2-D photography belongs to this group. Different forms of such devices that are based on optical signals have been listed in Chapter 1. A brief description of techniques of their acquisition and retrieval of information are given in the sections to follow.

4.3.1 Interferometry

Optical Interferometry can be construed as a form of signal processing optically. It is a phenomenon wherein a countable number of beams (two or more) interact with each other with a defined phase (caused by modulation of the signal by a measurand), where the field vectors have a finite non-zero projection on one another. For interference light signals must be coherent and of same frequency. However, superposition can easily be applied to get the effect under broadband condition provided coherence length and bandwidth are identical. The interference pattern observed visually (on a screen) or detected by photodetector is the steady state behaviour with respect to the frequency of light, in which degree of coherence controls the modulation depth.

Consider beams from two coherent sources P_1 and P_2 incident at some points on the screen OQ (Fig. 4.23) at the same instant of time. These two beams (electro magnetic fields) can be represented as

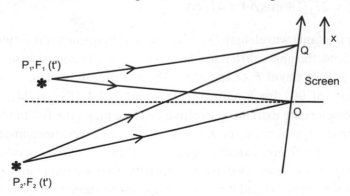

Fig. 4.23 Two source interference

$$E_1 = E_{01} \exp \ [i\,(\omega t + 2\beta_1 z_1)]$$

$$= E_{01} \cos(\omega t + \phi_1)$$

and $\quad E_2 = E_{02} \exp \ [i\,(\omega t + 2\beta_2 z_2)]$

$$= E_{02} \cos(\omega t + \phi_2) \tag{18}$$

where E_{01} and E_{02} are the amplitudes of the two beams (or two electric fields), z_1 and z_2 are the distances the light travels between the two sources and the screen, β_i =propagation constant = $2\delta n / \lambda_i$, n = refractive index of the medium, λ_i is the wavelength of the light, ω is the angular frequency of the light source, and ϕ_1 and ϕ_2 are the variation in phases of the two beams as a result of propagation through distances z_1 $(P_1 O)$ and z_2 $(P_2 O)$ (spatial coherence). The resultant electric field at any point, say at O, for two beam interaction is,

$$E = E_1 + E_2$$

or, $\quad E^2 = E_1^2 + E_2^2 + 2 < E_1.E_2 > \tag{19}$

For an ideal case when $|E_1| = |E_2|$ and $E_1.E_2 = |E_1|^2 = |E_2|^2$,

$$E^2 = \ 2E_1^2 + 2E_1^2 \cos\delta \ = 2E_1^2(1 + \cos\delta)$$

$$= 4E_1^2 \cos^2 \frac{\delta}{2} \tag{20}$$

where $\delta = \phi_1 \sim \phi_2$

Since light intensity $I \alpha E^2$,

$$I = 2I_1(1 + \cos\delta) = 4I_1 \cos^2 \frac{\delta}{2} \tag{21}$$

For the same wavelength ($\lambda_1 = \lambda_2 = \lambda$) and frequency ($\omega$) between the two sources, the phase difference between the two beams will depend on spatial path of travel $P_1 O$ and $P_2 O$. The two beams will have a phase difference of 90° for a path difference of $\lambda / 4$ (Fig. 4.24). Hence interferometers are extremely sensitive because they give full modulation for a spatial displacement of $\lambda / 4$. However, to give a meaningful and reliable measurement, external perturbation due to variables other than measurand has to be avoided. High sensitivity gives a comparatively small range of measurement and it needs a very good alignment of the reference arm.

Displacement sensitivity of the interferometer is

$$\frac{dI}{d\delta} = \frac{d\{2I_1(1+\cos\delta)\}}{d\delta} \quad \alpha \quad \sin\delta \tag{22}$$

Hence sensitivity to be maximum (constructive interference)

$$\delta = \pm m\pi \pm \pi/2 = \pm(2m+1)\frac{\pi}{2}$$

$$m = 0,1,2 \tag{23}$$

and for sensitivity to be minimum or zero (destructive interference).

$$\delta = \pm m\pi \quad (m=0,1,2 \ldots\ldots\ldots\ldots) \tag{24}$$

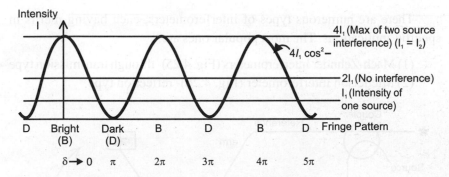

Fig. 4.24 Two source interference

The fringe pattern of an interferometer (alternate dark and bright circles or strips) can be observed visually, projected on a screen or recorded photoelectrically. Both intensity and photodetector current follow square law. A photo detector will produce a current i in the external circuit

$$i \alpha \, \eta_{qe} \, P_{abs} / h\upsilon \tag{25}$$

where P_{abs} = power absorbed by the photodetector,

η_{qe} = quantum efficiency = ratio number of the number of carriers generated to the number of photons absorbed, h, υ are Planck constant and optical frequency, respectively.

Since a photodetector is a square law detector

$$i \, \alpha \, P_{abs} \alpha \quad [E_1 \cos(\omega t + \phi_1) + E_2 \cos(\omega t + \phi_2)]^2$$

or, $i = \varepsilon \left(1 + \mu \cos\delta\right)$

where ì = fringe visibility = 1, if $E_1 = E_2$,

$$\delta = \delta(t) = \quad 2\beta \left| z_2(t) - z_1(t) \right|$$

ε = factor to take into consideration optical to electrical conversion.

In visual interference patterns, the distance between two bright and dark strips is due to effective spatial difference of $\lambda/2$ between the two beams. Any electrical detector (counter) used to count p, the number of changes from maximum to minimum gives the total path difference $= p.\dfrac{\lambda}{2}$. Piezoelectric transducers (PZT) are used in optical fibre based interferometers in the reference arm for phase compensation and setting the system at the maximum sensitivity condition.

There are numerous types of interferometers, each having merits in specific applications. The most popular ones are:

(1) Mach-Zehnder interferometers (Fig. 4.25)- through transmission type
(2) Michelson interferometer (Fig. 4.26)- reflection type

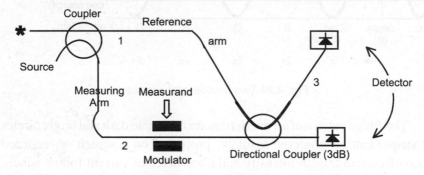

Fig. 4.25 Mach-Zehndar interferometer

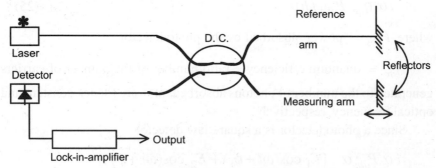

Fig. 4.26 Michelson interferometer

Source and detector can be on the same side. Modulation can be done by relative displacement between reflectors (Fig. 4.26) as well.

(3) Sagnac interferometer (Fig.4.15)- gyroscopic application.

(4) Fabry- Perot interferometer (Fig. 4.27)- multiple beam type used in laser construction.

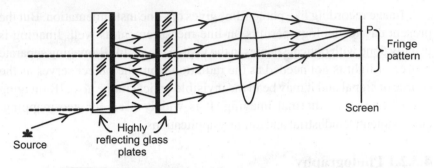

Fig. 4.27 Fabry Perot interferometer

Interferometric technique of signal processing is adopted in phase modulated sensors. Optical interferometer output is fed to a detector circuit (electronic) to determine the phase changes. A piezoelectric transducer (PZT) is used in the reference arm (of the waveguide) to extract the sign change in phase variation.

Lock-in-amplifier is used when the signal available is very weak. In some applications Photo-multiplier-tube (PMT) is used to capture very low signal. PMT sensitivity is very high so as to respond to a few photons even. Optical time domain reflector (OTDR) is used in specific applications like scattering loss modulated sensors.

4.3.2 Two Dimensional Optical Imaging

Conventional photography is the most common form of 2-D imaging in which identification of object and measurement of object parameters are done through experience. Different forms of 2-D optical imaging are:

- Photography
- Videography
- IR imaging
- Optical fibre imaging

Techniques of image based measurement differ in physical configuration as the technology involved vary from one to another. The steps involved (Fig 4.29) are:

- Configuring devices or apparatus for image acquisition
- Image recording

- Image processing
- Referencing
- Read out and interpretation

Image recording like photo films gives off-line instrumentation. But the present day technology permits on-line measurement as well. Imaging is usually done with optical sources in the visible range. In IR imaging separate source of light is not needed as the radiation from the object serves as the source of signal and it may be in the invisible optical range also. IR imaging is also known as thermal imaging. It is widely used for geo-mapping, environmental, industrial and military applications.

4.3.2.1 Photography

(a) Basic Triangulation method

It is a method to determine the height or distance of a remotely located object utilizing the property of triangles. In its simple version a source of light at an angle θ_1 emits beams to strike the object (Fig. 4.28). The reflected light is detected by a detector at an angle θ_2. If the distance between the source and detector is d,

$$h = \frac{d \sin \theta_1 \sin \theta_2}{\sin \{180^o - (\theta_1 + \theta_2)\}} \qquad (26)$$

and

$$\frac{\sin \theta_1}{r_2} = \frac{\sin \theta_2}{r_1} = \frac{\sin [180^o - (\theta_1 + \theta_2)]}{d} \qquad (27)$$

(b) Passive Triangulation method

It is also referred to as stereo or binocular vision which is biologically motivated because this is the way that humans and most animals perceive depth. Two imaging devices like two TVs, diode matrix or CCD cameras are analogous to two eyes in the biological system. To calculate the range r, from the cameras to a given point P on the object, both cameras are used to scan the scene and generate a picture matrix (Fig. 4.29). Thus, there will be two pixels representing the point P, each pixel is located at a given distance x_i ($i = 1, 2$) from the centre of its image. If one overlaps the two camera images, the two image points x_1 and x_2 will not coincide. The absolute value of the difference is called disparity between two image points. Range r from any or both camera such that the point is between the two cameras is

$$r = \frac{d\sqrt{f^2 + x_1^2 + x_2^2}}{(\text{mod } |x_1 - x_2|)} \tag{28}$$

where d = distance between the two camera lens centres,
f = focal length of the cameras

(c) Active Triangulation method

Active triangulation involves movement of the imaging device or the light source not both. Either the imaging device is stationary and the scene is scanned with a projected beam of light or illumination is constant and the camera is moved to generate different prospectives of the scene. This type of system can employ one of the two techniques for depth (or range) measurement :

(1)   Spot sensing
(2)   Light stripe sensing

In spot sensing a single beam light projected onto an object is reflected back into a camera positioned at known distance d, from the spot projector (Fig. 4.30). This creates a triangle between the projector, object and camera such that the range r of the object spot from the camera is

$$r = \frac{b \sin \theta_1}{\sin \{180^o - (\theta_1 + \theta_2)\}} \tag{29}$$

$$\theta_2 = \tan^{-1} \frac{f}{x} \tag{30}$$

Proper choice of wavelength of the light source would minimize the environmental interference. Laser light is practically useful and ratiometric measurement is adopted.

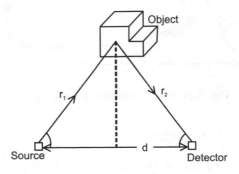

Fig. 4.28 Simple triangulation method

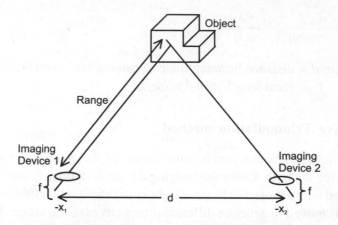

Fig. 4.29 Passive triangulation method

Light stripe sensing is an extension of spot sensing. Rather than projecting a spot, a stripe of light is projected on the scene. As a result, the stationary camera sees a line, or stripe of light. The stripe can then be divided into individual image point and range calculated for each point along the stripe. The range calculation is identical to that for spot sensing. The light stripe can be formed by passing ambient or infrared light through a slit on the projector.

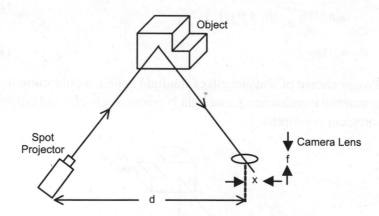

Fig. 4.30 Active triangulation with spot sensing

(d) High speed photography

It can be used for on-line measurement of parameters for moving objects.

4.3.2.2 Videography

For on line imaging and measurement thereby video imaging technique can be used. The image formed on photodetector arrays can be used for extraction of parameters of objects, stationary or moving. An advantage of video imaging is that image processing including ambient noise filtering can be effectively done using software.

4.3.3 Imaging and Image Processing [16]

When large objects or systems having varying characteristics in its different parts require simultaneous assessment of its parameters, discrete sensors or probes fail to give satisfactory results. For example, temperature mapping for environmental or pollution status of a region require an instrumentation technique different from single point sensing concept.

Optical imaging involves grabbing the image by an imaging device when the object is illuminated by radiation in the optical range or the object itself radiates optically (self illuminated). Thermal or IR imaging and imaging of luminescent objects belong to the latter category.

Image recording or storage is done by films or photo detector arrays.

Image processing is the analysis and manipulation of a (usually digitized) image, especially to improve its quality. It includes encoding and decoding of images for efficient transmission and less storage space, image enhancement from the bad quality, poor contrast and noisy images to good quality and high contrast images, segmentation of image as used in computer vision and robotics, image classification, and image restoration using mathematical model and digital signal processing. Generally special parallel computers are used for real time image processing.

Three dimensional optical imaging in measurement and instrumentation is needed for visual inspection, computer vision, robotic applications and night vision systems used in defence and security.

4.3.4 Optical Fibre Imaging

Image processing technique along with active triangulation method has been used to determine the distance between two objects under static and dynamic conditions [17]. The method has been successfully applied to monitoring of overhead equipment in railway traction. Optical fibre based biomedical instruments are widely used for diagnostics and laser surgery. Fibre bundles are used in endoscopy for visualization of internal organs of the human body, such as, ulcers, tumours. Subsequent developments in this area have resulted in gastroscopy, bronchoscopy, rectoscopy, laparoscopy, cystoscopy, and proctosigmoidoscopy (cancer in rectum and colon). Further more, recent developments have shown that configurations of rigid, straight or curved fibre optics can be used in medicine, such as a rigid endoscope and a hypodermic probe, to allow visualization of regions under the skin.

Fibre endoscope consists of a coherent bundles of fibres i.e. fibre position within the cross section is maintained over the length of the device so that optical images (intensity based) can be transmitted. Generally, a second bundle of fibres is incorporated into the endoscope to provide illumination. This bundle need not be coherent.

Application of endoscopy has been extended from diagnostic to therapeutic techniques. Using laser it is being used in microsurgery, namely, in ophthalmology, removal of tissues, cure of internal ulcers, etc.

4.3.5 Optical Signal Processing

In optical sensors signal processing is done optically and electrically. Electrical signal processing is adopted after the optical signal is converted into electrical signal (current).

The main objective in optical signal processing is to eliminate optical noise such as ambient noise, source noise or detector noise. It is also useful in integrated optics so that the signal can be processed inside the chip and whatever is the output, it is in the form of electrical signal. A few techniques adopted are:

(a) Optical Wheatstone Bridge
(b) Ratiometric Measurement
(c) Interferometric Technique

(a) Optical Wheatstone Bridge [11]

The output of a source may change because of voltage fluctuation due to temperature variation. Typically output of LED changes by about -0.7%°C. This technique is used for compensation for link and source fluctuations. It is useful to intensity modulated sensors. The device uses two sources S_1 and S_2, and two detectors D_1 and D_2 (Fig. 4.31). If L_1, L_2, L_3 and L_4 are transmission losses in dB (in the wave guide), R_1, R_2, R_3 are fixed attenuation in dB, and M is the output of the sensor (optical or optical fibre based), we have with one source at a time

$$D_{11} = D_1 \Big|_{\substack{S_1 = on \\ S_2 = off}} = S_1 L_1 M \, L_2 \qquad D_{12} = D_1 \Big|_{\substack{S_1 = off \\ S_2 = on}} = S_2 L_3 R_3 \, L_2$$

$$D_{21} = D_2 \Big|_{\substack{S_1 = on \\ S_2 = off}} = S_1 L_1 R_1 \, L_4 \qquad D_{22} = D_2 \Big|_{\substack{S_1 = off \\ S_2 = on}} = S_2 L_3 R_2 \, L_4$$

and

Set $\qquad U = \dfrac{D_{11}}{D_{12}} = \dfrac{S_1 L_1 M L_2}{S_1 L_1 R_1 L_4} = \dfrac{M L_2}{R_1 L_4}$ \hfill (31)

$$V = \dfrac{D_{12}}{D_{22}} = \dfrac{S_2 L_3 R_3 L_2}{S_2 L_3 R_2 L_4} = \dfrac{R_3 L_2}{R_2 L_4} \tag{32}$$

Hence $\qquad \dfrac{U}{V} = M \cdot \dfrac{R_2}{R_1 R_3}$ \hfill (33)

Typically source S_1 may be modulated at 1 KHz (M_1) and source S_2 at 10 KHz (M_2)

(b) Ratiometric Measurement

Optical noise due to ambient light, temperature effect, vibration etc. on sensor output is compensated by ratiometric technique. This is more suitable for sensors using wavelength modulation. Let $I(\lambda_1)$ and $I(\lambda_2)$ be the sensor output intensities at wavelengths λ_1 and λ_2 respectively. If we can choose $I(\lambda_2)$ to be independent of the modulation effect caused by the measurand, the variation in $I(\lambda_2)$ will be only due to variation in ambient noise. Assuming the ambient noise to affect both $I(\lambda_1)$ and $I(\lambda_2)$ equally, the output

$$\frac{I(\lambda_1)}{I(\lambda_2)} = \frac{I(Measurand, ambient\ noise)}{I(ambient\ noise)} \, \alpha \ Measurand\ only$$

In ratiometric measurement various optical and electrical components and devices like interference filters (chopper modulated), gratings, spectrum analyzers, monochromators, etc. are used.

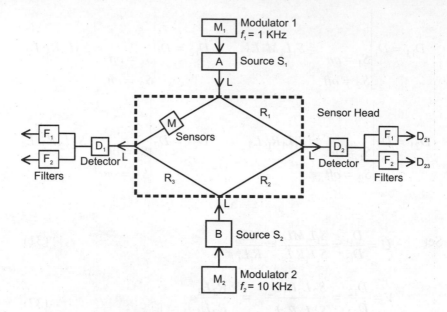

Fig. 4.31 Optical Wheatstone Bridge

4.3.6 Holography [18]

The term holography has been coined from Greek word 'holos' meaning 'whole' and 'graphein' meaning 'to write' and is being used to designate a class of 3-dimensional photography. The differences between two-dimensional photography, interferometer and holography are depicted in Fig 4.32.

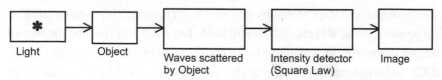

(a) Two-dimensional photography: one–step intensity recording

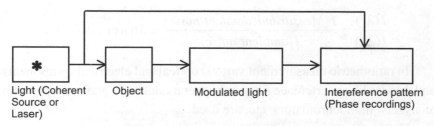

(b) Interferometry: one-step phase recording process

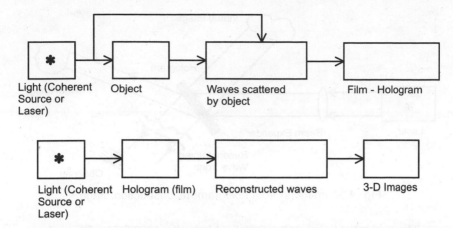

(c) Holography: three – dimensional photography – two-step image (amplitude and phase) construction.

Fig. 4.32 Comparison of 2-D photography, interferometry and holography

In the first step holography does not record directly the image but records the interference pattern created by the wavefronts scattered by the object and the reference beams (Fig.4.33). This is the first or the construction stage to record the hologram. In the second or reconstruction stage, the 3-D

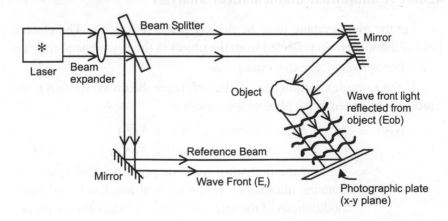

Fig. 4.33 Recording of a hologram: construction stage. Interference pattern i.e. hologram is recorded on the photographic plate if the two beams are perfectly coherent.

image from the hologram is restored by illuminating the hologram by a suitable coherent light source (Fig.4.34) that possesses spatial and temporal coherence with the reference beam used for recording the hologram.

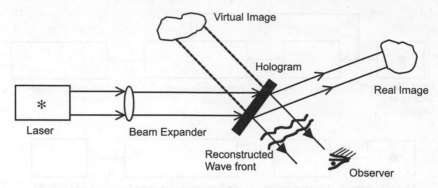

Fig. 4.34 Image(s) from hologram: reconstruction stage

The method of holography is applicable to all types of waves e.g., electron waves, x-rays, acoustical or ultrasonic waves, optical waves, microwaves and seismic waves, that is, wherever strong coherency and occurrence of interference are possible. Holograms are, in reality, photograph of the interference pattern of the object waves with coherent reference beam that can be unmodulated spherical or plane wave fronts. When the hologram is illuminated from the front, the observer sees the virtual image and when illuminated from the back, a real image is viewed.

4.3.6.1 A simplified mathematical analysis

Let the photographic plate be the x-y plane (Fig 4.33). The electric field of the wavefront reflected from the object in the x-y plane at time t is

$$E_{ob} = A_{ob} (x,y) \exp (- jwt) \tag{34}$$

If the complex amplitude of the reference beam is $A_r (x,y)$, the irradiance recorded in the photographic plate is

$$
\begin{aligned}
I (x,y) &= [A_o + A_r] = (A_o + A_r) (A_o^* + A_r^*) \\
&= (A_o A_o^* + A_r A_r^*) + (A_o A_r^* + A_o^* A_r) \tag{35} \\
&= \text{(sum of the individual irradiances)} + \text{(interference that} \\
&\quad \text{contains information in the form of amplitude and phase} \\
&\quad \text{modulations of the reference beam) (* denotes conjugate)}
\end{aligned}
$$

Let the plate be processed to form a transmission hologram and the transmission be a fraction K of irradiance function $I(x,y)$.

When it is illuminated by original reference beam, the original amplitude of transmitted light will be

$$
\begin{aligned}
A_t (x,y) &= A_r K I(x,y) \\
&= K [A_r (A_o A_o^* + A_r A_r^*) + A_r A_o^* + A_r A_r^* A_o] \tag{36}
\end{aligned}
$$

The first term (within the first bracket) of the above equation represents the direct beam and the second term Ar Ar* is constant and hence the last or fourth term is essentially Ao, the object wavefront amplitude (diffracted beam which is the reconstruction of the wavefront from the original object and forms the virtual image).The third term represents the other diffracted beam and forms the real (conjugate) image.

4.3.6.2 Applications

Holography is more widely used for off line and on line measurements of stress developed in materials, computer memories (theoretically can go upto 10^{10} bits / mm) with rapid access.

4.4 Spectroscopy [19]

Spectroscopy is the general term for the phenomena that deal with the interactions of various types of radiation with matter.The radiation signal can be in the form of light, radiant heat, gamma rays and X rays, ultraviolet, microwave and radio frequency radiation, and also acoustic waves and beams of particles such as ions and electrons. Optical spectrometry employs an optical source and a detector to ascertain the changes in the optical intensity as a result of interaction with matter. It is widely used for chemical analysis in scientific and industrial applications. It is also known as absorption spectrometry.

In absorption spectrometry based chemical analysis matter (chemical sample) absorbs radiation very selectively with respect to wavelength. Three such analyzers are

(a) UV analyzer

(b) IR analyzer and

(c) Near IR analyzer.

4.4.1 UV Analyzer

It consists of a radiation source (UV), optical filter, sample cell, detector and the end device. Quartz prism, Hg, H_2, Xe, Na are usual UV sources. Photo tubes, photomultiplier tubes and photo cells are used as detectors. Fig. 4.35 represents two types of UV analyzers. The opposed beam design is simple, low cost, moderately accurate instrument. In split-beam technique optical noise like source variation, sample turbidity, cell window dirt is eliminated. High sensitivity accuracy with low drift is achievable.

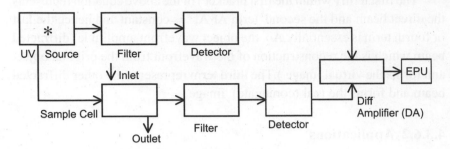

(a) Opposed Beam Technique

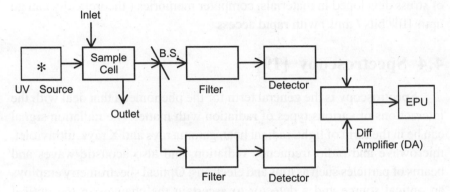

(b) Split Beam Technique
Fig. 4.35 Industrial UV analyzers

4.4.2 IR Analyzer

The basic components of an IR absorption spectrometer (Fig. 4.36) are: an infrared source, an optical system, a sample cell, a prism, a detector and a recording system. The source works usually in the 2 and 15 micron range because of hardware and technology limitations. The source may be a coiled Nichrome wire or Nernst filament (tubular element of rare oxides heated at 1800°C to 2000°C). The sample cell material is chosen depending on the wavelength of the source i.e. from quartz, calcium fluoride, sodium chloride etc. The detectors used are thermal such as thermopiles and bolometers (which measure IR energy emerging from the cell after absorption), or pneumatic (which causes expansion of gas volume in a gas-pressure cell by IR emerging from the cell after absorption in the sample gas), followed by a microphone. The output of the infrared detector is recorded electronically or using a stripchart recorder. Using suitable filters it is possible to have output at specific wavelengths (non dispersive) or at various wavelengths permitted by the prism and the cell (dispersive).

4.4.3 Near IR Analyzer

Near IR analyzers working in the range of 0.7 to 2.5 microns use simpler optical techniques and photocell as a detector (Fig. 4.37). Conventional glass (sample-cell window) transmits in this region and a standard tungsten filament lamp can be used. Two interference filters (very narrow band pass filters) are used for ratiometric measurement.

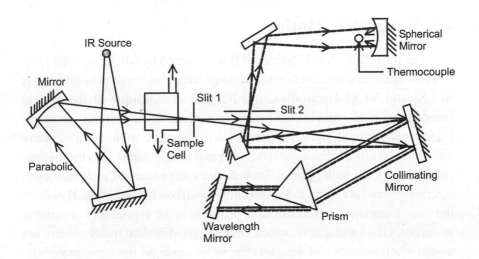

Fig. 4.36 IR Analyzer

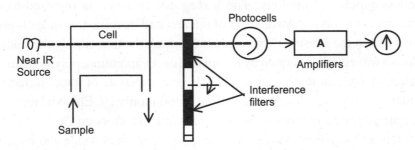

Fig. 4.37 Near IR Analyzer

4.4.4 Comparison of Absorption Spectrometers

UV is not as specific as IR, but it is a more sensitive technique. Aqueous samples can be analyzed using UV but not using IR. UV is generally quantitatively superior to IR in determination of complex organic structures. But UV analyzers can not be used to analyze mixture of metals as they are completely opaque to UV radiation. UV anlyzers are very suitable for materials that absorb UV such as nitrates, halogens, unsaturated hydrocarbons, ketones, benzene compounds, naphthalene, etc.IR anlyzer is useful for qualitative and quantitative analysis of a number of organic gases and liquids. But it is unsuitable for O_2, H_2,Cl_2 and inert gases that do not absorb infrared radiation. Near IR analyzer is used for measurement of moisture of organic streams.

4.5 Integrated Optics

The term 'Integrated Optics' (IO) was coined by Miller in 1969 [20], wherein optoelectronic circuits and systems were conceived to be analogous to LSI and VLSI electronics, and IO will incorporate all the optical components onto a single chip for performing necessary circuit functions. Functionally IO will have less thermal drift, be more immune to moisture and vibration effects, consume less power and more reliable than the discrete components (the bulk optics). Such devices are expected to reduce cost, size, weight and are likely to be compatible with batch fabrication. However, the main hindrance towards its rapid progress is the availability of suitable materials. Glass and quartz, which are very good optical materials, are not useful when sources and detectors are to be made on the same material.

The primary building block of an IO circuit is the waveguide whose successful fabrication by and large depends on thin-film techniques and semiconductor technology. The waveguide material is required to be transparent to optical wavelength of interest and its refractive index higher than that of the substrate, the medium in which the thin-film is embedded. Such a waveguide is formed by RF sputtering, evaporation, polymerization, diffusion and ion implantation methods on a substrate of lower refractive index, usually glass. The top of the film is commonly air. Epitaxial methods are already used in fabrication of sources and detectors on Ga As, Si, etc. Structures of diffrent waveguides are: planar, channel (ridge and buried), curved.

In view of the fact that progress made towards the goal of achievement of all optical IO is slower than it was predicted (IO was predicted to take the role of electronics and microelectronics by turn of 21st century), presently 'hybrid' IO, a combination of IO and bulk optics, is resorted to. A few IO components are already in the market and the work on others is being vigorously pursued. The successful ones are: lasers (laser diodes), optical fibre and optical switches. Considerable progress has been made towards implementation of fibre lasers (also known as fibre amplifiers), modulators, fibre gratings, etc. Directional couplers (replacement of beam splitters) presently available require more miniaturization and improvement. Other IO devices being pursued are: detectors, prisms, lasers, polarizers, etc. On the signal processing front the advancement in development of all fibre interferometers, multiplexers (using WDM, FDM, TDM, etc) is noticeable.

Application of hybrid IO systems can be cited in the following areas: RF spectrum analyzers for defence application [22], micro optic gyroscope [18], time integrated correlator for radar signal processing, etc. Since the light is coupled into a thin-film of much smaller cross section, its intensity gets enhanced. This enables interesting studies on nonlinear optics (harmonic generation, enhanced Raman spectroscopy, bistable optical switches, etc.). Commercial viability of IO circuits is expected to revolutionaize the technology influencing all walks of our life.

Acknowledgement: The author gratefully acknowledges the help received from Professor M L Mukherjee (Former Professor, Department of Physics and Meteorolgy, IIT Kharagpur) and Professor P K Dutta, (Professor, Department of Electrical Engineering, IIT Kharagpur), and his many students in writing the above chapter.

4.6 References

1. Ghatak, A. Optics. TMH Pub, New Delhi, 1994.

2. Ghatak, A. and Thyagarajan, K. Optical Electronics. Cambridge University Press, NY, 1989.

3. Hawkes, J. and Latimer, I. Lasers - Theory and Practice. Prentice Hall, London, 1995.

4. Verdyen, J. T. Laser Electronics. PHI, New Delhi, 1993.

5. Kingston, R.H. Optical Sources, Detectors and Systems— Fundamentals and Applications. Academic Press, London, 1995.

6. Wilson, J. and Hawkes, J.F.B. Optoelectronics— An Introduction. PHI, New Delhi, 1999.

7. Ghosh, M.K. et at. Optical device for measuring angular deviations / disposition of an object. Indian Patent No. 1985 73 dated July 19, 2002.

8. Kao, C.K. Optical Fiber Systems: Technology, Design and Applications. Mc Graw Hill, NY, 1982.

9. Senior, J.M. Optical Fiber Communications- Principles and Practice. PHI, New Delhi, 1996.

10. Culshaw, B. Optical Sensing and Signal Processing. Peter Peregrinus Ltd, IEE, 1984.

11. Culshaw, B. and Dakins, J. (Ed) - Optical Fiber Sensors- Principles and Applications. Vol I, II. Artech House, London, 1989.

12. Agarwal, D.C. Fiber Optic Communication. Wheeler Publishing, New Delhi, 1993.

13. Ghosh, M.K. et al. - An Optical Fibre Based Linear Displacement Meter. -Indian Patent No 163703 of 1985 dated May 1989.

14. Ghosh, M.K. and Mishra, A.- Device for Transmitting Pressure and Differential Pressure Signals. - Indian Patent No. 183428 of 1996 dated 7.7.2000.

15. Sanjiv, S. et al. - An Optical Fibre Based Low Temperature Sensor, Journal of The Institution of Engineers (India) 13-18, Val 87, July 2006.

16. Jain, A.K. Fundamentals of Digital Image Processing. PHI, New Delhi, 1997.

17. Ghosh, M.K. et al. - Apparatus for Measurement of Spatial Parameters of a Distantly Located Object Under Static and Dynamic Conditions, Indian Patent application No. 280/Cal/2000 dated 10.5.2000.

18. Hariharan, P. Optical Holography : Principles, Techniques and Applications. Cambridge University Press, 1987.

19. Liptak, B.G. (Ed)- Instrument Engineers' Hand Book. Vol. I and Supplement I. Chilton Book Co, Philadelphia.

20. Miller, S.E. An Introduction to Integrated Optics. Bells Systems Technology Journal, Vol. 48, pp 2059, 1969 and American, Vol. 28, pp 230, 1974.

21. Mergerian, D. et al. Operational Integrated Optical RF Spectrum Analyzer, Applied Optics, Vol.119, pp 3033,1980.

22. Lawrence, A. The Micro Optic Gyro, Report : Northorp Corp, Mass, 1983.

Chapter 5

Chemical Sensors

S. Basu
S. Roy

Abstract

Chemical sensors are capable of quantitative or qualitative recognition of atomic, molecular or ionic species. Based on the physics and operating mechanism, they are categorized into four general groups:

(1) Chromatography and spectrometry,

(2) Electrochemical sensors,

(3) Mass sensors, and

(4) Optical sensors.

Some of these sensors can be integrated into IC chips with electrical output.

5.1 Introduction

The microprocessor revolution has transformed our ability to provide electronic control of processes and products. Dramatic improvement in industrial process control, in the functioning and facilities of domestic devices and in the control of environmental pollution through, for example, vehicle exhaust emission controls, are already widely evident. The key factor that has helped achieving the improved control is the superior interface, owing to the recent advancement in sensor technologies, between the device and the environment to be monitored. So, in a nutshell sensors have become an integral part of modern technology.

Between the two broad classifications of sensors - physical and chemical, based on their operation principle, the chemical sensors find a gamut of prospective industrial, domestic and biomedical applications. Chemical sensors are devices, which are capable of a continuous monitoring of the concentration of molecules (or ions) in their ambient [1]. The molecules change a physical property of the device, like mass [2] temperature [3], current [4], resistance [5], light absorption [6], etc. Several sensing principles and types of devices have therefore been developed and many of them are also commercially available.

The need for chemical sensors is clearly defined, yet the specifications are highly demanding. Unlike the control electronics, the sensor must interact with, and often be exposed to, the environment. Even apparently benign atmospheres such as in domestic dwellings may contain many corrosive and contaminating species, which can seriously interfere with sensor functioning and make sensor design and development a painstaking and expensive business. Applications such as natural gas leak detection, non-invasive blood glucose monitoring [7], home indoor air quality [8], personal/ portable air quality monitors, LPG detector [9], and home food spoilage detectors [10] need a sensor that is small, efficient, accurate, sensitive, reliable, and inexpensive.

Connecting an array of these next generation chemical sensors to wireless networks that are starting to proliferate today creates many other applications. Asthmatics could preview the air quality of their destinations as they venture out into the day. HVAC systems could determine if "fresh air" intake was actually better than the air in the house. Internet grocery delivery services could check for spoilt food [11] in their clients refrigerators. City emission regulators could monitor various emission sources throughout the area from their desks and so on.

5.2 Categories of Chemical Sensors

The chemical sensors can be categorized into four general groups: (1) chromatography and spectrometry [12]; (2) electrochemical sensors; (3) mass sensors and (4) optical sensors. Categorization of these sensors is based primarily on the principle of physics and operating mechanisms of the sensor. For example, chromatography relies on separation of complex mixtures by percolation through a selectively absorbing medium, with subsequent detection of compounds of interest. Electro-chemical sensors include sensors that detect signal changes (e.g., resistance) caused by an electrical current being passed through electrodes that interact with chemicals. Mass sensors rely on disturbances and changes to the mass of the surface of the sensor during interaction with chemicals. Optical sensors detect changes in visible light or other electromagnetic waves during interactions with chemicals. Within each of these categories, some sensors may exhibit characteristics that overlap with other categories. For example, some mass sensors may rely on electrical excitation or optical settings.

5.2.1 Chromatography and Spectrometry

Chromatography is a method for the separation and analysis of complex mixtures of volatile organic and inorganic compounds [13, 14]. A chromatograph is essentially a highly efficient apparatus for separating a complex mixture into individual components. When a mixture of components is injected into a chromatograph equipped with an appropriate column, the components travel down the column at different rates and therefore reach the end of the column at different times. A detector is positioned at the end of the column to quantify the concentrations of individual components of the mixture being eluted from the column. Several different types of detectors can be used with chromatographic separation as discussed below.

5.2.1.1 Bench-Top Gas Chromatographs (GC)

A laboratory bench-top GC, which provides flexibility and performance, required for research and method development in industry applications. It is rugged and reliable, so it can be used for routine methods that require multiple columns or valves, specialty inlets or detectors, or a broad temperature range. This unit can be configured with a variety of columns or detectors and can be tailored to individual needs. Approximate dimensions of a typical Agilent 6890 model GC are 50 x 58 x 54 cm and it weighs approximately 49 kilograms. It should be operated in temperatures ranging from 15^0 C to 35^0 C in 5 to 95% humidity. The price for the bench-top model varies depending upon specifications but is typically in the range of $20,000 to $50,000.

Advantages: The bench-top GC can provide superior discrimination capabilities (relative to other device and sensors) with excellent precision, sensitivity, and reproducibility.

Limitations: Not portable. Expensive. Requires training to operate.

5.2.1.2 Portable Gas Chromatographs

Femtoscan (USA) has developed a new, hand portable Gas Chromatograph/Ion Mobility Spectrometer (GC/IMS). The instrument is called the Environmental Vapor Monitor II (EMV II) and is based on Ion Mobility Spectrometry (IMS) technology for sensitive detection of gas phase analytes with high speed. Automated Vapor Sampling Transfer Line Gas Chromatography sampling and separation capabilities have been developed be FemtoScan and the University of Utah, USA. The DEVM II is a sensitive and selective near-real-time vapor detector. A wide range of volatile and semi-volatile contaminants can be detected with EVM II, which can be operated from a 24 volt battery pack or from an external power supply.

Advantage: Portable. Reliable with good reproducibility. Real-time measurement (in seconds). Parts per billion (ppb) level sensitivity to vapors. Remote monitoring capability. No carrier gas required for operation. Wide range of volatile and semi-volatile components.

Limitations: Cannot be used in situ. Fairly expensive.

5.2.1.3 Micro-Chem-Lab (μ ChemLab) on a Chip

The μChemLab has been developed at Sandia National Laboratories, Albuquerque, New Mexico. This sensor is not commercially available at this time. Micro fabrication has been utilized to provide a miniaturized GC-type device that provides a fast response with an ability to utilize multiple analysis channels for enhanced versatility and chemical discrimination. The μChemLab is an autonomous chemical analyzer the size of a palm-top computer that incorporates a gas phase analysis system for detecting chemical warfare agents (e.g., sarin, soman, mustard gas) and a liquid phase analysis system for detecting explosives. The μChemLab improves the sensitivity and selectivity to individual chemicals by using a cascaded approach where each channel includes a sample collector/concentrator, a GC separator, and a chemically selective surface acoustic wave (SAW) array detector.

Advantages: Batch micro fabrication provides several advantages for the μChemLab. The low heat capacity of the thermal desorption stage allows it to be heated rapidly with low power. The rapid heating provides sharp chemical pulses that provide improved temporal separation in the GC column. Additionally, components for the μChemLab can be manufactured at low cost, which can open a variety of new markets where current chemical analysis systems are cost prohibitive. Finally, the extreme miniaturization of this device may open new markets where traditional GCs and portable GCs were too large or where other micro sensors cannot adequately discriminate the analytes of interest.

The μChemLab is still under development and requires additional research and testing before being deployed to in-situ settings.

5.2.1.4 Ion Mobility Spectrometry

The ion mobility spectrometer (IMS) [15] can be considered a subclass of chromatographic separators. The principle of every IMS is a time-of-flight measurement. After a gaseous sample has entered the spectrometer it will be ionized by a radioactive source, the resulting positive and negative charged species will be accelerated over a short distance and the time-of-flight will be determined. The IMS is different than the mass spectrometer in that it operates under atmospheric conditions and does not need large and expensive vacuum pumps. Because of this, IMSs can be easily miniaturized.

Bruker-Daltronics, Inc. has developed a hand-held chemical agent detector, designed for automatic chemical agent detection referred to as **Rapid Alarm and Identification Device (RAID).** It is suited for the screening of traces in gas and for the detection of toxic industrial compounds and chemical warfare agents down to the ppb-range. Equipped with automatic polarity switching, the instrument enables continuous monitoring. The built-in microprocessor evaluates the recorded ion mobility spectra. The results, identified substances and their concentrations, are shown on the display. An integrated alarm function responds according to programmed threshold values. The substance library can be updated at any time, as a special Teach-in function allows the integration of new substance data. Main application fields are on-site investigations and personnel protection, especially for fire brigades, rescue services and military use.

Advantages: Small packaging. Can be used to detect toxic industrial compounds and chemical warfare agents in ppb-range. Integrated alarm for threshold detection. Radioactive ionizing source.

Limitations: Cannot be used in situ.

5.2.1.5 Mass Spectrometry

The principle of the mass spectrometer is similar to the ion mobility spectrometer, except a vacuum is required. Sampled gas mixtures are ionized, and charged molecular fragments are produced. These fragments are sorted in a mass filter according to their mass to charge ratio. The ions are detected as electrical signals with an electron multiplier or a Faraday plate.

Low mass ions are displayed as a vertical line at the left end of a scale while heavy ions are displayed towards the right. The length of a line represents the quantity of that ion in the gas mixture.

A large number of commercialized vendors exist that sell mass spectrometers. A few of the vendors that sell portable units are listed below:

> http://www.geo.vuw.ac.nz/analytical/dycor.htm
> http://www.kore.co.uk/tcat.htm
> http://www.moorfield.co.uk/newprodqms2.htm

Advantages: The mass spectrometers have good discrimination capabilities and can detect a wide range of chemicals. Some of the mass spectrometers are portable enough to carry into the field.

Limitations: the units appear to be quite expensive (the ecoSys-P device is in excess of $40,000). Spectral overlaps can be a problem in detecting mixtures of unknown composition. Cannot be placed in situ.

5.2.2 Electrochemical Sensors

Electrochemical sensors are usually categorized into three groups: potentiometric (measurement of voltage) [16]; and amperometric (measurement of current) [17]; and conductometric (measurement of conductivity).

5.2.2.1 Conductometric Sensors

Three different types of conductometric sensors are presented in this section. The first is a polymer-absorption sensor that indicates a change in resistance in the conductive polymer electrode when exposed to chemicals. The second is the catalytic bead sensor, which requires elevated temperatures to burn combustible hydrocarbon vapors and change the resistance of an active element. The third sensor is the metal-oxide semiconductor sensor, which responds to changes in the partial pressure of oxygen and requires elevated temperatures to induce combustion of chemical vapors that change the resistance of the semiconductor.

5.2.2.1.1 Polymer-Absorption Chemiresistors

The concept of using polymeric absorption [18] to detect the presence of chemicals in the vapor phase has existed for several decades. These polymer-absorption sensors (chemiresistors) consist of a chemically sensitive absorbent that is deposited onto a solid phase that acts as an electrode. When chemical vapors come into contact with the absorbant, the chemicals absorb into the polymers, causing them to swell. The swelling changes the resistance of the electrode, which can be measured and recorded. The amount of swelling corresponds to the concentration of the chemical vapor in contact with the absorbent. The process is reversible, but some hysteresis can occur when exposed to high concentrations. Several companies and organizations have developed chemiresistors [19], but the specific attributes and types of absorbents, which are generally proprietary, vary among the different applications.

Cyrano Sciences™ (http://cyranosciences.com/technology) developed a hand-held "electronic nose" device that employs an array of chemiresistors. They use array of 32 chemiresistors that consist of polymer films as the absorbent. The large number of chemiresistors in the handheld unit is used to increase analytic discrimination. However, the unit must be "trained" for each analyte of interest. The cost of this hand-held device is ~$7000, but it is not currently amenable for in-situ sensing.

Adsistor Technology™ (www.adsistor.com) developed and patented a chemiresistor for the particular application of vapor detection of gasoline spills and leaks in the subsurface. This simple sensor consists of a metal leads connected by a conductive polymer (Fig.5.1). This company manufactures only the sensor itself, and they desire to team with vendors or customers to implement the sensors. Their focus appears to be on the passive detection of hydrocarbons like gasoline, and they do not appear to have developed integrated software packages for data acquisition and interpretation.

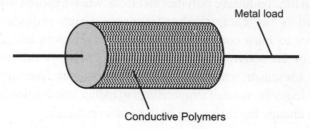

Fig. 5.1 Polymer-absorption sensor developed by adsistor Technology.™

Several companies have incorporated the Adsistor™ sensor into their own vapor sensing devices for use in subsurface sensing: For example

SiteSentinel ™ Hydrocarbon Vapor Sensor used by PetroVend ™ (www.petrovend.com)

Veeder-Root ™ Vapor Sensor (www.veeder.com)
VedoScan ™ Hydrocarbon Vapor Sensor
http://www.biorenewal.com/vadohydro.htm).
Cost = $200-$300

The first two companies provide services to petroleum stations ranging from automated fueling systems to monitoring systems. The third company focuses on applications in soil remediation and long-term monitoring. None of the companies exhibited information regarding quantification of the sensor data (e.g., discrimination of analytes, characterization of the contaminant plume, etc.).

Sandia National Laboratories has developed chemiresistors using polymer films deposited on microelectrodes. Rather than using a single electrode and conductive polymer, the chips used at Sandia can house an array of chemiresistors . The Sandia sensor-array chip has several advantages over the Adsistor ™

(1) an array of differing sensors can be used to identify different volatile organic compounds; a single Adsistor cannot; (2) the footprint of a cylinder is not conducive to temperature control and measurement like the Sandia chip, which is also much smaller; (3) the chip geometry and preconcentrator design allows Sandia to look for improvements in sensitivity in an integrated MEMs produced package.

Advantages: Chemiresistors are small, low power devices that have no moving parts and have good sensitivity to various chemicals. As a result, they are amenable to being placed in-situ in monitoring wells. Another big advantage for chemiresistors in comparison to the standard electrochemical sensors is that they do not require liquid water to work properly.

Limitations: May not be able to discriminate among unknown mixtures of chemicals. Some polymers react strongly to water vapor. Uncertain durability of polymers in subsurface environments; need to develop robust packaging. May need pre-concentrator to detect very low limits (for regulatory standards). Although reversible, signal may experience hysteresis and a shift in the baseline when exposed to chemicals.

5.2.2.1.2 Catalytic Bead Sensors

Catalytic bead sensors are low-power devices (50-300 mW) that have been used for many years in the detection of combustible gases, particularly methane in air. They are used widely in portable gas detection instruments. The catalytic bead sensor is comprised of a passive and active element, both made from an embedded coiled platinum wire in a porous ceramic (Fig. 5.2). The active element is coated with a catalyst such as platinum, and the passive element is coated with an inert glass to act as a reference element to compensate for environmental conditions. Both elements are heated to a prescribed operating temperature ranging from 300^0C to 800^0C. When a combustible gas such as methane comes in contact with the elements, the vapor combusts on the active elements and the active element increases in temperature. As a result, the resistance of the platinum coil changes. The two elements are connected to a Wheatstone bridge circuit, so the changes in resistance are measured as changes in voltage.

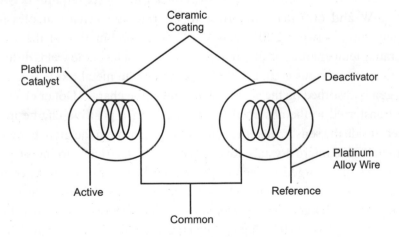

Fig. 5.2 Diagram of catalytic bead sensor.

GasTech™ (http://www.gastech-inc.com/) has a portable soil vapor monitor that uses the catalytic bead sensor for detecting combustible hydrocarbon gases. The benefits of this monitor are that it has the ability to eliminate methane from the readings. Methane occurs naturally in the subsurface, so it can provide false reading. This is a hand-held device that provides discrete readings of real-time gas concentrations. Sampling occurs through a probe, and an internal pump draws the sample. It is lightweight (~5 pounds) and has a built-in data logger. User-defined alarm points (visual and audible) can be programmed. The battery can last up to 20 hours. Cost ~$2000.

Advantages: This unit is very portable and can allow the operator to distinguish between methane and other volatile hydrocarbon vapors.

Limitations: This unit is not amenable for long-term in-situ operation. The catalytic bead sensor requires elevated temperatures for operation. Internal pump is required to sample gas. Sensitivity to aromatic and halogenated hydrocarbons is questionable.

5.2.2.1.3 Metal-Oxide Semiconductor Sensors

Metal-oxide thick film gas sensor

The metal-oxide semi-conductor (MOS) thick film sensor [20] is comprised of a metal oxide that is sintered on a small ceramic tube. A coiled wire is placed through the center of the ceramic tube to act as the sensor heater. Metal wires provide electrical contact between the tin oxide and the rest of the electronics. The MOS thick film sensor requires between 300 mW and 600 mW of power to operate the sensor at elevated temperatures between 200^0C and 450^0C. The combination of the sensor operating temperature and the composition of the metal oxide yields different responses to various combustible gases. When the metal oxide is heated, oxygen is absorbed on the surface with a negative charge. Donor electrons are transferred to the absorbed oxygen, leaving a positive charge in the layer. Inside the sensor, electrical current flows through the grain boundary of metal oxide micro crystals. Resistance to this electrical current is caused by negatively charged oxygen at grain boundaries. In the presence of a reducing gas, a surface catalyzed combustion occurs and the surface density of negatively charged oxygen decreases, thereby decreasing the resistance of the sensor. The relationship between the amount of change in resistance to the concentration of a combustible gas can be expressed by a power-law equation.

Figaro (http://www.figarosensor.com/) has developed thick film MOS sensors for detection of various gases. It requires a heater current of 42mA and it has a quick response time (seconds) when exposed to the target gas. The sensor shows a stable baseline for periods over a year. A range of popular Figaro gas sensors is enlisted below.

TGS 813-for the Detection of Combustible Gases

The TGS 813 has high sensitivity to methane, propane, and butane, making it ideal for natural gas and LPG monitoring. The sensor can detect a wide range of gases, making it an excellent, low cost sensor for a wide variety of applications.

The Fig. 5.3 below represents typical sensitivity characteristics, all data having been gathered at standard test conditions. The Y-axis is indicated as sensor resistance ratio (Rs/Ro), which is defined as follows:

R_s = Sensor resistance of displayed gases at various concentrations
R_0 = Sensor resistance in 1000ppm methane

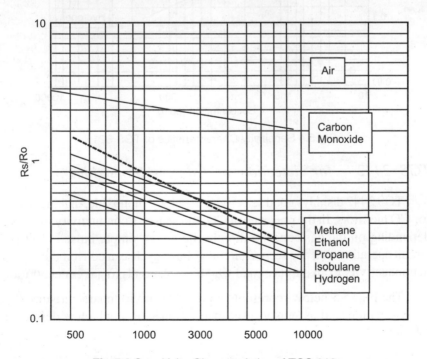

Fig 5.3 Sensitivity Characteristics of TGS 813

TGS 821 - Special Sensor for Hydrogen Gas

The TGS 821 has high sensitivity and selectivity to hydrogen gas. The sensor can detect concentrations as low as 50ppm, making it ideal for a variety of industrial applications.

The Fig. 5.4 below represents typical sensitivity characteristics. The Y-axis is indicated as sensor resistance ratio, which is defined as follows:

R_s = Sensor resistance of displayed gases at various concentrations
R_0 = Sensor resistance at 100ppm of hydrogen.

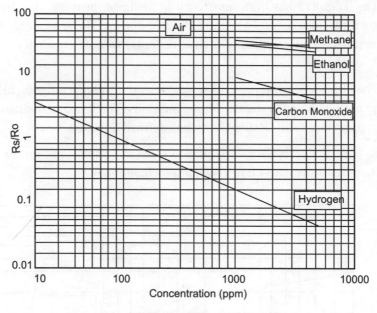

Fig.5.4 Sensitivity Characteristics of TGS 821

TGS 2442 - for detection of Carbon Monoxide

TGS 2442 displays good selectivity to carbon monoxide, making it ideal for CO monitors. In the presence of CO, the sensor's conductivity increases depending on the gas concentration in the air. A simple pulsed electrical circuit operating on a one second circuit voltage cycle can convert the change in conductivity to an output signal which corresponds to gas concentration.

The Fig. 5.5 below represents typical sensitivity characteristics. The Y-axis is indicated as sensor resistance ratio (R_s/R_0), which is defined as follows:

R_s = Sensor resistance of displayed gases at various concentrations
R_0 = Sensor resistance in 100ppm CO

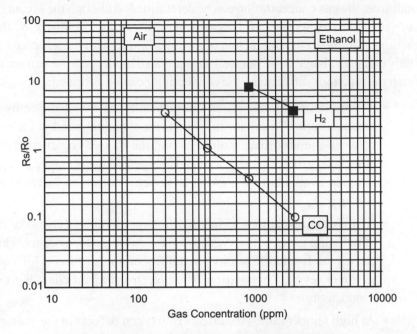

Fig.5.5 Sensitivity Characteristics of TGS 2442

Advantages: The MOS sensors have high sensitivity to combustible gases (e.g,. hydrogen, carbon monoxide, methane, ethane, propane, alcohols, etc.). They are compact and durable. The cost is also relatively low.

Limitations: U.S. EPA (1995) performed tests on some Figaro MOS sensors and found that they had more drift during exposure to xylene than the polymer-absorption sensors. The MOS sensor has a fair amount of sensitivity to water humidity, which may be problematic in subsurface environments. Sensitivity to aromatic and halogenated hydrocarbons is questionable.

Metal-oxide thin film Sensor [21]

The metal oxide is prepared in the form of a thin film on a substrate, which ensures close contact between the detector material and the gas phase. The electrical resistance of the gas-sensitive thin film is measured using electrodes prepared from precious metals. An electrical heater heats this chip to its operating temperature in the range of several hundreds of degrees, at which point a chemical reaction between the gas and the metal oxide changes its electrical conductivity. The measured resistance of the oxide is a function of the temperature and gas concentration. For a constant temperature, the function can be determined inversely, and under favorable conditions, the gas concentration can be determined. Although the layout of a metal oxide gas sensor can be easily described, the chemistry of the reactions that impart the interaction of the metal oxide with the gas phase is highly complex. Three basic but well substantiated models can be named, to which the reactions influencing the conductivity can be assigned:

- At low temperatures, gases interact with the metal oxide by a 'chemisorption' mechanism: a covalent chemical bond is formed between the absorbate molecules and the oxide. The chemically absorbed particles receive a partial charge and the opposite charge is made available to the oxide as a free electron to increase its conductivity.

- At higher temperatures, surface reactions predominate: an oxidation of reducing gases with oxygen from the oxide surface occurs. This causes the formation of oxygen defects on the surface, which act as donor levels, i.e. contribute free electrons-again changing the conductivity.

- At high temperatures, the effects of oxygen defects in the volume of the metal oxide predominate. A balance in the concentration of the oxygen defects in the volume of the metal oxide with the oxygen content of the surrounding gas atmosphere is reached. This results in the existence of a variable density of donors in the volume and therefore variable charge-carrier density and electrical conductivity.

Siemens have developed metal-oxide (Ga_2O_3, $SrTiO_3$, WO_3) based thin film gas sensors operating at temperatures up to 1000^0 C. to manufacture the sensor chips, ceramic base substrates (Al_2O_3) were provided with structured layers using thin-film processes (sputtering, photo-lithography, ion etching), The circuit strictures were applied using the thin-film process when fine structures (10 to 100ìm) were used. Response characteristics are shown in Fig. 5.6.

Advantages: As referred to its thick film version the metal oxide thin film gas sensors exhibit a series of fundamental advantages. The fault-prone influence of grain boundaries on conductivity does not occur at higher operating temperatures (500 to 1000°C). The mobility of the charge carriers is no longer dependent upon the grain boundaries but upon the crystal lattice. The conductivity is controlled in a stable and reproducible manner directly by the change in charge-carrier density. The high operating temperature also ensures an always-clean surface that can react in a defined manner. Extremely short response times of 5 ms can be achieved with layers based on perovskite $SrTiO_3$ at operating temperatures of 1000°C.

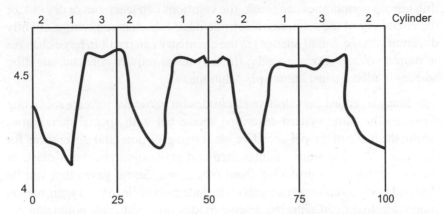

Fig.5.6 Response time of acceptor-doped SrTiO3 in the exhaust flow at engine test rig: cylinders 1 and 3 are lean, cylinder 2 is rich

Limitations: Same as its thick-film counterpart.

5.2.2.2 Potentiometric and Amperometric Sensors

Potentiometric and amperometric sensors [22] employ an electrochemical cell consisting of a casing that contains a collection of chemical reactants (electrolytes or gels) in contact with the surroundings through two terminals (an anode and a cathode) of identical composition. For gas sensors, the top of the casing has a membrane which can be permeated by the gas sample. Oxidization takes place at the anode and reduction occurs at the cathode. A current is created as the positive ions flow to the cathode and the negative ions flow to the anode. Gases such as oxygen, nitrogen oxides, and chlorine, which are electrochemically reducible, are sensed at the cathode while electrochemically oxidizable gases such as carbon monoxide, nitrogen dioxide, and hydrogen sulfide are sensed at the anode. Potentiometric measurements are performed under conditions of near-zero current. Amperometric sensor are usually operated by imposing an external cell voltage sufficiently high to maintain a zero oxygen concentration at the cathodic surface; therefore, the sensor current response is diffusion controlled.

A common application for potentiometric and amperometric sensors is for water analysis. The most common is the pH sensor system. The basic principle of these devices is that they require two separated carefully controlled liquid reservoirs with two different chemically stable electrodes (called reference electrodes), for example a silver wire with a coating of silver chloride. The pH is measured by the voltage difference between the two reference electrodes, so the unknown sample must be in electrochemical connection with both solutions through a glass membrane. However, these thin porous membranes can break, the solutions can leach out or dry out, or the chemistry of the reference electrode itself can change giving a slightly different voltage. Small changes in the chemistry can result in large changes in output voltage. Consequently, these systems require constant attention and calibration against known pH solutions.

Many so-called ion selective electrodes for particular ions are sold using basically the same system described above but with special membranes taking the place of the pH-sensitive glass that give potential differences for different ions. The same maintenance and calibration problems exist, as well as interference problems from other ions. Some gases that can be detected using potentiometric methods include carbon dioxide, oxygen, carbon monoxide, hydrogen, chlorine, arsenic oxides, and oxidizable pollutants.

Potentiometric CO_2-Sensor (Fig. 5.7)

The most important characteristics of this device are:

- the reference electrode does not need to be sealed
- the signal is independent of oxygen partial pressure
- the sensor signal is thermodynamically well defined and thus does not drift with time
- response-times are in the order of seconds

Principle of a Potentiometric Sensor for Gases is explored below with response characteristics for CO_2 in Fig. 5.8(a), (b).

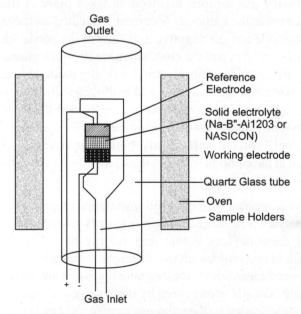

Fig. 5.7 Schematic of Potentiometric CO_2 - Sensor

Cell Reactions

Working Electrode:

$$Na_2CO_3 \rightarrow 2Na^+ + CO_2 + 1/2O_2 + 2e^-$$

Reference Electrode:

$$2Na^+ + 1/2\ O_2 + 2e^- + 6TiO_2 \rightarrow Na_2Ti_6O_{13}$$

Overall Cell Reaction

$$Na_2CO_3 + 6TiO_2 \rightarrow CO_2 + Na_2Ti6O_{13}$$

$$EMF = -\frac{\Delta_r G^O}{2F} - \frac{RT}{2} In \frac{P_{co_2}}{P^O}$$

with:

EMF = "electromotive force" (voltage between reference electrode and working electrode)

$\Delta_r G°$ = standard free reaction enthalpy of the overall cell reaction

F = Faraday's constant

R = universal gas constant

T = operating temperature in K

P_{co_2} = partial pressure of CO_2

P° = standard pressure (1000mbar)

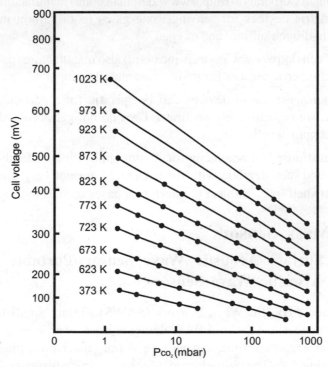

Fig. 5.8(a) Measured Signal as a Function of CO_2 - Partial Pressure.

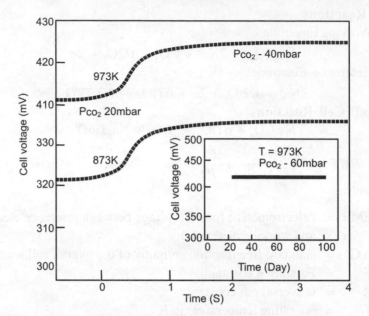

Fig. 5.8(b) Typical Measured Signal as a Function of CO_2 Operating
Temperature

Delphian Corporation (http://www.delphian.com/) manufactures several potentiometric devices for sensing toxic gases (e.g., carbon monoxide, chlorine, hydrogen sulfide) and oxygen.

GasTech (http://www.gastech-inc.com/) also manufactures devices that use potentiometric sensors for toxic gases and oxygen.

Advantages: These devices can be specific for a particular gas or vapor and are typically very accurate. They do not get poisoned and can monitor at ppm levels.

Limitations: Not amenable for in-situ applications. Membranes are sensitive and may degrade with time. Devices are not very durable and have short shelf lives. Subject to interfering gases such as hydrogen.

5.2.3 Mass Sensors

5.2.3.1 Surface Acoustic Wave Sensors/Portable Acoustic Wave Sensors

Surface Acoustic Wave Sensors (SAWS) [23]are small miniature sensors used to detect gases. A SAW device consists of an input transducer, a chemical absorbent film, and an output transducer on a piezoelectric substrate (Fig.5.9). The piezoelectric substrate is typically quartz. The input transducer launches an acoustic wave which travels through the chemical film and is detected by the output transducer. The device runs at a very high frequency 100MHz. The velocity and attenuation of the thin film which can allow for the identification of the contaminant. Heating elements under the chemical film can also be used to desorb chemicals from the device. A signal pattern recognition system that uses a clustering technique is needed to identify various chemicals. SAWS have been able to distinguish organophosphates, chlorinated hydrocarbons, ketones, alcohols, aromatic hydrocarbons, saturated hydrocarbons, and water.

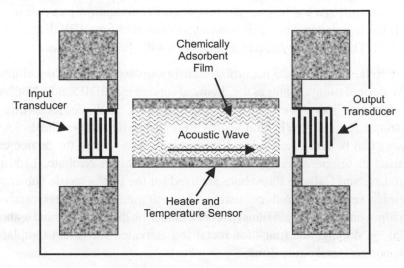

Fig. 5.9 Schematic of SAW device.

Sandia National Laboratories has developed and tested a six-SAW device array and have been able to identify 14 different individual organic compounds over a wide range of concentrations with 98% accuracy (http://gaas6.mdl.sandia.gov/1315.docs/ sawarray.html). They have also been able to identify 21 different binary mixtures of 7 of the compounds with 96% accuracy. They also claim that with random calibration errors of 25%, they were still able to identify individual chemicals with 92% accuracy. They were able to isolate chemicals in the following classes:

Organophosphates, chlorinated hydrocarbons, ketones, alcohols, aromatic hydrocarbons, saturated hydrocarbons, and water.

Sandia National Laboratories has developed an integrated GaAs SAW sensor (http://gaas6.mdl.sandia.gov/1315.docs/sawgaas.html). This microsensor is a SAW device that has been integrated with a microprocessor on a single 5mm × 5mm chip. GaAs is used as the piezoelectric material instead of quartz. This device has the advantages of being small, having reduced power consumption, and having simple packaging.

Sandia National Laboratories has developed a hand held SAW system (http://gaas6.mdl.sandia.gov/1315.docs/paws.html).This system is 5.5"× 3.3"×1.5" (27 in^3) which includes batteries for portable or field operation. This device contains a single SAW sensor. The system uses a pump to pull air samples into the test cavity and data is fed into a computer for control, real-time display, data storage, and pattern recognition. A down-hole monitor has also been developed. Also developed with Texas A&M, New Mexico Institute of Mining and Technology, and industrial partners.

PNL (http://www.technet.pnl.gov/sensor/chemical/project/es4cwsen. html) has developed a portable sensor to detect real-time CW dispersal in the field. They have developed a portable SAW chemical sensor.

ORNL (http://htm29.ms.ornl.gov/diffgroup/zeol.html) is developing SAW devices using zeolites as the chemical capture material. Sandia National Laboratories has developed high temperature SAW devices (http://gaas6.mdl.sandia.gov/1315.does/hightempacoustic.html). These SAW sensors can be fabricated using different materials so that the device can be used at temperatures as high as 525°C. Lithium Niobate, Lithium Tantalate, and Gallium Phosphate are used for the piezoelectric substrate. Chemical sensing materials consisting of pure or mixed noble metal catalytic thin films, binary oxide thin films (zirconia, titania, tin dioxide) with and without metal ion doping, and transition metal ion activated surfactant-templated mesoporous metal oxide films.

Advantages: Small, low power, no moving parts other than the high-frequency excitation, good sensitivity to various chemicals. Can detect chemicals in very low concentrations.

Limitations: May not be able to discriminate among unknown mixtures of chemicals. Some polymers react strongly to water vapor; uncertain durability in subsurface environments.

5.2.3.2 Microcantilever Sensor

Microelectromechanical system (MEMS) can be composed of multiple micron-thick cantilevers (visualize miniature diving boards) that respond by bending due to changes in mass. Appropriate coatings are applied to the cantilevers to absorb chemicals of interest. This particular technology has been used for developing infrared sensors to "see" images in darkness (http://www.sarcon.com/), but commercial devices using microcantilever sensors to detect volatile organic chemicals were not found. Active research in this area is being performed by Oak Ridge National Laboratory (http://lsd.ornl.gov/babs/thundat/Thundat.htm).

5.2.4 Optical Sensor

5.2.4.1 Fiber Optic Sensors

Fiber optic sensors [24] are a class of sensors that use optical fibers to detect chemical contaminants. Light is generated by a light source and is sent through an optical fiber. The light then returns through the optical fiber and is captured by a photo detector. Some optical fiber sensors use a single optical fiber while others use separate optical fibers for the light source and for the detector. There are three general classes of fiber optic sensors (Fig.5.10). the first type is completely passive. A spectroscopic method can be used to detect individual types of contaminants. This method involves sending a light source directly through the optical fiber and analyzing the light that is reflected or emitted by the contaminant. The refractive index of the material at the tip of the optical fiber can be used to determine what phases (vapor, water, or NAPL) are present. A second class of fiber optic sensors consists of a fiber optic sensor with a chemically interacting thin film attached to the tip. This film is formulated to bind with certain types of chemicals.

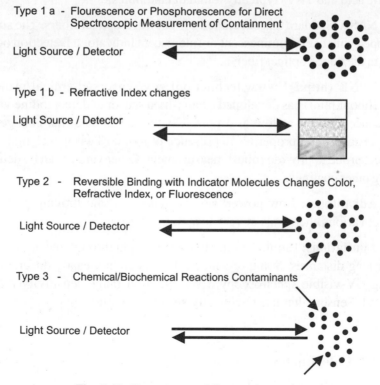

Fig. 5.10 Three types of fiber-optic sensor

Contaminant concentration can be found by measuring the color of the thin film, the change in refractive index, or by measuring the fluorescence of the film. The third type of fiber optic sensors involves injecting a reagent near the sensor. This reagent reacts either chemically or biologically with the contaminant. The reaction products are detected to give an estimate of the contaminant concentration.

LLNL developed an optical fiber sensor for TCE, using pyridine as the reagent, and has performed both lab and field tests. This system uses separate light source/detector fiber optic lines. The estimated error for measurements between 25-500 ppb was 10%.

EIC Laboratories (http:/ /www.eiclabs.com/resspeccone1.htm) has developed two types of optical fiber sensors that are encased inside a cone penetrometer. One sensor detects the refraction index changes and is used to detect the presence of DNAPL contaminants. This probe has been tested in the field. The other is a Raman fiber optic probe that identifies DNAPLS by measuring the raman scattering pattern generated from chemical contaminants that have adsorbed on the surface of specially treated metals. This probe has been used in the field to detect nitrate-based aromatics (picric acid and TNT) typically found in landmines.

NASA Goddard has developed sol-gel filled optical fibers. The sol-gel is doped with chemiluminescent or fluorescent indicators. Detection occurs within the doped optical fiber.

PNNL (http:/ / www.technet.pnl.gov/sensors/chemical/projects/ es4i2fibopt.html) has developed a hand-held sensor to detect iodine vapors for concentrations as low as 5 ppm. Coating developed that changes the optical absorption properties in presence of iodine. Two-wavelength fiber optical probe to provide robust measurement. Other vapors can be detected using other coatings.

Advantages: Low power; several types have no moving parts; can detect various chemicals at very low concentrations.

Limitations: Limited ability to transmit light through the optical fiber over long distances. Some organic pollutants are not easily differentiated using UV-visible spectroscopy. Concentration range sensitivity may be limited. Sensors that use chemically sensitive coatings may degrade with time.

5.2.4.2 Colorimetry

Pocket colorimeter test kits can be used to measure trace levels of contaminants. They work by analyzing the color of contaminated water that has been mixed with a particular chemical reagent. Hach sells pre-measured unit-dose reagent that react with water samples. To test water samples, the pocket colorimeter compares a reacted sample with a sample blank and yields results in concentration units.

Hach (www.hach.com) makes a variety of calorimetry kits. The only chemical of interest on the list of available kits is total petroleum hydrocarbons (TPH in water and in soil). TPH levels in water can be found between 20-200 ppm in soil and 2-10 ppm levels in water. Most of the kits available are focused on metals (chromium, copper, iron, etc.) or for parameters of interest for drinking water (chlorine, fluoride, hardness, pH, etc.). Products include hand-held devices and strips that can be dipped in water.

American Gas & Chemical Co. (http://www.amgas.com/ttpage.htm) also makes color strips to detect a variety of toxic gases (e.g., carbon monoxide, hydrogen sulfide, chlorine). These strips are intended to be worth for visible detection of toxic gases.

CHEMetrics, Inc.sells the RemediAid kit (http://www.chemetrics.com/TPH.html) that will determine total petroleum hydrocarbons across a wide range of soil types and petroleum products, so it is suited to use in the field. The system enables the user to run 10 tests concurrently, providing the potential to run 25 tests in one hour.

RemediAid can also be calibrated to measure quantitative amounts of specific petroleum products including: BTEX, PAH, diesel fuel, leaded and unleaded gasoline, weathered gasoline, brent crude and lubricating oil. The LED based calorimeter gives ppm hydrocarbons. The cost of the unit is ~$800.

Advantages: Portable, simple to use. Visual evidence of gas detection event. Not prone to interferences.

Limitations: Limited chemical sensitivity to individual VOCs; needs actual water samples (cannot be used in situ); most kits do not meet U.S. EPA method requirements and may not be used for compliance monitoring. Requires visual inspection and is not amenable to long-term in-situ applications.

5.2.4.3 Infrared Sensors

Infrared sensors [25] can be used to detect gases, which, in general, have unique infrared absorption signatures in the 2-14 μm range. The uniqueness of the gas absorption spectra enables identification and quantification of chemicals in liquid and gas mixtures with little interference from other gases. These devices are typically comprised of a source of infrared radiation, a detector capable of seeing the infrared radiation, and a path between the detector and source that is exposed to the gas being detected. When gas in the path absorbs energy from the source, the detector receives less radiation than without the gas present, and the detector can quantify the difference.

GasTech (http://www.gastech-inc.com) has developed an FX-OIR Infrared Single Gas Transmitter. This device uses a dual infrared technology to detect hydrocarbon gases (e.g., hexane (0-1000 ppm), methane (0-5000 ppm), propane (0-2000 ppm)). Cost ~$1200.

Ion Optics (http://www.ion-optics.com/MEMSGasSensorDetailed.htm)

has developed an infrared sensor that uses a silicon element that acts as both the infrared emitter and detector. The emitted radiation and a lower temperature will exist, which is detected.

Advantages: These devices can be made to identify specific gases; they require less calibration than other sensors; good durability with minimal maintenance.

Limitations: They can only monitor specific gases that have non-linear molecules; they can be affected by humidity and water; they can be expensive; dust and dirt can coat the optics and impair response, which is a concern in in-situ environments.

5.3 Conclusions

In this article an overview of a sequential development of sensors for detecting the chemical agent has been presented. A comparative discussion on the application of chemical sensors for qualitative and quantitative measurement has also been presented in a detailed manner. From the very first development of chromatography and spectrometry as the laboratory equipments for the detection of chemicals to the ppm levels, with the modern advancement of microelectronics, microsensors with extremely high sensitivity are available for automotive emissions and for monitoring environmental pollutions. Of late, the technology of micro-electro-mechanical-systems (MEMS) has brought a revolution in the field of sensors in general and in the area of chemical sensors in particular. The silicon planar technology has made possible the development of the latest sensor arrays with complete electronic circuitry in the form of monolithic sensors. Parallely, hybrid sensor technology is also creating enormous interest among the scientists and engineers for large-scale commercial applications. The area of chemical sensors in respect of the materials development and sensor devices is an extremely important and interesting topic of R & D. Finally the outcome of intensive research has given birth to so called intelligent sensors or smart sensors. In order to keep pace with the global development of chemical sensors it is very much necessary to develop a module on education for chemical sensor technology in the high school and college levels.

Some of the chemical sensors which have been developed in the Material Science Centre of I.I.T Kharagpur are:

(a) Room temperature hydrogen sensors using thick and thin films of ZnO deposited on different substrates like glass, quartz, alumina and Si by a modified chemical vapour deposition method,

(b) Conducting polymers like poly - aniline and poly - pyrrole,

(c) High temperature hydrogen sensors using ZnO, TiO_2 and SiC, and

(d) Odor sensors as a part of the electronic nose to verify the quality of food products using pure high resistive ZnO thin film deposited on optically flat glass and quartz substrates by the modified CVD method. The sensor can detect (as a chemo - resistor) dimethylamine (DMA) and tri methylamine (TMA) vapours.

5.4 References

1. Yu.G. Vlasov, J.Anal. Chem..47(1), (1992), 80

2. Wieslaw P. Jakubik, Marin W. Urbanczyk, Stanislaw Kochowski and Jerzy Bodzenta, Sensors and Actuators B: Chemical 82(2-3), (2002) 265-271.

3. Tadashi Takada, Toru Maekawa and Naganori Dougami, Sensors and Actuators B: Chemical, 77(1-2) (2001) 307-311

4. C.K. Kim, J. H. Lee, Y. H. Lee, N. I. Cho and D. J. Kim, Sensors and Actuators B: Chemical, 66(1-3), (2000) 116-118

5. K. Galatsis, Y. X. Li, W. Wlodarski, E. Comini, G. Sberveglieri, C. Cantalini, S. Santucci and M. Passacantando, Sensors and Actuators B: Chemical 83(1-3), (2002) Pages 276-280.

6. R. Uhl, T. Reinhardt, U. Haas and J. Franzke, Spectrochimica Acta Part B: Atomic Spectroscopy, 54(12) (1999) 1737-1741.

7. Uemura, T; Nishida, K; Sakakida, M; Ichinose, K; Shimoda, S; Shichiri, M, Frontiers of Medical and Biological Engineering: the International Journal of the Japan Society of Medical Electronics and Biological Engineering, 9(2), (1999), 137-153.

8.   J. Frank and H. Meixner, Sensors and Actuators B: Chemical, 78(1-3), (2001), 298-302.

9.   M. H. Madhusudhana Reddy and A. N. Chandorkar, Thin Solid Films, 349 (1-2) (1999), 260-265.

10. P. Mielle and F. Marquis, Sensors and Actuators B: Chemical, 76(1-3) (2001), 470-476.

11. P. -M. Schweizer-Berberich, S. Vaihinger and W. Gopel, Sensors and Actuators B: Chemical, 18 (1-3), (1994), 282-290.

12. Journal of Chromatography A, 926(2), (2001), 291-308.

13. W. Fiddler, R. Doerr, R.A. dan Gates, J. Assoc. off. Anal. Chem. 74 (1991) 400-403.

14. R.C. Lundstrom, L.D. Raciot, J. Assoc. off. Anal. Chem,. 66(5) (1983) 1158-1163.

15. Ching Wu, Wes E. Steiner, Pete S. Tornatore, Laura M. Matz, Wiliam F. Siems, David A. Atkinson and Herbert H. Hill, Jr, Talanta, 57(1), (2002), 123-134.

16. Susumu Nakamaya, Ceramics International, 27(2), (2001), 191-194.

17. Jing-Shan Do and Wen-Biing Chang, Sensors and Actuators B: Chemical,72 (2), 2001, 101-107.

18. Frank Zee and Jack W. Judy, Sensors and Actuators B: Chemical, 72 (2), (2001), 120-128.

19. Frank Zee and Jack W. Judy, Sensors and Actuators B: Chemical, volume 72 (2) (2001), 120-128.

20. S. M. Sze, Semiconductor Sensors, John Willey & Sons, Inc.p. 399.

21. K. Galatsis, Y.X.Li, W. Wlodarski and K. Kalantar-zadeh, Sensors and Actuators B: Chemical, 77(1-2), (2001), 478-483.

22. Kuo-Chuan Ho and Wen-Tung Hung, Sensors and Actuators B: Chemical, 79 (1) (2001), 11-16.

23. S. M. Sze. Semiconductor Sensors, John Willey & Sons, Inc. p.124.

24. Sheila A. Grant, Joe H. Satcher Jr. and Kerry Bettencourt, Sensors and Actuators B: Chemical, 69 (1-2), (2000), 132-137.

25. A. Chtanov and M. Gal, Sensors and Actuators B: Chemical, 79 (2-3), (2001), 196-199.

26. S. Basu and A.Dutta, Materials Chem. Phys. 47(1997) 93.

27. S. Roy, S. Sana, B. Adhikari and S. Basu, Journal of Polymer Materials 20 (2003) 173.

28. S. Roy, C. Jacob and S. Basu, Sensors & Actuators B 94 (2003) 298.

29. S. Roy and S. Basu, J of materials Science: materials in Electronics 15 (2004) 321-326.

| Chapter 6 |

Biochemical Sensors

B. Adhikari

Abstract

Biochemical sensors are sensors containing a sensing biological component, an immobilized agent (enzyme, tissue cell, etc.) and a transducer to detect or measure a chemical compound (analyte). An interaction of the immobilized agent with the molecule to be analyzed is transduced into an electronic signal. Different aspects of principles and design of biochemical sensors and their applications are presented in the chapter.

6.1 Introduction

Biochemical sensors have evolved from the combined activities of chemical substances and bio-signals. It is known as a bioelectronic analytical device containing a sensing biological component, an immobilized agent, and a transducer to detect or measure a chemical compound (analyte). An interaction of the immobilized agent with the molecule to be analyzed is transduced into an electronic signal. The immobilized agent, e.g., an enzyme, sequence of enzymes, organelle, whole cell, tissue slice, antibody or binding protein etc., remains in intimate contact with the transducer which can be one for measuring weight, electrical charge, potential, current, temperature or optical activity. Biochemical sensors are simple to use, provide single-step measurement, reagentless devices, disposable and compatible with conventional data-processing technologies.

The purpose of using biochemical sensors is to analyze and estimate variety of substances in biological samples. Originally such analytical devices were based on the withdrawal of samples from the body and were the normal practices till late eighties. Later online analysis and monitoring of the bioanalyte concentrations attracted attention of people. Not only that a high degree of selectivity of the modern biochemical sensors eliminated prior separation of analyte component from a multicomponent sample in natural matrix, e.g., in body fluids like blood, urine etc.

Biochemical sensors must possess good biocompatibility, high selectivity and specificity, resistance to degradation or fouling in the biological environment. To achieve these characteristics, polymers are frequently required and used in fact. A typical biochemical sensor contains three primary elements: a physical component (measurement), a membrane component and a biological (biospecific) component [1]. Some polymers might be used in the physical component (e.g., optical fibers, piezoelectric elements, insulators). The membrane component may be one or more polymeric biomaterials. This is used for protecting the physical element, for providing biocompatibility, for immobilizing the bioactive element, and for offering transport selectivity. The bioactive component typically contains enzymes, antibodies, or environmentally sensitive dyes. An important concern for biochemical sensors and other implanted electronic devices is the material needed to protect the device from the biological environment and to protect the recipient from the device. Silicone elastomers, polyurethanes, plasma-deposited polymeric films, polyimides and a variety of metals such as stainless steel and titanium are used for such protection. Some important characteristics of biochemical sensors are:

- Require simple or no sample preparation steps.
- Provide high selectivity and specificity in direct measurements.
- Have a detection limit of minimum concentration of the analyte.
- Continuously analyze one or more analytes within the body.
- Have better accessibility for measurements within body.
- Provide continuous and reversible response.
- Helps measurement with minimum perturbation of the sample.
- Should possess good biocompatibility.
- Scope for miniaturization and portability.
- Low cost and use convenience.
- Should be resistant to degradation or fouling on the electrode in the biological environment.

6.1.1 Immobilization of Bioselective/Biospecific Component

In order to have a proper biosensing response, the chemically or biologically selective components must be intimately associated with an energy transducer. In other words, molecular recognition agents of either biotic or abiotic origin must be immobilized for selectivity or specificity towards many types of analyte.

Established methods of immobilization of various biologically active substances on a transducer are based on adsorption, entrapment, crosslinking, and covalent bonding. For example, biosubstance is immobilized in between porous membrane, which is attached onto an electrode or can be covalently bound to the electrode through spacer arms or immobilized on beads. Covalent or entrapment immobilization in a conductive polymer such as polyaniline or polypyrrole is also a good method. Isolated, soluble enzymes, cells and tissues may be held to a surface by simple containment behind a permeable, protective membrane such as dialysis tubing or micro porous polymers.

Adsorption of enzymes and antibodies are also followed in sensor design, e.g., protein macromolecules adsorb on metal/metal oxide transducer surfaces as well as on carbon and glasses. Increasing use of spontaneously adsorbed species to develop organized thin films (OTF) has done a revolutionary job to control and design interfaces, e.g., self-assembled monolayers (SAMs) of organic thiols on gold (Au-S-R-X) and L-B films [2-4].

Another immobilization strategy is the entrapment and cross-linking of biochemically selective agents in polymer network produced on the transducer surface by electrochemical, photochemical or plasma polymerization [5-9].

Covalent immobilization particularly applied to proteins, for enzymes and antibodies, are normally bound through involvement of the amino, carboxyl, sulfhydryl or aromatic side chains of the amino acids in the macromolecule [10-15]. Covalent immobilization provides a better surface structure, which enhances the amount of reagent on the surface. Furthermore, covalent binding produces a more durable surface modification. Covalently bound compounds do not leach as badly from a reaction layer as some entrapped species do along with a long-term stability under severe experimental conditions such as exposure to flowing samples, stirring, washing etc.

6.1.2 Biocompatibility

In addition to the engineering problems, such as encapsulation, reliability, durability, reproducibility, need for recalibration encountered in biosensors etc., problem of unwanted effects in the body calls for a proper attention while designing and making a biosensor. In other words, it must be biocompatible, so as not to cause toxication, thrombosis or inflammatory effects due to irritation of blood vessels or tissue.

Meyerhoff has discussed in a review [16] intraarterial measurements of blood gases and electrolytes. Normal ranges of analyte, device sizes, flow-rate effects, stability and sterilization requirements are described. For non-hazardous performance, biocompatible polymers such as polyurethanes or silicone rubber are used. Poly (2-hydroxyethyl methacrylate) (poly-HEMA) encapsulation worked well in a potentiometric device [17]. For implantable biosensors both blood and tissue compatibility should be given due consideration. Encapsulating membrane must have adequate permeability for the sensing mode such as the permeability of the ions for accurate and stable signals.

6.1.3 Sterilization

Development of nonfouling and sterilizable biochemical sensors for bioreactors such as human body or fermentors is a very difficult and complex task. They need to be produced either under sterile conditions or the biological macromolecules are added in sterile form. Some details of sterilization are discussed in the literature [18-20].

6.1.4 Chemical Selectivity and Molecular Recognition

Since the concentrations of biological analytes are very low, e.g., concentration of some blood proteins may be as low as few micrograms per liter, their specific estimation requires a discrimination ratio or selectivity of

10^7-10^8. So molecular recognition is achieved through affinity pairs, i.e., molecules to react selectively with the other member of the pair, e.g., antibody-antigen, lectin-carbohydrates etc.

Chemical selectivity is achieved by chemical process and molecular recognition relates to very specific interactions implying both bonding complementarity and biological function. In the biosensor devices immobilized biotics such as enzymes, antibodies, cells and tissues exhibit their biological function derived from molecular recognition through binding specificity on to a substrate through their active sites. In these sensors biological molecular recognition is utilized in the chemical transduction step. A general classification of molecular recognition components based on biotic chemical transduction mode is presented in Fig. 6.1. In other sensors, with selectivity based on abiotic chemical transduction, partially selective metal-ligand binding, partitioning between phases, and size or charge exclusion are examples of selectivity without true molecular recognition [21].

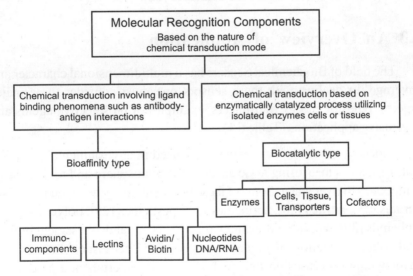

Fig. 6.1 General categories of molecular recognition components based on biotic chemical transduction mode.

6.2 Measuring Principles in Biochemical Sensor Design

In biochemical sensor device a transducer is combined with a biologically active material. The transducer, remaining in intimate contact with the biologically sensitive material, measures parameters like weight, electric charge, potential, current, temperature or optical activity. Biologically active species used in the device may be an enzyme, a multienzyme system, a bacterial cell or other whole cell, antibody, antigen, a receptor, mammalian or plant tissue. Such devices can specifically determine sugars, amino acids, alcohols, lipids, nucleotides etc. The major principles used to detect these substances are electrochemical, optical transducers, thermal and piezoelectric. Electrochemical principle is applied in oxygen sensor, glucose sensor, bilipid layer sensor for acetylcholine, human IgG detection on protein A. Optical transducers are used to detect fibrinogen-antifibrinogen (ellipsometry), albumin and IgG (reflectometry), measurement of CO_2 from the urease reaction with urea for urea detection (photoacoustic) and glucose sensor with matrix-bound peroxidase immobilized on a photodiode (luminescence). Thermistors are used for sensing cholesterol with cholesterol oxidase as the immobilized biocatalyst. In piezoelectric principle, a surface acoustic wave (SAW) device is used for assay of human IgG. Some details of energy transduction modes are available [22].

6.3 An Overview of Application

The field of Biochemical Sensors has a multidimensional character and coverage. In dealing with these sensors one must consider three major aspects: types of energy transducers, examples of bioselective agents and biosensing applications (Table 6.1).

Biochemical sensors are extensively used in environmental pollution control such as measuring toxic gases in the atmosphere and toxic soluble compounds in river water. Some of these pollutants are heavy metals [23,24], nitrates [25], nitrites [26], herbicides [27], pesticides [28], polychlorinated biphenyls [29], etc. In food manufacturing process and assessment of food quality the estimation of organic compounds is necessary. In industrial fermentation processes on-line analysis of raw materials and products is also necessary. The use of enzyme sensor can help direct measurement of such compounds including organic pollutants for the maintenance of natural environments. Hydrogen peroxide because of its ability to oxidize, bleach and sterilize is used in food, textile and dye industries. Enzyme sensors as per the following equation can directly measure it and the liberated oxygen can be detected by oxygen electrode.

$$H_2O_2 \xrightarrow{\text{Catalase}} H_2O + \tfrac{1}{2}O_2$$

This technique using enzyme sensor is faster and more convenient than the classical techniques like colorimetric and volumetric methods.

Assessment of freshness of food products is essential. Various compounds such as amines, carboxylates, aldehydes, ammonia, sulfur compounds and carbon dioxide are liberated during putrefying of fish, meat and green vegetables. Measurement of these compounds by conventional methods involves complicated operations such as extraction, centrifugation, steam distillation, and titration all of which seriously lack reproducibility and continuity. On the contrary an enzyme sensor can measure the freshness of the food products by estimating those putrefied products in a more easy and simple manner. Karube et al. [30] have developed a monoamine oxidase-collagen membrane based enzyme sensor and an oxygen electrode for the determination of monoamines in food analysis including the determination of monoanines in meat pastes.

A rapid and simple amperometric method for the determination of choline over an enzymatic method which required long incubation time, more reagents, and expensive enzymes has been developed using choline oxidase immobilized on a partially aminated polyacrylonitrile membrane and an oxygen probe [31]. The consumption of dissolved oxygen by the following enzymatic reaction was measured amperometrically.

$$\text{Choline} + 2O_2 + H_2O \xrightarrow{\text{Choline oxidase}} \text{Betaine} + 2H_2O_2$$

Gargullo and Michael [32] prepared amperometric microsensors by immobilizing horseradish peroxidase and choline oxidase onto carbon fiber microcylinder (7 or 10 pm dia and 200-400 pm long) electrodes with a cross-linkable redox polymer for the detection of choline in the extracellular fluid of brain tissue. When microsensors were operated for the detection of cholin at applied potential of -0.1 V vs SCE, the electrode did not detect ascorbate and other easily oxidizable interferant molecules present in brain tissue. Since ascorbate can interfere with the response to choline by acting as a reducing agent in the enzyme-containing polymer film, a Nafion overlayer was applied in order to reliably detect choline in presence of physiologically relevant concentrations of ascorbate ($\sim$200 μM). The Nafion-coated microsensors showed a detection limit of $\sim$5 μM choline and gave a linear response beyond 100 μM. Amperometric sensors were prepared by Garguilo et al. [33] for determination of hydrogen peroxide, choline, and acetylcholine. They immobllized horseradish peroxidase, (HRP), choline oxidase, and acetylcholinesterase in a cross-linked redox polymer deposited on glassy carbon electrodes. Peroxide sensors, prepared by immobilization of HRP alone, showed 10 nM detection limit and 1 mM linear response. HRP and glucose oxidase were coimmobilized to establish the feasibility of highly efficient bienzyme sensors at low substrate levels. Choline oxidase in place of glucose oxidase produced sensors with submicromolar detection limit and a lInear response up to 0.8 mM. A choline electrode was designed by Leca et al. [34] using photocrosslinkable polymer coating of the transducer tip through in situ polymerization of the sensing layer. They used poly(vinyl alcohol)-bearing styrylpyridinium groups (PVA-SbQ), which was mixed with choline oxidase (ChOD) in a definite ratio. A small quantity of this solution was deposited on a platinum electrode followed by drying and photopolymerization. They reported the influence of the degree of polymerization of the PVA backbone and the quantity of styrylpyridinium on the sensitivity of the resulting electrodes. The electrode had a detection limit of as low as 2.5×10^{-9} M and a response time of as short as ~30 s.

A lactate electrode was developed for the determination of L-lactate in blood serum samples in flow systems. Commercially available lactate oxidase from *Mycobacterium smegmatis* was immobilized on a nylon net, which was fixed on an oxygen probe to provide a simple L-lactate sensor. The serum sample was diluted 20-50 times with 0.1 M citrate buffer (pH 6.0) and passed through the flow cell at 1 ml/min. The oxygen was consumed by the following reaction

$$\text{Lactate} + O_2 \xrightarrow{\text{Lactate oxidase}} \text{Acetate} + CO_2 + H_2O$$

The decrease in dissolved oxygen in the vicinity of the electrode surface causes decrease in current, which was measured. The high activity of the immobilized enzyme permits the use of only 20-100 μl of plasma diluted with citrate buffer [35]. Chaube et al. [36] immobilized lactate dehydrogenase (LDH) on electrochemically polymerized polypyrrole-polyvinylsulphonate (PPy-PVS) composite films, followed by cross-linking using glutaraldehyde, for application to lactate biosensors. They measured the activity of LDH both in solution and immobilized in PPY-PVS films by UV-visible spectrophotometry. These PPy-PVS-LDH electrodes possess a detection limit of 1×10^{-4}M, a response time of about 40 s, and a shelf-life of about 2 weeks and suitable for L-lactate estimation from 0.5 to 6 mM. These electrodes are also stable upto about 40^0C after which the enzyme starts denaturation. A reagentless, disposable biochemical sensor [37] was developed for lactic acid based on a screen-printed carbon electrode containing Meldola's Blue and coated with enzyme L-lactate dehydrogenase (LDH) and its cofactor nicotinamide adenine dinucleotide (NAD) using a cellulose acetate (CA) membrane cast in situ. Baker and Gough [38] prepared an implantable sensor for continuous monitoring of lactate in the bloodstream or tissues. The sensor responds specifically to lactate over a broad concentration range and the response is stable for more than 1 week during continuous in vitro operation at body temperature. The sensor device consists of a lactate electrode based on immobilized L-lactate oxidase coupled with a potentiostatic oxygen electrode and an oxygen reference electrode to account for local oxygen. Enzyme microelectrode array strips were prepared by combination of thick-film technology and laser micromachining procedures to yield effective amperometric test strips for glucose and lactate [39]. Such enzyme microelectrode array strips offer fast and sensitive measurements from small sample volumes and can be used as reliable enzyme-based diagnostic strips.

Rahni et al. [40] have constructed an oxalate oxidase immobilized enzyme electrode for the determination of oxalate in urine. The partial pressure of oxygen, consumed in the following reaction catalyzed by the enzyme, was amperometrically monitored.

$$(COOH)_2 + O_2 \xrightarrow{\text{Oxalate oxidase}} 2CO_2 + H_2O_2$$

The responses of the enzyme electrode to oxalate in urine samples from 14 patients were compared to those obtained by a standard spectrophotometric method. The authors have claimed this method has high sensitivity and specificity to oxalic acid with almost no loss of relative activity due to interference studied. No sample pretreatment is required and they claimed that the method can be modified to be equally useful for the determination of oxalic acid in any biological or nonbiological sample.

Sugawara et al. [41] have reported some ion-channel chemical sensor that mimics biological ion channels for sensitive detection of analytes such as Ca^{++}, ClO_4^-, K^+ ions in solution. A glassy carbon electrode was coated with synthetic lipid multilayer membranes by the Langmuir-Blodgett technique. The membrane functions as an ion channel which can be switched on by the analyte. The analyte-stimulated on/off switching of ion channels was ascribed to a conformational change in the lipid membranes and/or electrostatic interaction between the lipid molecules and marker ions, which are allowed to permeate across the membrane.

Development of accurate, portable, relatively inexpensive and easy to use biochemical sensors in medical diagnostic and monitoring devices has become a high priority all over the world. Some bioanalytical sensors used in clinical assay are shown in Table 6.2. Table 6.3 shows the applications of some selected biosensors. These sensors are based on continuous amperometric or potentiometric estimation of enzyme reaction products without any pretreatment such as extraction, dialysis or filtration, which are normally required in the conventional spectrophotometric method. Enzyme sensors have found extensive application in clinical analysis as well as in process and environmental analyses. They normally deal with the following classes of substances:

Carbohydrates	Hydrogen peroxide
Amino acids	Sucrose
Organic acids	Monoamines
Alcohols	Hypoxanthine
Lipids	Choline
Urea	Inorganic ions
Enzyme activities	
Creatinine	

Diabetes is now a serious global problem that has attracted continuous interest for the development of efficient glucose sensor. Artificial pancreas has come to a reality for dynamically responding to glucose level and control insulin release based on the sensor's response. Clark and Lyons [56] first developed enzyme sensor for glucose analysis and Updike and Hicks [57] used glucose oxidase immobilized in polyacrylamide gel and an oxygen electrode (Fig.6.2). Enzymatic reaction consumes oxygen and decreases the concentration of dissolved oxygen around the enzyme membrane resulting marked decrease in electrode current until a steady state is reached. A Clark-type oxygen electrode for sensing the decrease of dissolved oxygen in contact with enzyme is shown in Fig. 6.2. In this type of electrode the sensing is based on the reduction of oxygen, a cosubstrate for glucose oxidase contained in solution behind the protective membrane. Zhang et al. [58] developed a glucose biosensor based on immobilization of glucose oxidase in electropolymerized o-aminophenol film on platinized glassy carbon electrode, that offered excellent characteristics, including high sensitivity, long-term stability, very short response time, and significantly reduced interference. Jung and Wilson [59] developed a miniaturized, potentially implantable amperometric glucose sensor with a sensing area of 1.12 mm^2 based on the enzymatically catalyzed reaction used by Clark and others. Sangodkar et al. [60] have reported a biosensor array based on polyaniline. They have fabricated polyaniline based microsensors and microsensor arrays for the estimation of glucose, urea, and triglycerides. Microelectronics technology was used to produce gold interdigitated microelectrodes on oxidized silicon wafers. Polymer deposition and enzyme (glucose oxidase, urease, and lipase) immobilization was done electrochemically. They have claimed that an analyte containing a mixture of three substances can be analyzed by a single measurement using a single sample of a few microliters and also claimed this as a general technique, which can be extended to other enzyme-substrate systems, eventually leading to the development of an "electronic tongue".

The estimation of sucrose in food and fermentation industries is necessary. The principle of this sucrose sensor is based on the monitoring of the decrease in dissolved oxygen due to consumption in the following enzymatic reactions [61]:

$$Sucrose + H_2O \xrightarrow{\text{Invertase}} \alpha - D - Glu\cos e + D - Fructose$$

$$\alpha - D - Glu\cos e \xrightarrow{\text{Mutarotase}} \beta - D - Glu\cos e$$

$$\beta - D - Glu\cos e + O_2 \xrightarrow{\text{Glucose oxidase}} D - Glucono - \delta - lactone + H_2O_2$$

On dipping the sensor into a sucrose solution, the current resulting from a reduced amount of dissolved oxygen decreases very rapidly. An amperometric flow injection analysis enzyme sensor for sucrose using a tetracyanoquinodimethane (TCNQ) modified graphite paste electrode has been reported by Filho et al. [62]. Tobias-Katona and Pecs [63] reported a multienzyme-modified ion-sensitive field effect transistor for sucrose measurement.

Brain glutamate is essential for the functioning of the nervous system. Berners et al. [64] studied on-line measurement of brain glutamate in brain microdialysate using poly (o-phenylenediamine) coated tubular electrode where glutamate oxidase was immobilized within the polymer coating. Cosford and Kuhr [65] developed a novel capillary biosensor for the analysis

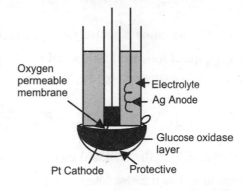

Fig. 6.2 Clark type oxygen electrode.

of glutamate. This allows biological transduction of glutamate signal during transport of analyte from the sampling site to the detector using laser-induced fluorescence. The authors [65] reported the successful attachment of glutamate dehydrogenase to the inner wall of a small diameter fused silica capillary maintaining the enzymatic activity. The high sensitivity of this biosensor and fast response time claim its suitability for *in vivo* application to neurochemical purposes.

6.4 Some Selected Biochemical Sensors

Major techniques for biochemical sensors based on electroanalytical principles are:

Potentiometry
Amperometry
Hydrodynamic voltammetry
Cyclic voltammetry
Small amplitude pulse techniques (differential pulse and square wave voltammetry)
Chronoamperometry and chronocoulometry

Potentiometric method is simple and good for ions and ionizable dissolved gases in solution for bioanalytical problems. The principles of potentiometric measurements are represented in the Nernst equation stating the fundamental relationship involving potential difference between electrodes and the activity of the species responsible for the response. When a metal M is immersed in a solution having its own ion M^{z+} an electrode potential is developed between the metal surface and the ions as per the following equilibrium:

$$M^{z+}_{(aq)} + ze^- \rightleftharpoons M_{(s)}$$

The electrode potential (E_{elec}), which depends on the activity ($a_{M^{z+}}$) of the metal ions in solution and the number of electrons involved in the process, is expressed by the Nernst equation

$$E_{elec} = E^0_{elec} + \frac{RT}{zF} \ln a_{M^{z+}}$$

where, E^0_{elec} is the standard electrode potential at $a_{M^{z+}} = 1$, R is the gas constant, T is the absolute temperature, F is the Faraday constant and z is the number of electrons involved in the process.

Potentiometric biochemical sensors have been designed with the concept of polymer membrane-based ion selective electrodes (ISE) by doping the membranes with ion-exchangers, neutral or charged carriers. Moody et al [66] have developed a calcium selective polymer-membrane electrode based on a liquid ion exchanger in a poly (vinyl chloride) (PVC) matrix that is still the basis for most modern devices. Initially because of more effectiveness, cation selective devices were given more emphasis than anion selective electrodes, that is to say, for easy selectivity, control and optimization.

Presently both cation and anion selectivity have been developed. Meyerhoff et al. [67] have prepared polyion-sensitive membrane electrodes for biomedical analysis. A chloride ion-selective polymeric membrane electrode based on a hydrogen bond forming ionophore was developed [68]. For neutral and charged carriers, the selectivity depends not only on the partitioning phenomena but also on the ion-carrier complexation constants [69].

Electropolymerized films were also used for ion sensing. For example, a sodium-selective electrode based on a calixarene ionophore in a PVC membrane with $NaBF_4$-doped polypyrrole (PPy) solid contact exhibited essentially no oxygen sensitivity in contrast to a Pt/PVC coated-wire electrode (CWE) [70]. Impedance measurement indicated that the PPy layer definitely lowered the charge transfer resistance to facilitate the necessary ionic to electronic conductivity transition. An implantable microsensor having ion-selective device with Ag/AgCl contact was made with polyHEMA layer between the contact and the ion-selective membrane [17]. In a report on some thick-film ion sensors Pace and Hamerslag [71] described the design and characterization of a multianalyte device for determination of blood electrolytes.

Field-effect transistor (FET) has proven to be an effective device for potentiometric measurements in the area of biochemical sensors, e.g., development of durable K^+-selective chemically modified field effect transistors with functionalized polysiloxane membranes [72] and sensors for environmental control [73]. The classical Severinghaus-type sensor, which is basically an ion-sensing electrode equipped with a gas permeable membrane, has undergone much variations in its original design. The Telting-Diaz device [74] is one of such modified versions having a dual lumen, implantable catheter for simultaneous potentiometric determination of pH and CO_2. In an *in vivo* test for using the device in non-heparinized animals, the catheter was coated with a biocompatible tridodecylammonium-heparinate complex although the coating had an adverse effect on the pH function, but not on the CO_2 function of the sensor.

Amperometric sensors are based on the measurement of current flow as the primary transduction phenomenon. Numerous amperometric sensors have been developed for various purposes. Different types of electrodes used in amperometric sensor designs are:

> Clark-type electrode: immobilized biocomponents (enzyme, cells or tissue) in gel with protective membrane.

> Needle type electrode with a configuration for an implantable amperometric device.

Interdigitated array (IDA) - a type of multiple electrode devices.

Carbon-fiber electrode - an ultramicroelectrode (UME) used for *in vivo* brain studies.

Miniaturized, disposable amperometric biochemical sensors employing electropolymerized permselective films for determination of creatinine, the product of cyclization of creatine by lactam formation, in human serum are described in a communication by Madaras and Buck [75]. The multienzyme system (creatininase, creatinase, sarcosine oxidase) is immobilized on top of the permselective layer using cross-linking of the proteins with glutaraldehyde. These sensors are reagentless, provide fast response time with a detection limit of 10-20 µM and are applied in a differential setup for creatinine assay in control and hospital human serum samples. The inert Pd metal-modified, enzyme-doped carbon-Ormosil (organically modified silica) ceramic biochemical sensors constitute a new type of renewable amperometric surface biochemical sensors with a controllable size reaction layer for determination of lactate, glucose and L-phenylalanine [76]. How the properties of these sensors, including stability and rate-limiting step, can be altered by minor changes in their preparation protocol are demonstrated. Pd- and redox enzyme-loaded carbon powder percolates through the porous silicate network. The biochemical reaction occurs in the outermost section of the electrode. Some amperometric sensors with recent state-of-the-art design for bio-analytes are shown in Table 4.

6.5 Current Trends in Biochemical Sensor Research

The state-of-the- art in biochemical sensor research indicates that over the years multiple sensors have been developed for the analytes, e.g., glucose oxygen, hydrogen peroxide, Ca^{++}, K^+, lactate etc. in terms of either sensor arrays designed for multiple analytes or for specific analytes in specific sensing environments. But only in few cases accurate, reliable, selective, sensitive, rapid, miniaturized, reagentless and stable devices have been achieved. So an extensive research is necessary to meet the need of the day. Following trends are notable for shaping the future research:

1. Improvement through chemical modifications in the immobilization strategy of biocomponents, through fine tuning of molecular recognition for selectivity and the use of new materials for both transducers and chemical transduction strategies.

2. Exciting innovations for multianalyte sensors, sensing arrays and chemometric approaches for non-selective or partially selective

sensing, e.g., an implantable multianalyte sensor array for sensing pH and K^+ in a beating heart [82].

3. Miniaturization and integration of components, for example, an amperometric microsensor array with 1024 individually address-able elements for two dimensional concentration mapping have been developed [83].

4. Studies of biochemical phenomena in non-aqueous media will open more biosensing opportunities [80].

A significant progress in chemical modification of biochemical sensor devices has been observed in four major areas: molecular design of chemically selective and biospecific agents, improvement in immobilizations for molecular recognition, assessment of the actual interfaces between sample and energy transducer and designing the energy transducers. Although arrays for multicomponent biochemical analysis have been developed, but with miniaturization of devices, improvements in immobilization methods and use of chemometrics, the development of arrays for all energy transduction modes should accelerate.

Overall, for biochemical sensors the following applications are emerging:

- Clinical diagnostics including patient monitoring and bedside analy-sis. Generally these are based on the measurements of species such as ions, gases, and hormones. In surgery, implantable biosensors provide real-time measurements in critical condition with efficient feedback system. In order to obtain quick analysis/tests for pregnancy, cholesterol, urine and blood glucose, and Helicobactor in the gastric mucose biosensors are most suitable.

- In environmental analysis including control of waste-water and water pollution biosensor can be used to monitor the presence of pesti-cides, herbicides, chemical fertilizers, and several other water pol-lutants.

- In quality control of food for purity and freshness.

- Biochemical sensors are used for monitoring the presence of heavy metallic ions, e.g., Hg, Pb, Cd, Cr(VI) etc.

- Biochemical sensors are used for on-line testing and process con-trol in chemical and pharmaceutical industry.

- Biochemical sensors are being used for the detection of drugs.

- In agriculture and horticulture, biosensors are being used in the measurement of the degree of ripeness of several products.

Table 6.1. Energy transducers, bioselective agents and sensor applications.

Energy Transducers	Bioselective Agents	Biosensing Applications		
		Medical/ Biochemistry	Environmental	Bioprocessing Technology
Optical	Enzymes	Cocaine	Polyaromatic	pH
Thermal	Immunochemicals	Glucose	Hydrocarbons	Sugars
Electrical	Nucleic acids	Blood gases	(PAH)	Dissolved gases
Acoustic wave	Cells and tissues	Acetylcholine	Polychlorinated	Recomb DNA
Light-addressable potentiometric sensor(LAPS)	Abiotics	DNA sequence	Biphenyls (PCB)	process
		pH, electrolytes	Pesticides	Alcohols
		Neurotransmitters	Herbicides	Amino acids
Biomagnetic		Metabolites	Heavy metals	Glucose
		Enzymes	Nitrates & nitrites	Monoclonal
		Drugs	Trichloroethylene	antibodies
		Immunocomponents,e.g., Hepatitis B antigen Herpes virus AIDS virus antigen		

Table 6.2. Some analytes in blood and corresponding transduction modes in healthcare applications.

Analytes	Normal Range	Chemical Transduction	Energy Transduction
Na^+	136-143 mmol/L	Ionophore complexation	Potentiometric
K^+	3.6-5.0 mmol/L	Ionophore complexation	Potentiometric
Ca^{++}	1.15-1.31 mmol/L	Ionophore complexation	Potentiometric
Cl^-	98-107 mmol/L	Ionophore complexation	Potentiometric
pH	7.31-7.45 (blood) 5-8 (urine)	Ionophore PH sensitive chromophore	Potentiometric Optical
O_2	80-104 Torr	Gas permeable membrane Perylene or Ru fluorophore	Amperometric Optical (quenching)
CO_2	33-48 Torr	Gas permeable membrane with bicarbonate solution	Potentiometric
		Membrane with bicarbonate solution and pH chromophore	Optical

Glucose	65-105 mg/dL	Enzymatic catalysis	Amperometric Potentiometric Optical Thermal
L-Lactate	3-7 mg/dL	Enzymatic catalysis	Amperometric Potentiometric Optical
Urea	7-18 mg/dL	Enzymatic catalysis	Potentiometric Thermal Optical
Pyruvate	$10-10^3$ mg/L	Enzymatic catalysis	Amperometric
Uric acid	$10-10^3$ mg/L	Enzymatic catalysis	Amperometric
L-Amino acid	$5-10^2$ mg/L	Enzymatic catalysis	Amperometric
Cholesterol	$10-5 \times 10^3$ mg/L	Enzymatic catalysis	Amperometric
Phospholipid	$10^2-5 \times 10^3$ mg/L	Enzymatic catalysis	Amperometric

Table 6.3. Some selected biosensors and their sensing modes

Substance measured	Biospecific component	Polymers/ immobilization	Sensing mode	Reference
Lactate ion	Lactate oxidase	Nylon	Oxygen electrode	35
Cholesterol	Cholesterol esterase	Covalent	Platinum electrode	42
Phospholipid	Phospholipase	Covalent	Platinum electrode	43
Penicillin	Penicillinase	Entrapment	pH electrode	44,45
Pyruvate	Pyruvate oxidase	Adsorption	Oxygen electrode	44
L-Glutamine	Glutaminase	Adsorption	Ammonium ion elect.	44,46
CO_2	—	Silicone rubber	pH electrode	47
IgG	Anti-IgG, protein A	Polyacrylamide	Piezo crystal	48, 49
Benzo(a) pyrene(BaP)	Anti-BaP	—	Fluorescence	50
Urea	Urease	Poly(vinyl alcohol) Crosslinked	Ammonia electrode	51
Uric acid	Uricase	Nitrocellulose	Oxygen electrode	44
DNA	Enzyme-DNA	—	Luminescence Voltammetry	52
Ca^{++}	Lipid membrane	Langmuir-Blodgett	Glassy Cerbon Elec.	41
Oxalate	Oxalate oxidase	Natural tissue	Oxygen electrode	40
Glucose	Glucose oxidase	Amine hydrogel, silicone rubber	Volume change Oxygen electrode	53 54
Coagulation proteins	Antibodies	Poly(ethylene oxide)	Fluorescence	55

Table 6.4. Selected amperometric sensors and analytes.

Substance measured	Biospecific component	Important features	References
Glucose	Electrically-wired Gox on Au-SAM (self-assembled monolayers), reconstituted holoenzyme on surface Ferrocene-containing polyacrylamide based redox gel with enzyme	Highest current density reported to date One step polymerization	77 78
Sucrose	TCNQ mediator with Gox and invertase in carbon paste electrode (CPE) for flow-injection analysis (FIA)	Cheap, works with high concentration of sucrose without dilution	62
Glutamate, dopamine	Glutamate oxidase, polyphenol oxidase immobilized electrochemically using amphiphilic polymer	Promotes immobilization of negatively charged enzymes	79
Peroxides, Phenols, Bilirubin	Organic phase enzyme electrode, organohydrogel	Works with variety of sensing media and analyte solubilities, tested with horse-radish peroxidase (HRP)	80
Human luteinizing hormone (hLH)	Immunosensor using enzyme channeling design	Disposable, one-step, separation-free	81

6.6 References

1. Ross, P, Bio Techniques, $\underline{1}$, 204(1983).

2. Zhong, C.J and Porter. M. D, Anal. Chem., $\underline{67}$ 709A(1995).

3   Finklea, H. O in Electroanalytical Chemistry Bard. A. J and Rubinstein. I (eds.), Vol. 19, Dekker Marcel, New York, 1996.

4. Sackmann. E, Science, $\underline{271}$, 43 (1996).

5. Shul'ga, A. A, Koudelka-Hep. , M, Rooij. N. F. de and Netchiporouk. L. I., Anal. Chem., $\underline{66}$, 205 (1994).

6. Pantano, P- and Walt, D. R- Anal Chem., $\underline{67}$, 481A (1995).

7. Bartlett, P. N- and Cooper, J. M-, Electroanal, J- Chem., $\underline{362}$, 1 (1993).

8. Ohara, T. J-, Rajagopalan, R- and Heller, A-, Anal. Chem., $\underline{65}$, 3512 (1993).

9. Garguilo, M. G-, Huynh, N-, Proctor, A- and Michael, A. C-, Anal. Chem., $\underline{65}$, 525 (1993).

10. Weetal, H.H- Appl. Biochem. Biotech., $\underline{41}$, 157 (1993).

11. Ong, S-, Cal, S-J, Bernal, C-, Rhee, D-, Qiu, X- and Pidgeon, C. Anal. Chem., 66, 782 (1994).

12. Pidgeon, C-, Ong, S-, Choi, H- and Liu, H- Anal. Chem., 66, 2701 (1994).

13. Guilbault, G. G- and Mascini, M- (eds), Use of Immobilized Biological Compounds, Kluwer, Dordrecht, The Netherlands, 1993.

14. Iwuoha, E. I-, Adeyoju, O-, Dempsey, E-, Smyth, M. R-, Liu, J- and Wang, J- Biosens, Bioelec., 10, 662 (1995).

15. Iwuoha, E. I-, Leister, I-, Miland, E.-, Smyth, M. R- and Fágáin, C. Ó- Anal. Chem., 69, 1674 (1997).

16. Meyerhoff, M. E- Trends Anal. Chem., 12, 257 (1997).

17. Cosofret, V. V-, Erdosy, M-, Johnson, T. A-, Bellinger, D. A-, Buck, R. P-, Ash, R. B- and Neuman, M. R- Anal. Chim. Acta, 314, 1 (1995).

18. Clarke, D. J-, Calder, M. R-, Carr, R. T. G-, Blake-Coleman, B. C-, Moody, S. C- and Collinge, T. A-, Biosensors, 1, 213 (1985).

19. Mullen, W. H- and Vadgama, P. M- J. Appl. Bacteriology, 61, 181 (1986).

20. Enfors, S. O- and Nilsson, H- Enzyme Microbe Technol., 1, 260 (1979).

21. Cunningham, A. J- Introduction to Bioanalytical Sensors, John Wiley & Sons, New York, 1998, p.28.

22. Cunningham, A. J- Introduction to Bioanalytical Sensors, John Wiley & Sons, New York, 1998, p.16.

23. Lin, Z-, Burgess, L. W- Anal. Chem. 66, 2544 (1994).

24. Turyan, I- and Mandler, D- Anal. Chem. 69, 894 (1997).

25. Cosnier, S-, Innocent, C- and Jouanneau, Y- Anal. Chem. 66, 3198(1994).

26. Strehlitz, B-, Grundig, B-, Schumacher, W-, Kroneck, P.M.H-, Vorlop, K-D- and Kotte, H- Anal. Chem. 68, 807(1996).

27. Preuss, M- and Hall, E.A.H- Anal. Chem. 67, 1940(1995).

28. Bauer, C.G-, Eremenko, A.V-, Ehrentreich-Forster, E-, Bier, F.F-, Makower, A-, Halsall, B.H-, Heineman, W.R- and Scheller, F.W- Anal. Chem. 68, 2453(1996).

29. Roberts, M.A- and Durst, R.A- Anal. Chem. 67, 482(1995).

30. Karube, I-, Satoh, I-, Araki, Y-, Suzuki, S- and Yamada, H- Enzyme Microb. Technol. 2, 117 (1980).

31. Matsumoto, K-, Seijo, H-, Karube, I- and Suzuki, S- Biotechnol. Bioeng., 22, 1071 (1980).

32. Gargullo, M. O- and Michael, A. C- Anal. Chem. 66, 2621(1994).

33. Garguilo, M. G-, Huynh, N-, Proctor, A- and Michael, A.C- Anal. Chem. 65, 523-528(1993).

34. Leca, B-, Morelis, R.M-, Coulet, P.R- Sensors and Actuators B 26-27, 436(1995).

35. Mascini, M-, Moscone, D- and Palleschi, G- Anal. Chim. Acta, 157, 45 (1984).

36. Chaube, A-, Gerard, M-, Singhal, R-, Singh, V. S- and Malhotra, B. D- Electochimica Acta, 46, 723 (2001).

37. Sprules, S. D-, Hart, J. P-, Wring, S. A-, Pittson, R- Anal. Chim. Acta 304, 17(1995).

38. Baker, D. A- and Gough, D. A- *Anal. Chem.* 67, 1536(1995).

39. Wang, J- and Chen, Q- Anal. Chem. 66, 1007 (1994).

40. Nabi Rahni, Md. A-, Guilbault, G. G- and Olivera, N. G. de- Anal. Chem., 58, 523 (1986).

41. Sugawara, M-, Kojima, K-, Sazawa, H- and Umezawa, Y- Anal. Chem., 59, 2842 (1987).

42. Karube, I- Hara, K- Matsuoka, H- and Suzuki, S- Anal. Chim. Acta 139, 127 (1982).

43. Karube, I- Hara, K- Satoh, I- and Suzuki, S- Anal Chim. Acta 106, 243 (1979).

44. Karube, I- in Biotechnology, Vol. 7a, Vol. Ed. Kennedy, J. F- VCH, 1987,Germany, Chap.13.

45. Nikolelis, D. P- and Siontorou, C. G- Anal. Chem. 67, 936 (1995).

46. White, S. F- Turner, A. P- Bilitewski, U- Bradley, J- and Schmid, R. D- Biosens. Bioelec., 10, 543 (1995).

47. Opdycke, W. N- and Meyerhoff, M. E- Anal. Chem., 58, 950 (1986).

48. Thompson, M- Arthur, C. L- and Dhaliwal, G. K- Anal. Chem., 58, 1206 (1986).

49. Muramatsu, H- Dicks, J. M- Tamiya, E- and Karube, I Anal. Chem., 59, 2760 (1987).

50. Vo-Dinh, T- Tromberg, B. J- Griffin, G. D- Ambrose, K. R- Sepaniak, M. J- and Gardenhire, E. M- Appl. Spectrosc., 41, 735 (1987).

51. Hsiue, G. H- Chou, Z. S- Hsiung, K. P- and 'Yu, N- Polym. Mater. Sci. Eng., 57, 825 (1987).

52. Downs, M. E. A- Warner, P. J- Turner, A. P. F- and Fothergill, J. C- Biomaterials (Guildford, Engl.), 9, 66 (1988).

53. Kost, J- Horbett, T. A- Ratner, B. D- and Singh, M- J. Biomed. Mater. Res., 19, 1117 (1985).

54. Gough, D. A- Armour, J. C- Lucisano, J. Y- and McKean, B. D- Trans. Am. Soc. Artif. Intern. Organs, 32, 148 (1986).

55. Andrade, J. D- Herron, J- Lin, J. N- Yen, H- Kopecek, J- and Kopeckova, P- Biomaterials (Fuildford, Engl.), 9, 76 (1988).

56. Clark, L. C- and Lyons, C- Ann. N. Y. Acad. Sci., 102, 29 (1962).

57. Updike, S. J- and Hicks, G. P- Nature, 214, 986 (1967).

58. Zhang, Z- Liu, H- and Deng, J- Anal. Chem., 68, 1632 (1996).

59. Jung, S-K- and Wilson, G. S- Anal. Chem., 68, 591 (1996).

60. Sangodkar, H- Sukeerthi, S- Srinivasa, R. S- Lal, R- and Contractor, A. Q- Anal Chem. 68, 779 (1996).

61. Satoh, I- Karube, I- and Suzuki, S- Biotechnol. Bioeng. $\underline{18}$, 269 (1976).

62. Filho, J. L. L- Pandey, P. C- and Weetall, H. H- Biosens. Bioelec., $\underline{11}$, 719 (1996).

63. Tobias-Katona, E- and Pecs, M- Sensors and Actuators B, $\underline{28}$, 17 (1995).

64. Berners, M.O.M- Boutelle, M.G- and Fillenz, M- Anal. Chem. $\underline{66}$, 2017 (1994).

65. Cosford, R. J. O- and Kuhr, W. G- Anal. Chem. $\underline{68}$, 2164 (1996).

66. Moody, G. J- Oke, R. B- and Thomas, J. D. R- Analyst, $\underline{95}$, 910 (1970).

67. Meyerhoff, M. E- Fu, B- Bakker, E- Yun, J-H- and Yang, V. C- Anal. Chem. $\underline{68}$, 168A (1996).

68. Xiao, K. P- Bühlmann, P- Nishizawa, S- Amemiya, S- and Umezawa, Y- Anal. Chem., $\underline{69}$, 1038 (1997).

69. Bakker, E- Meruva, R. K- Pretsch, E- and Meyerhoff, M. E. Anal. Chem., $\underline{66}$, 3021 (1994).

70. Cadogan, A- Gao, Z- Lewenstam, A- Iraska, A- and Diamond, D- Anal. Chem. $\underline{64}$, 2496 (1992).

71. Pace, S. J- and Hamerslag, J. D- in Edelman, P. G- and Wang, J. - (eds), Biosensors and Chemical Sensors: Optimizing Performance through Polymeric Materials, ACS Symposium Series #487, American Chemical Society, Washington DC, 1992.

72. Reinhoudt, D. N- Engbersen, J. F. J- Brzozka, Z- van den Vlekkert, H. H- Honig, G. W. N- Holterman, H. A. J- and Verkerk, U. H- Anal. Chem. $\underline{66}$, 3618 (1994).

73. Karube, I- Nomura, Y- and Azikawa, Y- Trends Anal. Chem., $\underline{14}$, 295 (1995).

74. Telting-Diaz, M- Collison, M. E- and Meyerhoff, M. E- Anal. Chem., $\underline{66}$, 576 (1994).

75. Madaras, M. B- and Buck, R. P- Anal. Chem. $\underline{68}$, 3832 (1996).

76. Sampath, S- and Lev, O- Anal. Chem., $\underline{68}$, 2015 (1996).

77. Willner, I- Heleg-Shabtai, V- Blonder, R- Katz, E- Tao, G- Bückmann, A. F- and Heller, A- Amer, J-Chem. Soc. $\underline{118}$, 10321 (1996).

78. Bu, H- Mikkelsen, S. R- and English, A. M- Anal. Chem., $\underline{67}$, 4071 (1995).

79. Cosnier, S- Innocent, C- Allien, L- Poitry, S- and Tsacopoulos, M- Anal. Chem. $\underline{69}$, 968 (1997).

80. Guoy, Y- and Dong, S- Anal. Chem., $\underline{69}$, 1904 (1997).

81. Invitsky, D- and Rishpon, J- Biosens. Bioelec., $\underline{11}$, 409 (1996).

82. Cosofret, V. V- Erdosy, M- Johnson, T. A- Buck, R. P- Ash, R. B- and Neuman, M. R. Anal. Chem., $\underline{67}$, 1647 (1995).

83. Harmes, Bühner, M- Bü cher, S- Sundermeier, C- Dumschat, C- Börchardt, M- Cammann, K- and Knoll, M- Sens. Actuators (B), $\underline{21}$, 33 (1994).

Chapter 7

Nucleonic Sensors

M.K. Ghosh

Abstract

Application of radioisotpes in scientific research, biomedical diagnostics and industrial measurements is not new. However, in view of its versatility, cost effectiveness, non invasive sensing principle and ability to offer solutions to tricky problems, they find a special place in industrial measurements. Today's petrochemical, chemical, offshore industries, heavy solid handling plants like quarrying and cement industries, steel plants and paper mills employ nucleonic sensors for strategic measurements. Principles of nucleonic measurements, sources and detectors, signal processing are discussed in this chapter along with the issues related to special care to be taken for handling the sources and likely health hazards. The horizon for application of nucleonic sensors seems to be unbounded.

7.1 Introduction

Nucleonic sensors are one of the most versatile classes of sensors used in non-invasive techniques of industrial measurements including identification. They are also known as radio isotope sensors or nucleonic gauges. Though emission of radiation signal arising out of disintegration of nucleus of the nucleonic sources is random in nature (follows Poisson distribution), because of the very nature of measurement, the sensors are considered to make deterministic measurements

Nucleonic gauges basically work either (i) on ionization technique, or (ii) on absorption (nucleonic energy absorption by measurement environment causing modulation of its intensity) technique. The basic principle of a nucleonic gauge is presented in chapter 1 (Fig. 1.3).

7.2 Nucleonic Sources

All nucleonic sources emit more or less all of the common types of radiations: α, β, γ, neutron. However, based on predominance of any one type of radiation, it acquires the specific nomenclature.

α-sources emit predominantly a flux of α- particles equivalent of a *He* nucleus composed of two protons and two neutrons (4 *He* or $^2_2He\,4$). Naturally decaying elements like Ra liberate α- particles, e.g.

$$^{138}_{88}Ra226 \rightarrow\ ^{136}_{86}Rn222 + \ ^2_2He4 \rightarrow 222\,Rn + \alpha \tag{1}$$

α radiation has relatively low energy and very little penetration property (air - 11 cm, fabric 0.1 mm, Al foil 0.006 cm). But it has a very high ionization capability.

β -rays are a flux of electrons or positrons appearing in the courses of nuclear disintegration of β -emitting isotopes as a result of neutron-proton transformation.

$$\begin{aligned}
&^{26}_{33}Fe59 \rightarrow\ ^{27}_{32}Co59 + \beta^- \\
&^{26}_{26}Fe52 \rightarrow\ ^{25}_{27}Mn52 + \beta^+
\end{aligned} \tag{2}$$

Radiation energy of β particles (2-3 *MeV*) is medium, ionization property medium (causes ionization when colliding with atoms of air), penetration property appreciable (10 m in air, 10-12 mm in fabrics, 5-6 cms in Al and 1 mm in *Pb*).

γ rays are emitted when during a reaction the nucleus attains a higher state of energy level and returns to the stable state by emitting γ rays, i.e.,

$$Co * 59 \rightarrow Co\ 59 + \gamma \tag{3}$$

γ rays are electromagnetic radiations similar to X rays $(0.01\overset{o}{A} < \lambda < 1.4\ \overset{o}{A})$. γ rays are much more penetrating than β rays. Energy level is high. Neutron radiation, such as,

$$^{26}_{26}Fe\ 52 + \beta^- \rightarrow\ ^{25}_{27}Mn\ 52 + Neutron \tag{4}$$

has maximum penetration power (concrete several metres), very high energy and has no charge.

For industrial applications the radioisotopes are chosen based on:

- Half-life
- Activity
- Principles of sensing
- Cost

Half life

The isotopes change from one state to a more stable state by emitting charged particles at a rate given by

$$\frac{dn}{dt} = -\lambda n \tag{5}$$

where n = number of nuclei and λ is the decay constant. Half life t_{half} indicates the time by which the radiation intensity reduces to half. Thus,

$$t_{half} = \frac{0.639}{\lambda}$$

Activity

Activity is measured by Courie, where $1\ mc = 3.7 \times 10^7$ disintegrations per second. Energy content of a source is kept low for safety reasons such that the activity is of the order of 1000 mC.

Principles of Sensing [1]

Nucleonic gauges working on the principle of ionization of the medium (gas) uses α sources. Other sources based on penetration technique of measurement are either β, γ or sometimes neutron sources. For these sources range of emission is an important factor. Intensity of radiation

$$I = I_0 \exp(-\mu_L d) \tag{6}$$

where

μ_L = linear absorption coefficient = $\rho \, \mu_m$

ρ = density of the medium,

μ_m = mass absorption coefficient of the medium, and

d = thickness of the medium

Half-distance d_{half} is a measure of range of emission. For $d = d_{half}$, $I = I_0 / 2$. It is dependent on material as well as energy content of radiation.

Cost

When several different radioisotopes can be used for a particular application, the choice will depend on the cost except for technical questions of penetration and half life.

Almost all (about a thousand) radioisotopes are created artificially in linear accelerators, synchrotrons, radiation channels of nuclear reactors. Approximately 150 of them are used in industries. The popular ones are $60\,Co$ (t_{half} = 5.25yrs,β,γ),137Cs(t_{half} = 30 years,β,γ),192 Ir (t_{half} = 74.4 days,β,γ),170Tu (t_{half} = 127 days, β,γ) sources.

Sources are shielded so as to force the radiation to pass in the form of a narrow parallel beam through a narrow slit. This makes the source nearer to an ideal point source and eliminates radiation leakage from to the source to the detector.

7.3 Nucleonic Detectors [2]

For detection of radioactivity, the phenomena that are utilized are:

- ionization
- luminiscence
- specific characteristics of materials in presence of nucleonic radiation.

The various detectors used are:

(a) Photographic films
(b) Ionization chambers
(c) Proportional counters
(d) Geiger Müller counters
(e) Scintillation counters
(f) Semiconductors

Photographic films are sensitive to electromagnetic radiation emitted by α,β,γ sources and hence they are universal detectors. The intensity of radiation exposure is studied using a microscope or a densitometer. However, it cannot be used as an on-line detector and is used only in personnel badges for record of radiation exposure undergone by them.

Gas discharge counters employ gas filled under vacuum pressure inside a chamber with an electric potential applied across the electrodes (Fig.7.1). The ionization current is

$$i = e \int_{V_0} n_o (x, y, z) dx\, dy\, dz \qquad (7)$$

where e = charge, v_o = active volume, n_o = ion pairs/unit volume. Hence, i is a measure of the radiation falling on the detector (through its window). The casing of counters is of glass metal or of metal for γ radiation and of Al for β radiation.

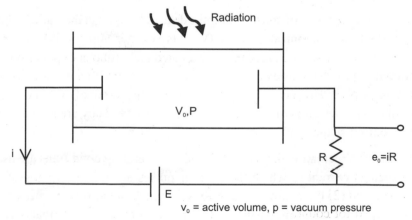

v_0 = active volume, p = vacuum pressure

R=High resistance., E-Supply voltage, e_0 = output voltage

Fig. 7.1 A general gas discharge counter

The characteristics of the gas discharge counters vary based on applied voltage E (Fig. 7.2). Ionization chambers operate in the ionization region (zone II). In the proportional region (zone III), the primary ionization process undergoes internal amplification due to avalanche ionization. Proportional counters operate in this region. They are suitable for measurements of low energy radiation (α particles) or neutrons. They are generally used in pulse type operation and rarely in control equipment.

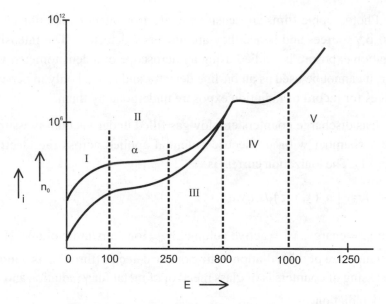

Fig. 7.2 Operation of Gas Discharge Counters

When pulses of constant amplitude are observed at the output whatever be the value of ionization (zone IV), it is a Geiger Müller (GM) counter. A GM tube consists of a metal or metal coated glass tube or cylinder with a thin wire of tungsten as electrode. The tube is filled with a gas mixture of 90% A (or Ne or He) and 10% organic vapour (ethyl alcohol). GM counters are universal detectors and have a higher sensitivity. They are available in the required shapes and sizes.

In instrumentation devices, two types of operating conditions are used (1) average current in which the counter pulses are fed to an integrating network and (2) the counting conditions in which the pulses are amplified and shaped for counting purposes. Phosphors (organic like anthracene, or inorganic like Na (Tl), Zn S (Ag) used in thin layers, produce tiny flashes of light (duration $\sim 10^{-8} s$) when nuclear radiation is incident on them. The light from this scintillation falls on a photo sensitive cathode of a photo multiplier tube (PMT). The assembly is known as a Scintillation Counter (Fig. 7.3) whose output is a measure of radiation. The counter has a very high gain, it is very sensitive and can detect all types of radiation.

Thermoluminiscent materials like Felspar (Ca F$_2$) absorb radiation and when heated liberate it [3]. However, they are not suitable for use in instrumentation.

Charged particles passing through silicon or germanium crystals of semiconductor detectors produce electron hole pairs. These detectors have low response to radiation and can be used as α-detectors. They are unaffected by magnetic fields, possess a good resolution and linearity. They can be made in any shape, but the noise present in the system requires careful processing.

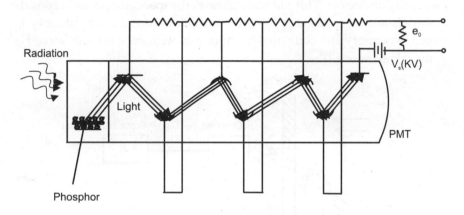

Fig 7.3 Scintillation Counter schematic

Scintillation counters are usually chosen for industrial instrumentation. GM counters are used for radiation leakage detection and other safety purposes. Photographic films are used for keeping record of radiation exposure undergone by operators.

7.4 Signal Processing: The Measuring Circuit [1,2]

Most of the industrial nucleonic gauges are required to generate control action along with display. Since the output of the industrial detectors is electrical, signal processing adopted is conventional. However, it is common practice to compensate for the measurement errors concurrently (dynamic error compensation). The sources of error are: (1) the uncalibrated nucleonic source (important for sources with small half life), (2) radiation leakage from a non-ideal point source, (3) change in detector characteristics and noise in electronic circuits.

Precalibrated sources and properly designed source guide eliminate errors due to the source. Noise in detector and electronic circuits are taken care of through hardware and software.

An instrument servomechanism for display and control along with dynamic compensation is shown in Fig 7.4, where error due to detection and signal processing are eliminated. The signal processing circuit (1) consists of an integrated network, amplifier, pulse shaping circuit, an amplitude detector and a comparator. The wedge (2) is moved up and down all the time by a motor (3) so that the intensity of standard flux I_{ref} reaching the detector varies continuously with a saw tooth waveform. The instant (t_o) at which compensating and measured fluxes are equal is registered by (1) and the measuring device (4). This method increases the speed of operation considerably. It uses α modulator (5) and the modulated radiation intensity is stored in memory and compared continuously with the sawtooth wave I_{ref} at the comparator (3).

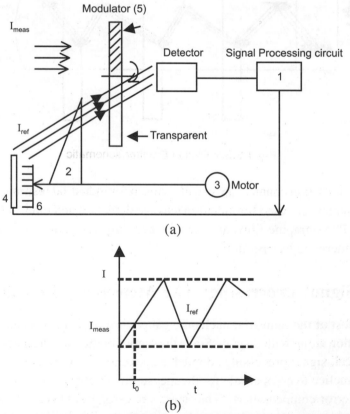

(a)

(b)

Fig. 7.4 (a) Measurement using dynamic compensation (b) Flux wave form

7.5 Nucleonic Sensors

Radioisotopes are used in biological, archaeological, geological, agricultural applications. The methods of application of radioisotopes in sensing are varied. Radioisotopes today are used as industrial tools in such applications as

(a) Non-destructive inspection and gauging for quality control of materials, measurement and control of process variables such as level, density, thickness, weight, flow rate, etc.

(b) Determination of process parameters like residence time mixing rates, wear characteristics of mobile parts, detection of gas leakage, etc. using tracers.

(c) Commercial production of radiation-based products such as wood - plastic composites, industrial polymers, vulcanizing of rubber latex, etc.

7.5.1 Sensing based on Ionization

Ionization chamber (Fig. 7.1) has an output

$e_o = \phi$ (gas, pressure p, E, intensity of radiation)

Intensity of α radiation can be measured by keeping other variables constant. Alphatron is a vacuum pressure transducer in which an α-source is kept inside the chamber which is connected to the gas line at vacuum pressure. The same scheme can be used for identification of the gas at a known vacuum pressure. In one of the applications accurate measurement of high resistances is done by replacing R in the external circuit by the resistances to be measured.

7.5.2 Sensing based on Intensity Modulation

In this group of sensors, radiation flux is modulated by the measurand and the change is interpreted as a function of the measurand. Two schemes are adopted :

(a) penetration of radiation into the medium thereby effecting modulation by the properties of the medium, and

(b) modulation of the distance between the radiation source and detector

(a) Thickness measurement

Three methods for thickness measurement are shown in Fig. 7.5. Let us consider the direct transmission technique (Fig 7.5a). If the nominal (desired) thickness of the sheet is d_o and the actual thickness is $d=d_o + \Delta d$, and μ_L= linear absorption coefficient of material,

$$I = I_o \exp(-\mu_L d) = I_o \exp[-\mu_L(d_o + \Delta d)$$
$$\approx I_o e^{-\mu_L d_o}(1-\mu_L.\Delta d +.......), \text{for } \mu_L.\Delta d \ll 1 \tag{8}$$
$$\therefore \quad \Delta I = I_o\, e^{-\mu_L d_o}(-\mu_L.\Delta d)$$

$$\frac{\Delta I}{I_o e^{-\mu_L d_o}} = -\mu_L.\Delta d \tag{9}$$

$$\Delta I \propto \Delta d.$$

The above schemes are the basis for measurement of any parameter that has a dimension of thickness (length). It is possible to measure, in a similar way, thickness of a substance deposited as a coating during manufacture; thickness of corrosion in a vat or pipe, especially when the interior is inaccessible; thickness of carbon deposit in the cylinders of internal combustion engines; measurement of very thin layers of materials such as the printing ink (3 to 7μm) on the rollers; spray painting thickness, etc. The method can also be used in measuring thickness of coal seams. Nucleonic thickness gauges are very popular in rolling mills (speed 60 to 100 kmph, temperature 850° C).

In Fig. 7.5 below, (a) depicts a Direct transmission technique while (b), (c) depict a Back scattering technique.

S - Dource, D - Detector

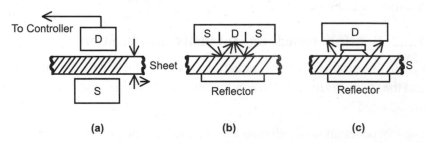

Fig 7.5 Thickness measurement configurations

(b) Level Measurement

A few methods of level measurement using radioisotopes are shown in Fig. 7.6. along with the measurement characteristics. The most accurate methods are (c) and (d). A method for measuring (and automatic filling) level in tinned condensed milk, pastes in tubes is shown in Fig. 7.7. The accuracy of measurement is as small as 0.14 mm for a liquid of density 1 gm/ml [4]. The method is successfully employed for measurement of levels in gas cylinders, aircraft fuel tank, etc.

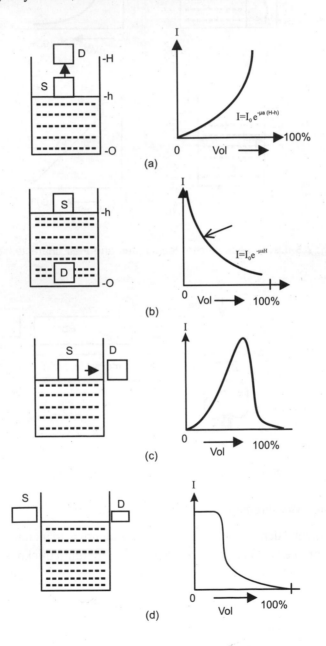

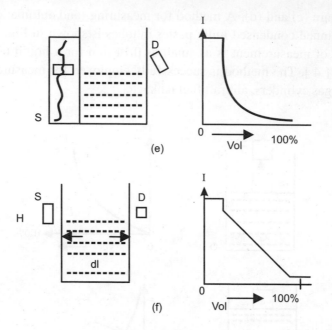

Fig. 7.6 Level measurement techniques and their characteristics

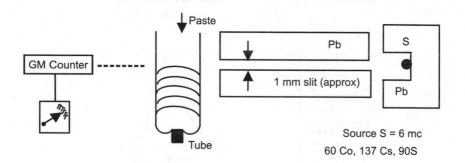

Fig. 7.7 Level measurement in paste tubes

(c) Density Measurement [5]

Incremental density ($\Delta\rho$) measurement is identical to incremental thickness measurement. Since $\mu_L = \mu_m . \rho$, keeping medium thickness d (in this case liquid filled tank or pipe line) constant, we have

$$\Delta I \propto (-\mu_m (\Delta\rho) d_o)$$
$$\propto \Delta\rho \tag{10}$$

where μ_m = mass absorption coefficient of material.

Continuous and accurate density measurements of liquid, granular material, soil in pipes, bins, tanks, hoppers or other types of containers can be made by placing the source and the detector diametrically opposite to each other (direct transmission) or by back scattering method. The method is applicable to all fluids, irrespective of their high temperature, high pressure, toxicity, inflammable or corrosive nature. It is also possible to detect physical behaviour of the fluid. Thus, with 100 mC of 60 Co, the concentration of gas in water-gas emulsion flowing in a pipe of 60 cm in diameter is possible. Oil and sugar industries use this method with 137 Cs and with an accuracy of 0.3%. Fig. 7.8 shows block diagram of a densitometer adopting instrumentation servomechanism and dynamic compensation. The γ ray source fixed to disc (1) revolves at a constant speed. The radiation passes alternatively through the object (2) and compensation wedge (3). The pulses from the output of the counter (4) are processed by integrating network (5), amplifier (6) and P.S.D. (7). The servosystem (8) moves the wedge until balance is established. The wedge is linked to a pointer (9) for indication.

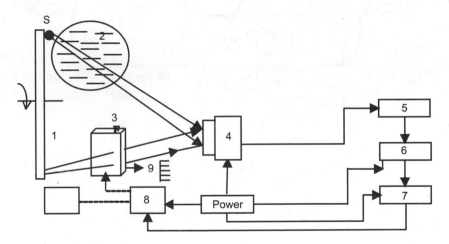

Fig. 7.8 Densitometer with dynamic compensation

(d) Flow measurement [6]

In the method based on variation in flux intensity due to movement of liquid, a thin, elastic lamina or sheet with a radiation source is placed with one end fixed in the stream (Fig. 7.9). Its deflection and hence the radiation received by the detector varies with the liquid flow rate. Fig. 7.10 gives scheme based on frequency method. Source S is fixed to one or more blades of the propeller (1) and the measuring device (2) is a frequency meter counting the number of fluxes. By placing two electrodes (at a difference in potential between them) in the gas stream and detecting the rate of movement of ions, the gas flow rate can be measured. This method is suitable for relatively low displacement speeds (<90 m/min).

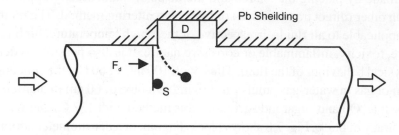

Fig. 7.9 Flow measurement using bending of plate by drag force

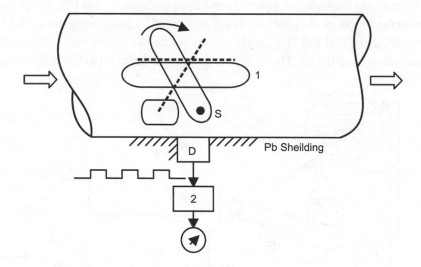

Fig. 7.10 Frequency type flow meter

(e) Other applications

Radioisotopes are used for gas analysis, gas identification, smoke detection, velocity measurement, temperature measurement, etc. The horizon of their applications seems to be unbounded.

7.6 Health Hazards and Protective Measures [4]

A fear psychosis predominates among personnels from industry as regards the dangers involved in handling of radioisotopes and appliance. They are, in fact, no more dangerous to handle than the most violent natural poisons, provided always appropriate protective measures are adopted. The chief effect of radioactivity on living tissues is to cause ionization. Single intense irradiation is less harmful than repeated less intense ones. Hazards of a radioisotopes are greater with those of nuclides of long half-lives with greater ionizing powers and with those which are absorbed selectively in the body (bones, blood forming organs).

Protection against radioactivity is achieved by taking sufficient care and training the personnel involved in handling the devices or coming under exposure of radioactivity. Protection is to be taken against direct irradiation and contamination. Norms have been established as to the safeguard to be adopted regarding premises housing radioactive sources, with clothing of personnel, equipment in premises, storage and handling, disposal of effluents, medical supervision and education of management and staff are also important.

National and International Monitoring and Licensing Authorities are responsible for evaluation of protection of health and environmental standards.

7.7 References

1. Shumilovskii, N.N. and Mel'ttser, L.V. - Radioactive Isotopes in Instrumentation and Control, Pergamon Press, 1964.

2. Knoll, G. T. - Radiation Detection and Measurement, 3rd ed, Wiley, NY, 2000.

3. Ghosal, D.- Design and and fabrication of Thermoluminiscent Nucleonic Detector, B. Tech Project, E E Dept, IIT Kharagpur, May 1999.

4. Piraux, H - Radioisotopes and their Industrial Applications, Philips Technical Library, 1964.

5. Khazan, A. D. - Transducers and Their Elements, Prentice Hall, 1994.

6. Desa, D. O. J. - Instrumentation Fundamentals for Process Control, Taylor & Sons

Chapter 8

Microwave Sensors

A.K. Mallick

Abstract

This chapter presents an exhaustive and comprehensive current account of modern microwave sensors. The chapter has been divided into number of sections, such as, passive and active sensors that may have imaging and non-imaging provisions. Microwave sensors have been brought into light in respect of different applications as well, namely remote-sensing, industry/commerce and biomedical etc. At the beginning, basics of microwave principles have been discussed in brief for easy understanding of the review that follows.

8.1 Introduction

Microwave engineering has led to various useful technological and engineering applications in industry, commerce, medicine and remote sensing, especially in climatology, oceanography, cryospheric data-collection and many others. In these applications, microwave-sensing technology plays an important role in monitoring, imaging and mapping of various physical parameters of variations useful in daily life.

The advantages of electromagnetic waves in microwave-frequency ranges for there applications arise from the following:

A. A number of microwave windows exists that allows the wave to propagate through the atmosphere with insignificant attenuation.

B. Presence of spontaneous thermal microwave emissions from various geophysical objects without being illuminated by any radiation.

C. Availability of suitable detectors with adequate sensitivity and reliability in adverse ambient conditions.

D. Presence of pronounced absorption features that can be used to monitor atmospheric properties in microwave ranges.

E. Microwaves can penetrate through the atmospheric hindrances like cloud, smoke, fog etc. unlike optical / IR signals.

In order to clarify the last point above, one may note that for years, microwave sensors have been used individually and with optical remote sensing aids to study the earth and its environment. It has been noticed that clouds frequently obscure observations of earth from space at visible and infrared wavelengths. However, microwave radiation, by virtue of its having much longer wavelengths, can 'see through' most of the clouds, mist, fog etc. and can also penetrate the top layer of the land's surface. The unique characteristic features of microwaves are used to measure a wide range of the geophysical properties.

Microwave radiation interacts with water in its various states e.g. gaseous, liquid and solid, in distinctly different ways. This property can be used to study efficiently the water cycle and the energy balance of the earth. For instance, microwave techniques are used to detect humidity, rain and hail in the atmosphere as well as soil moisture, snow and permafrost on the land. Over the ocean, microwave sensors are capable of measurement related to sea-surfaces, wind speed and direction, Arctic and Antarctic sea ice; e.g. its boundary, age and thickness, and sea surface temperature and salinity. Microwave sensors are already in routine operational use on satellites. Space-borne microwave sensors for study of oceanography and glaciology now routinely measure, for example, sea ice extent and sea surface wind speed and direction, as well as ocean surface topography. The meteorological community has come to rely heavily on satellite-based microwave radiometers for all-weather measurements of atmospheric temperature and humidity profiles globally every 12 hours for their weather forecast needs.

The first passive microwave sensor to record sea ice images is the Electronically Scanning Microwave Radiometer (ESMR) [1] that went into orbit in 1973. This first generation sensor recorded images at 25 kilometres of spatial resolution. The range of measurement that a passive microwave sensor can make is as follows: 1) temperature and moisture sounding, 2) rainfall, 3) sea ice, 4) snow/ice cover, 5) soil moisture, 6) oceanic wind speed etc.

In active microwave sensors, the information on the measurand is derived from the backscattered signal from the target due to pulses of radiation emitted from the sensor itself. Out of many active microwave sensors, some of the sensors include synthetic aperture radar, scatterometer, altimeter etc.

Besides, in industries, commerce, biomedical and many other useful areas, microwave sensors are successfully used with preference.

8.2 Basics of Microwaves Propagation

Wavelength is an inherent characteristic feature associated with the radiation or propagation of radio waves through any transmitting medium. The value of the wavelength depends entirely on the frequency of the propagating wave, as the velocity of the wave propagation through a medium is a constant factor. The entire electromagnetic spectrum is divided into number of regions based on wavelength or frequency of the propagating wave. Table-8.1 shows the different designations of the electromagnetic spectrum starting from very low frequency to extremely high frequency in which microwave takes a place in the centimetric range.

In a similar fashion, the microwave spectrum from 3 to 30 GHz is spread out conveniently in different bands which is as follow (Table-8.2):

It may be noted that microwaves occupy a greater part of the electromagnetic spectrum than visible and infrared radiations. Passive sensors sense microwaves of selected frequencies, while active microwave sensors transmit and receive at fixed frequencies. The bands noted in Table-8.1 and Table-8.2 represent standards used for governments and commercial applications. The most common bands for active satellite remote sensing today are C and L Bands.

Table-8.1: Electromagnetic Bands Spectrum

Designation	Frequency Range	Wavelength Range
Long waves	30 Hz - 300 kHz	10 Km - 1 K m
Medium waves	300 kHz - 3 MHz	1 Km - 100 m
Short waves	3 MHz - 30 MHz	100 m - 10 m
Very high freq.(VHF)	30 MHz - 300 MHz	10 m - 1 m
Ultra high freq.(UHF)	300 MHz - 3000 MHz	1 m - 0.1 m
Microwaves (MW)	**3 GHz - 30 GHz**	**10 cm - 1 cm**
Millimetre waves (mm)	30 GHz - 300 GHz	10 mm - 1 mm
Sub-millimetre waves	300 GHz - 3000 GHz	1 mm - 0.1 mm
Infrared (including Far Infrared region)	3000 GHz - 416 THz	$10^{-4}\mu m$-0.72 μm

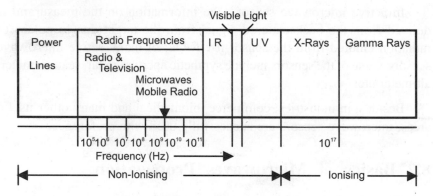

Table-8.2 Microwave Bands

Designations	Frequency Ranges
S Band	2.60 GHz - 3.95 GHz
C Band	3.95 GHz - 5.85 GHz
XC Band	5.85 GHz - 8.20 GHz
X Band	8.20 GHz - 12.4 GHz
Ku Band	12.4 GHz - 18.0 GHz
K Band	18.0 GHz - 26.5 GHz
Ka Band	26.5 GHz - 40.0 GHz

Electromagnetic waves can be divided into two categories, e.g., ionising and non-ionising radiation. Ionisation is a process by which electrons are stripped off from atoms and molecules producing molecular changes or damages in biological tissues including some remarkable effects on DNA. X-rays and Gamma rays often have sufficiently high levels of energy that produce ionisation on biological materials. So this region is classified as ionisation radiation. On the other hand, radiations associated with microwaves or lower radio frequencies do not have energy levels that may cause ionisation of atoms in biological bodies. This region of the electromagnetic spectrum is, therefore, termed as non-ionisation radiation.

8.2.1 Polarization and Modes

As is apparent from Table-8.1 and Table-8.2, the microwave energy spreads over a band of frequencies (3 GHz to 30 GHz). Not only it occurs in divergent frequencies, its fields may have different polarisations as well. Fig. 8.1 shows such an example of the linear polarisation (1) vertical and (2) horizontal field-components of microwave energy. Sometimes, in complex situation, the polarisation could be, besides linear polarisation, circular or elliptical polarisation with right-handed or left-handed rotation. As has been illustrated in Fig. 8.1, in free space propagation, the linear electric and magnetic vectors are mutually perpendicular and orthogonal to the direction of propagation along z direction, i.e., the mode of propagation here does not have any longitudinal component of electric and magnetic fields, or in other words, $E_z = H_z = 0$. This mode of propagation is known as Transverse Electric and Magnetic (TEM) Mode. It will be worthwhile to mention, here, that TEM modes are prevalent in free space, twin-wire lines, coaxial cables etc. Besides TEM modes, other type of possible modes is either TE (transverse electric) or TM (transverse magnetic) normally found to propagate through hollow metal wave-guides, as they are to satisfy certain particular boundary conditions. In TE modes, invariably, $E_z = 0$ and $H_z \neq 0$, and in TM modes, $H_z = 0$ and $E_z \neq 0$. Because of peculiar complex boundary conditions to be satisfied in a two-layer dielectric structure of core and cladding in optical fibres, the propagating modes present in the optical fibre are hybrid in nature where both TE and TM modes are supported simultaneously in the form of HE or EH modes. In literatures, they are conveniently expressed as Linearly Polarised (LP) modes.

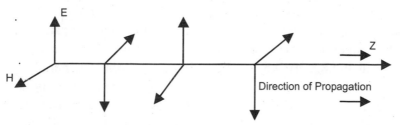

Fig. 8.1 Linear polarisations of propagating electromagnetic wave. Electric field vector, E and Magnetic field vector H are vertically and horizontally polarized respectively at z = 0. These polarizations change with propagation along z direction.

8.2.2 Antennas

In microwave sensors, whether active or passive, imaging or non-imaging, antennas play an important role in an effective and efficient remote sensing system. In fact, antennas act as a perfect transition device or matching unit interfacing between free space and circuitry. They convert photons to electrons and vice versa. In transmitting antennas, the electrical signal from an electrical circuit produces electron in the antenna to oscillate and sets up an electric current distribution on the conductive surface of the antenna structure. In case of time-varying surface electric current density present on the antenna surface, the antenna will radiate a time-harmonic electromagnetic field in free space. Conforming the rule of reciprocity, the same nature of behaviour is observed in receiving antennas in a reverse way i.e., the conductive surface of the receiving antenna on interception of electromagnetic field from the free space experiences an electric current flow that, ultimately, is thrown to the circuitry for signal processing.

The basic equation of radiation may be expressed as $\dot{I}L = Q\dot{v}$ where, $\dot{I}$ = time-changing current, (As^{-1}), L = length of current element, (m), Q = charge, (C), and $\dot{v}$ = time-change of velocity which equals the acceleration of the charge, (ms^{-2}).

Before various antenna structures used as in the microwave sensors are put into record, it will be worthwhile to discuss various antenna parameters that are relevant to the direct operation of microwave sensors. These are as follows:

8.2.2.1 Radiation Pattern

The angular distribution of the relative field strength of the radiated field of an antenna, at a fixed distan from the antenna, in any one of the orthogonal principal planes, either on the vertical or on the horizontal plane is known as radiation pattern or sometimes called antenna pattern. When plotted, either on polar coordinates or in rectangular coordinates, it is as shown in Fig.8.2. However, the radiation pattern is generally three-dimensional. The pattern, normally, posses a main beam with its peak value at 0 dB and a few numbers of side-lobes that are normally designed to be insignificant with respect to its main lobe in remote sensing applications and hence ignored. The presence of side-lobes always produces false alarms and increases the noise-temperature of the antenna, which is not desirable at all for remote sensing.

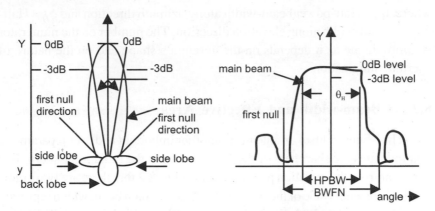

(a) Radiation Pattern of an antenna
     in a polar coordinate system. Y
     represents radiated power in dB
     along the radial direction of the
     Polar coordinate system. Angle
     q_H is the -3dB beam-width.

(b) Radiation Pattern of an antenna
     in a Cartesian coordinate system.
     Y is the radiated power in dB.
     HPBW= half-power or -3dB
     Beam-width. BWFN = Beam-
     Width w.r.t the First Null angle
     of the pattern

Fig. 8.2 Typical schematic presentation of the Radiation Pattern of an antenna plotted on (a) polar coordinates and (b) cartesian coordinates.

8.2.2.2 Antenna Gain

Being absolutely passive in nature, how an antenna can have gain? This is a valid question that instantly peeps into the mind of a man, when he encounters with this parameter of the antenna. However, in order to clarify this, it can be said that it is nothing but a comparison of responses of two antennas at a point when they are fed with the same RF energy - one being the antenna under consideration whose gain is to be defined, and the other is the reference antenna, say, an isotropic antenna which radiates equally in all direction. To be more precise, the antenna gain $G(\theta, \phi)$ can be defined as the ratio of the power per unit solid angle radiated in a particular direction by a directive antenna to the power per unit solid angle emanating from an isotropic antenna with 100 per cent efficiency. Here, gain G expressed in dB_ϕ is a function of θ, the azimuth angle and also a function of ϕ, the elevation angle of a spherical coordinate system. Physically, the gain of an antenna is the measure of its ability to focus the RF energy fed into it in a particular direction with respect to a reference antenna like an isotropic radiator. The physical parameters controlling the gain of an antenna is shown in the expression (1) below:

$$G(\theta,\phi) \approx \frac{41,253}{\theta_H \phi_H} \tag{1}$$

where, θ_H = Half-power beam-width along azimuth direction and ϕ_H = Half-power beam-width along elevation direction. The number on the numerator is approximate as it depends on the antenna's shape and on the nature of aperture illumination.

8.2.2.3 Beam-width and Effective Area

The beam-width of an antenna is another important and useful parameter that indicates the sharpness of the major beam of the radiation pattern. It is apparent from the radiation pattern diagram Fig.8.2 that there are two points available on either side of the peak power level of main beam which represent the half-power level (-3 dB) with respect to the peak-value at 0 dB level at 0° angle. The angular distance between these two points at -3 dB levels is known as the half-power beam-width (HPBM) or 3 dB beam-width of the antenna. Some times, the beam-width may also be defined as the angular distance between the first nulls of the antenna pattern; hence, known as Beam-Width between First Nulls (BWFN). The pattern, a three-dimensional solid figure in general, is measured, usually, on either of the two orthogonal principal planes. These are E- plane and H-plane. In E- plane (elevation plane), the beam-width is designated as ϕ_E, and in H-plane (azimuth plane), it is θ_H. However, the beam-width depends on the operating wavelength λ and the antenna aperture dimension D in the corresponding principal plane in a manner shown below:

$$\theta_H = \lambda / D \tag{2}$$

Beam-width, typically, equals to 65 λ / D, when the first side-lobe level is 25 to 28 dB below the peak value of the major-beam. This leads to the fact that the narrower beam-width corresponds to larger antenna-size and vice-versa at a fixed operating frequency.

It may be interesting to note that depending upon application the shape of the beam may change. For example, it may be either pencil beam or fan beam. The beam-widths of a pencil beam are identical in both the azimuth as well as elevation planes, i.e., $\theta_H \approx \phi_H$. Generally, it is a few degrees; typically about a degree or so. In case of fan beams, the azimuth beam-width θ_H is normally very small, typically one degree or so, whereas the elevation beam-width ϕ_H is much larger, say, about four to ten times the azimuth beam-width.

It is to be mentioned that the power gain G defined earlier is named as directive gain or directivity of the antenna as the gain is referred to its direction of maximum response. It is related to the effective area of the aperture A_e (also called as effective aperture) as given below:

$$G = \frac{4\pi A_e}{\lambda^2} = \frac{4\pi \eta_a A}{\lambda^2} \qquad (3)$$

where, λ = wavelength, η_a = antenna aperture efficiency and A = physical area of the antenna.

8.2.2.4 Radiation Resistance and Bandwidth

Radiation resistance of an antenna is defined as the ratio of the power radiated by the antenna to the square of the current at its feed point. The bandwidth of an antenna refers to the frequency range over which operation is satisfactory and is taken between half power points in respect of input impedance.

In addition to these features of antennas, there are other equally important characteristics that may include radiation efficiency, bandwidth, antenna losses etc.

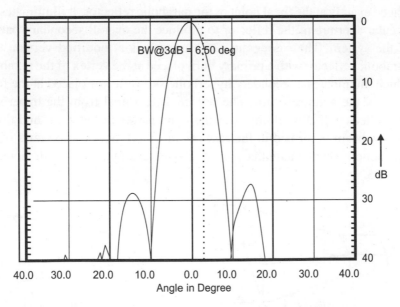

Fig. 8.3 Graphical representation of the radiation pattern of a remote sensing antenna in the Cartesian coordinates system [4].

As an example, Fig.8.3 shows that the 3 dB beam-width (HPBW) of the radiation pattern on the elevation plane of RADPAT system [4] is 6.5 degree and side-lobe level is 27 dB below the peak value of the main-beam. Y-axis represents relative radiated power in dB normalized to 0 dB at the peak-value. The frequency of operation is 9.9 GHz and polarization is horizontal. The first null points on either side of the peak are at 10 degree.

8.2.2.5 Types of Antennas

For microwave sensors, antennas, normally used are the following: (i) Horn antenna, (ii) slotted antenna, (iii) parabolic reflector, (vi) cassegrain antenna, (v) phased array antenna and the like. Some of the configurations of these antennas are displayed in Fig. 8.4.

In a pyramidal horn antenna, both the dimensions (a, b) of a uniform rectangular waveguide are flared up into new rectangular dimensions (a, b) so as to have a gracefully taper-transition for the purpose of proper matching between the space and the waveguide.

In conical horn antenna, a circular waveguide is similarly transformed into a circular aperture of larger dimension. In sectoral horn, either E-plane or H-plane is flared up while other dimension remaining unchanged. In parabolic reflector, horn antenna is normally used as a primary feed horn placed exactly at the focal point of the parabolic reflector. It illuminates the circular aperture of the reflector to produce the overall secondary pattern of the system. The cassegrain [2,3] antenna is a modified version of a parabolic reflector with a primary horn placed at the vertex of the parabola, which illuminates a secondary hyperbolic sub-reflector placed at the focal point of the main reflector. The reflected RF signal from the hyperbolic sub-reflector, finally, illuminates the main parabolic reflector and thus a narrow beam is formed in the forward direction. Schematic views of the parabolic reflector antenna and of Cassegrain antenna are depicted in Fig. 8.5.

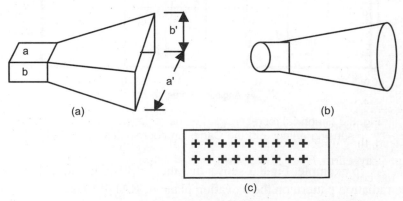

Fig. 8.4 Microwave antennas (a) pyramidal horn (b) conical horn (c) slotted array antenna

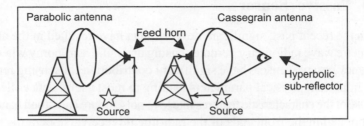

Fig. 8.5 Schematic view of parabolic reflector antenna and Cassegrain antenna used in microwave sensing system

8.3 Classification of Microwave Sensors

The broad classification of remote sensors is shown Fig. 8.6. The interest of the paper lies in the microwave sensors only, both passive as well as active. The family of remote sensors, in general, can broadly be classified into a number of groups - active and passive. Further, these can be either imaging sensors that produce two- dimensional images or non-imaging sensors, which do not produce images, but measure some important parameters, like altitude, phase, velocity, acceleration etc. The presentation, here, focuses on some features of both the passive sensors like scanning microwave radiometers and active sensors like microwave altimeters, synthetic aperture radars etc as sensing elements.

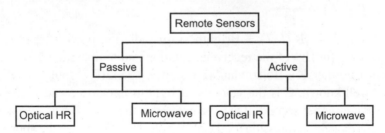

Fig. 8.6 Classification of Remote Sensors of which interest lies with microwave sensors only.

It may be noted that microwave passive sensors and active sensors can be either image producing sensors or non-image producing sensors. Passive sensors do not possess any radiating system of its own for detection. Instead, they directly receive the signal emanating spontaneously from the target, say, earth. Even, sometimes, radiation emanates beneath the ground. On the contrary, active sensors always acquire radiating systems and the detection process depends very much on the reflected signal received from the target. It is due to a-priori known transmitted radiation from the sensor itself.

8.3.1 Passive Sensors

In the recent past, significant development have resulted in the areas of millimetre wave radiometry, remote sensing and radio astronomy where high frequency passive sensors have significant contribution for sensing remotely some important physical parameters leading to most useful data collections. Because of the characteristic features like low-noise figure and broad bandwidth incorporated in the front ends of the radiometric receivers operating at high frequencies, a spectacular and significant improvement in sensitivity for detection of thermal radiation has been achieved. This improved version of radiometric receivers finds successful applications in remote sensing of the environment and radio astronomy [5-9]. Besides, these receivers are also used in many other fields such as missile thermal guidance [10], detection of ships and vehicles [11], measurement of targets and terrain signatures [12]. Other applications include thermography [13] and navigation [14].

The basic principle for passive microwave sensing is similar to longwave thermal IR-sensing, only the wavelengths of energy are longer and more variable depending on the source material. However, the governing relation [15] of the power received, known as Brightness B, by a passive microwave sensor in terms of relevant parameters like frequency of radiation, (f) and temperature of the source radiation, (T) may be expressed through the well-known Rayleigh-Jeans Radiation Law which is as follows:

$$B(f,T) = \frac{2kTf^2}{c^2} \qquad (4)$$

In eqn. (4), B (f, T) = Brightness or radiation power dependent on frequency f (in Hz) of the received radiation and temperature T (in degree Kelvin). Further, K = Boltzmann's constant = 1.38×10^{-23} Joules per Kelvin. As is well known, C is the velocity of light in free space and is closed to 3×10^8 metres per second. It may be noted that the eqn. (4) has been derived from the Planck's expression for the blackbody radiation under the appropriate assumption of $hf \ll kT$ for the ambient temperature closed to 300K and in the frequency range of microwaves / millimetre waves [15-17].

The emissivity of snow, for instance, depends on the rate of accumulation of the snow, and increases with wetness. Passive sensors are quite useful in distinguishing sea ice edges in oceans, where there is a significant distinction between emissivity of open water and sea ice. Sea ice has a wide range of emissivity over different microwave wavelengths because of its complex composition e.g. density, salinity, air content, age etc. and there are even significant differences between the microwave properties of Arctic and Antarctic sea ice. These many differences complicate analysis, but also allow for improved processing once accurate classification algorithms are developed.

However, it is clear from Fig. 8.7 that the passive microwave sensors installed in the spacecraft receive microwave signals from various objects around it, conceptually, just like long-wave thermal- / IR-sensing, as stated earlier. Only difference is the frequency of energy that is being smaller and more variable depends on the source materials. As with thermal sensing, blackbody radiation theory becomes key to the conceptual understanding of passive microwave sensing. All objects in the natural environment emit microwave radiations of some magnitude, but the amounts are generally very small and faint. Passive microwave sensors simply measure the amount of radiation emitted or reflected by the targets. Everything on earth emits varying amounts of energy across the electromagnetic spectrum, including microwave energy (wavelengths from 1 mm to 0.8 m). Snow and ice are no exception. Fig.8.7 surveys the possible microwave sources of radiation for the microwave passive sensors on board and depicts them in the diagram that are as follows:

(1) Extra terrestrial body (the Sun, Stars etc.),
(2) Reflection from the ground
(3) Radiations from the ice-caped mountains
(4) Ocean-water
(5) Foliage/vegetation
(6) Target ground surface/subsurface
(7) Buildings
(8) Clouds etc.

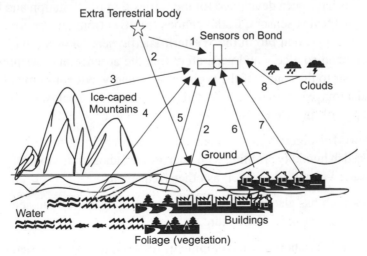

Fig. 8.7 Microwave radiations caused by different self-emanating sources even from within the Ground (ray no. 6) received by passive sensors mounted on the Satellite.

One example of a passive microwave sensor that collects data on the cryosphere is the Special Sensor Microwave/Imager (SSM/I). The SSM/I is a four band passive sensor that collects data at 25 km of spatial resolution over 1400 km of swath. The sensors record target microwave emissions, then processors calculate target temperature and brightness and derive environmental data. The data records calculate various parameters of ocean, land, atmosphere and ice. The four ice parameters derived from the data are ice concentration, ice age, ice edge and cloud water content over the ice.

The next passive microwave sensor in orbit is the Advanced Microwave Scanning Radiometer (AMSR), carried by both the Japanese Advanced Earth Observing Satellite-II (ADEOS-II) and US Earth Observing System (EOS) PM-I satellite. With a larger antenna allowing improved spatial resolution (from 50 km to 5 km, depending on frequency) and a swath width of approximately 1600 km.

AMSR provides the best passive images of any satellite sensor. AMSR features eight bands, two more than in the SSM/I, it collects data on the same four parameters as the SSM/I does e.g. sea ice concentration, ice age, ice edge, and water vapour content of clouds over ice.

Along with these activities, it is worth mentioning that the NASA centres have successfully developed the following research programmes [18] in course of time. In this regard, the research milestones achieved are given below:

For the use in the Airborne radar and radiometric measurements suitable algorithms have been developed for the retrieval of precipitation and latent heat from TRMM sensors. Radar echoes (in dBz) from an airborne-radar flying over Typhoon Flo in the Western Pacific near Okinawa have been photographed so nicely and accurately that the absence and the presence of rain near to the centre of the typhoon are clearly evident in the picture. The radar image is seen to be reflected by the sea surface below the airplane so that part of 'inverted image is also clearly visible.

Powerful algorithm and measurement Technique have also been developed in humidity profiling with microwave radiometers around the 183-GHz water vapour line as well as other frequencies.

Research has also been carried out on microwave remote sensing of snow, vegetation and soil moisture.

In passive radiometer development, research has been carried out on soil-moisture measurements with multifrequency microwave radio-meters at L, C and Ka bands in junction with the Shuttle Imaging Radar.

Millimetre-wave Imaging Radiometer (MIR) is an instrument, which has been developed as an airborne scanning radiometer having five bands at frequencies of 90, 150, 183, 235, and 320 GHz. It is being used aboard NASA's high-altitude (21 km) ER-2 to study atmospheric water vapour, clouds, rain and other forms of precipitation. Fig. 8.8 shows the block diagram of a radiometer on board explaining its functions.

Electronically Steerable Thinned Array Radiometer (ESTAR) is the synthetic aperture microwave radiometer. It employs an interferometric technique to synthesise high resolutions from small antennas. It operates at 1.4 GHz and has been deployed on the C-130 and P-3 aircrafts. It is supposed to measure soil moisture and ocean salinity. The high brightness temperature indicates dry soil.

The AMSU-A and B systems are at cutting edge of space-borne microwave radiometer technology. They utilise the most advanced narrow-band low-noise microwave radiometer receiver and antenna technologies. AMSU-A is designed to measure atmospheric temperature profiles up to 40 km, and AMSU-B will provide humidity profile using channels at the 183 GHz water vapour line. They require complex, state-of-the art, RF and antenna designs.

The TMI is a five frequency, dual-polarised, microwave imager ranging from 10 to 89 GHz.

8.3.2 Active Sensors

In Active Microwave Sensors, a transmitter is employed, which emits microwave radiations along the direction of expected target. The radiations hit the targets in view and the radiated energy gets scattered in various directions. A small part of the scattered energy directed along the backward direction i.e., towards the receiver located at the sensor. Thus, the sensing receiver, which is sensitive enough, picks up the back-scattered energy and the received signal is processed for getting back the information carried out by the microwave signal.

On RF amplification and detection by the receiver, the required information is finally extracted in a proper fashion and displayed on a suitable monitor. In fact, this is the basic principle of radar. A radar operating at microwave bands is basically a microwave sensor sensing various physical parameters of the targets (e.g., range, bearing, elevation, speed, velocity, temperature and many more).

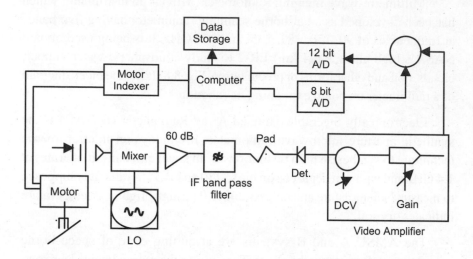

Fig. 8.8 Block diagram of the Millimeter-wave Imaging Radiometer (MIR) [18].

From these data, many other useful parameters of the hitting targets may be obtained employing suitable relationships and corresponding algorithms. Just to name a few, they are imagery, surface elevation, oceanic surface wind vectors, sea-state, speed, infiltrators etc.

Radar sensors may be of various types depending on applications and uses. It includes Pulse Radar, FM-CW (Frequency Modulated Continuous Wave) Radar, Pulse Doppler Radar, MTI (Moving Target Indication) Radar, Weather forecasting Radar, Tracking Radar, Fire-Control Radar, SAR (Synthetic Aperture Radar) and the like. Basic principles of these active sensors are described below:

8.3.2.1 Principle of Pulsed Radar

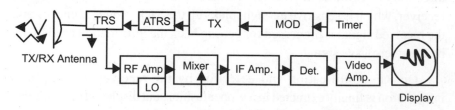

Fig. 8.9 Block diagram of a pulsed-radar.

Consider the block diagram displayed in Fig. 8.9. In this case, a train of short pulse modulated RF at microwave frequency is generated in the transmitter and leaves the sensor for the targets placed at a distance R_{max}. The back-scattered RF signal received carries all the information regarding the targets e.g. range, velocity, acceleration, direction etc, which may be processed in the receiver in a suitable manner and displayed in the indicator. The range information of the target is given below:

$$R_{max} = \left[\frac{P_t G^2 \lambda^2 \sigma}{(4\pi)^3 S_{min}} \right]^{0.25} \tag{5}$$

In eqn. (5), P_t = peak transmitted power, G = power gain of the antenna, λ = wavelength, σ = radar cross-section of the target and S_{min} = minimum detectable signal required by the radar sensor. The waveform in radar display consisting of transmitted and corresponding received pulses will be as per the diagram shown in Fig. 8.10.

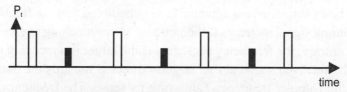

Fig. 8.10 RF pulses radiated out of the radar sensor ☐ and corresponding received pulses ▪

The sensor basically measures the time of return pulses from emitted ones and senses the range of the target.

8.3.2.2 CW Radar Sensors

In this sensor, instead of pulse modulated RF, the transmitter generates a continuous unmodulated oscillation of frequency f_0, which is radiated by an antenna. A similar backscatter signal returns after hitting the target and is received by the receiver. As it does not have any modulation, it cannot measure the range. In case the target has a relative (radial) motion with respect to the sensor, due to the Doppler effect, the received signal will have frequency other than f_0. The frequency transformation that will take place is $(f_0 \pm f_d)$, where f_d is the Doppler frequency shift. Once the sensor measures the Doppler frequency, the radial velocity of the target can be determined through the relation: $f_d = 2 v_r / \lambda$, losing the sense of the velocity i.e., approaching or receding target. v_r = relative velocity of the target and λ is the wavelength of the free space.

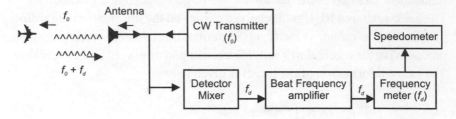

Fig. 8.11 Basic Block diagram of a CW radar sensor.

Consider the block diagram in Fig. 8.11. Basic principle of this non-imaging CW radar sensor lies on the generation of continuous wave sinusoid signal without any external modulation. Let its frequency be f_0. The signal is fed to a suitable antenna for radiation in space. When the radiated signal is encountered with radially directed moving target, the reflected signal from the moving target will have a transformed frequency $f_0 \pm f_d$, which will receive be by the receiving antenna. It may be noted that the frequency of the incoming signal increases if the target is an approaching one i.e., $f_0 + f_d$; on the contrary, the frequency decreases if the target is a receding one i.e., $f_0 - f_d$. The received signal at this stage is directed towards the detector to extract the Doppler frequency f_d missing its sense. The frequency is read and converted into velocity through the relation $f_d = 2v_r / \lambda$. The velocity meter enjoys certain advantages e.g.,

(i) Very good accuracy of measurement
(ii) Simple circuit used
(iii) Low transmitting power
(iv) Low power consumption
(v) Comparatively small equipment size
(vi) Can ignore all stationary targets
(vii) Ability to operate even at zero range
(viii) Capable to measure a large range of target speeds quickly & accurately
(ix) With some additional circuitry, it can measure the range as well.
Among its disadvantages are the following
(i) Can not measure the range
(ii) Misses the sense of the target speed.
CW radar sensors have a number of applications. These are as follows:
(i) Speed measurement for aircraft navigation
(ii) Rate-of-climb meter for vertical -take off planes
(iii) Speedometer for police radar sensors.

8.3.2.3 Altimeter

It is again a CW radar sensor where the continuously generated microwave signal is frequency modulated in a linear, or in otherwords, triangular manner. It measures the target altitude. Equipped with very precise timing devices, the sensors transmit microwaves to Earth's surface, and based on the timing of the return signal, the sensors can determine target altitude. Such sensors are extremely useful when the data is combined with other imagery, and help provide accurate maps of remote polar region. In literature, the device is known by various names, such as FM-CW radar, terrain avoidance radar, altimeter and many others.

The block diagram of an altimeter is given below (Fig. 8.12):

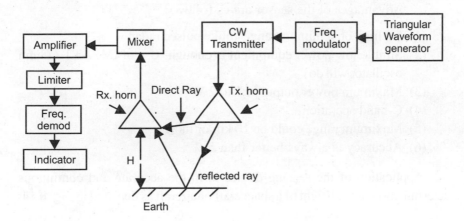

Fig. 8.12 Block diagram of an FM-CW radar sensors.

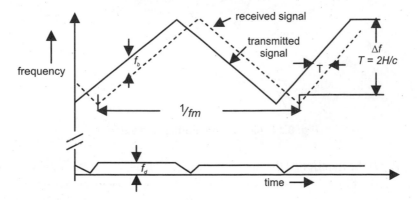

Fig. 8.13 Frequency-time plot of the sensor showing the direct and reflected rays received by the receiver. The beat frequency f_b is measured as a function of time. The altitude of the target is obtained from eq. 6.

The principle of the FM-CW sensor is explained in the accompanying Fig. 8.13. Here, the expression for beat frequency, f_b is given in eqn. (b) 6 as follows:

$$f_b = 4 H f_m \, \Delta f \, / \, c \tag{6}$$

Where, H　　= height (altitude) or range of the target, m

　　f_m　　= rate of frequency modulation, Hz

　　Δf　　= frequency excursion, Hz

　　c　　= velocity of wave propagation in free space, m/s

Thus, the measure of average value of f_b determines unambiguously the altitude of the spacecraft. As in this case, the target itself is the Earth; the question of Doppler shift is not coming into the picture.

Advantages of the sensor are as follows:

(1) No limit on minimum range unlike pulsed radar
(2) Simple low power equipment is enough. Use of IMPATT / Gunn oscillator will do
(3) Maximum power output of 1 to 2 W
(4) C-band operation
(5) Maximum range could be 10 km or more
(6) Accuracy is always better than 5 %

Application of the sensing device is in the accurate and continuous determination of the height of a spacecraft or aircraft as depicted in Fig. 8.14.

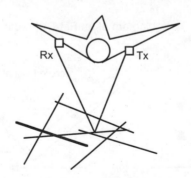

Fig. 8.14 Aircraft measuring its altitude.

It may be mentioned that the operation the radar sensor explained above through the Fig.8.11 & 8.12 is valid only for stationary target whose range or altitude is not changing. In case the target is moving, the Fig.8.12 will be modified where the Doppler frequency shift f_d will be superimposed on the FM range beat note f_b as depicted in Fig.8.15.

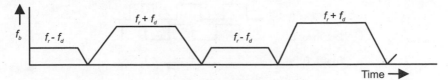

Fig. 8.15 Beat frequency, f_r as a function of time for a moving target with Doppler frequency f_d in a FM-CW radar sensor.

If, with the help of suitable averaging circuit, the two frequency levels $f_r - f_d$ and $f_r + f_d$ are separately measured by switching a frequency counter in every half modulation cycle, then beat frequency f_r and the Doppler frequency f_d can be separated and exact values of these frequencies are obtained. The beat frequency f_r gives the range and the Doppler frequency f_d supplies the speed of the target.

However, satellite radar altimeters are used mainly for oceanographic studies. In fact, these altimeters can measure the dynamic sea surface elevation or in other words the sea surface topology.

8.3.2.4 Rate-of-Climb Meter

The following block diagram explains the operation of the rate-of-climb meter used in the vertical take-off aeroplane. It determines the velocity of the aircraft with respect to the ground during take-off and landing. The received signal is divided in two channels A and B and fed into separate detectors. A portion of the transmitted signal is fed directly into the detector of channel A. In channel B, the reference from the transmitter is delayed by 90 degree. The sign of the phase shift determines the direction of motion as shown in Fig.8.16.

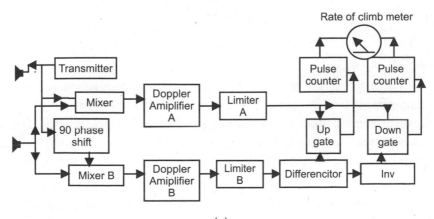

(a)

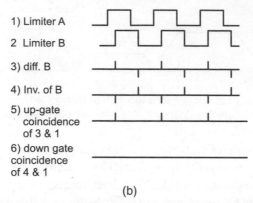

(b)

Fig. 8.16 Measurement of Doppler direction as used for VTO aircraft by rate-of-climb meter. (a) Block diagram, (b) Signals at different stages.

8.3.2.5 Synthetic Aperture Radar (SAR) Sensor

Synthetic aperture radar sensors are perhaps the most powerful and most prevalent space-borne sensors monitoring the cryosphere. Unlike passive microwave sensors, radar sensors send out microwave energy pulses, and then measure the strength and polarity of the return signal or backscatter.

Table- 8.3

Polarisation Combination (Transmission - Reception)
VV
HH
VH
HV

Table-8.3 illustrates different polarities transmission/reception combinations for SAR sensors. The latest SARs take advantages of these different combinations of transmission and reception polarimetry. These polarimeters can vary the polarisation of their signals, thereby taking advantages of the unique reflectance signature of different target materials.

In real aperture radar, backscatter is received from the same location as the initial transmitted microwave signal. A synthetic aperture radar sensor sends out a signal, but the forward motion of the satellite to receive the backscatter over a distance, thus simulating a large aperture. This simulated aperture and the more sophisticated signal processors allow for higher resolution than standard aperture sensors. In cryosphere, the strength, or brightness, of the return signal is based on salinity, surface roughness, geometry, density and target wetness. Typical spatial resolution is 18 to 30 metres, with swath width of 75 to 100 km. This resolution is similar to some high-resolution visible/IR sensors, but SAR sensors can collect data independent of weather or light conditions. With such a fine resolution, SAR is the sensor of choice for zooming in on small areas of interest.

8.4 Industry / Commerce Applications

In industry as well, the sensors used in microwave ranges are also found in abundance. The type of sensor applications in this field can be classified as either monostatic or bistatic depending on the configuration of the sensor antennas [19]. In monostatic sensors, both the transmitting as well as receiving antennas are either one and the same or located in the same place, where as in bistatic case, they are not only different but also separated wide apart. Being small in size due to high frequency application, the microwave sensors are extremely portable, cheap and can easily be installed at any odd place and made operational for convenient applications. In fact, the application areas of the monostatic microwave technology include zone monitoring, microwave fencing, portal monitoring and portable applications.

It is a very common problem, in an industry, to safeguard certain vulnerable areas that are susceptible to theft and pilferage. Being sensitive to security hazards, it is, rather, difficult to man the entire area in an industrial complex for keeping away the unwanted trespassers from outside. For a reliable fencing for 24 hours a day and 7 days per week, it is advisable to install a microwave protection system either using monostatic sensor or bistatic sensor. In case of monstatic one, there two or more are trans-receivers (monostatic sensors) placed at different points in the protected zone. These trans-receivers focus their beams towards a reflecting surface as shown schematically in Fig.8.17 so that these R F beams, both transmitted and received ones, form the required barrier wall around the protected area. Once a stealthy intruder cuts any one of these beams, he is going to change the received R F signal strength and the normal operation of the R F circuit is interrupted. It causes a trigger to operate the alarm in the control-room alerting the security personels to take immediate action.

A monostatic sensor normally operates in X-band or in K-band frequency range. It may also be used to detect people, vehicles and other large objects in motion. Among the advantages, easy installation, portability, low-power requirements, suitability for monitoring portal areas are worth mentioning.

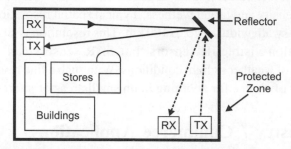

Fig. 8.17 Microwave fencing for protecting a vulnerable zone in industry using monostatic microwave senso has been depicted schematically in this diagram. Reflector positioned as indicated reflects RF signal from the Tx backed to the corresponding Rx formed as closed microwave fencing to protect the zone. Any body crossing the broader of the protected zone has to interrupt physically the radiation paths and gets detected.

The principle of bistatic microwave sensors arrangement is basically same as that of the previous case of monostatics system. The only difference lies in the fact that the transmitters and receivers are not located in the same place, but scattered in a particular fashion as illustrated in Fig.8.18. This system enjoys the same advantages as of monostatic case and is equally attractive for security applications.

In defence, the detection of buried land mines is a very demanding task. For the safety of the defence personnel, it is essential to locate the buried land mines and remove them from the ground on road. In case the land mines are made up of metals, metal detectors can detect them easily. But the metal detector fails when the land mines are made of plastics, use of which is the practice of the day. A hand-held mine detector based on the principle of passive microwave sensor [20] can sense the inherent microwave radiation naturally and spontaneously emitted by a land mine, whether made up of metal or plastic, buried beneath the soil. After identifying and locating the mines by sensors, they are physically removed or destroyed.

The sensing of speed of a moving object is, very often, an essential demanding task and requirement in an industrial ambience. In the detection and measurement [21-22] of radial velocity of high speed moving objects like motor cars, aeroplanes, jet-rockets, missiles, ammunition, speeding cricket-ball etc., the CW radar microwave sensors are very well-suited. The operating principles of sensing the velocity of these moving objects are to employ the FM-CW radar, which has already been discussed before.

The sensor not only measures the exact value of the velocity of the moving objects, it also evaluates the range of the objects as well. One of the popular applications of the CW radar sensor is the Police radar that checks the speed limit of rushing vehicles on the highway even in the midst of large clutter echoes. It is also used to monitor the docking speed of a large ship. Besides, the detection of radial speed or velocity of a moving target, the CW Doppler radar sensor are used in the synthetic aperture radar and in the inverse synthetic radar sensor for producing images of targets.

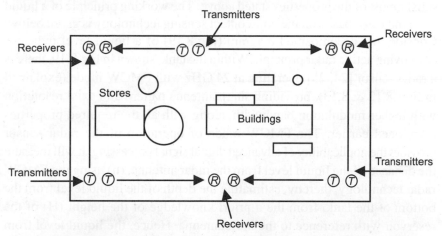

Fig. 8.18 A schematic view of a typical bistatic microwave sensor installation for protecting industrial zone. In the case, a few sets of transmitters ⓣ and receivers ⓡ combinations are so arranged that the R F beams form a closed microwave fencing for keep away the stealthy intruders.

In meteorological radar, the technique is well suited for the measurement of cloud velocity and the prediction of the weather. The sensor finds appropriate applications in the measurement of railroad-freight-car velocity during humping operations in marshalling yard and in use as detection device to the maintenance personnel an advance alarm of approaching trains. In industry, the principle of this sensing radar has been advantageously and successfully employed in monitoring the turbine-blade vibrations, the peripheral speed of grinding wheels and the vibrations in the cables of suspension bridges. The development of intruder-alarm, burglar alarm, fire alarm, etc for safety of industrial arena can be potentially based on this principle as well. The sub-surface radar imaging finds applications to help reveal the invisible, e.g., land-mine detection, identification of weak points in bridge structures, measurement of the glacier-thickness, identification and location of voids and cavities within coal mines, and under ground archaeological sites etc.

Microwave sensors, essentially, are well suited for robust non-contact measurement of distance and velocity of moving objects. In fact, the sensor applications employing radar technology (active sensors) are steadily increasing [23] in these areas. The measurement of liquid level in a reservoir has significantly high potential requirements in the industrial market. Truly speaking, the process control industry requires badly the monitoring of liquid levels in tanks of power plants, chemical plants, paper mills, oil refineries, cement factories and the like. The essential properties that are expected in these plants are as follows: high reliability, repeatability, good accuracy, easy installation, easy operation and maintenance etc. Microwave sensors satisfy most of the properties stated above. The working principle of a liquid level indicator based on the non-contact sensing technology is given below. Consider the accompanied diagram (Fig.8.19) of a liquid level indicator employing active radar principle.Within the tank, shown in Fig.8.19, there is a radar sensor [24, 25] operating at 24 GHz with FMCW mode (explained in Figs.8.12 & 8.13), providing sharp antenna pattern and axial resolution with higher modulation bandwidth, along with all the required properties mentioned earlier. The FMCW mode of operation of the radar sensor provides the applicability of advanced digital signal processing. It will measure the distance of the liquid level from the radar antenna, (h) using the FMCW radar technology, thereby, estimating the depth of the liquid level from the bottom of the tank, from the a-priori knowledge of the height (H) of the reservoir with reference to the horn antenna. Hence, the liquid level from the bottom of the tank is equal to H-h. A preset value of the liquid levels will operate the valves that control the liquid outlet and inlet maintaining a required constant liquid level within the reservoir automatically. It has got tremendous industrial market.

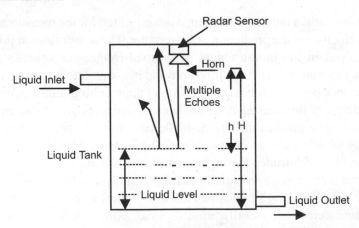

Fig. 8.19 Schematic view explaining the working method of radar sensor liquid level indicator

Besides industry, microwave radar sensors based on Doppler principle are also popular in the vehicular technology especially for the short-distance sensor functions in cars, e.g., for lane-change aid, park distance control, pre-crash detection, blind-spot sensing, side-crash detection and stop & go distance control [26] . 24 and 77 GHz are being used in the module installed in cars. When properly designed, it is, also seen to be used for the purposes of detection with accuracy, the seed of off-road vehicles and proximity for collision avoidance [27].

8.5 Biomedical Applications

In the biomedical applications area, microwave sensors are equally important in detection of various physical disorders, e.g. tumour within the human body using non-invasive diagnostic methods. Not only as an adjunct to surgery diathermy to coagulate, prevent excessive bleeding and seal of traumatized tissues and for localized heating for the cancer therapy [28], the use of microwave sensors radiating rays inside the human body is being made use of monitoring of heart-beats, lung water detection [29], breast cancer etc.

Presently, the early detection and treatment of breast cancer are considered to be the most important to the doctor as, in many countries, this disease is the main course of women death. Specialists believe that early detection of breast cancer is a key to the survival. At present, the standard and established life-saving method of such detection is the X-ray mammography that provides an inexpensive and simple approach to the problem. However, despite immense progress, the technology is still limping because of its relatively high rate of false-negative and false-positive diagnoses, especially in young women. The mammography technique is, sometimes, found to miss a significant number of cancerous tumours.

On the other hand, the microwave imaging technique [30], when applied for early diagnosis of breast cancer, provides an excellent result with satisfactory accuracy that is confirmed by biopsy tests. The best part of microwaves is its ability of identifying extremely small-sized tumours; say of the order of a few millimetres in size and even less. This feature of the technique, thus, makes it more attractive and valuable for the purpose of early diagnosis and treatment of bread cancer. Further, the technique is capable of distinguishing between a malignant and benign tumour helping women to avoid trauma of unnecessary biopsy tests.

In fact, a low-power microwave signal in pulsed form is thrown into the body where the affected organ is present with the help of a transducer having an ultra-wide band microwave phased array antenna. Being non-ionising, the electromagnetic wave at microwave frequency penetrates into the body and interacts with the tissue according to the water content present within the tissue itself. The same phased array antenna placed in the transducer receives the backscattered signal from the affected tissue. By changing the relative phase difference between the elements of the phased array antenna located within the transducer, the beam of the signal is allowed to steer on the entire affected domain of the breast. The received signal is processed to extract the required information based on the fact that the larger is the water content in the tissue, the stronger will be the backscattered signal received. The sensor is controlled by computerized software that can develop a 3-D image of the breast tumour showing clearly the size and the depth of the cancerous tissue. It can very well distinguish between the normal tissues and cancerous tissues of the breast. However, attempt is, still, being carried out to improve upon accuracy of its results.

There is another interesting method of detecting the breast cancer with equal effectiveness. Here the method of detection is based on accurate measurement of the intensity of the electromagnetic radiation automatically and spontaneously generated from the patient's internal tissue by microwave radiometer at microwave frequencies. The intensity of the radiation is proportional to the temperature of the affected tissue. Thus, with the help of suitable software, the 3-D distribution of internal temperature measured by the microwave radiometry is displayed on the monitor of the computer. Further, the specific heat generated in the tumour is dependent linearly with the grow-rate of the tumour. So, first growing tumours being hotter, they are more distinguishable than the normal tissues of the breast in thermograms [31, 32, 33]. Thus, the radiometric method at microwave frequencies is unique for diagnosis of fast-growing tumours in the breast of young ladies. The method can distinguish mastopathy and fibroadenoma with proliferation from those with non-proliferation. Of obvious reason, the technique is absolutely safe and harmless both to the patients as well as to the medical personnel.

The microwave radiometry is not only effectively used in mammology, it is employed in urology for diagnostic prostate cancer, gynaecology, etc as well.

8.6 Conclusions

Over the past thirty years there has been an explosion of data collected on the cryosphere-glaciers, the Greenland and Antarctic ice sheets, sea ice and snow. The rapid advance of remote sensing technology mostly employing microwave sensors is largely responsible for this data explosion. No single sensor is responsible for data collection. Instead, a group of imaging and non-imaging sensors using visible, infrared and microwave energy provides detailed data on many parameters of the cryosphere. These sensors have contributed to our understanding of the composition and physical processes of these often-remote corners of the globe. Today these satellites play a critical role in monitoring changes in the sensitive cryosphere-changes that may be indicative of regional climate change.

In the field of industry/commerce, the use of microwave sensors are popular and well liked. Some typical examples with their basic principles are described with illustrations. In fact, there are many more areas where microwave sensors are very much ahead of their competitors. In diagnosis of breast cancer for women and live-saving land mine identification for defence personnel, microwave sensors play a very important role possibly not found before by traditional and conventional means. Besides, this type of sensors has plenty of beautiful applications in industry/commerce/ biomedical and other areas. In this review, an attempt has been made to identify all possible concepts involved in this emerging technology.

8.7 References

1. 'Passive Microwave Sensors', Obtained from the web site h t t p : / www.personal.psu.edu/users/m/r/mr/cryosphere/passive_microwave_ sensors.htm, 2002.

2. Skolnik, M.I., Introduction to Radar Systems', 3rd Edition, Tata McGraw-Hill Publishing Company Limited, New Delhi, pp. 556-558, 2001.

3. Byran Edde, 'Radar-Principles, Technology, Applications', Pearson Education Inc., Delhi, pp. 438, 1995.

4. Augustin, E. P. et. al., 'RADPAT: A New Portable Digital Radiation Pattern Recording System', obtain from the web site http:// www.technicalsystems.net/PAPERS/RDAPAT.PDF - an Internet Publication,

5. Cog, H. I, A. R Kerr and R. J. Mattauch, 'The Low-Noise 115 GHz Receiver On the Columbia GIISS 4 ft. Radio Telescope', **IEEE Trans. Microwave Theory Tech.,** Vol.MTT-27, No.3, March 1979, pp. 245-248.

6. Ulich, B., et. al, 'Absolute Brightness Temperature Measure-ment at 3.5 mm Wavelength', **IEEE Trans. Antenna Propag.,** Vol. AP-28, No. 3, May 1980, pp. 367-375.

7. Tolbert, C. W., A. W. Straiton and L. C. Krause, 'A 16-foot Diameter Millimeter Wavelength Antenna System, Its Characteristics and its Applications', **IEEE Trans. Antenna Propag.,** Vol.AP-13, No. 2, March 1965, pp.225-229.

8. Townes, C. H., 'The Challenge of Astronomy to Millimetre Wave Technology', **IEEE Trans. Microwave Theory Tech.,** Vol. MTT-24, No.11, November 1976, pp.709-711.

9. Kollberg, E. L. and P. T. Lewin, 'Travelling Wave Masers for Radio Astronomy in the Frequency 20-40 GHz', **IEEE Trans. Microwave Theory Tech.,** Vol.MTT-24, No. 11, November 1976, pp. 718-725.

10. Seashore, C. R., J.E. Miley and B. A. Kearns, 'MM-wave Radar and Radiometer Sensors for Guidance Systems', **Microwave journal,** Vol. 22, No, 8, August 1978, pp.47-51.

11. Richer, K. A, 'Environmental Effects of Radar and Radiometric Systems at Millimetre Wavelength', **Proc. Polytechnic Inst. Brooklyn Symposium on Sub-millimetre Waves,** Polytech. Press, Polytechnic Institute of Brooklyn, New York, April 1970, pp. 533-542.

12. Schuchardt, J. M., et. Al., 'The Coming of mm-wave Forward Locking Imaging Radiometers', **Microwave Journal,** Vol. 24, No. 6, June 1981, pp. 45-62.

13. Edrich, J., 'Centimetre and Millimetre-Wave Thermography - A Survey on Tumour Detection', **J. Microwave Power,** Vol. 14, No. 2, June 1979, pp. 95-104.

14. Moor, R. P., C. A. Hawthrone and M. C. Hoover, 'Position Updating with Microwave Sensors', **IEEE NAECON-76 Proc.,** 1976, pp. 13-19.

15. Bhartia, P. and I. J. Bhal, 'Millimetre Wave Engineering and Applications', A Wiley- Interscience Publication, John Wiley & Sons, Inc. NY, 1984, Chapter 10, pp. 660-662.

16. Griersmith, D. C. 'Radiative Transfer Equation / Microwave Sounding and Rain Detection" in Resources in Earth Observation', http://ceos.cnes.fr:8100/cdrom- 98/ceos1/science/dg/dg21/23.htm, An Internet Publication, May 2002.

17. Murphy, R. et. al., 'Earth Observing System Volume IIe: HMRR High-Resolution Multi- frequency Microwave Radiometer', Published by NASA, Goddard Space Flight Centre, Greenbelt, Marryland 20771, USA, 1987, pp.59.

18. Peter Hildebrand, 'Microwave Sensors', obtained from the web site http://Neptune.gsfc.nasa.gov/mmicrowave.html, Microwave Sensor Branch, Code 975, NASA Goddard Space Flight Centre, Greenbelt, Maryland 20771, USA. pp.1-5.

19. 'Monostatic & Bistatic Microwave Intrusion Detection System' from the web page published by CMC Technology Coordinator, Sandia National Laboratories, Albuquerque, New Mexico, Sepember 1999.

20. Johnson, Joel. T., and M. Zhang, 'Microwave to Reveal Ocean Weather, Locate Land Mines', obtained in the web site at http://www.acs.ohio-state.edu/researchnews/archive/microwav.htm

21. Merriman, R. H and J. W. Rush, Microwave Doppler Sensors, **Microwave Journal**, vol.17, pp. 27 - 30, July, 1974.

22. Whetton, C. P., Industrial and Scientific Applications of Doppler Radar, **Microwave Journal**, vol. 18, pp. 39 - 42, November 1975.

23. Heide, P., Commercial Microwave Sensor Technology - An Emerging Business, **Microwave Journal,** vol.42, no. 5, pp.348-352, May 1999.

24. Kielb, J. A., et. al., Application of a 25 GHz FMCW Radar for Industrial Control and Process Level Measurement, **1999 IEEE MTT-s** Int. Microwave Symposium, Anaheim, USA, pp.   281-284.

25. Vossiek, M, et. al., Novel 24 GHz Radar Level Gauge for Industrial Automation and Process Control, **SENSOR 99,** Int. Fair and Congress for Sensors, Transducers & System, Nurnberg, Garmany, pp. 105-110.

26 Kerssenbrock, T.V. and Heide, P., Novel 77 GHz Flip-Chip Sensor Modules for Automotive Radar Applications, **1999 IEEE MTT-s,** Int. Microwave Symposium, Anaheim, USA, pp. 289-292.

27. Heyward, S. W., Microwave Motion Sensors for Off-Road Vehicle Velocity Data aand Collision Avoidance, Web Site, http://www.sensorsmag.com/articles/1299/index.htm, December 1999.

28. Short, J.G., and Turner, P.F, Physical Hyperthermia and Cancer Therapy, **Proc. IEEE**, vol. 68, pp. 133-142, 1980.

29. Iskander, M. P., and Durney, C. H., Electromagnetic Techniques for Medical Diagnosis: A Review, **Proc. IEEE**, vol. 68, pp. 126-132, 1980.

30. Hagness, S., Confocal Microwave Imaging for Breast Cancer Detection, obtained from the web page of the University of Wisconsin-Madison, **Perspective**, vol. 25, Spring 1999.

31. Barret, A., Myers, P. C., and Sadowsky, N. L., Detection of Breast-Cancer by Microwave Radiometry, **Radio Sci.** vol. 12, pp. 167-171, 1977.

32. Carr, K. L., Microwave Radiometry: Its Importance to the Detection of Cancer, **IEEE Trans. Microwave Theory & Tech.**, vol. MTT-37, no. 12, pp. 1862-1869, 1989.

33. Burdina, L.M., et. al., Detection of breast cancer with microwave radiometry, **Mammology,** no. 2, pp.3-12, 1998.

Chapter 9

Silicon Micromachining, Microsensors and MEMS

S. Kal

Abstract

The present chapter provides an overview on design, fabrication and applications of MEMS devices. The major emphasis is on fabrication techniques of different microstructures. The bulk and surface micro machining techniques for fabrication of microstructures have been discussed in detail. The fabrication techniques of few micro sensors and micro actuators developed by the author and his research group have also been reported.

9.1 Introduction

The emergent field of microsensors has grown rapidly during the last decade. Microsensor is now used to describe a miniature device that converts a nonelectrical quantity, such as pressure, temperature, vibration, gas concentration, fluid flow, etc. into an electrical signal. Although low cost microelectronic components are available now-a-days, but sensors with performance/price ratios comparable to that of microelectronic circuits are much in demand. This article reports on: (a) the recent developments of miniaturisation of a sensor to produce microsensor, (b) the integration of a microsensor, a microactuator, and their microelectronic circuitry to produce a microsystem. Silicon microelectronics is being employed today to fabricate microsystems and they are called Micro-Electro-Mechanical Systems or MEMS, in short.

MEMS is an innovative new technology with possible applications in wide range of industries including telecommunications, medicine, automobiles and information technology. Most of the development work in the area of MEMS and Microsystems is currently carried out in university and other research labs in different countries, and as a result, some of MEMS devices have already found their way into marketplace.

Micromachining is a technique for fabricating microstructures that are used to develop various kinds of microsensors [1-5]. Silicon etching is a key technology for micromachined sensors. If the micromachined structures are to be integrated with the electronics it is necessary to design new process lines in addition to the standard process (CMOS or bipolar). It is important to establish a strategy in terms of how the additional steps required for micromachining can be inserted into the standard process. In the 1980's the micromachining process was introduced. Polysilicon and SiO_2 layers were used as mechanical and sacrificial layers, respectively. Both surface and bulk micromachining have merits and demerits. Micromachining techniques are elaborately discussed in section 3.

Single crystal silicon is being used as the unique microelectronic material for VLSI. Because of excellent mechanical properties, silicon is also recommended as a mechanical material for fabricating micro-mechanical sensors and actuators. Silicon processing technology for ICs has been suitably modified to fabricate microstructures for realising accelerometers, gyros and other inertial and non-inertial sensors.

MEMS devices are now less than 1 mm^3 in size, including both electronic and mechanical systems. Examples of MEMS devices that are currently drawn large attention include microsensors used in automobiles, process control and monitoring in industrial applications, ink-jet nozzles used in office printers, tinny nozzles in nuclear reactors etc. Researchers are working on MEMS that would be "micromachines on a chip", such as arrays of cantilever micromirrors that could operate as switching systems in the optical fibres of telephone systems. MEMS devices have lot of promise in biological applications. Tiny MEMS instruments might be able to clean out arteries or conduct microscopic neurological operations through the use of catheter-based microsurgical instruments [3, 5].

The two major advantages to MEMS are - (a) capability of operating in a very small confined area and (b) potentially very inexpensive to produce, once appropriate manufacturing technology has been perfected. Micromachined microsensors permit diverse uses as back up sensors in automobile fuel systems. Arrays of micro actuators could continually adjust shape of an aircraft wing during flight. Thus, continued MEMS research is to be optimally directed to make cost-effective analysis of potential uses of MEMS.

The four major direction of R&D efforts for the development of sensors and actuators are:

- Miniaturisation
- Combination of sensors with signal processing in a minimum amount of space (intelligent smart sensors and actuators)
- Interconnection of several sensors and actuators for systems
- Development of new material for selective sensors, in particular chemical and biological measurements

New product quality depends on the developments in the above four areas of Microsystems solutions. Components used in microsystem technology are system knowledge converted into hardware, knowledge that comes from highly different areas of technology, production and application. The basis of microsystem technology is to interlink different technologies like glass fibre technology, micromechanics and VLSI fabrication technologies. The development of CAD, simulation and testing tools as well as standards and norms is of decisive importance for the entire field of microsystem technology.

Some more fascinating application areas of intelligent sensors and microsystem technology are [1]-

- Intelligent, independent equipment for the detection and storage of measured data, e.g. for long-term field measurements of air and water pollution in the field of environment technology.
- Low-cost, reliable, intelligent, motor-vehicle microsensors, micro-

actuators and microcontrollers that can be mass produced and incorporated in comprehensive system concepts.

- Multi-sensor-controlled assembly tools in production engineering.
- Positioning drives and micro-metric accuracy in manufacturing technology.
- Adaptive aerials in communication technology.
- Navigation systems for unguided micro-measuring instruments, micro-tools, and micro-dosing devices in medical technology.

The manufacturing technologies for miniaturised sensors and actuators compatible with microelectronics are derived from the areas of micromechanics, integrated optics, fibre optics, ceramic technologies, thin film and semiconductor technologies. Some of the recent developments of microsystem are used in industrial manufacturing processes by small and medium size companies, while other technologies are still at the basic research stage. The evolution of microsensors and MEMS is discussed in section 2. Few micromachined microsensors are described in section 4. Microsystems technology and MEMS are elaborated in section 5. Section 6 highlights the future trends.

9.2 Evolution of Microsensors and MEMS

The advancement of VLSI technology has resulted in a decreasing device size which is currently fabricated with minimum feature size of about 200 nm. The total number of transistor in an IC has risen from about 100 in 1970 to 100 million in 2000. This is equivalent to a doubling of the number per chip every 18 months. The clock speed of microprocessor has increased from 100 kHz to 100 MHz by the year 2000. In the same time the capacity of a standard DRAM chip has increased from 1 kB in 1970 to 512 MB in 2000.

The microelctronics revolution has enabled manufacture of complex signal processing chips. Furthermore, sensors and actuators are now combined with these processors to make an information processing triptych as shown in Fig. 9.1. A sensor converts a nonelectrical quantity into an electrical quantity and an actuator is the converse. It has been observed till 1980s that the price-to-performance ratio of both sensors and actuators had fallen woefully behind integrated processors. Consequently the measurement systems tended to be more expensive. Work has therefore started to use microelectronic technologies to make silicon sensors, the so-called microsensors. This work was inspired by the vision of microsensors being manufactured in volumes at low cost.

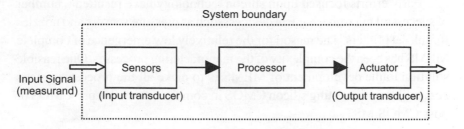

Fig. 9.1 Schematic of an information processing system

Major advantages of making microsensors with microsystems technology include,

- Batch production of wafers for high volume
- Production of miniature sensors
- Product of less bulky and much lighter sensors
- Employment of well-established technology
- Integration of processors

The ambitious goal is to fabricate monolithic chips that can not only sense with microsensors but also actuate with microactuators, that is to develop a microsystem that encampasses the information processing system. The technology employed to make such a microsystem is commonly referred to as Micro-System Technology (MST). Figure 9.2 provides an overview of MST along with some application areas. Work has been started in this direction since late 1980s, and there has been tremendous effort to fabricate Micro-Electro-Mechanical Systems (MEMS) using MST.

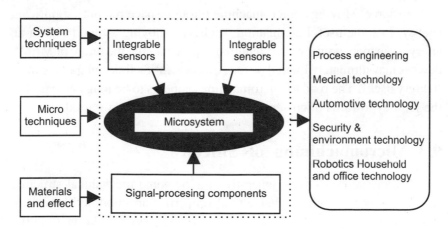

Fig. 9.2 Various elements of a MEMS chip along with
some application areas

Early efforts focused upon silicon technology have resulted a number of successful micromechanical devices, such as pressure sensors and nozzles for ink-jet printers. The reason for the relatively low emergence of complete MEMS has been the complexity of the manufacturing processes. One feasible solution to the development of MEMS is to make all the processing steps compatible with existing silicon CMOS in combination with a pre-CMOS or post-CMOS MST.

The present steady MEMS market mainly consists of some simple optical switches for the communication industry, pressure sensors, and inertial sensors for the automotive industry. Recent report on world market for MEMS devices [1] indicates the major growth areas of MEMS such as microfluidics, photonics and communications. However, there has been some exciting developments in methods to fabricate true 3-D structures on micron scale. The technique of microstereolithography can be used to make variety of 3-D microparts, such as microsprings, microgears, micromotors, micoturbines etc. Two major challenges we are facing today are (a) to develop methods that will manufacture microparts at high volume at low cost and (b) to develop microassembly techniques.

The concept of micromachines is now emerging and microrobots, microcars, microplanes, microsubmarines and microstellite are now at a conceptual stage. Micromachines will need sophisticated microsensors so that they can determine their location and orientation in space and proximity to other objects. They should also be able to communicate with a remote operator and hence will require a wireless communication link. Another major problem of microsystem research is to solve miniaturisation of suitable power source. Moving a micromachine through space requires significant energy. In some useful application, such as removing a blood clot in an artery, even more power will be required. Consequently, the future of MEMS devices may ultimately limited by the communication link and the size of its 'battery pack'! The road to micromachines appears to be long and hard but the journey towards microsensors and MEMS have been started.

9.3 Micromachining of Silicon

The miniature devices generally employ some mechanical structure whose one or more physical/electrical properties are modified by one or more ambient environment variables. Single crystal silicon is being increasingly used in a variety of such new products not because of its well-established electronic properties, but rather of its excellent mechanical properties. Large number of techniques have been derived to exploit the mechanical properties of silicon and have been grouped under the common heading of *micromachining* of silicon, permitting mechanical structures and relevant circuitry to exist side by side on the same wafer. Micromachining of silicon entails the manufacturing of micromechanical structures either in or on the silicon substrate. It is used to fabricate a variety of mechanical microstructures including beams, diaphragms, grooves, orifices, springs, gears, suspensions, and a great diversity of other complex mechanical structures. These mechanical structures has been used successfully to realize a wide range of microsensors and microactuators. The emergence of silicon micromachining has enabled the rapid progress in the field of MEMS.

Processes involved in micromachining are very simple, and the accuracy, speed and repeatability of manufacture are phenomenal compared with ordinary machining. Although new techniques continually being developed for realizing micromechanical structures, the most powerful and versatile tool continues to be etching.

9.3.1 Silicon as a Micromachining Material

Silicon is the major material used for micromachining and microengineering because it has excellent properties including a tensile strength and young's modulus comparable to steel, a density less than that of aluminum and a low thermal coefficient of expansion. Silicon micromachining is of great importance for the development of inexpensive, batch fabricated, high performance sensors which can be interfaced with microprocessors easily.

The properties of silicon that have made micromachining feasible are indicated below:

1. Exposing it to steam or dry oxygen can readily oxidize silicon. It allows silicon wafers to be masked selectively during chemical etching.

2. Single crystal silicon is brittle and can be cleaved like diamond but it is harder than most metals. It is resistant to mechanical stresses and the elastic limit of silicon is greater than that of steel.

3. Single crystal silicon can withstand repeated cycles of compressive and tensile stresses.

4. The rate of chemical etching in certain etching solutions is dependent on the crystal orientation of single crystal silicon wafer. This is important in creating various structures.

9.3.2 Bulk and Surface Micromachining

The two distinctly different approaches of micromachining for realizing microsensors and actuators are bulk micromachining and surface micromachining. Isotropic and anisotropic etching of silicon have been used for realizing micromechanical parts from bulk silicon wafer and form the basis of *"bulk micromachining"*. While bulk micromachining silicon, the backside of the wafer is conventionally protected against the etchant with an oxide or nitride layer in which windows are opened where the micromechanical structures are to emerge. An accurate alignment of the etch windows is essential to obtain the structure at a proper position with respect to the photolithographic patterns at the front. In another approach for micromachining called *"surface micromechining"*, the silicon substrate is primarily used as a mechanical support upon which the micromechanical elements are fabricated. The bulk of the silicon wafer itself is not etched in surface micromachining. In surface micromachining a sacrificial layer is deposited on the silicon substrate which may be coated first with an isolation layer. Windows are opened in the sacrificial layer and the micro structural thin film is deposited and etched. Selective etching of the sacrificial layers leaves a free standing micromechanical structure.

9.3.2.1 Bulk Micromachining

Bulk micromachining emerged in the early 1960s and has been used since then in the fabrication of many different microstructures. Bulk micromachining is utilized in the manufacture of the majority of commercial devices-almost all pressure sensors and silicon valves and 90 percent of silicon acceleration sensors. The microstructures fabricated using bulk micromachining may cover the thickness range from sub-microns to the thickness of the full wafer (200 to 500 μm) and the lateral size range from microns to the full diameter of a wafer (75 to 200 mm).

Bulk Micromachining Process

As mentioned earlier, etching is the key technological step for bulk micromachining. The etch process employed in bulk micromachining comprises one or several of the following techniques [1, 10]:

(a) Wet isotropic etching

(b) Wet anisotropic etching

(c) Plasma isotropic etching

(d) Reactive ion etching (RIE)

(e) Etch-stop techniques

The etching process is used immediately after photolithography to etch the unwanted material from the wafer. This process is not selective and that is why the pattern has to be traced onto the wafer using photoresist. There are two main methods of etching, namely wet etching and dry etching.

Wet etching is done with the use of chemicals. A batch of wafers is dipped into a highly concentrated pool of chemical etchant and the exposed areas of the wafer are etched away. Wet etching is good in that it is fairly cheap and capable of processing many wafers quickly. Dry etching refers to any of the methods of etching that use gas instead of chemical etchants. Bulk micromachining of silicon uses wet and dry etching techniques in conjunction with etch masks and etch steps to make micromechanical devices from silicon substrate.

9.3.2.2 Isotropic and Anisotropic Etching of Silicon

Selective etching of silicon can be carried out by using both isotropic etchant and anisotropic etchant. Isotropic etchant under-etch large area in lateral direction than the area defined by mask opening. On the other hand, anisotropic etchant, which are also known as orientation-dependent or crystallographic etchant, etch the silicon surface at different directions in the crystal lattice at a different rate. They can form well-defined shapes with sharp edges and corners.

Wet chemical etching and some forms of dry etching are examples of isotropic etching. The problem of isotropic etching is that the etchant material will etch lateral as well as vertical direction. Anisotropic etching occurs in most forms of dry etching. In this process there is no lateral etching and an exact representation of the pattern is obtained onto the wafer.

One common MEMS (Micro-Electro Mechanical Systems) fabrication technique is the anisotropic etching of crystalline silicon, where etch rate is a function of orientation. Complex microsystems can be generated using the anisotropic properties of single-crystal silicon in an orientation dependent dissolution reaction. V-groove structures for example, useful for the passive alignment of optoelectronic devices, are easily fabricated using an anisotropic etchant like KOH or tetramethylammonium hydroxide (TMAH). A fundamental understanding of these processes is required to achieve a well-defined anisotropic ratio and a good surface finish. Mostly used technology for bulk structuring of silicon is the anisotropic etching with KOH. For sensor applications (100) oriented silicon is mostly used.

The property of orientation dependent wet etchants is to dissolve a given crystal plane of a semiconductor much faster than other planes. In diamond and zinc-blende lattices, the (111) plane is more closely packed than the (100) plane and, hence, for any given etchant, the etch rate is expected to be slower in (111) plane. Anisotropic etchants of silicon, such as Ethelene Diamine Pyrocatecol (EDP), KOH and hydrazine are orientation dependent. This means that they etch different crystal orientations with different etch rates. Anisotropic etchants of silicon etch the (100) and (110) crystal planes significantly faster than the (111) crystal planes. The etch rate for (110) surface lies between those for (100) and (111) surface. A commonly used orientation-dependent etchant for silicon consists of a mixture of potassium hydroxide (KOH) in water and isopropyl alcohol. The etch-rate is about 2.1 μm/min for the (110) plane, 1.4 μm/min for the (100) plane, and only 0.003 μm/min for the (111) plane at 80°C; therefore, the ratio of the etch rates for the (100) and (110) planes to the (111) plane are very high at 400:1 and 600:1, respectively. Figure 9.3 demonstrates the basic concepts of bulk micromachining by anisotropic etching of a (100) silicon substrate. For example, for a (100) silicon substrate etching proceeds along the (100) planes while it is practically stopped along (111) planes. Since the (111) planes make a 54.75° angle with the (100) planes, the slanted walls are formed as shown in Fig.9.3.

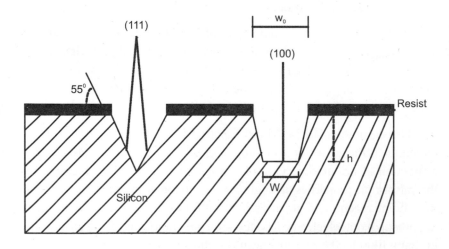

Fig. 9.3 Basic concept of bulk micromachining by anisotropic etching of (100) Silicon

Silicon dioxide, silicon nitride and some metallic thin films provide good etch masks for typical silicon anisotropic etchants. These films are used to mask areas of silicon that are to be protected from etching and to define the initial geometry of the regions to be etched. Fig.9.4. Shows orientation-dependent etching of (100) oriented silicon through patterned silicon dioxide (SiO_2), which acts as a mask. Precise V-grooves, in which the edges are (111) planes at an angle of approximately 55° from the (100) surface, can be realized by the etching. If the etching time is short, or the window in the mask is sufficiently large, U-shaped grooves could also be realized. The width of the bottom surface, W, is given by

$$W = W_0 - 2h \, \coth (55°) \quad \text{or} \quad W = W_0 - 1.4h \qquad (1)$$

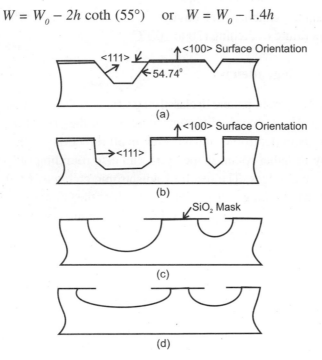

Fig. 9.4 Orientation dependent etching of (100)- oriented silicon through patterned silicon dioxide (SiO_2)

where W_0 is the width of the window on the wafer surface and h is the etched depth. If (110)-oriented silicon is used, essentially straight walled grooves with sides of (111) planes can be formed as shown in Fig. 9.4.

The anisotropic etch rate in the (100) direction of monocrystalline silicon of (100) oriented wafers was investigated focusing on the dependence of temperature and concentration of potassium hydroxide. The rate was found to be directly proportional to the KOH concentration and temperature. This difference in etch rate can drastically change the evolving shape. The different rates are due to the crystal structure, with the (111) planes usually being slowest, (100) and (110) being intermediate and higher order planes such as (311) and (522) being fastest. Because of the differences in rates, some planes grow, while others disappear.

Etchants can be categorized by using the following characteristics:

a. direction dependency (isotropic or anisotropic)
b. etch rate (0.25 to 40 μm/min), and its variability
c. anisotropic etch rate ratio (only for anisotropical etchants, 1:1 to 400:1 for (100)/(111)-planes
d. dopant dependence / selectivity
e. temperature of etching (20 to 100°C)

(a) Direction dependency

The most important feature in classifying silicon etchants is their ability to have different etch rates in different directions of the crystal lattice that is exposed to them. Isotropic etchants etch in all directions with the same rate, resulting in rather round shaped pits, and also rounding off previous sharp corners and edges. The result of anisotropic etchants, on the other hand, is different, looking perpendicular on etch of the crystal planes. This makes it possible to fabricate sharply formed structures or narrow gaps, whose borders have to coincide with the crystal planes. Depending on what kind of structure is desired, the proper etchant type (isotropic or anisotropic) has to be chosen.

(b) Etch rate

The etch rate can vary with temperature, mix of ingredients, concentration of etch solution and sometimes optical circumstances (light intensity). Fast etch rate sometimes results in rough etch surface. This variability of etch rate can either be wanted or not. Thus etch rate is to be chosen depending on the desired results.

(c) Anisotropic etch rate ratio

If an anisotropic etchant is chosen, the ratio of etch rates concerning the different crystal planes, can vary in a wide range. Again, the desired result influences the choice of etchant, since this ratio from 1:1 to 400:1 are possible, if one compare the (100) and the (111) planes.

(d) Dopant dependence (selectivity)

Another, very important attribute is the dopant dependency of etchants. Some etchants are very selective on the material that they are exposed to, so that a doped layer of different material can be used as an etch stop or a direction of a much higher etch rate.

(e) Etching temperature

In general, lower temperature is better than higher ones, since higher temperature induces stress concentrations. In addition, the occurrence of hazardous gases is lower at low temperatures.

9.3.2.3 Etch Chemicals for Micromachining of Silicon

The following chemicals are of particular interest due to their versatility:

a. EDP (ethylene diamine, pyrocatechol, and water)
b. KOH and Isopropyl alcohol / water
c. HNA (acetic acid CH_3OOH)
d. TMAH

(a) EDP

EDP has three properties that make it attractive for silicon micromachining:

- It is anisotropic (important for special structures that otherwise would be impossible to fabricate)
- It is highly selective
- It is dopant dependent (stops on highly boron-doped layers).

(b) KOH and water

The main advantage of KOH is that it is orientation dependent with a much higher (100):(111) etch rate ratio than EDP, therefore useful for groove etching on (100) wafers.

The etch rate of silicon in KOH solution is studied at different temperatures and at different concentrations. The etchant, a solution of KOH and water was continuously stirred and kept at the selected temperature by means of a hot – water thermostatic system. The completely processed wafer, ready to be etched, is mounted on the holder, immersed in the solution and the etching starts. The samples are etched for one hour each to find the etch rate. The etch depth is found from the DEKTAK surface profiler (resolution of 25 Angstroms). The etch rate can then be

found from the measured values of etch depth and etch time. Fig. 9.5 shows
the variation of etch rate of silicon with temperature at a fixed concentration
and Fig. 9.6 shows the variation of etch rate with concentration at a fixed
temperature of 60°C. From the results indicate that the etch rate increases
with concentration and as well as with temperature. The KOH solution is
standardized to 44 gms in 100 ml of water at 80°C to get the etch rate of
0.92 µm/min.

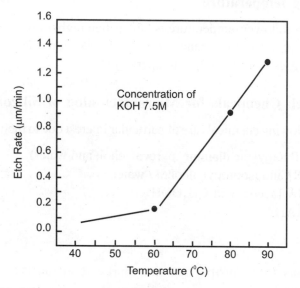

Fig. 9.5 Variation of etch rate of silicon in KOH solution with temperature

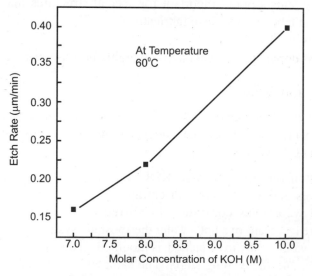

Fig. 9.6 Variation of etch rate of silicon in KOH solution with KOH molar
concentration

(a) HNA

The HNA system is highly variable in its etching rates and the etch characteristics depends on

- Silicon dopant concentration
- Mix ratios
- Degree of etchant agitation

A major disadvantage is that SiO_2 mask is etched for all the mixtures of HNA, so that it only can be used for short etching times, otherwise, Si_3N_4 or Au are to be used as etch mask.

(b) TMAH [13]

Tetra-methyl ammonium-hydroxide (TMAH), is an isotropic silicon etchant that is gaining considerable interest in silicon micromachining due to its excellent silicon etch rate, etch selectivity to masking layers (even with aluminum film), degree of anisotropy, and relatively low toxicity.

9.3.2.4 Etch-Stop Techniques

Different chemical etchants for silicon are discussed in previous section. The properties that make some of these etchants indispensable to micromachining of 3-D structures are selectivity and directionality. As etching process in polar solvents depends on charge transport phenomena, the etch-rate may be dopant-type-dependent, dopant-concentration-dependent, and bias-dependent. Etch process thus can be made selective by the use of dopants - heavily doped regions etch more slowly – or even halted electrochemically when observing the sudden rise in current through an etched n-p junction. A region where wet (or dry) etching tends to slowdown (or halt) is called an etch-stop. There are several ways in which an etch-stop region can be created, namely, Doping-selective etching (DSE), Bias-dependent etching (BSE).

Silicon membranes are generally fabricated using the etch-stop phenomenon of a thin heavily boron-doped layer, which can be epitaxialy grown or formed by the diffusion or implantation of boron into a lightly doped substrate. This stopping effect is a general property of basic etching solution such as KOH and EDP. Because of heavy boron-doping, the lattice constant of silicon decreases slightly, leading to highly strained membranes that often show slip planes. Boron etch-stop properties for (a) KOH and (b) EDP are shown in Fig.9.7. In case of EDP, an early study showed that the Si etch rate falls to 0.015 μm/min from 0.75 μm/min when boron concentration is raised to a critical value of 7×10^{19} cm^{-3}. However this technique is not suited to stress sensitive microstructure. The main advantages of the high boron etch-stop are the independence of crystal orientation, the smooth surface finish. On the other hand, the introduction of electrical components for sensing purposes into these microstructures, such as implantation of piezoresistors, is inhibited by excessive background doping. This consideration limits the applicability of the high boron dose etch stop. Consequently bias-dependent selective etching (BSE), known as an *electrochemical etch stop* is currently the most widely used etch-stop technique.

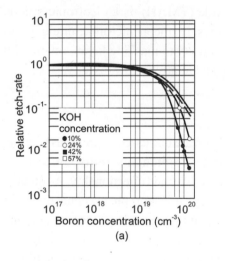
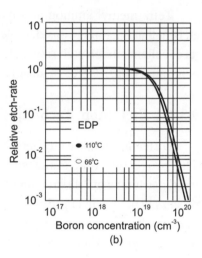

(a)   (b)

Fig. 9.7 Boron etch stop properties for (a) KOH and (b) EDP etchants

   In electrochemical etching of silicon, a voltage is applied to the silicon wafer (anode) with respect to a counter electrode (cathode) in the etching solution. The fundamental steps of the etching mechanism are as follows:

1. Injection of holes into the semiconductor to raise it to a higher oxidation state Si$^+$
2. Attachment of negatively charged hydroxyl group , OH−, to the positively charged Si
3. Reaction of the hydrated silicon with the complexing agent in the solution
4. Dissolution of the reaction products into etchant solution

Fig. 9.8 shows a set up for electrochemical etching of silicon in 5% HF solution. The cathode plate used is made of platinum. Holes are injected into the silicon electrode and they tend to reside at the Si surface where they oxidize Si at the surface to Si^+. The oxidized silicon interacts with incoming $OH-$ that are produced by dissociation of water in the solution to form the unstable $Si(OH)$, which dissociates into SiO_2 and H_2 gas. The SiO_2 is then dissolved by HF and removed from silicon surface. The current density of the electrochemical cell is very much dependent on the type and the resistivity (doping level) of Si. This dependence on the type and resistivity is the property that is utilized in the electrochemical etch stop phenomenon.

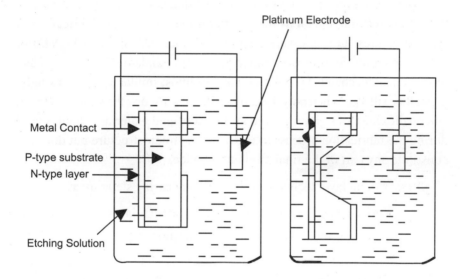

Fig. 9.8 The electrochemical etch –stop technique used to form a diaphragm.

9.3.2.5 Surface Micromachining

Since the beginning of the 1980s, much interest has been directed towards micromechanical structures fabricated by a technique called surface micromachining. The dimensions of these surface-micromachined structures can be an order of magnitude smaller than the bulk - micromachined structures. The main advantage of surface-micromachined structures is their easy integration with integrated circuit (IC) components, as the same wafer surface can also be processed for IC elements. However, as miniaturisation immensely increased by silicon surface micromachining, the small sizes of masses that are created are often insufficient for viable sensors and, particularly, for actuators. In a particular case of accelerometers, the sensitivity reduces due to lower mass of the proof mass. There are several approaches to make microelectromechanical system (MEMS) devices using surface micromachining, namely (a) the sacrificial layer technology for the realization of mechanical microstructures, (b) IC technology and wet anisotropic etching and (c) plasma etching to fabricate microstructures at the silicon wafer surface.

Sacrificial layer technology

Sacrificial layer technology uses, in most situations, polycrystalline rather than single-crystal silicon as the structural material for the fabrication of microstructures. Low-pressure chemical vapour deposition (LPCVD) of polysilicon (poly-Si) is most common process in standard IC technology and has excellent mechanical properties that are almost similar to those of single-crystalline silicon. When polycrystalline silicon is used as the structural layer, silicon dioxide (SiO_2) is employed as the sacrificial material, which is used during the fabrication process to realize some microstructure but does not constitute any part of the final miniature device.

In sacrificial layer technology, the key processing steps are as follows:

a. Deposition and patterning of a sacrificial SiO_2 layer on the substrate

b. Deposition and definition of a poly-Si film

c. Removal of the sacrificial oxide by lateral etching in hydrofluoric acid (HF), that is, etching away of the oxide underneath the poly-Si structure

Here, we refer to poly-Si and SiO_2 as the structural and sacrificial materials, respectively. In almost all practical situations this is the preferred choice of material combination.

A Simple Surface Micromachining Process

The simplest of surface micromachining process involves just one poly-Si layer and one oxide layer. This process is a one-mask process and is illustrated in Fig. 9.9 in which it is designed to form a poly-Si cantilever anchored to a Si substrate by means of an oxide layer. The oxide scarificial layer is deposited first (Fig. 9.9(a)). The poly-Si structural layer is then deposited on top of the oxide. Next, the poly-Si layer is patterned, forming both the cantilever beam and the anchor region (Fig. 9.9(e)). Following the poly-Si patterning step, the cantilever beam is released by laterally etching the oxide in an HF solution. The oxide etch needs to be timed so that the anchor region is not etched away (Fig. 9.9(d)).

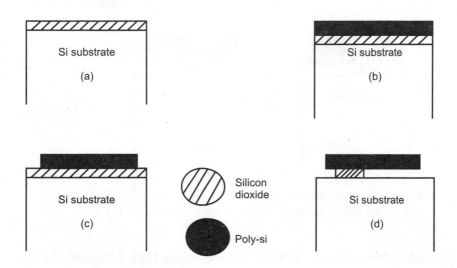

Fig. 9.9 Process flow for the fabrication of polysilicon cantilever using surface micromachining technology (one mask process)

To implement successfully the process described in the preceding paragraph, the release etch must be very carefully controlled. If the release etch is continued for too long a period, the anchor region will be completely cut, resulting in device failure. However, to avoid such a failure, the process may be extended to a two-mask process in which the poly-Si cantilever is directly anchored to the substrate. This two-mask process is shown in Fig. 9.10. In this process, the deposited oxide (Fig. 9.10.(a)) is patterned for an anchor opening by the first mask (Fig. 9.10.(b)). This is followed by a conformal deposition of poly-Si and subsequent patterning of the poly-Si cantilever beam using the second mask (Fig. 9.10.(c)). The cantilever is then released by a lateral oxide etch in HF solution (Fig. 9.10.(d)). Because the anchor region in this case is made out of poly-Si, the oxide release etch poses no threat of device failure.

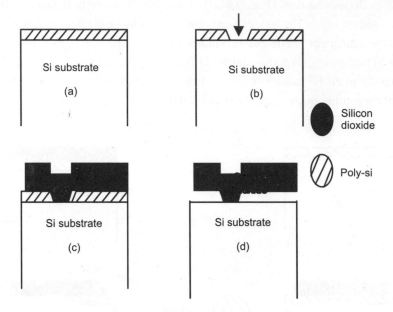

Fig. 9.10 Process flow for fabrication of polysilicon cantilever anchored directly on silicon substrate using surface micromachining technology (two mask process)

9.3.2.6 Fabrication of Micro-structures using Bulk Micromachining

In bulk micromachining process, the vertical dimensions of the fabricated structures are limited by the wafer thickness (200 - 500 µm), which also depends on the wafer diameter, for example the thickness of a 4-inch wafer will be approximately half a millimeter. The minimum feature depends on the precision of the lithographic process used, but in many cases the feature size is process dependent and is in the order of the wafer thickness. Bulk micromachining is frequently followed by wafer-to wafer or wafer-to-glass bonding which extends the possibilities, allowing for more complex three-dimensional electro-mechanical structures.

The micromechanical structures are designed using the relation between the oxide window and thickness of the wafer (eqn.(1)). These structures are fabricated using wet anisotropic etching in KOH/TMAH/EDP solution. The realization of various micromechanical structures such as nozzles, cantilever beams, pits, membranes, mesh, stensil mask are now described.

The cross-section of anisotropic etching of silicon is shown in Fig. 9.11. The micromechanical structures are designed by using the formula as given in Eqn.(2).

$$W_0 = W_R - \frac{2\,t_{si}}{tan\ \theta} \tag{2}$$

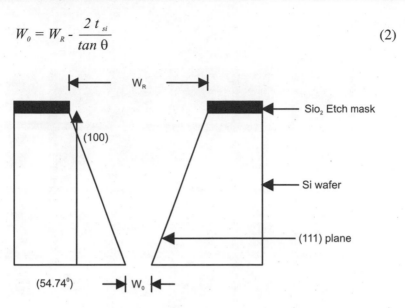

Fig. 9.11 Designn of micro-nozzles

where W_O is the nozzle size, W_R is the mask opening, t_{Si} is the silicon wafer thickness, θ is the angle between (100) and (111).

Nozzles

These simple structures etched in a silicon wafer can be designed and utilized to provide solutions in unique and varied applications. Some of the application of silicon micro-nozzles are (i) generation of very high precision moulds for micro miniature structures, (ii) for high speed, high quality ink-jet printing. Forcing pressurized ink through an array of nozzles forms fine jets of conducting ink. The jets break up into uniform streams of droplets by vibrating the nozzle array at a fixed frequency by means of a piezoelectric transducer. In one method of printing, unwanted drops are selectively charged and deflected electrostatically from the main streams, and the uncharged drops are allowed to strike the paper to form the required character.

As shown in the Fig. 9.11 the geometry of the pyramidal hole in (100) silicon can be adjusted to completely penetrate the wafer, the square hole on the bottom of the wafer forming the orifice for an ink jet stream. The size of the orifice depends on the wafer thickness and mask dimension on top surface. Using the equation (2) the mask dimensions are calculated to different sizes of nozzles. Fig. 9.12 shows the microphotograph of some array of nozzles fabricated in the Microelectronics Laboratory at IIT

Kharagpur.

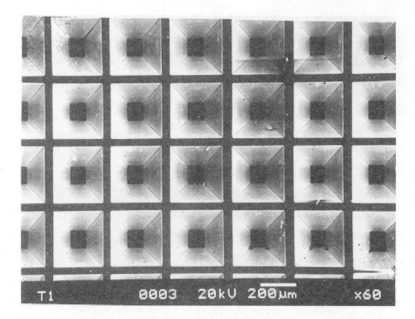

Fig. 9.12 Photomicrograph of array of nozzles fabricated using
bulk micromachining technique

Cantilever beams

This section outlines the design of SiO_2 cantilever beams oriented in
(100) direction by etching (100) Si wafer. Micromachined silicon dioxide
cantilever beams are simple structures commonly used in microsensor and
microactuator designs, electromechanical switches, IR detectors. The etchant
can remove the silicon from backside of a silicon wafer consisting of layered
structure of Si and SiO_2.

By applying elasticity theory, it is not possible to obtain a stable oxide
film if the ratio of its thickness to the window dimension t/w is made too
small. The boundary condition is given as -

$$0.52\sqrt{\Delta\alpha\ \Delta T} < t/w < 1.2\sqrt{\Delta\alpha\ \Delta T} \tag{3}$$

where $\Delta\alpha$ is the difference between the coefficients of thermal expansion
for the SiO_2, and ΔT is the change in temperature. For our process, the
oxide growth was carried out at 1100°C so ΔT is roughly 1075°C. For Si
and SiO_2, $\Delta\alpha$ is 2 x 10⁻⁶, so that $\sqrt{\Delta\alpha\ \Delta T}$ = 0.046. The differing thermal

expansions for Si and SiO$_2$ are responsible for bending in composite beams. When a cantilever of Si and SiO$_2$ is formed, it can be expected to bend downward as the tension in the Si and compression in its oxide seek accommodation. However, cantilevers formed of one homogeneous material should not bent in either direction unless there is some reconfiguration caused by the release of stored strain energy.

The responsivity or sensitivity depends on the geometry of the device. For a cantilever beams of width w, length l, and thickness t, the responsivity is proportional to l/w. Hence the responsivity benefits from a large length/width ratio. The cross section of cantilever beams fabricated in the author's laboratory is shown the Fig. 9.13.

Fig. 9.13 SEM photograph of silicon cantilever beam fabricated using bulk-micromachining technique

Membranes

Membranes are made by etching most of silicon substrate from the backside of the wafer. A simpler way to achieve a thin membrane would have been to etch a wafer for a period just short of what it would take to etch through it. This technique is known as time-etch stop technique. But it is difficult to make membrane of uniform thickness by this time-etch method, because the thickness of the membrane is determined by that of the wafer and wafer thickness varies typically by 10μm. The cross section of the membrane is shown in the Fig. 9.14.

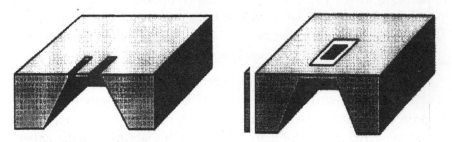

Fig. 9.14 Cross-section of silicon membrane using bulk-micromachining technique

9.4 Silicon Microsensors

The field of microsensors has grown rapidly during the past two decades because of tremendous advances in the technologies originating in the semiconductor industry. The term microsensor is now commonly used to describe miniature device that converts a non-electrical quantity, such as temperature, pressure, vibration, or gas concentration into an electrical signal. During the past 10 years, a sharp increase in sensor market has taken place. Here, we focus on the main types of microsensors, which have accelerated this sensing revolution, together with some of the emerging new designs [4].

This section reports on the recent developments in

(a) the miniaturisation of a sensor to produce a microsensor

(b) the integration of a microsensor and the corresponding micro-electronic circuitry for signal processing to produce a so called smart sensors

(c) the integration of microsensors, a microactuator, and their microelectronic circuitry to produce a microsystem.

Silicon microsensors, working in the fascinating interface between electronics and physics, is an attractive and interesting area of research, especially when it is carried out within the framework of integrated circuit technology, with its possibilities and novelties.

A sensor may be simply defined as a device that converts a non-electrical input quantity $\bar{E}$ into an electrical output signal E; conversely, an actuator may be defined as a device that converts an electrical signal E into a non-electrical quantity $\bar{E}$. In contrast, a processor modifies an electrical signal (e.g. amplifies, transforms) but does not convert its primary form. Basic input-output representation of sensor, actuator and processor is shown in Fig. 9.15. A transducer is a device that can be either a sensor or an actuator. Some devices can be operated both as a sensor and an actuator. For example, a pair of interdigitated electrodes lying on the surface of a piezoelectric material can be used to sense surface acoustic waves (SAWs) or to generate them. This device is referred to as an interdigitated transducer (IDT).

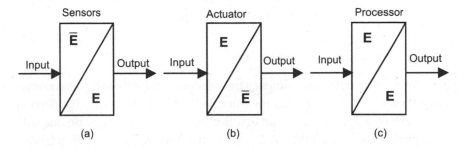

Fig. 9.15 Basic input-output representation of sensor, actuator and processor in terms of their energy domain

It has been proposed by Middelhoek [4] that a sensor or an actuator can be classified according to the energy domain of its primary input-output (I/O). There are six primary energy domains and the associated symbols are as follows:

- Electrical   E
- Thermal   T
- Radiation   R
- Mechanical   Me
- Magnetic   M
- Bio(chemical)   C

Sensors and actuators may also be classified more precisely in terms of the electrical principle employed.

In this article we are concerned with miniature sensors, so-called *microsensors*, which are fabricated using predominantly the bulk- and surface- micromachining technologies. Some sensing devices have a part or all of the processing functions integrated onto the same silicon substrate. We refer to these devices as '*smart*' sensors. We reserve the label of '*intelligent*' for devices that have in addition some biomimetic function such as self-diagnosis, self-repair, self-growth, and fuzzy logic. We discuss some of microsensors in the following sections.

9.4.1 Thermal Sensors

Thermal sensors are sensors that measure a primary thermal quantity, such as temperature, heat flow, or thermal conductivity. Other sensors may be based on a thermal measurement; for example, a thermal anemometer measures fluid flow. The most important thermal sensor is the temperature sensor. Conventionally, the temperature of an object can be measured using a platinum resistor, a thermistor, or a thermocouple. Resistive thermal sensors exploit the basic material property that the bulk electrical resistivity ρ, and hence resistance R, varies with absolute temperature T.

9.4.1.1 Silicon Microbolometer

An infrared detector is simply a transducer of radiant energy. It converts radiant energy into some other measurable form. An infrared is the portion of the electromagnetic spectrum that lies between visible light on one side and microwave on the other. A large spectrum of emitted radiation from a hot body lies in the infrared region. In thermal sensor, the energy of radiation absorbed in the membrane raises the temperature of the detecting element on the membrane. This increase in temperature will cause a change in the temperature dependent properties of the detector. Monitoring one of these changes enables the radiation to be detected. Two fundamental types of infrared detectors available today are thermal detectors and photon detectors. In thermal detectors, the radiation is absorbed by the material, generating phonons and causing heating of the lattice. This change in the lattice temperature is then converted into a change in the electrical properties of the structure. In photon detectors, the radiation is absorbed by the material, resulting in a direct modification of its electrical properties. The two-step transduction process of thermal detector is considerably slower than the single-step process associated with photon detectors. However, the thermal detectors have a broader spectral response, low cost, ease of operation and insensitivity to ambient temperature as compared with highly wavelength sensitive photon detectors. The compact and low cost thermal detectors are useful for low power thermal imagers in civilian as well as military applications. There are mainly two types of thermal detectors i.e., (1) thermopile detectors, (2) bolometer detectors. The working principle of thermopile detector is based on Seebeck effect. The emf developed across a thermopile, which is a series combination of a large number of thermocouples, is proportional to the temperature difference due to IR radiation. On the other hand, the bolometer is sensitive radiation detector, which indicates the presence of radiation by the process of changing electrical resistance due to the change in temperature of the sensing elements caused by the absorption of the radiant energy. Bolometer has an advantage in IR detection in that they generally have a faster response time than thermopile detector [6, 8].

Design of Bolometer Detector [9]

In a bolometer, the increase of temperature due to absorption of IR causes a change in resistance of the sensing element of the bolometer and can be characterized by certain figure of merit such as temperature coefficient of resistance (TCR), responsivity (S) and detectivity (D). A most convenient and generally useful IR detector would have the following properties: low noise equivalent power, fast response, wide spectral range at room temperature operation and ease of fabrication. The small size of detector results in reduced noise equivalent must be composed of a material, which has an appreciable temperature coefficient of resistance (TCR). The essential component of a bolometer is the sensing element that absorbs the IR, supported by a thin isolated membrane and its associated electrical contact, and also a suitable package fitted with a transmitting window. The schematic diagram of the cross-section of a silicon microbolometer is shown in Fig. 9.16. The figure shows the thin insulating silicon membrane, supported by bulk of silicon substrate, fabricated by micromachining technology. A thin layer of bismuth metal strip was formed at the center of the membrane that acts as the IR absorbing element. The contact leads of the metal strip are extended up to the bulk of silicon.

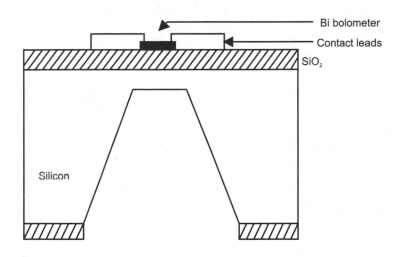

Fig. 9.16 Schematic diagram of a silicon microbolometer

For small current through the element the signal output voltage is proportional to the bias current. As the bias current is increased, the power input to the element in the form of Joule heating increases to the point where this quantity has serious influence on the dynamic behavior of the detector. For a bolometer biased at a dc voltage V_b, the change in voltage ΔV across the bolometer due to the incident radiation is $\alpha_T V_b \Delta T$, where α_T is the temperature coefficient of resistance.

The voltage responsivity S is the ratio of ΔV to the incident radiation with power ΔP and is given by

$$S(f) = \Delta V / \Delta P = (\alpha_T V_b) / (G_t + j2\pi fC) \qquad (4)$$

where G_t is the thermal conductance of the element, C is the heat capacity of the sensor. The speed of response of sensing element is given by

$$\tau = C / G_t \qquad (5)$$

where τ is the time constant. For maximum response of detector, the value of thermal conductance G_t is minimized by reducing the conduction losses to supporting membrane and surrounding atmosphere to the least possible value consistent with mechanical stability. It is also necessary to minimize the heat capacity per unit area of the detector. This can be achieved by making the sensing strip very thin.

The detector is generally called metal bolometer if the material is a metal and a thermistor if the material is semiconductor. Metals have positive TCR. Higher values of α_T increase the bolometer responsivity and suppress the contribution of Johnson noise and amplifier noise effectively increasing the bolometer bandwidth. The responsivity of the detector varies inversely with the thickness of IR sensing element.

Experiments [11]

Realization of micro-bolometer is essentially a two step process. Initially a thin insulating membrane was fabricated on the silicon wafer followed by formation of bolometer sensing element on the membrane. In the present study, bolometer detector was realized using bismuth material. Bismuth metal was chosen because of its large TCR value ($0.0043/^0C$), and low electrical and thermal conductivity. Photolithography and wet chamical etching technique cannot be implemented to pattern bismuth film due to rapid deterioration of bismuth in most acids. Lift-off and ion milling are common techniques to define this material. In the present case, the bolometer detector was fabricated by depositing bismuth metal through a specially designed silicon vacuum mask. In general silver was used as contact leads for the metal strip to get ohmic contact. However, in the present case gold metal was used for ohmic contact and the contact area was made very large compared to metal strip for minimum contact resistance.

Silicon membrane has been fabricated by anisotropic etching of single crystalline silicon wafer using KOH solution. Bismuth metal was deposited on the oxidized membrane by thermal evaporation technique through the silicon vacuum mask. To achieve the best performance of the bolometer detector the array of holes was aligned at the centre of the membrane. High purity bismuth metal was deposited in a vacuum chamber on oxidized silicon wafer through vacuum mask by thermal evaporation technique at a pressure of 2×10^{-5} Torr. The substrate temperature was around 80^0C during deposition.

The resistance was measured at different temperatures for different dimensions of bolometer sensing element. A wide-band IR source was used to heat the bolometer keeping the source at a fixed distance. The temperature of the sensor was measured using a copper-constantan thermocouple placed very near to the sensing element. The change of resistance vs. temperature is shown in Fig.9.17. for different dimensions of bismuth metal lines. It was observed that the resistance of bolometer sensing element increases linearly with the increase of IR radiation. The maximum percentage change of resistance was around 70 % at 150^0C for a Bi metal strip of width 47 μm and length 1100 μm, the TCR value calculated from experimental data was 0.0037 $/^0$C which is very close to the theoretical value.

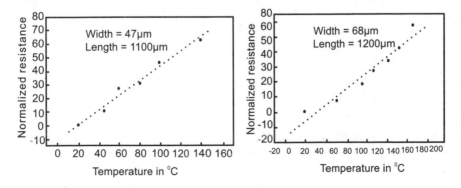

Fig. 9.17 Plot of normalised resistance of thin film bismuth bolometer with temperature for different dimension of the bolometer

9.4.1.2 Silicon Thermopile IR Detector / Temperature Sensors

The use of infra-red detectors, both thermal and photon is not confined to research and development laboratories, but has many applications in industry, medicine, meteorology, astronomy and defence. In fact without touching the object, IR technology can determine its existence, its shape, its temperature and its composition. A temperature sensor can be integrated in a silicon process to become either a temperature sensor or part of a silicon-based MEMS devices. A silicon thermocouple can be made in an IC process with doped polysilicon and a standard metal contact, for example aluminum. The advantage of thermocouple is that it comes to thermal equilibrium quite rapidly in the system because of its small mass [6].

Thermopile-type smart sensors on ultra-thin silicon membranes have relatively higher sensitivity and large bandwidth compared to conventional thermal sensors. These sensors are compact and may be combined with monitoring IC chip, and have the capability of direct interfacing with a microcomputer. These IR detectors have immense application in moving target detection, night vision, process automation etc.

Although two-step transduction process of thermal detector is considerably slower than the single-step process associated with photon devices, however, in applications where high speed is not of primary importance, the thermal detectors have a number of advantages, including a board spectral response, low cost, ease of operation. Among thermal detectors, two of the most common detectors are thermopile and pyroelectric devices. Thermopile detectors require no bias voltage and are used over a wider temperature range and easily interfaced with monolithic circuit. The thermopile detectors make use of Seebeck effect which arises from the fact that the density of charge carriers (electron in metal) differs from one conductor to another and depends upon the temperature.

The basic Si thermopile IR detector/thermal sensor structure is shown in Fig.9.18. It consists of a series of thermocouples whose hot junctions are supported by a thin Si membrane (window) and whose cold junctions are on the thick chip rim. Since the boron-doped p^+ membrane is a [8] good electrical conductor, the couples must be insulated from the membrane. Thermally grown SiO_2 is used as a insulated layer. The metal A and B constitute the thermocouple. In some study, doped poly-Si and Al are used as metal A and metal B, respectively. The membrane area is generally coated with a thin layer of an absorbing material, such as bismuth black, so that energy incident on the chip is absorbed efficiently over a broad spectral range from visible to far infrared. The thermal conductivity of the Si membrane is a relatively strong function of doping concentration. It is evident that the Si membrane has the lowest thermal resistance and cannot be easily loaded with any other material. For a temperature difference, ΔT between the hot and cold junctions, a voltage will be developed across the thermopile which is equal to

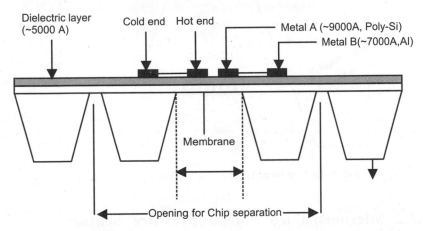

Fig. 9.18 Basic structure of silicon thermopile IR detector / thermal sensor

$$V = N\alpha_s \, \Delta T \tag{6}$$

where α_s is the Seebeck coefficient of the two materials and N is the total number of couples. In addition to the dc response, it is useful to compute the transient response of the device. The time constant of a thermopile radiation detector is defined as

$$\tau = R_{th} \, C_{th} \tag{7}$$

where C_{th} is the thermal capacitance of the window and R_{th} is the thermal resistance between the appropriate tap point in the resistor string and the rim. A figure of merit for a thermocouple material is defined as

$$Z = (\alpha^2 \sigma)/ k \tag{8}$$

where α is thermoelectric power, σ is electrical conductivity and k is thermal conductivity. The materials for thermocouple is selected such that they can be deposited and patterned easily on Si and SiO_2, have higher figure of merit and low time constant. Polysilicon and Al are preferred material for micro-thermocouple because (i) they are compatible to each other as far as adhesion and thermo-emf are concerned, (ii) they are easily deposited and patterned, (iii) they produce thermo-emf in the millivolt range, and (iv) they have moderate responsivity and speed.

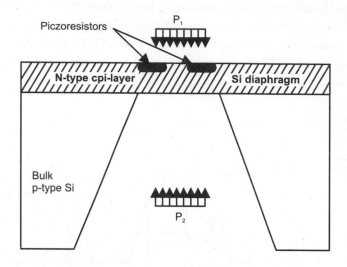

Fig. 9.19 Typical structure of a MEMS pressure sensor

9.4.2 Micromachined Silicon Pressure Sensor

One of the most needed sensors for next generation electronic systems is the silicon pressure sensor. Applications range from health care to transportation. More than a dozen applications for pressure transducers on the automobile have been identified and silicon thin diaphragm piezoresistive sensors appear very promising for many of these applications. The instrumentation and robotics require lot of high precision pressure sensors. The microcomputer is creating high volume markets for such sensors and may provide sufficient motivation to develop low-cost high-performance devices needed in future systems.

The pressure sensor design incorporates a four-arm Wheatstone bridge diffused into a silicon membrane/diaphragm for maximum sensitivity and frequency response. The piezoresistive pressure sensors consists of a thin diaphragm, which flexes in response to differentially-applied pressure. The stress associated with deflection alters the carrier mobilities in the resistors, giving rise to their pressure sensitivities. It is noted that piezoresistive and capacitive pressure sensors are quite different devices in that the former measure the point stresses while the latter measures average deflection. While deflection and stress are directly coupled for uniformly-applied pressure, they may not be in other situations (e.g., due to temperature). The cross-section schematic of a typical MEMS pressure sensor is shown in Fig. 9.19.

9.4.3 Micromachined Silicon Accelerometer

A miniature accelerometer fabricated in silicon using MEMS technology integrates a delicate micromechanical structure consisting of a proof mass, a frame and cantilevers or flexures for suspension, tiny sensing elements and electronics circuitry for converting the sensor output to the desired electrical output for interface with control instruments / computers. A typical diagram of micromachined silicon accelerometer is shown in Fig. 9.20.

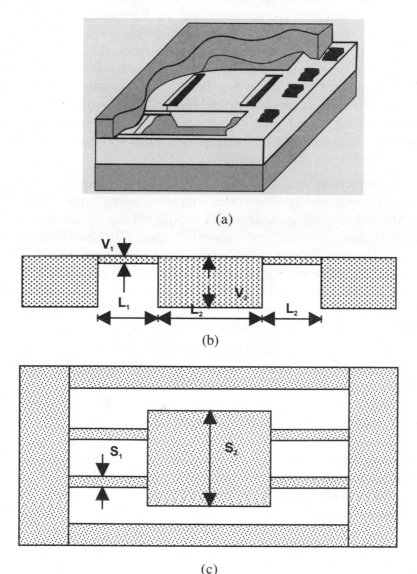

(a)

(b)

(c)

Fig. 9.20 Structure of a typical micromachined silicon accelerometer :
(a) Isometric view, (b) Cross-section view, (c) plain view

Design of Accelerometer

The MEMS accelerometer requires the following parts to design:

(i) Design and simulation of micromechanical part,

(ii) Design of sensing element,

(iii) Electronics design.

MEMCAD simulator may be used for the design and simulation of the mechanical structure of the accelerometer. The critical design parameters are - proof mass geometry and dimension, the cantilever thickness and width, the gap between the top and bottom cap layers etc. While the minimum resonant frequency, static sensitivity and range of acceleration are governed by the proof mass and cantilever geometry, the maximum frequency of response is sensitively controlled by the damping. In the proposed structure, the damping is provided by the thin air gap on both sides of the proof mass. For aircraft motion sensing, heavy damping is necessary because the aircraft vibration frequency may be quite low.

The structural simulation using MEMCAD will give the dominant resonant frequencies, the lowest due to vibrations normal to the proof mass, and the two other frequencies due to rocking vibrations around an axis perpendicular to the flexures and torsional vibrations. The natural frequencies and the Q-values (related to damping) are to be decided and simulated appropriately for the given application. The design will produce data for the device geometry, the natural frequencies and frequency response with critical damping and the stress and strain distribution over the flexures for given values of acceleration along the normal to proof mass.

Different sensing elements can be used, such as piezoresistors, piezocapacitors, tunneling, resonant vibrations etc. In the present development we have used piezoresistors because of easy fabrication and controllability, Piezoresitive accelerometer is becoming more popular although the devices will be temperature sensitive. MEMCAD software offers accurate simulation of piezoresistors. Resistor geometry, dimension and doping profile etc are simulated by process simulator SUPREM. Hence the process parameters like mask geometry, pre-dip/ implant and drive-in diffusion parameters (temperature, time, dose, etc) can be determined. Several iterations may be necessary to achieve design specifications.

Fabrication and wafer Level Testing

This accelerometer consists of three layers of silicon: (a) middle layer containing the proof mass suspended from a frame via four cantilever flexures, containing eight piezoresistors and interconnection pattern forming a Wheatstone bridge, (b) top capping layer providing over-retardation mechanical stop and air friction for damping and (c) bottom capping layer providing over-acceleration mechanical stop and air friction for damping. After fabrication, the individual layers are tested at the wafer level. Finally the chips are bonded, packaged and assembled to form accelerometer.

9.4.4 Micromachined Silicon Flow Sensors [12]

Flow sensors are used in a variety of applications including environmental monitoring (water flow, wind flow, air conditioning and exhaust control) biomedical applications (arterial flow, blood flow) and process control (liquid cooling system, micro-chemical reaction). These sensors are used to measure the movement of a fluid per unit time. Fluid flow can either be a gas or liquid flow. The measurement of flow can be considered as mass flow, average velocity or volumetric flow. To detect flow sensor rate, a device needs to translate the signal of interest from physical domain to electrical domain. MEMS technology is highly suitable for realizing such sensors.

MEMS flow sensors based on thin film anemometer is discussed in this section. A thermal anemometer refers to sensors that measure total heat loss. In this technique a thin film resistor is put inside liquid. Due to flow of liquid, temperature of resistor changes and consequently its value is changed. The change of resistance will be a measure of liquid flow. The 3D view of flow sensor is shown in Fig.9.21.

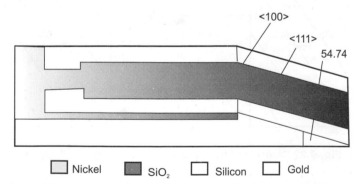

Fig. 9.21 Schematic 3D view of flow sensor based on thin film amenometer

A micromachined silicon beam is used as the base for thin film resistor. On the micromachined silicon cantilever beam nickel film is deposited as a resistor material by vacuum evaporation and subsequently gold is deposited as contact pad material. A thin film resistor is designed using the relation:

$$R = (L / W) \times R_S \tag{9}$$

where, R = Resistor value, L = Length of the thin film resistor, W = Width of the thin film resistor, R_S = Sheet resistance of the nickel film.

The sheet resistance of the deposited nickel film was found to be 5Ω / sq. Width and length of thin film nickel resistor are adjusted to obtain a $15\ \Omega$ resistance (design value).

Experimental Procedure

Nickel metal is used for the resistor because pure nickel has the highest resistance versus temperature sensitivity. Pure nickel is not oxidized easily in ambient condition. A planar thin film nickel resistor may be used to measure the sheet resistance and resistivity for resistance calculation. The TCR value of the nickel film may also be determined using the planar device. The diagram of the planar device is shown in Fig.9.22. Planar nickel resistor was fabricated in author's laboratory and TCR measurement was performed. In planer structure silicon (100) is oxidised and nickel metal was deposited. Using photolithography nickel film is patterned to get resistor structure and subsequently gold metal was deposited and patterned for contact pads.

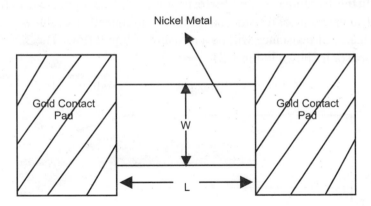

Fig. 9.22 Diagram of planar device of thin film nickel resistor

The Temperature Coefficient of Resistance (TCR) was measured using the relation

$$TCR = [(R_F - R_I) / R_I (T_F - T_I)] \times 10^6 \text{ PPM} / {}^0C \tag{10}$$

where, R_F = Final resistance (Ω), T_F = Final temperature $({}^0C)$, R_I = Initial resistance (Ω), T_I = Initial temperature $({}^0C)$

Thin film nickel resistor is measured in the temperature ranges from 25^0C to 75^0C. The resistance variation is linear in 40^0C - 75^0C temperature range. A plot of ΔR / R Vs temperature change (ΔT) calculates the TCR value of the nickel film as 0.01507 Ω / Ω / ^0C.

9.5 Microsystems Technology and MEMS

In the early days of microsystems engineering the attention has been focused exclusively on silicon micromechanics because microsystem would have to be monolithically integrated on silicon with all elements (mechanical, optical, fluidic and electronic) in one chip. Due to increased complexity in monolithic integration of all elements, it has been substituted with a more realistic way of looking at things, namely hybrid approach. The hybrid solution will probably succeed as a technological middle course with the components obtained in different processes joined to a system on a substrate. This seems to constitute the compromise between a challenging performance profile and favorable costs of manufacture. However, to achieve an economic breakthrough on a large scale, it is of high importance to present the spectrum of the potential of microsystem engineering in full width [3, 14].

MEMS devices can be considered as smart devices because they integrate sensors with actuators (i.e. a smart microsystem), but the degree of integration can vary significantly. Many MEMS devices are commercially available today, but two most exciting ones are used in optical and chemical instrumentation. The former is sometimes referred to as micro-opto-electomechanical system (MOEMS), and is being driven by the optical telecommunications and biotechnology industries.

The individual subsystems of a microsystem are - sensors, actuators, signal processing, intrfaces, packaging and interconnection technology. The possibility of integrating sensors into a microsystem as arrays is probably the principle advantage of a microsystem. It improves the reliability of a system by orders of magnitude. As microstructures can be integrated on a substrate at favorable costs and in a high packing density, it is possible to make actuator subsystems with redundancy feature. This possibility is of high interest in making safety related systems in medical engineering, process engineering and traffic engineering. The tasks in microsystem impose stringent requirements of signal processing. Sometimes tasks are to be solved in real time during operation. New concepts of signal processing must be used such as processing of fuzzy information and application of neural nets.

The interface plays an important role in a microsystem, both those between the components within a microsystem and those coupling it to the macroscopic environment. In microsystem technology interfaces among electric, thermal, fluidic, mechanic, acoustic and optical interfaces are involved. A mature technology for developing such interfaces is still in its infancy. Great efforts in this respect must be made in future in order to provide methods and optimize them with a view to manufacture in large scales.

Packaging and interconnect technology in microsystems play a key role in commercialization on industrial scale. Important tasks in this technology include (i) Si-Si bonding, (ii) anodic bonding of Si on glass, (iii) electric connections (wire bonding) over wide gaps and with great difference in height and (iv) development of further innovative techniques of joining components. Techniques have been developed not only to improve the chemistry of bonding, but also to design microsystems in such a way that bonding components or subsystems is feasible without sophisticated dosage of the adhesive. One of the unsolved problems in microsystem engineering is the dicing microcomponents manufactured batch-wise on a large substrate. Here, the dicing methods used in microelectronics cannot be applied because many mechanically sensitive packages would not survive the rude scribe and break operation used to break silicon wafers.

The spectrum of potential applications of microsystems technology is extremely vast. In a first estimate four areas of application can be recognized: (a) general instrument and control, (b) telecommunications, (c) environmental engineering and (d) medical engineering. Besides these fields of application, complete new fields will certainly emerge in the future.

9.6 Future Trends in MEMS

Looking to the future, it seems that there are two areas that will develop further. The first is that we will be able to make increasingly sophisticated micromachines that parallel the human senses, such as the microgrippers, micronoses, microtongues etc. The second area is the way in which we communicate with these microdevices and specially micromachines, such as microrobots, microcars, microplanes. The human-machine interface will probably become a limiting step and there will be a need to communicate remotely, perhaps via speech, with these intelligent micromachines. Perhaps we will even see the day in which we implant these microdevices in our body to augment our own limited sense, that is, we could have infrared vision, detect poisonous and toxic biological agents, and feel magnetic and electric fields. We could also have speech translators implanted in our ears, and so understand foreign languages, and perhaps, attach devices that monitor our health and automatically warn us of imminent problems. It is hard to say how rapidly the development of technology will take us in the next few years. We all hope that it will be beneficial to our quality of life and is not misused.

9.7 References

1   J. W. Gardner, V. K.Varadan and O.O. Awadelkarim, *Microsensors MEMS and Smart Devices*, John Wiely, 2001

2. J. W. Gardner, *Microsensors: Principles and Applications*, John Wiley &Sons Ltd, 1994

3. T. Fukuda and W. Menz, Eds., *Micromechanical systems: Principles and Technology*, Elsevier, 1998

4. S. Middelhoek and S. A. Audet, *Silicon Sensors*, Academic Press, New York, 1989

5. S. Fatikov and Rembold, Microsystems Technology and Microrobotics, Springer, New York, 1997

6. S. Kal, S. Das and S.K.Lahiri, "Silicon Membranes for Smart Silicon Sensors", Defense Science Journal, vol. 48, pp423-431,October, 1998

7. K.E.Peterson, "Silicon as a Mechanical Material", Proc. IEEE, Vol 70, pp420-457, 1982

8. G. R. Lahiji, K. D. Wise, A Batch Fabricated Silicon Thermopile Infrared Detector, IEEE Trans. Electron Devices, 29(1), pp14-22, 1982

9. S. P. Madhavi, "Relization of Micromechanical Structures and some studies on Microbolometer IR Detector" MS Dissertation, IIT Kharagpur, July, 2000

10. S. Kal, S.Das, S.K.Lahiri, "Micromachining of Silicon", Proc. Of MAERO-98, 1998

11. S. P. Madhavi, S.Das and S. Kal, "Silicon Microblometer IR Detector by MEMS Technology", Proc. ICCCD-2000, pp 247-250, IIT Kharagpur, 14-16 Dec., 2000

12. S. Kal, D.K.Maurya, K.Biswas, S.Das, R.K.Singh and S.K.Lahiri, "Design and Development of Bulk Micromachined Flow Sensor" Proc. IWPSD - 2001, pp537-539, New Delhi, 11-15 Dec., 2001

13. S.Das and S.Kal, "Formation of Silicon MEMS using Doped TMAH Solution" Proc. IWPSD - 2001, pp 534-536, New Delhi, 11-15 Dec., 2001

14. I. Fujimasa, Micromachines: A New Era in Mechanical Engineering, Oxford University Press, Oxford, 1996

Chapter 10

Design of MEMS Capacitive Accelerometers

K. Biswas
S. Sen
P.K. Dutta

Abstract

In the present chapter, different stages of design of a MEMS device have been described with particular reference to a MEMS capacitive accelerometer. A typical case study has been taken up to elaborate the design of a comb-structured accelerometer. The step-by-step methodology of design and optimization of the performance of the accelerometer have been illustrated.

10.1 Introduction

With progress in VLSI technology, the prospective of replacing conventional electro-dynamic devices by micron sized ones appeared feasible. Fully assembled devices and systems would perform the same function as the macro devices do, but at a considerable lower cost, size and weight. This gave birth to the field of Micro-Electro-Mechanical-Systems (MEMS). The fact that the mechanical strength of silicon is comparable with that of steel and density is nearly one-third of steel, added extra impetus to this thought [1]. Moreover both MEMS and VLSI may be fabricated using the same foundry and process technology [2].

Although it was expected initially that the advancement of VLSI technology would provide a ready-made solution for MEMS, in practice, it was observed that the MEMS process technology is much more complex. Moreover unlike VLSI system, the design process is not as structured. In MEMS the 2-D planar technology alone is not sufficient to model the electromechanical behaviour of the system. In fact, MEMS design methodology is still in its infancy. Design is very much application dependent and at the design stage itself one has to take care of most of the aspect of its realization, like fabrication technology, packaging [3, 4], signal conditioning etc. Besides, MEMS design is interdisciplinary, involving fields comprising electronics, material science, fluid mechanics, optics, chemistry and structural engineering [5]. Since MEMS incorporates multiple energy domains (e.g. electrical, mechanical, chemical, etc.), analysis capabilities are needed to simulate system dynamics in these domains efficiently, including their interaction [6, 7].

Design of a MEMS device has to be carried out in several stages. After the basic structure is conceived, the dimensions are to be chosen carefully, based on factors such as specification, available fabrication technology, type of signal conditioning etc. The next stage is the optimization of the design where minimization of specified design objective function(s) is carried out under a set of constraints.

In the present chapter, the design methodology for MEMS devices is discussed with particular reference to a MEMS capacitive accelerometer. It discusses design of the sensing element as well as the signal conditioning circuit. A brief outline of the design steps and the optimization procedure has been illustrated for a MEMS capacitive accelerometer.

10.2 Design Methodology

The design of a MEMS unit starts from a specification of desired function and leads to an optimized fabrication of a MEMS device. Since MEMS design is not standardized, different design flows exits [2] for the realization of a MEMS device. However, the flow can typically be represented as shown in Fig.10.1 [8].

Based on a set of performance specifications, in the first phase, the identification of the structural element is needed. Next, the functional schematic for both the MEMS device and the signal processing system are simulated at the block level using standard software such as PSpice, Matlab, etc. and subsequent high level design modifications are done [9, 10, 11].

Next, mask layout design is carried out to form a device. The mask design is dependent on the available foundry technology and the designer must formulate the particular custom fabrication process using the available foundry for the MEMS device. During the process construction, the designer should take into consideration details of material incompatibilities, layer etchants, and thin-film materials appropriate for design [12]. This requires deep understanding of the fabrication process and is a major time consuming stage in the design cycle.

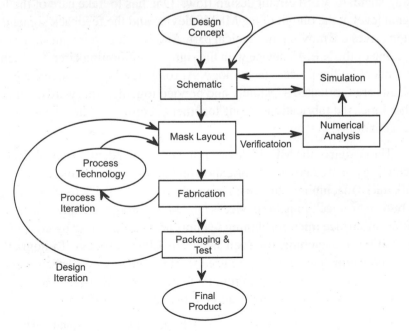

Fig. 10.1 Flow chart for Design

Along with the custom fabrication process, a mask set must be generated to form a device. This is done using a conventional layout editor for VLSI. There is no formal guideline for the design of mask set, for except some heuristics rules to avoid undesirable effects of undercut and stringers. Typical practice is to borrow some process CAD tools developed for semiconductor processing (such as SUPREME, TANNER) [2, 5], which help to determine the shapes assumed by diffused and reactively grown layers. But these tools do not provide sufficient information regarding stress effects which is one of the major causes of device failure.

In the next phase, a solid model of the actual microstructure is constructed and simulated using conventional FEM tools such as ANSYS and MEMCAD [2]. If the dynamic behaviour is acceptable, the designer can further optimize its performance by adjusting physical parameters such as mask features or thickness. However, if the simulation results are not satisfactory, the whole process must be repeated including design of the custom fabrication process and mask set design.

In the system, the thin film mechanical structure must interact with the electrical subsystem. For example in a capacitive accelerometer the displacement has to be converted to the capacitance change. Therefore, the mechanical dynamics of the MEMS device is interlinked with the signal conditioning circuit. The design of signal conditioning circuit is performed using standard VLSI circuit design flow. One has to take care of the low signal level at the output of the MEMS device and the feedback signal that often drives the moving mechanical part. Another important question is the bonding of the MEMS device with the signal conditioning circuit, which is dependent on the process technology. At times the MEMS device and the circuits are fabricated separately and bonded together afterwards, while at other times the fabrication is done together according to the availability of the technology.

The design of the MEMS device is heavily dependent on the fabrication technology. Three types of fabrication technologies are available; namely bulk micro machining, surface micromachining and the LIGA process [12]. In bulk micromachining the device is fabricated through etching from a bulk wafer. In surface micromachining features are built up layer by layer using deposition and etching for each layer. In LIGA process, Lithography, Electroforming and Molding are used to build the device. The process of micromachining determines minimum feature size and aspect ratio of any part of the MEMS device that can be fabricated using the process.

MEMS components have relatively large surface areas, but very small stiffness. At the same time the gap between the substrate and the moving object is only few microns. The combinations of these factors makes MEMS device very susceptible to the surface forces. Strong attractive capillary forces develop during dehydration causing collapse. The collapse can also develop when the device is exposed in humid condition leading to capillary condensation. Adhesion related failures also occur if the suspended member is placed in contact with its substrate by external forces (accidental shock). The above phenomenon is termed as stiction and remains a major concern in MEMS fabrication [13, 14]. These failures can be minimized by designing more stiff structure and also adopting fewer fabrication steps [7, 12]. But stiffness of the structures is inversely related to the sensitivity and complicated structures require more number of fabrication steps. Hence, tradeoff between these has to be done during optimization.

Bulk micromachining is less prone to stiction but complicated structures can not be built using it. The LIGA process can fabricate structures with high aspect ratios but is much more costly than the other processes.

Packaging of a MEMS device is needed to protect the device from working environment at the same time it is required to expose them in the environment for measurement. If the packaging issue is not taken care of properly in the design step, the cost of the device shoots up. The design requirements are application specific and include aspects such as [12]

- Expected environmental effects such as temperature, humidity and chemical toxicity.
- Design margin for mishandling and accidents.
- Choice of material.
- Minimum electrical feedthrough and bonding
- Cost in manufacturing, assembly and packaging.

So the design considers the formation of the electrodes as well as electrical connections with minimum wiring so that fabrication can be achieved in minimum number of steps and does not introduce parasitic effects.

The design methodology described here is far from complete. In this section we have intended to give the reader an overview of the different aspects of design of a MEMS device only. Some of these aspects have been further elaborated in the next section while describing the design steps for a MEMS Capacitive Accelerometer.

10.3 MEMS Capacitive Accelerometer

The MEMS Capacitive Accelerometer is the second most popular MEMS device (coming next to MEMS pressure sensor) [7, 15]. It finds several important applications, e.g. vibration measurement for machines, acceleration measurement for inertial vehicles (automobiles, spacecrafts etc.), inclination measurement etc.

The main advantage of the MEMS Capacitive Accelerometer is that the sensing element can easily be incorporated in a feedback loop to convert the output voltage into electrostatic force [16, 17]. Like other MEMS devices it has two distinct parts. The first part is the mechanical sensing element and the second part is the electronic signal conditioning circuit. The input to the sensing element is acceleration that changes the position of the proof mass. The change is sensed by a capacitance pickoff circuit. The block diagram [18] of the accelerometer may be represented by Fig.10.2.

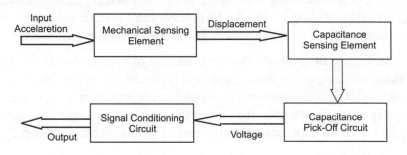

Fig. 10.2 Functional Block Diagram

Typical accelerometer specifications are sensitivity, resolution, frequency response, full scale nonlinearity, maximum operating range, offset, off-axis sensitivity, shock survival, etc. Different structures of the mechanical sensing element and mode of capacitance sensing have been experimented with by the researchers [16, 19, 20, 21]. Some of them are shown in the Fig.10.3.

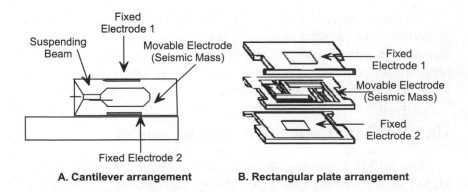

A. Cantilever arrangement   **B. Rectangular plate arrangement**

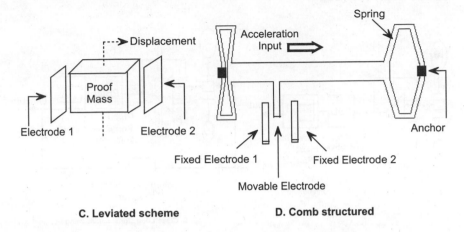

C. Leviated scheme   **D. Comb structured**

Fig. 10.3 Different sensing mechanisms used in MEMS capacitive accelerometer

Among the above structured arrangements, comb fingers and rectangular electrodes are most common. The ADXL series of accelerometers use the comb structure, where the gaps between moving and fixed fingers vary in a push-pull manner. Electrostatically levitated schemes have also been adopted to avoid the stiffness [21] constraints in their design.

In the electrical domain we have to design both for the sensing element and the signal conditioning circuit. In the sensing element part we have to design for the electrodes that will form the capacitance bridge. The nominal capacitance, the capacitance sensitivity, bandwidth etc. are also important design parameters.

The signal conditioning circuit can be of two types, close loop or open loop. The basic block diagrams of the signal conditioning circuits [16, 18, 22] are shown in Fig.10.4. The charge amplifier converts the charge accumulated in the electrode/electrodes due to the movement of the proof mass in the sensing element, into a voltage signal. The phase sensitive detector detects the direction of movement of the proof mass and its output is fed to a low pass filter to get the demodulated output. In close loop operation the output voltage is used to exert electrostatic force on the sensing element electrodes to bring back the moved proof mass to its nominal position [16, 17, 23]. The open loop configuration is simple and more stable whereas in the close loop configuration the sensitivity, bandwidth and noise performance can be improved [22].

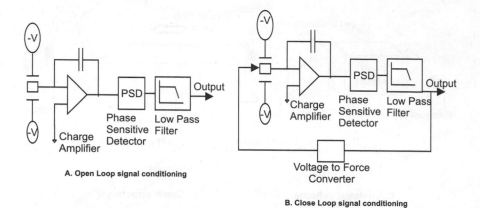

Fig.10.4 Basic signal conditioning scheme

10.3.1 Design of a MEMS Capacitive Accelerometer

In this section, the design of a typical MEMS capacitive accelerometer in the range 0 - $\pm$ 10g is presented. The sensing element is essentially a comb structured one, but unlike conventional accelerometers (e.g.ADXL50), the capacitance sensing is based on variable overlapping area [24, 25] in a differential mode, as evident from Fig.10.5 and 10.6. Here two sets of fixed fingers, with an offset of one pitch are formed on the substrate. The proof mass is suspended above the fixed electrodes.

The major advantages of this scheme are better linearity, less Brownian noise [25], and less susceptibilites to operational stiction. Better linearity is achieved due to the fact that the capacitance changes (ΔC) due to the change in the overlapping of the areas between the sensing electrodes, where change in capacitance is linearly related with area change.

Less Brownian noise is achieved since the damping phenomenon involved here is lateral film damping [16, 24]. In contrast, in most of the comb structured sensing methods, it is the distance between the electrodes which changes with the movement of the proof mass. This causes nonlinear capacitance change and involves squeezing of viscous fluid between the small gap of the sense fingers which offers high Brownian noise. The variable gap sensing methods are also more prone to operational stiction due to the change in the narrow gap between the sensing electrodes.

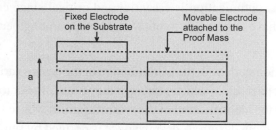

Fig. 10.5 The electrode arrangement

Fig. 10.6 The moving electrode arrangement

10.3.2 Design Steps

As mentioned already, the design starts with a preconceived structure and a preliminarily chosen set of dimensions. Based on these dimensions, the performance analysis of the device is carried out. This has to be done in two stages. The first one is based on available analytical expressions for determination of spring constant, damping, sensitivity, etc. Simultaneously, simulation has to be carried out with design tools (like MEMCAD, ANSYS etc.), since analytical expressions are not available for all the performance parameters. If the preliminary design is able to meet the desired specifications, then one has to go for the next step - *optimization* [19, 26]. Here we specify a certain performance index that has to be minimized (e.g. volume, area etc.) with a set of equality and inequality constraints. For this, we require mathematical expressions relating the performance parameters to the dimensions and a suitable algorithem for nonlinear optimization. At the end, one has to simulate the overall performance again using MEMCAD and similar software [27].

10.3.3 Optimization

It has been mentioned earlier that MEMS device operates in multiple interacting energy domains. Hence, for proper functioning of the device in all the domains optimization is essential and critical since careful design compromises have to be made to obtain a balance among performances in each domain.

Optimization starts with the identification of the design variables (length, width, input voltage, etc.), which represent the accelerometer topology. This is followed by the identification of the main variables whose values is to be chosen for optimum design. A design space is defined by setting minimum and maximum values of these variables. This design space may be further limited by the design constraints that are imposed by consideration of manufacturability or application specific specifications. Such design constraints may be geometric or functional [19].

The next step is to choose appropriate objective functions that are to be optimized; e.g. sensitivity, area, gain-bandwidth-product, etc. The objective functions are expressed in terms of the design variables, and optimum values for the design variables are searched for in the specified design space.

In the present design we have chosen the following seventeen design variables with the preliminary chosen dimensions as given in Table 10.1. The geometric dimensions are shown in Fig. 10.7.

Table 10.1: Design Variables

Var.	Description	Value	Var.	Description	Value
l	Total length of the sensing element	440µm	l_s	Length of the U-spring	318 µm
w	Total width of the sensing element	650µm	w_s	Width of the U-spring	6µm
t	Thickness of the sensing element	3µm	t_s	Thickness of the U-spring	3µm
l_b	Proof mass beam length	303 µm	l_c	Length of the comb fingers	120µm
w_b	Proof mass beam width	48µm	w_c	Width of the comb fingers	3µm
t_b	Proof mass beam thickness	3µm	N	Number of fingers	50
l_t	Length of the truss beam	20µm	d_0	Gap between the sense fingers	2µm
w_t	Width of the truss beam	3µm	w_0	The overlapping width between the sense electrodes	1.5µm
V	Voltage amplitude	1.0V			

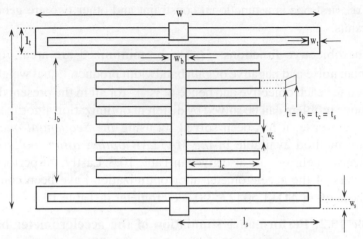

Fig. 10.7 The geometric dimensions of the sensing element

With these dimensions, the performance analysis of the accelerometer has been carried out, both analytically and using the MEMCAD design tool and given in Table 10.2.

The table provides an idea about the types of performance specifications normally used to judge the merit of a design. But this design may be improved further by adjusting the dimensions within permissible limits So, the next step is design optimization.

In the present case, nonlinear constrained optimization problem can be formulated as [28, 29]

$$\min_{\underline{u}} \quad Z = \sum_{i=1}^{k} w_i * f_i(\underline{u}) \tag{1}$$

subject to $\quad h(\underline{u}) = 0, g(\underline{u}) <= 0, u <= U_p$

where $\underline{u}$ is the vector of independent design variables; $f(\underline{u})$ is a set of objective functions; w_i is the scale weights to balance competing objective. $h(\underline{u})$ and $g(\underline{u})$ are the geometric and functional constraints and U_p in eqn.1 is the set of allowable values of $\underline{u}$. The elements in $\underline{u}$ with upper and lower bounds are listed in Table-10.3.

Two inequality constraints and one equality constraint are chosen as follows:

$sensitivity \geq 5 \times 10^{-9} \, m/g$;

$\dfrac{l_s}{d_0} \geq 100$;

$l_s = l_c$.

The, first one is a functional constraint and other two are geometric constraints.

The objective functions $f(\underline{u})$ to be minimized are area, volume, Brownian noise and negative of gain bandwidth product. Equal weights are assigned for each objective and hence w_i is '1' for all in the present design. The above problem can be solved by different optimization algorithems. In the present case, it has been solved by using *the Sequential quadratic function* method available in *the MATLAB optimization toolbox*. The optimized dimensions are also given in Table 10.3. Lastly, the performance parameters of the accelerometer after optimization have been compared with those of an ADXL50 [27, 38] acelerometer in Table 10.4.

Table 10.2: Performance simulation of the accelerometer based on preliminary design [8, 25]

Specifications	Analytical		Simulated with CoventorWare software	
Mechanical Spring constant along sensitive axis [30, 31]	$k_s = \dfrac{2Ew_s t_s^{\,3}}{l_s^{\,3}} = 1.71 N/m$		1.75 N/m	
Electrical Spring constant along sensitive axis [15, 32]	$K_E = -\dfrac{1}{x}\dfrac{\partial E}{\partial x} \approx -\dfrac{2C_0 V^2}{d_0^2} = 0.048$ N/m for V= 1.0V		*	
Displacement Sensitivity (Y-Axis)	$\dfrac{x}{a} = \dfrac{m}{k_s} = 4.46 nm/g$		4.40 nm/g	
Cross-axis Sensitivity (X-axis) Cross-axis Sensitivity (Z-axis) [19, 33]	*		$\sigma_{yx} = 10.23\%$ $\sigma_{yz} = 0.19\%$	
Natural Frequency of Oscillation	$fr = \dfrac{1}{2\pi}\sqrt{m/ks} = 7.46 kHz$		7.58 kHz	
Modal Analysis (3-Modes) 1st Mode 2nd Mode 3rd Mode	Freq (Hz) 7.46 kHz	Mass (Kg) 7.69×10^{-10}	Freq (Hz) 7.58×10^3 1.34×10^4 2.53×10^4	Mass (Kg) 7.69×10^{-10} 7.74×10^{-10} 2.83×10^{-11}
Nominal Capacitance	$C_0 = \varepsilon_0 \dfrac{A}{d_0} = 89.94 fF$		79 fF	
Capacitance sensitivity	$\dfrac{\Delta C}{g} = 0.266 fF$		0.201 fF/g	
Lateral film damping constant [16, 34, 35]	$b = \dfrac{1}{2}\dfrac{\mu_{eff} A}{d_0} = 4.48\times10^{-7}$ N/(m/s)		*	
Brownian Noise [36]	11.73μg/√Hz		*	
Mechanical bandwidth [29]	7.78 kHz		*	
Linear Range	± 10g		± 10g	
Shock level survival [30,37]	± 500g		*	

* could not be carried out due to the non availability of the software support or suitable analytical expression.

Table 10.3: Constraints chosen in the optimization routine and optimized values

Variables	Minimum	Maximum	Optimized Value
Length of the spring beam: l_s	$100\,\mu m$	$1000\,\mu m$	$456.18\mu m$
Width of the spring beam: w_s	$2\,\mu m$	$50\mu m$	$2.0\mu m$
Proof mass beam length: l_b	$100\,\mu m$	$1000\,\mu m$	$199\mu m$
Proof mass beam width: w_b	$20\,\mu m$	$50\mu m$	$49\mu m3$
Length of the comb fingers: l_c	$100\,\mu m$	$1000\,\mu m$	$456.18\mu m$
Width of the comb fingers: w_c	$2\,\mu m$	$20\mu m$	$2.0\mu m$
Thickness of the Proof mass: t	$1\,\mu m$	$3\mu m$	$2.0\mu m$
Gap between the electrodes: d_0	$1\,\mu m$	$5\mu m$.	$2.0\mu m$
Length of the truss beam: l_t	$10\mu m$	$30\,\mu m$	$9.99\mu m$
Width of the truss beam: w_t	$2\,\mu m$	$5\mu m$	$5\mu m$

Table 10.4: Comparison of performance of the designed accelerometer after optimization with ADXL150 accelerometer

Performance Criteria	Optimized Design	ADXL50
Sensitivity	8.18×10^{-9}	2.76×10^{-10}
Area	2.24×10^{-7}	2.23×10^{-7}
Volume	8.97×10^{-13}	7.36×10^{-13}
Brownian Noise	2.63×10^{-5}	1.40×10^{-4}
Gain-bandwidth Product	9.04×10^{-5}	1.66×10^{-5}
Sensitivity/Area	3.64×10^{-2}	1.23×10^{-3}
Sensitivity/Volume	9.11×10^{3}	3.75×10^{2}
Brownian Noise/Gain-bandwidth product	2.91×10^{-1}	6.03
Spring Constant	6.21×10^{-2}	5.31×10^{-1}

10.4 Conclusion

The design of a MEMS device is quite involved and is an area of interdisciplinary research. In the present chapter, we have presented a general outline for the design steps of a MEMS capacitive accelerometer. The design and optimization methodology for a comb-structured accelerometer with open loop configuration has been discussed. It has been shown that starting from the initial assumed dimensions, the performance can be improved through optimization. The performance criteria, after optimization, are comparable and sometimes better in some respects than a standard commercial accelerometer. However, the design procedure discussed here is incomplete. Several important stages, like mask layout, fabrication process flow, design of the signal conditioning circuit, its layout and integration, and finally packaging have to be taken into consideration before the device can qualify for fabrication.

10.5 Reference

1. N.Yazdi, F. Ayazi and K. Najafi, "Micromachined inertial sensors", Proc. Of IEEE, vol. 86, no. 8, pp.1640-1659, August ,1998.

2. Final Workshop Report, *Structured Design Methods for MEMS*, National Science Foundation, California Institute of Technology, November 12-15, 1995.

3. Doughlas R. Sparks, S. Massoud-Ansari and Nader Najafi, "Chip-Level Vacuum Packaging of Micromachines Using NanoGetters", IEEE Transactions on Advanced Packaging, vol. 26, no. 3, pp. 277-282, August 2003.

4. Seong Joon Ok, Chunho Kim and Daniel F. Baldwin, "High Density High Aspect Ratio Through- Wafer Electrical Interconnect Vias for MEMS Packaging", IEEE Transaction on Advanced Packaging, vol. 26, no. 3, pp. 302 - 310, August 2003.

5. Oliver Nagler, Michael Trost, Bend Hillerich and Frank Kozlowski, "Efficient design and optimization of MEMS by integrating commercial simulation tools", Sensors and Actuators A 66, pp. 15-20, 1998.

6. Hao Luo, Gang Zhang, L. Richard Carley and Garry K. Fedder, "A post-CMOS micromachined lateral accelerometer", Journal of MicroElectroMechanical System, vol. 11, No. 3, pp. 188-195, June, 2002.

7. James M. Bustillo, Roger T. Howe and Richard S. Muller (1998), "Surface Micromachining for Microelectromechanical Systems", Proceedings of the IEEE, vol. 86, no. 8, pp. 1552-1574, August, 1998.

8. Karabi Biswas, Siddhartha Sen, Pranab Kumar Dutta, "Design Methodologies for MEMS Devices", International Symposium on Advanced Materials and Processing, pp. 804-814, I. I. T. Kharagpur, 6-8th December, 04.

9. Prabhat Kumar, Karabi Biswas, Siddhartha Sen, Pranab Kumar Dutta, "Modeling and Simulation of MEMS Capacitive Accelerometer", Proceedings of National System Conference, pp. 442-447, I. I. T. Kharagpur, 17-19 December, 2003.

10. Pradeep Kumar Veeramalla, "Design and Simulation of Capacitive MEMS Digital accelerometer", M. Tech. Thesis in Instrumentation, Electrical Engineering Department, Indian Institute of Technology, Kharagpur, May, 2004.

11. Prabhat Kumar, "Design and Simulation of MEMS Capacitive Accelerometer", B. Tech Thesis in Electrical Engineering Department, Indian Institute of Technology, Kharagpur, May, 2004.

12. Tai-Ran Hsu, *MEMS and Microsystems Design and Manufacture,* Tata Mcgraw Hill Edition, New Delhi, India, 2002.

13. C. H. Mastrangelo and C. H. Hsu, "Mechanical Stability and Adhesion of Microstructures under Capillary Forces- Parts {I} and {II}", Journal of Micromechanical Systems, vol.2, pp.131 - 136, March, 1993.

14. Dongmin Wu, Nicholas Fang, Cheng Sun and Xiang Zhang (2002), "Adhesion force of polymeric three-dimensional microstructures fabricated by microstereolithography", Applied Physics Letter, vol. 81, no. 21, pp. 3963-3965, November, 2002.

15. Kyu-Yeon Park and Chong-Won Lee and Hyun-Suk Jang and Yongsoo Oh and Byeoungju Ha, "Capacitive type surface-micromachined silicon accelerometer with stiffness tuning capability", Sensors and Actuators A: Physical, vol. 73, pp. 109-116, March, 1999.

16. Mark Lemkin, Bernard E. Boser, "A Three-Axis Micromachined Accelerometer with a CMOS Position-Sense Interface and Digital Offset-Trim Electronics", IEEE Journal of Solid-State Circuits, Vol. 34, No. 4, pp. 456- 469, April 1999.

17. Analog Devices, *Monolithic Accelerometer with Signal Conditioning*, Datasheet ADXL50, Norwood, MA, U. S. A.

18. E. O. Doeblin, *Measurement System Application and Design*, Mcgraw Hill, Singapore, 1990.

19. G. K. Fedder and T. Mukherjee (1996), "Physical Design for Surface-Micromachined MEMS", Proc. 5th ACM/SIGDA Physical Design Workshop, Reston, VA, April 1996.

20. M. Kraft, C. Lewis, T. Hesketh and S. Szymkowiak, "A novel micromachined accelerometer capacitive interface", Sensors and Actuators A: Physical, vol. 68, pp. 466-473, June 1998.

21. R. Houlihan, A. Kukharenka, M. Gindila and M. Kraft, "Analysis and design of a capacitive accelerometer based on a electrostatically levitated micro-disk", Proc. SPIE Conf. on Reliability, Testing and Characterization of MEMS/MOEMS, San Francisco, pp. 277-286, 2001.

22. Larry K. Baxter, *Capacitive Sensors,* IEEE Press, U. S. A, 1997.

23. O. Ludtke, V. Biefeld, A. Buhrdorf, J. Binder, "Laterally driven accelerometer fabricated in single crystalline silicon", Sensors and Actuators 82. pp. 149-154, May 2000.

24. Arjun Selvakumar, Farrokh Ayazi and Khalil Najafi, "A High Sensitivity Z-Axis Torsional Silicon Accelerometer, International Electron Devices Meeting", San Francisco, 8-11 December 1996.

25. Karabi Biswas, Siddartha Sen and Pranab Kumar Dutta, "Design and Simulation of Variable-Area MEMS Capacitive Accelerometer", Proc. of 15th Micromechanics Europe Workshop, Leuven, Belgium, pp. 61-65, 5-7 September 2004.

26. Jeongheon Kim, Seonho Seok, Byeungleul Lee, Songyi Kim, Jeongleul Park, Innam Lee, Gilho Yoon and Kukjin Chun, "The optimized design and fabrication of a capacitive torsional accelerometer", Proc. International Conference on Electrical Engineering, Korean Institute of Electrical Engineering, pp. 1307-1310, 2002.

27. M. H. Zaman, S. F. Bart, J. R. Gilbert, N. R. Swart and M. Mariappan, "An environment for design and modeling of electromechanical microsystem", Journal of Modeling and Simulation of Microsystems, vol. 1, no. 1, pp. 65-76, 1999.

28. S. S. Rao, *Optimization theory and application,* 2nd edition, New Age International (p) Limited, India, 1995.

29. Yong Zhou, "Layout Synthesis of Accelerometers, Master Thesis", Department of Electrical and Computer Engineering, Carnegie Mellon University, 1998.

30. S. P. Timoshenko and J. M. Gere, *Theory of Elastic Stability*, 2nd edition, McGraw-Hill, 1994.

31. Alexander D. Khazan, *Transducers and Their Elements,* Prentice Hall, Englewood Cliffs, NJ, 1994.

32. R. P. Van Kampen, M. Vellekoop, P. Sarro and R. F. Wolffenbuttel, "Application of electrostatic feedback to critical damping of an integrated silicon accelerometer", Sensors and Actuators A: Physical, vol. 43, pp. 100-106, 1994

33. T. Mukherjee and Y. Zhou and G. K. Fedder, "Automated Optimal Synthesis of Microaccelerometers", Proc. IEEE MEMS Conference, Orlando, Florida, 1999.

34. Young-Ho Cho, Albert P. Pisano, Oger. T. Howe, "Viscous Damping Model for Laterally Oscillating Microstructures", Trans. ASME, Journal of MicroElectroMechanical System, vol. 3, no. 2, pp. 81-87, June 1994.

35. Timo Veijola and Marek Turowski, "Compact damping models for laterally moving microstructures with gas-rarefaction effect", Journal of MicroElectroMechanical System, vol. 10, no. 2, pp. 263-273, June 2001.

36. Crist Lu, Mark Lemkin and Bernhard E. Boser, "A Monolithic Surface Micromachined Accelerometer with Digital Output", IEEE Journal of solid state circuits vol. 30, no.12, pp. 1367-1373, December 1995.

37. W. T. Thomson, *Theory of Vibration,* 3rd edition, Allen & Urwin, New Zealand, 1988.

38. Wolfgang Kuehnel, "Modelling of the Mechanical behaviour of Differential Capacitance Acceleration Sensor", Sensors and Actuators A 48, pp. 101-108, 1995.

PART C
Signal Conditioning and Processing

Chapter 11

Signal Conditioning and Processing

S. Sinha

Abstract

Signal conditioning and processing are important for rationalizations of the output from a physical transducer. The present day developments are system-on-chip circuits for the necessary signal conditioning. Beyond that point, the processor takes over.

The chapter introduces the necessary dc and ac signal processing circuits for passive transducers converting parameter changes to electrical signals. Later, various analog and digital signal conditioning circuits like amplifiers, filters, non-linear elements and data acquisition modules are described. The chapter rounds off with a discussion on wiring, noise considerations and grounding techniques.

11.1 Introductory

10.1.1 Introduction

Signal conditioning (SC) and signal processing (SP) circuits are so many and so diverse in nature, that it is impossible to try to catch them in a single net or within a small compass. What has been attempted in the following few pages, is to understand the underlying philosophy in different kinds of SC and SP circuits and unifying them. This in turn may excite one to think about and arrive at solutions that are immediate, satisfying and long lasting.

Way back in 1946, A E Kinnard of General Electric Co. tried to analyze all measurement systems in terms of three basic blocks, namely, a primary transducer, an intermediate means and the end device (Fig.11.1).

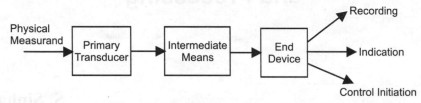

Fig. 11.1 Measurement system analysis

The primary transducer converted the physical measurand into an electrical signal. Today sensor technology is a huge subject covering mechanical, electrical, chemical and optical means and integrated circuit (IC) technology.

The intermediate means mainly consisted of the developing electronic circuits. For a direct current (DC) signal, e.g., thermocouple sensing temperature, it could be DC amplification only before display. For an alternating current (AC) signal, it might be a bridge rectifier or a phase sensitive detector, before being indicated on a meter. Today, a host of analog and digital signal processing circuits and techniques come into picture. Computer compatibility of instruments and data acquisition are the next steps.

The end device, which at the time was mainly a DC permanent magnet moving coil (PMMC) instrument, was used for indication; recording and / or control initiation were the other purposes. The end device today could be a digital panel meter, the instrument could have a general purpose (GPIB) interface for data acquisition. Data logging and control are initiated through computers.

Some kind of analyzing and a macro-level breakdown into blocks, however, helps in developing an 'object oriented' design of SC-SP circuits. In the next section, we look into a few transducers. The signal processing requirements are analyzed later.

11.1.2 Different Kinds of Transducers

Transducers can be broadly classified into two classes: active and passive. Whereas *active* transducers generate a voltage, charge or current signal, however small, proportional to the change in the physical quantity, a passive transducer induces parameter changes like resistance, inductance or capacitance of the transducer element with the physical quantity being measured.

An active transducer at best requires amplification while interfering and modifying inputs and noise are to be kept at bay. A passive transducer requires separate excitation, a bridge or other kind of circuitry, for translating the parameter change into measurable electrical signal. Wheatstone's bridge circuit is widely used in *DC signal conditioning* circuits.

For AC signal conditioning, a transformer type bridge or Blumlein bridge, is preferred for high sensitivity. *AC signal conditioning* circuits additionally has to be direction sensitive. Normally, a phase sensitive demodulator (ΦSD) followed by a low pass filter (LPF) achieves this objective. Alternatively, we could use a 'lock in' amplifier.

Before going into the signal conditioning circuits, we need to know the transducers and their operating principles. The table-11.1 is taken from R C Weyrick's book [1]

Table 11.1 Physical Variables and Transducers

Measurand	Transducer	Operating Principle
Acceleration (absolute)	Seismic mass	Accn. causes displacement of the mass relative to the case
Attitude (spacecraft)	Gyroscope	Change in orientation causes angular displacement with respect to fixed axis of rotating wheel of an axle
Displacement (translational / rotational)	Potentiometer	Sliding contact on a resistance element produces change in output voltage
	Synchro pair	Movement of the rotor from position of *correspondence* produces change in ac voltage

Fluid flow	Linear variable differential transformer (LVDT)	Change in magnetic coupling with core position causes change in mutually induced voltages
	Orifice plate	Pressure differential across a small opening in a plate in the flow path is proportional to the flow rate
	Venturi tube	Pressure differential across constriction is proportional to flow rate
	Turbine	Bladed rotor turns at speed proportional to flow rate
	Electro-magnetic (EM) flow meter	Electrically conductive fluid in magnetic field has electromotive force induced that is proportional to flow rate
Humidity	Resistance hygrometer	Change in ambient relative humidity produces variation in resistance of sensing material
Light	Photovoltaic cell	Illumination of junction between two dissimilar metals produces output voltage
	Photo-conductive cell	Resistance of semiconductor material changes with incident light level
	Photodiode or transistor	Resistance across junction in semiconductor device changes with light level
Liquid level	Float	Position of float on liquid surface is mechanically coupled to a displacement transducer
	Resistance probe	Resistance between probes changes with level (in electrically conductive liquid)
	Capacitive probe	Variable dielectric length with level change causes capacitance change between electrodes
	Optical	Interaction with liquid surface with light beam is detected by light sensor
Nuclear radiation	Ionization chamber	Incident radiation generates ion pair in gas or solid material which produces ionization current
	Scintillation Detector	Radiation generates light emission
Pressure	Bourdon tube	Internal pressure causes straightening deflection in curved or twisted tube sealed at one end

	Bellows	Axial deflection produced in thin walled tube formed in convolution
	Diaphragm	Deflection produced in circular plate fastened continuously around edges
	Manometer	Displacement of liquid column in U-shaped tube
	Piezoelectric Crystal	Voltage produced by mechanical distortion of crystal
Strain	Strain gauge	Deformation of sensing material causes change in electrical resistance
Temperature	Resistance thermometer	Resistance of sensing material changes with temperature
	Thermocouple	Electromotive force proportional to temperature produced at junction between two dissimilar metals
Velocity	Tachometer	Armature rotating in magnetic field generates voltage proportional to speed

The above list is not exhaustive. It is a representative list, however. Many of them require secondary pick-ups (a second transducer, e.g., a differential pressure transmitter for flow measurement or a displacement transducer for pressure measurement by Bourdon tube), a conditioning circuit like Wheatstone's bridge (for say, strain measurement) or a ΦSD-LPF combination (e.g., for the LVDT), or an instrumentation amplifier for amplifying the thermo-emf.

Some mathematical operations have also to be undertaken, like square rooting for connecting the flow variable to the differential pressure. Instrumentation engineers often prefer analog processing compared to digital handling of the signals. In addition 'buffering' or isolation, are important considerations. Besides, electromagnetic (EMI) and electrostatic interferences and *radiative noises* are to be guarded against by proper *grounding* and *shielding*.

11.2 Signal Conditioning Circuits

11.2.1 DC Signal Conditioning Circuit

The transducer or sensor, which is passive in nature, e.g., strain induced resistance change in a resistance strain gauge (RSG), is built into the Wheatstone bridge as one or more of the arms. The physical quantity to be measured, in this case strain, changes the resistance of the arm and the bridge balance is destroyed and there is an unbalance voltage. This unbalance voltage is *linearly* proportional to the resistance change, as long as the strain induced resistance change is small, that is, $[\Delta R / R] \leq 0.1$.

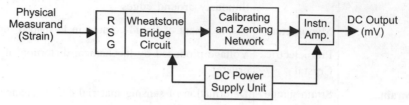

Fig. 11.2 DC Signal Conditioning Circuit

Normally a four-arm resistance strain gauge bridge (Fig. 11.2) is used for temperature compensation. Of the four RSGs, two may be in tensile mode in opposite arms, and the other two in compressive mode in the rest two arms. Equal resistance arms ensure maximum sensitivity of the bridge output and independence of the positions of the supply and indicator.

As the RSGs have equal *nominal* resistances (sets of metal gauges of 120 ohms, 350 ohms or 1000 ohms are used), their actual resistances may differ giving rise to an unbalance voltage under 'no strain' condition. This is taken care of by the potentiometer or pot in short, shown in parallel to the bridge (Fig. 11.2). The pot along with the two fixed resistances of comparatively large value, constitutes the *zeroing* network.

Calibration is carried out by switching on the calibrating resistance in parallel with one of the arms. This produces a measured quantity of unbalance. If the indication on the meter does not go to full scale value, the pot in series with the battery is adjusted to give full scale value, failing which the battery need to be replaced.

The instrumentation amplifier used to amplify the small unbalance voltage of the bridge forms the subject of a separate topic.

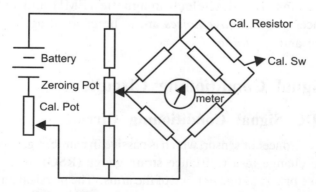

Fig. 11.3 Zeroing and Calibration incorporated

11.2.2 AC Signal Conditioning Circuit

AC signal conditioning systems are used with variable inductance or capacitance type transducers. They are used with resistance transducers where telemetering (metering at a distance) is necessary. These systems are better known as carrier systems, the carrier (AC excitation) frequency ranges from power frequency (50 Hz) to audio frequency (20 kHz) in case of resistance and inductance transducers and in the range of 500 kHz or above, for capacitance transducers. The bridge of which the transducer forms one or more of the arms (e.g., in a push pull type inductance transducer two of the arms are used) could still be of the Wheatstone bridge configuration with other two arms resistive, *alternatively* they could be transformer type bridges.

The ac signal conditioning circuit (Fig. 11.4) consists of a carrier oscillator (for excitation and use as reference for the phase sensitive detection), an AC amplifier replacing the DC amplifier, a phase sensitive detector (ΦSD) followed by a low pass filter (LPF). The last two are important additions required to detect the direction of movement (positive or negative) of the physical quantity being measured and render the final output DC.

The carrier oscillator and AC amplifier, phase sensitive demodulator, the reference line phase shifter, output low pass filter are often made integral in a *'lock-in amplifier'*, which we would take up separately. Because the lock-in amplifier acts like a narrow band detector, spurious signals outside the frequency band as also mains frequency pick up, are kept in check. However, because of modulation of the carrier signal by the measured signal or *message*, and subsequent demodulation for detection, the signal frequency bandwidth is limited to only about one tenth of the carrier frequency.

In both DC and AC signal conditioning circuits, the output analog meter could be replaced by a digital voltmeter (DVM) with numeric display and additional provisions for overload indication and auto-ranging. The output signal may also be converted into digital words using suitable analog-to-digital converters (ADCs) and inputted to a computer for the purpose of data logging and control.

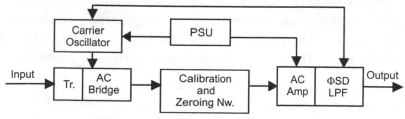

Fig. 11.4 AC Signal Conditioning System

11.2.3 Transformer Type Bridge

Transformer bridge or Blumlein bridge [2] as it is called, has two arms made of 1 : 1 tightly coupled (coefficient of coupling, k =1, iron cored coils) transformer primary and secondary coils joined at one vertex of the bridge. The other two arms are made of a push-pull type inductive or capacitive transducer. Redrawing the closely coupled inductive ratio arms of the transformer as an equivalent T network, the bridge circuit reduces to Fig. 11.5.

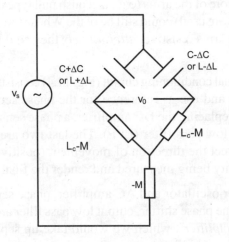

Fig. 11.5 Transformer Type Bridge Equivalent Circuit

The unbalance voltage for push pull connected inductance is given by, $v_o = v_s (\Delta L / L) [(4 L_c/L) / (1+ 2 L_c/L)]$, assuming $(\Delta L / L)^2 \ll 1$. When $L_c/L \gg 1$, last factor becomes 2. This is a four-fold increase in sensitivity than when $k = 0$.

The same for capacitance is given by $v_o = v_s (\Delta C / C) [(4 w^2 L_c C) / (2 w^2 L_c C - 1)]$ having a resonance peak at $w = 1/\sqrt{2 L_c C}$.

11.2.4 Instrumentation Amplifier

One of the basic requirements of the signal processing circuit is to amplify a signal so that it becomes suitable for indication, data log ging or control initiation. The amplifier should be free from drift and show long term *thermal* stability. The instrumentation amplifier fulfils this requirement.

A single Op-Amp could be used either in inverting or non-inverting configuration to amplify signals. Combining the two as shown in the circuit Fig. 11.6(c) will produce a differential amplifier with suitable differential mode gain and a high common mode rejection ratio (CMRR). CMRR is defined as the ratio of differential mode gain (A_D) to that of common mode gain (A_{CM}) expressed in decibels, that is, $20 \log_{10} (A_D / A_{CM})$. High CMRR helps attenuate the effects of ground loop, AC power line pick up and noise induced common mode error signals.

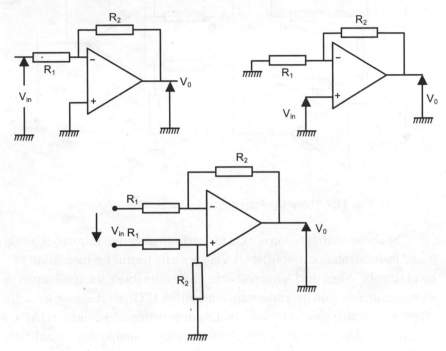

Fig. 11.6 (a) Op-Amp in Inverting Configuration, Gain = - R_2 / R_1
(b) Op-Amp in Non-inverting Configuration, Gain = 1+ R_2 / R_1
(c) Single Op-Amp based Differential Amplifier, Gain = R_2 / R_1

However, the above simple differential amplifier (DA) have two basic shortcomings: one, any gain change requires ganged variation of the resistance pair R_2 or R_1 (over and above they have to be matched pairs), and two, the input impedance presented to the signal source is $2R_1$ only.

The above two difficulties are obviated in the following (Fig.11.7) three Op-Amp based instrumentation amplifier (IA). The final stage of the amplifier is same as the single Op-Amp based DA but has a fixed gain parameter. If one thinks of a virtual ground somewhere in the middle of resistance R_3, we get an initial amplification through a pair of Op-Amps in non-inverting configuration. This ensures high input impedance and the gain is continuously variable through R_3 pot,

$$\text{Gain} = R_2 / R_1 \, (1 + 2 \, R_4/R_3)$$

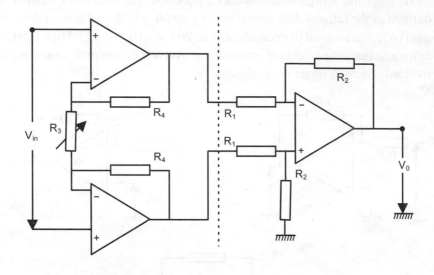

Fig. 11.7 Three Op-Amp based Instrumentation Amplifier

The above configuration is used in nearly all integrated circuit (IC) chip based instrumentation amplifiers. IA is specially useful for measuring low level signals. When the IA has software programmable gain, it is known as a programmable gain instrumentation amplifier (PGIA). Because virtually all programmable gain amplifiers in data acquisition and control (DA&C) systems are IAs, they are simply called programmable gain amplifiers (PGA).

IA amplifies DC as well as AC signals up to a certain bandwidth, depending on the Op-Amps used and thus is used as a *signal processor* for active transducer as well. Actually, the dividing line between signal conditioning and processing is rather thin and the two terms are used interchangeably. We would say we have left the domain of passive transducers and signal conditioning circuits and are now moving into general signal processing circuits.

11.2.5 Lock-in Amplifier

A lock-in amplifier [3] is used to recover low-level *alternating* (not necessarily sinusoidal) signals. Low-level signals can be detected in a very noisy environment, that is, where the signal-to-noise ratio (SNR) is poor. This is possible because of *a priori* knowledge of the signal frequency and (cross-) correlation principle is used to fish out the signal from noise.

The lock-in amplifier is shown in block diagram form in Fig. 11.8. It has two inputs: one, the signal to be measured and the second, a reference that has the same frequency as the input but a phase shift with reference to the input. The signal input is pre-filtered after amplification and is fed to a phase sensitive detector. The reference signal is also filtered and after passing through a 0 to 360° continuously variable phase shifter, is fed to the phase sensitive detector through a driver. The output of the ΦSD is low pass filtered and after further amplification becomes the output of the lock-in amplifier. The phase shifter is adjusted to maximize the output signal.

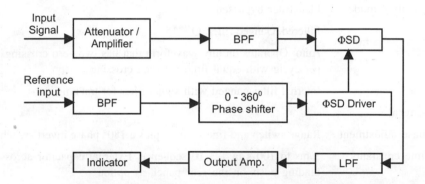

Fig. 11.8 Block Diagram of a Lock-in Amplifier

Noise that has frequency components very close to or at the same frequency as the signal are not easily filtered out, but can produce a low frequency beat at the output. By having a time constant as large as 100 seconds in the low pass filter, the system will attenuate noise components farther than 0.0016 Hz away from the signal frequency. The noise component is all but eliminated and the system provides high signal-to-noise ratio. The following specifications of a lock-in amplifier will tell more about it.

Specifications of a lock-in amplifier:

Signal Channel

Frequency range	1.5 Hz to 150 kHz in 5 ranges. Continuous tuning, plus vernier
Input impedance	Differential mode: 10 MΩ ‖ 20 pF
	Direct: 100 kΩ ‖ 25 pF
Input sensitivity	Diff.: 1 mV to 10 mV in decade steps, plus linear vernier
	Direct: 100 mV to 1.0 V in decade steps, etc.

Internal noise	Diff. : Noise figure at 1 kHz, less than 3 dB for source resistances 20 k to 5 Meg. Equivalent noise voltage with input shorted, less than 50 nanovolts at detector.
	Direct: Equiv. noise voltage, 100 nV.
Filter characteristics	Single tuned, Q = 25

Reference Channel

Frequency range	1.5 Hz to 150 kHz in 5 ranges. Cont. tuning, plus vernier
Input impedance	100 k minimum
Input sensitivity	20 mV Minm., Cont. gain control
Operating mode	Flat, filter bypassed
	Filtered, Ganged filter, Q = 15
	Auto: Operates on any waveform that has two zero crossings per cycle with equal time between crossings
	Internal filter ganged with signal filter for tracking

General

Phase adjustment	Range switch and fine vernier, plus a 180° phase invert switch
Time constants	3 ms to 100 s in 1-3-10 sequence. External capacitor across binding posts on the rear panel, as option
Monitor switch	Meter monitors signal and reference inputs to detector, 100 mV, 10 mV and 1 mA outputs. 10 X switch multiples meter readings when desired.
Outputs	Direct signal, Reference and phase outputs, Scope-servo output, 10 mV recorder output, 1 mA recorder output, etc.
Zero suppression	Ten-turn pot provides up to 300% full scale suppression.

11.2.6 Charge Amplifier

When stressed, a piezoelectric crystal generates charge across its faces. Left to itself, this charge gradually leaks away. Connected through a cable and finite measuring impedance, the charge leaks away faster, thus rendering steady state measurement impossible. An ideal *charge amplifier* [4] consists of a single Op-Amp with a capacitor in the feedback path and the crystal connected to the inverting input of the amplifier.

However, this assumes that the Op-Amp is ideal and has zero offset current. Otherwise the offset current leads to steady charging of C_f and the amplifier is driven to saturation. To overcome this problem, a practical circuit incorporates a feedback resistance, R_f, in parallel (Fig. 11.9). Analysis of the circuit gives

$$[e_o / x_i] (s) = k \, s\tau / (s\tau + 1),$$

where $k = k_q / C_f$ volts/mm and $\tau = R_f C_f$ sec. The measurement is *quasi-static* only with the cable resistance and capacitance being replaced by more specific R_f and C_f.

A variation of the same circuit is used for passive variable capacitance transducers in the forward path and with a high frequency excitation in place of the charge generator. The output is a modulated AC signal. As we have commented before, variation of the same circuit is used for an active transducer (e.g., a piezoelectric crystal) as well as a passive variable capacitance transducer. The borderline is *thin*.

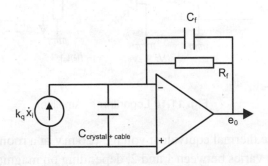

Fig. 11.9 Charge Amplifier

11.2.7 Some other Useful Analog Signal Processing Circuits

Mathematical operations are possible to be performed by Op-Amps. *Summing* and *integration* are basic operations which constitute the basis of analog computers or analog manifolds. More complicated operations like multiplication (and squaring), square rooting, etc. are also performed using Op-Amps.

For multiplication, we use a combination of log and antilog amplifiers. In the logarithmic domain, multiplication / division is reduced to addition / subtraction which part is easily achieved by using Op-Amp based summers. The *crux* of the log and antilog amplifier design lies in intelligently using a diode as the forward or feedback element in the Op-Amp circuit [5]. Fig. 11.10 shows the circuitry of the logarithmic amplifier. The exponential diode characteristic helps in realizing the logarithmic operation.

In the logarithmic amplifier shown, the output voltage is proportional to the log of the input voltage. The forward transfer characteristics of the diode is given by

$$I_F = I_s \exp [V_F / \eta V_T],$$

where I_s is the diode saturation (leakage) current, V_F is the forward voltage drop across the diode,

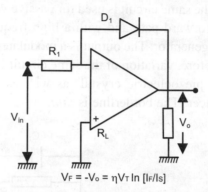

$$V_F = -V_o = \eta V_T \ln [I_F/I_s]$$

Fig. 11.10 Logarithmic Amplifier

V_T is the thermal equivalent voltage (26 mV at a room temperature of 20°C) and η varies between 1 and 2, depending on magnitude of I_F. For I_F near I_s value (less than 1 mA), $\eta = 2$. The diode could be substituted by a transistor with collector-base short.

In an antilog amplifier, the feedback diode and the input resistors are interchanged. An analog multiplier is made up using log and antilog amplifiers and a summer. Fig. 11.11(a) shows the multiplier symbol and Fig. 11.11(b) gives the multiplier in block diagram.

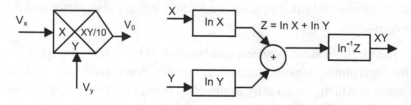

Fig. 11.11 (a) Symbol of Analog Multiplier, (b) Block Diagram
Representation of the same

Squaring circuit and a square rooting circuits could be made up as shown in Figs.11.12.(a) and (b).

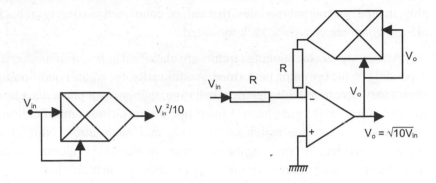

Fig. 11.12 (a) Squaring Circuit (b) Square rooting Circuit

Going into further details of these circuits is not necessary as all these are now available as ICs or packages, and sometimes the integration progresses much further. They are available as *devices* with the vendors. Mention of *Catalogs* of two such vendors have been included in the *reference* [9, 10].

We briefly go over the basic analog signal conditioning circuits and follow it up with digital signal conditioning circuits that helps interface the instruments with the computer.

11.3 Analog Signal Conditioning

11.3.1 Analog Signal Conditioning Devices

Slightly modifying the ideas of the first section, an instrument (or measuring sub system) consists of an input transducer, followed by a signal conditioner that interfaces with the data processing and display unit. Signal conditioners should preferably be linear in their amplitude (and phase) and have adequate bandwidth to handle the signal. The signal conditioners accept the signal from the transducers and produce a signal appropriate for feeding to the remainder of the instrument.

A wide variety of analog building blocks for signal conditioning are available in either modular or integrated circuit (IC) form [6]. Such building blocks include operational amplifiers, instrumentation amplifiers, isolation amplifiers, comparators, analog multipliers / dividers, log / antilog amplifiers, RMS-to-DC converters, trigonometric function generators, etc. Also available are complete signal conditioning subsystems consisting of various plug-in input and output modules, that can be connected via universal back planes which are chassis or rack mounted.

Analog signal conditioning circuits are classified as *linear* or *nonlinear* depending on the operation performed. Additionally, the signal conditioning circuit may have to provide electrical isolation, reference for phase detection and excitation for the transducer. Linear operations include amplification / attenuation, impedance matching, filtering and modulation. Nonlinear operations may require obtaining the root mean square (RMS) value involving squaring and square rooting, log-antilog operation as in a multiplier.

11.3.2 Amplification / Attenuation and Impedance Matching

Amplification is, in *general,* achieved by using an operational amplifier, an instrumentation amplifier or an isolation amplifier. We have not said much about the *operational amplifier* as a unit. An *ideal* Op-Amp is characterized by infinite gain, infinite input impedance, zero output impedance, infinite bandwidth and infinite slew rate. In differential mode it will have infinite differential input resistance and infinite common mode rejection ratio. It also should have zero bias currents, zero input offset current for differential input, zero input offset voltage, etc., etc. Real Op-Amps fall far short of this ideal.

Important parameters to consider while selecting an Op-Amp are: DC voltage gain, A_o, gain-bandwidth product, f_t, slew rate which is a large signal parameter. Other parameters like finite input impedance, non-zero output impedance, noise, offset voltage, input offset current, etc. are to be taken care of for particular applications and through selection of special Op-Amps.

The following special kinds of Op-Amp are listed in the catalogs besides the 'garden variety': Low noise Op-Amps (for use with particularly low level signals), Chopper-stabilized Op-Amps (where extreme DC stability is required), Fast Op-Amps (handling large slew rates and requiring large gain bandwidth products), Power Op-Amp (when output currents greater than a few milliamps are required to be delivered to the load), Electrometer Op-Amp (with very high input resistances $>10^{13}$ ohms).

Instrumentation amplifiers are used to provide high input impedance, low output impedance, stable gain, high common mode rejection ratio and relatively low offset and drift. They are suited for amplification of low level signals in the presence of high level common mode voltages. IAs are available in inexpensive IC form. Some IAs have digitally programmable gains, whereas gains for the others are set either by interconnecting resistors internal to the IA through external pins, or by connecting external resistors to designated pins.

Isolation amplifiers are used where a voltage or current signal is to be measured in the presence of high common mode signals safely. They are useful where DC or line frequency leakage current safety is to be ensured, as in biomedical applications. These amplifiers have three sections: an input stage, an output stage and a power circuit. The input stages are galvanically isolated from the output stage. Communication between input and output stages is by modulation / demodulation.

Two-port isolation takes place where there is a DC connection between power circuit and output stage and three-port isolation takes place when the power circuit is isolated from the output stage as well. Isolation amplifiers are also available in modular form.

Attenuation is the opposite of amplification. Most signals to be attenuated are voltage signals and attenuation is accomplished either through voltage dividers or through voltage transformers. Voltage dividers could be resistive or capacitive, depending on whether the elements in the chain are resistors or capacitors. Capacitance voltage dividers are applicable at high voltages and frequencies. Inductive dividers are also available. Most common form of inductive divider is the *autotransformer*. Inductive dividers work only up to a frequency of few kHz.

Voltage transformers are used on AC and they are included in the class, instrument transformers and are more generally known as potential transformers, or PTs in short.

Current signals are scaled down either by using shunt resistances or by using current transformers. Whereas current shunts are used for DC and low frequency applications, current transformers are used for alternating currents at power frequencies and above. Current transformers, or CTs as they are known in short, also come under the class of instrument transformers. CTs consist of a specially constructed toroidal core, upon which the secondary winding is wrapped and through which the primary winding is passed. A single turn primary (bar primary) or multi-turn primaries are used. CTs are operated with their secondary windings under near short circuit conditions to counterbalance the omni-present primary ampere-turns.

Besides, *attenuator pads* which in addition to voltage scaling serves the purpose of impedance matching (image impedance in passive filter circuits) are used in L, T and π, balanced or unbalanced, configurations.

Impedance transformation is used in electronic circuits for maximum power transfer between preceding and succeeding stage. Output transformers are used for impedance matching of load.

In most cases it is sufficient to provide buffering that presents very high impedance to the input stage, a very low impedance to the succeeding stage, and unity voltage gain. A voltage follower (Fig. 11.13), a single Op-Amp connected in non-inverting configuration, constitutes a good buffer.

Matching transformers, passive matching networks such as attenuator pads and unity gain buffers are standard means of accomplishing impedance transformation. Unity gain buffers are available in IC form.

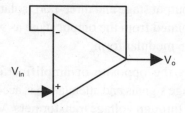

Fig. 11.13 Voltage Follower

11.3.3 Passive and Active Filters

Linear filtering: Filters are used within analog signal conditioners to reduce power frequency hum, limit signal bandwidth, as part of the demodulator, or if the signal is to be sampled, to limit its bandwidth to prevent *aliasing*. These filters could be made of passive components, or they could be active filters based on single or multiple Op-Amps, or 'biquad' circuits to be exact.

Fig. 11.14 (a) shows a single-ended, double pole *passive* filter circuit to attenuate 50 Hz noise. The filter has a -6 dB cutoff at about 1 Hz while attenuating 50 Hz, about 40 dB (200 times). The same in the differential mode is shown in Fig. 11.14 (b).

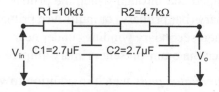

Fig. 11.14 (a) 1 Hz Low Pass Filter (single-ended, 2-port)

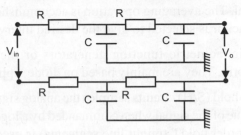

Fig. 11.14 (b) Low pass Filter (differential, 2-port)

Passive filters assume that the source impedance is many (at least 10) times less than R, and the load impedance, similarly many times greater than R.

Monolithic ceramic type capacitors have been found useful for filter circuits, as they possess high package density, low leakage current and they are non-polar. Values up to 4.7 mF at 50V are quite common.

For more exacting applications and where source loading, etc. are considerations, *active* filters are useful [7]. They form a class of their own and are kept outside the purview of the present discussion. We move to nonlinear analog signal processing circuits next.

11.3.4 Nonlinear Signal Processing Circuits

Nonlinear signal conditioning circuits are many and varied. Most of them are, however, available as ICs. Among them are included Comparators, Schmitt trigger, Multiplier / Divider, Squarer / Square-rooter, Log / Antilog amplifiers, Trigonometric function generators, Sample-and-hold and Track-and-hold circuits, precision diode based circuits such as Half-wave Rectifiers, Absolute Value circuits, precision Peak Detectors, and precision Limiters.

A comparator is a two-input device whose output voltage takes on two stable values depending on whether one particular input is greater or less than the other. A Schmitt Trigger (ST) is a comparator with hysteresis. A comparator may be realized from a single Op-Amp connected in differential fashion. ST is constructed from the comparator by applying positive feedback.

The multiplier circuit has already been worked out. In case of a divider, a summer is used after taking log of the divisor, rendering a sign change. Rest of the circuit is same as that of a multiplier. Squarer and square rooter circuits have already been discussed as also the logarithmic and anti-logarithmic amplifiers. A true RMS-to-DC converter computes the square root of the average, over some interval of time, of the instantaneous square of the input signal. The averaging operation is accomplished by a low pass filter whose capacitor is selected to give the desired interval.

Standard trigonometric function generators or their inverses, are available in IC form. They are mainly based on diode 'spline' generators.

Sample-and-hold (S/H) circuits samples the analog signal and holds the instantaneous value of the signal when commanded by a logic control signal. In 'track' a first order hold (straight line segments) is used rather than the zero order hold used in S/H circuits. We would come back to them again when we do the digital signal conditioning circuits.

Precision diode based circuits such as half-wave rectifiers, absolute-value circuits, precision peak detectors, and precision limiters are easily implemented based on Diodes and Op-Amps. All these nonlinear circuits are standard building blocks and fairly widely used in signal processing applications in instrumentation.

One of the reasons for using the analog circuits is to operate in *real time*, though averaging etc. compromises this requirement to some extent. The alternative is to use the power of the computer to do the operations. But the operation may no longer be in real time.

11.4 Data Acquisition and Control

11.4.1 Digital Signal Conditioning Circuits

We have nearly come back full circle in that we want to rewrite the introductory section in a slightly different fashion in today's world. Computer compatibility of the instruments is an often used term and is important when a computer do the data acquisition and control (DA&C). Fig. 11.1 now takes on a different configuration as below (Fig. 11.15).

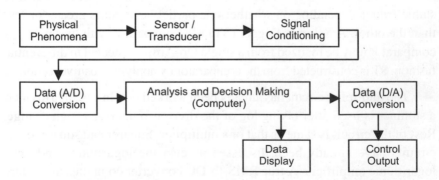

Fig. 11.15 Data Acquisition and Control Flow Diagram

The DA&C systems bridge the gap between the pervasive digital computer and the real world. The personal computer (PC) is the fastest growing 'engine' for the new DA&C system design. Data loggers and programmable controllers are quite implementable through the ubiquitous personal computer.

11.4.2 Analog Input Systems

The fundamental function of the analog input system is to convert the analog signal into digital words *or bit strings*, which are acceptable to the computer. Analog-to-digital converter (A/D) chips do it. In addition, preceding the A/D converter, analog input subsystems can include a (programmable) amplifier, a sample/hold circuit, a multiplexer and various analog signal conditioning elements - linear or non-linear.

Most widely used *analog-to-digital converters* are successive approximation, integrating (dual ramp) and parallel (flash) type converters. Flash converters are the fastest but the most expensive. Complexity generally limits this converter to low resolution (8 bits or less). When high speed is not required, integrating A/D converters can easily give 12-, 14- or even 16-bit resolution at low cost. Sampling speed is about 3 to 50 conversions per second. Data acquisition usually requires about 12- bit or more resolution and for speeds above 100 samples / second the successive approximation type A/D converter is the preferred one.

Some sensors have wide dynamic ranges, that is, there is a wide span between the lowest detectable value and the full (scale) value. Dynamic ranges are measured in decibels (dBs) and a 0.5% accurate transducer with a dynamic range of 80 dB, might require a resolution of at-least 12 bits. Sometimes pre-amplification is also necessary. A signal change below 2.44 μV level will not be seen by a 10-V range ADC with 12-bit resolution. Pre-amplification by a factor 1000 could reduce the resolution to 2.44 μV level. These are finer points of input system design.

Multiplexer (MUX) is an electronic switch that allows many input channels to be serviced by a single A/D converter. In general analog signals are continually changing with time. Successive approximation A/D converters require that the input signal amplitude does not change during the conversion cycle. A sample and hold (S/H) circuit 'grabs' the present value of the signal at the beginning of the conversion cycle and holds it constant during the cycle. A number of such S/H circuits connect to a single MUX/ADC combination as shown in Fig. 11.16.

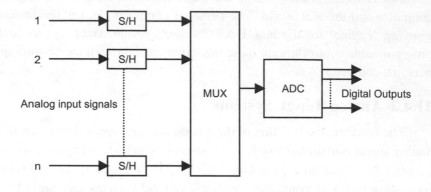

Fig. 11.16 Time multiplexing of many signals

Time multiplexing requires that the multiplexed signals follow the desired sample rate (above Nyquist frequency, that is, twice the highest signal frequency) and is matter of some attention. Ideally all signal channels should be sampled uniformly, but non-uniform sampling (in integral multiples of sampling time only) could easily be accommodated by suitable programming. In addition, the input signal subsystems might incorporate signal conditioning.

Analog Output Systems: Analog output signals are required, say, to drive chart recorders, to provide control signals to controllers / actuators and variety of other devices. Analog output ranges are ±5 V, 0-10 V and 4-20 mA. In voltage output mode most digital-to-analog (D/A) converter can supply up to 5 or 10 mA of load current. When large loads such as positioners, valves, lamps and motors are to be controlled, power amplifiers or current boosters are required.

11.4.3 Digital Inputs and Outputs

Digital inputs could be single or multiple bit *status* signals, could be output of a direct digital transducer, or a keyboard generated *set point* data. Similarly, outputs could be alarm annunciations, *on-off control* signal, etc. Most data acquisition systems are able to accept and generate TTL level voltage signals (0 to 5 volts). However, applications often require an interface for other discrete voltage levels. Higher voltages and current outputs are required to control devices such as solenoids, motors and relays.

Other applications require that devices be turned on or off for precise time periods. All these can be provided by counter / timer (C/T) circuits. The system's counter / timers are optimized for pulse applications including frequency measurement and time-base generation. For low frequency (below

10 Hz) measurement, a clock is counted for the period of the unknown input signal. For high frequency measurement (up to 8 MHz), the method is to count cycles of the unknown input signal for a fixed time interval. The counting methods are mostly software driven.

11.4.4 Software for Data Acquisition and Control

Whenever one is using a computer for data acquisition and control, availability of proper software packages is equally important. Data acquisition is a general term which includes in addition to acquiring data, a host of other functions like data processing, data logging, display and printing. Actually, many of the special kind of analog circuits developed for mathematical operations like squaring, square rooting, etc. become redundant as they could be handled by writing computer programs, the problem of real time operation notwithstanding. Similarly, filters could be implemented digitally.

There is already a significant amount of standardization in the matter of interfacing and a large family of hardware / software tools and application packages are available with which an instrumentation engineer today is able to implement a custom DA&C system without consuming much time and money.

11.5 Diverse Topics

11.5.1 Wiring and Noise Considerations

Digital signals are relatively immune to noise because of their discrete nature. In contrast, analog signals are directly susceptible to even relatively low-level noise. The noise transfer takes place through conductive, inductive (magnetic) and capacitive coupling. A radiative coupling at high frequencies is also possible.

The switching on (or off) of high current loads in the vicinity can induce noise signals by magnetic coupling (transformer action). Signal wires running close to AC power cables can pick up 50 Hz noise by capacitive coupling. Allowing more than one power or signal return path can produce ground loops that inject errors by conduction. Conduction involves current flow through ohmic paths. Interference via capacitive or magnetic coupling usually requires that the disturbing source be close to the affected circuit. At high frequencies, however, radiated electro magnetic signals propagate over long distance. The noise level induced will depend on several factors: signal source output impedance (at the transducer), signal source load impedance (at the input of the data acquisition system), lead wire length, shielding and grounding, proximity to the noise source(s), signal and noise amplitude.

There are few other facts to be taken into consideration. Transducers that can be modeled by a current source are less sensitive to magnetically induced noise pick-up than voltage driven devices. When the transducer appears as a voltage source, the magnetically induced noise errors add directly to the signal source without attenuation. Errors due to capacitive coupling affect equally both current and voltage transducer circuits. The error signal induced is proportional to $2\pi f RC$, where R is the load resistance, C is the coupling capacitance and f, the interfering frequency.

Many of the noise problems are solved by simply giving attention to a few grounding and shielding principles [8]. Besides, minimizing wiring inductance and ground currents, limiting number of antennas and their sizes and maintaining balanced networks wherever possible help in reducing noise problems. We would take up grounding first.

11.5.2 Grounding Techniques

We must first understand that the ground path and the return path are not the same. A ground wire connects equipment to earth for reasons of safety, to prevent accidental contact with dangerous voltages, or to put something to a potential where there is no ground capacitance. Grounds do not normally carry current. Return lines are an active part of the circuit, carrying power or signal currents. Return paths should have lowest possible impedance. Current takes the path of least resistance (or impedance). Return impedance is usually dominated by path inductance. Do not unnecessarily create ground loops by mixing the two.

Three different grounding techniques are suggested in Fig. 11.17(a), (b) and (c). In Fig. 11.17(a) the signal return line is grounded at each chassis. If a difference in potential exists between two grounds, a ground current flows. This current multiplied by the wire impedance results in an error voltage, e. The voltage applied to the amplifier is no longer V, but [V + e]. This may be acceptable where the signal voltage is much greater than the error voltage.

When the signal level is small and a significant difference in ground potentials exists, the connection of Fig. 11.17(b) is desirable. The return wire is not grounded at the amplifier and ground current is excepted. Any difference in ground potential appears, to the amplifier, as a common mode voltage.

If cost is not a limitation, Fig. 11.17(c) offers the highest performance.

Installing an isolator into the signal path faithfully conveys the signal to the amplifier while interrupting all direct paths. In this configuration, multiple ground paths can be tolerated along with several hundred volts between the input and output circuits. The isolator could be an isolation transformer or an opto-isolator.

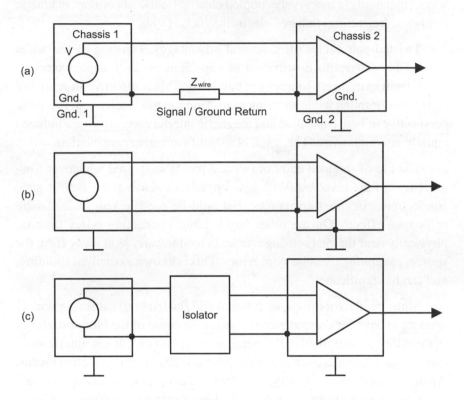

Fig. 11.17 (a) A single-ended connection, (b) a differential connection and (c) input-output isolated circuit

11.5.3 Different Kinds of Wires and Cables

There are four kinds of wires or cables, basically. They are: the plain pair, shielded pair, twisted pair and coaxial balanced. All but the coaxial (coax) wires, are said to be balanced. Coax differs from the others in that the return line surrounds the central conductor. In contrast a shielded pair is surrounded by a separate conductor (properly called a shield), that does not carry signal current. Coax offers a very different capacitance between each of its conductors and ground. Not only does the outer conductor surround the inner, but it is also connected to ground. Thus, coax is intended for single-ended applications only.

One method of reducing errors, due to capacitance coupling, is to employ a shield. The shield blocks the interfering current and directs it to ground. Depending upon how complete the shield is, attenuations of more than 60 dB are attainable. When using shielded wire, it is very important to connect one end of the shield to ground. A shield can actually work in three different ways, 'bypassing' capacitively coupled electric fields, 'absorbing' magnetic fields and 'reflecting' radiated electromagnetic fields.

Twisted-pair cables offer several advantages. Twisting of the wires insures a homogenous distribution of capacitances. Both capacitances to ground and extraneous sources are balanced. This is effective in reducing capacitive coupling while ensuring high common mode rejection. From the perspective of both capacitive and magnetic interference, errors are induced equally into both wires. The result is a significant error cancellation.

The use of shielded and / or twisted-pair is suggested whenever low-level signals are involved. With low impedance sensors, the largest gage connecting wires that are practical should be used to reduce lead-wire resistance effects. On the other hand, large connecting wires that are physically near thermal sensing elements tend to carry heat away from the source, generating measurement errors. This is known as thermal shunting, and can be significant.

Multi-conductor cables, both round and flat (ribbon) cables are widely used these days for computer connections. Because of the close proximity of the different pairs, they are susceptible to 'cross talk', that is, interference caused by inadvertent coupling of signals via capacitive or inductive means. Again, twisted pairs are very effective. Alternatively, connect alternate wires as return paths or run a ground plane under the conductors or use a full shield around the cable.

Still another source is the tribo-electric induction. This refers to the generation of noise voltage due to friction. All commonly used insulators can produce a static discharge when moved across a dissimilar material. Special low noise cables are available that employ graphite lubricants between the inner surfaces to reduce friction.

The key to designing low-noise circuits is recognizing potential interference sources and taking appropriate preventive measures. After proper wiring, shielding and grounding techniques have been made use of, input filtering can be used to further improve the signal-to-noise ratio. However, filtering can never be relied upon as a *fix* for improper wiring or installation.

11.6 Conclusion

The above by no means purports to be a complete text on signal conditioning and of processing circuits. As has been stated before, signal conditioning and signal processing encompasses host of other things and are by themselves full-fledged areas of study. This has only been a very humble attempt to give glimpses of signal conditioning and processing as they have been envisaged by an instrumentation engineer. Rest of the things will have to be looked into as and when they matter.

11.7 References

1. R C Weyrick, Fundamentals of Automatic Control (bk.), McGraw-Hill KogaKusha, International Student Edition, Tokyo, 1975.

2. B E Jones, Instrumentation, Measurement and Feedback (bk.), Tata McGraw-Hill Pub., New Delhi, India, TMH Edition, 1978.

3. C S Rangan, G R Sarma and V S V Mani, Instrumentation Devices and Systems (bk.), Tata McGraw-Hill Pub., New Delhi, 1988.

4. E O Doebelin, Measurement Systems: Application and Design, McGraw-Hill Kogakusha, International Student Edition, 1966.

5. D J Dailey, Operational Amplifiers and Linear Integrated Circuits, Theory and Applications (bk), McGraw-Hill International, 1989.

6. S A Dyer, Signal Conditioning, Section 139 in Engineering Handbook, pp 1501-1505, Part-B, Ed. R C Dorf, Jaico Pub., Mumbai, 1996.

7. Alok Barua and Satyabroto Sinha, Computer Aided Analysis, Synthesis and Expertise of Active Filters (bk.), Dhanpat Rai & Sons, Delhi, India, 1994.

8. The Handbook of Personal Computer Instrumentation for Data Acquisition, Test and Control, Burr-Brown, Intelligent Instrumentation Products, Tucson Az, 1998.

9. Analog Devices Data Book, Vol. 1 Integrated Circuits, Vol. 2 Modules Subsystems Norwood Mass, 1984.

10. Burr-Brown Product Data Book, Tucson Az, July 1984.

| Chapter 12 |

Telemetry and Data Communication

S. Sinha

Abstract

Telemetry is no different from communication systems. This is one field where the recent developments are tremendous and rapid and the area is in a state of flux. Whereas there are old telemetering systems surviving the ravages of time, it is a new look approach today embracing multimedia capabilities, hybridization of analogue and digital systems and data networking. Gradually, a shift is taking place from hardware to software and through miniaturization, less power consumptive circuits are taking over daily.

12.1 Introduction

12.1.1 Elements of Data Communication Systems

Telemetry is the science and technology of metering, that is, measurement at a distance. The term is similar to tele-vision and tele-phone and evolves to tele-control of systems. The data handled could be analog or digital in nature. In not so very distant past, the data generated by a transducer or sensor, could have been other than electrical in kind. But to-day, data is mostly electrical.

If a physical quantity is to be measured, it is converted into an electrical signal, preferably in linear proportion, by a transducer. Examples of common transducers are: thermocouples for temperature measurement, differential pressure transmitters (DPTs) for flow measurement, a float moving on a potentiometer for liquid level measurement, etc. Alternatively, transducers are also called sensors because they sense change in physical measurands, particularly as more and more of today's transducers are getting miniaturized. As for example, the touch sensor on a robot hand could be a small semiconductor strain gauge.

Except Geiger-Muller counter for measuring radioactivity, almost all transducers in the world are analog in nature, that is, they give out analog signals. Digital signals are at most status signals, that is, it gives the status of a device like whether a contactor is on or off, a valve is open or closed, a solenoid is excited or not, etc. But they also constitute important signals, particularly in tele-control for the purpose of command and acknowledging commands.

Today, computer compatibility of instruments is important for the purpose of data acquisition, monitoring and alarm annunciation and control action generation. Usually, the standard analog signals of 1-5 volts, 4-20 mA in current mode, is converted to digital words by analog-to-digital converters (ADCs) for either computer compatibility or digital data transmission.

Transmission brings in the concept of distance. The signal, analog or digital, is modulated and is usually made to ride a carrier and then transmitted over the transmission link which could be a pair of wires, power lines, co-axial cables, fibre optic cables, radar or satellite links and then demodulated at the receiving end for signal recovery. The distance may vary from a few meters to millions of kilometer.

The transmission link has its inherent limitations. The link is at best noisy; noise comes in the form of interfering signals, modifying signals due to environmental parameter changes and random noise, which is omnipresent, e.g., the static from the atmosphere. The signal also gets distorted because no transmission link is ideal, is of 'finite' bandwidth and / or of limited channel capacity.

Performance of a telemetering system in the face of interference, random noise (characterised by signal-to-noise ratio, or bit error rate), bandwidth requirement and transmitter power are important considerations and require careful study for various available analog and digital communication systems. Fig. 12.1 shows a telemetry / communication system block diagrammatically.

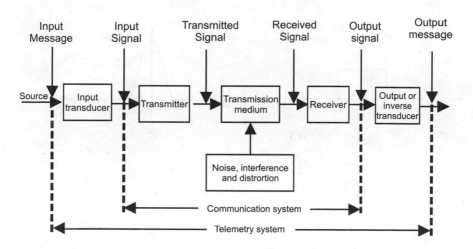

Fig. 12.1 Elements of a Telemetry/Communication System (after Carlson [1])

12.1.2 Signal Description in Time and Frequency Domain

A communication engineer should be able to describe a signal both in the time and frequency domain with equal ease. If the time domain description is needed for study of signal waveforms, frequency domain description is necessary for bandwidth determination, frequency division multiplexing (FDM) and design of filter banks.

All signals, when described in time domain, can be classified into three categories, namely, periodic or repetitive signals, *aperiodic* or signals transient in nature, and random signals which can only be statistically described. Whereas probability density function is a more accurate description of a statistical signal, statistical moments like mean value, variance, etc. are accepted characteristic description of the signal.

Periodic Signal: Composite periodic signals are analyzed by Fourier series method to separate them out into their component frequencies. The frequency spectrum for a periodic signal is a line spectrum with characteristic lines representing the frequencies contained in the signal. It is customary to use the exponential Fourier series.

Aperiodic signal: The time and frequency domain descriptions of the *a periodic* signal form a Fourier transform pair. In the frequency domain, the signal is continuous in frequency. A square pulse in the time domain will become a sinc function (=sinx/x) in the frequency domain, as shown in Fig.12.2(a) and (b). The spectrum will always be positive as shown in Fig.12.2(c).

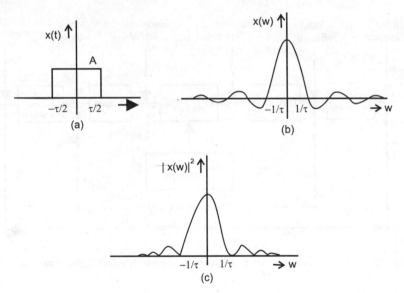

Fig. 12.2 (a) A square pulse in time-domain, (b) corresponding frequency domain representation and (c) energy spectrum of a square pulse

The Fourier transform pair has interesting properties. One of them is the reciprocal spreading effect; that is, if the pulse is constricted in time, made sharper, the frequency domain representation flattens out meaning more and more high frequency components are contained in the main lobe of the frequency domain description of the signal. Alternatively, if the signal is stretched in time, the frequency domain description peaks up. In the limit when T tends to infinity, the time domain description of the signal is a steady direct current quantity, whereas the frequency domain description is a spike at w = 0, that is, zero frequency.

Random Signal: If the signal is a random signal and is only statistically describable, things become much more complicated. A common noise signal having a Gaussian amplitude density distribution function with mean value, m, and a variance, σ^2, is represented as N (m, σ^2). To describe this signal in the time domain, the signal is *correlated* with itself in the time domain. If the random signal is stationary and ergodic in character, the correlation of the signal with a time shifted version of itself is given by the time auto-correlation function (ACF).

An auto correlation function is a good description of signal in the time domain. Using Wiener-Khintchine theorem, it could be shown that ACF and power spectral density (PSD) form a Fourier transform pair. That is, power spectral density is considered a good enough description of a random signal in the frequency domain.

Now, consider white noise as an example of a random signal. It contains, theoretically, all frequency components extending from -∞ to +∞ in the w-scale with equal measure. That is, it approximates a dc power in the spectrum domain. If we take the inverse Fourier transform of the same, we should get a spike centred at $\tau = 0$ and zero elsewhere. The spike is representative of the ACF of the white noise, that is, it is correlated with itself when there is no shift. Otherwise, it is completely uncorrelated. The PSD and ACF of a white noise signal are shown in Fig. 12.3(a) and (b) respectively.

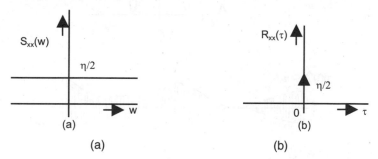

(a)   (b)

Fig. 12.3 (a) Power spectral density of white noise signal and
(b) auto-correlation function of the same

Further, the signal spectrum gives the power or energy, as the case may be, contained in the signal straightaway. Whereas the periodic signal is described by signal power after time averaging, *aperiodic* signals are transient in nature and are described by their energy content. The power content of the random noise signal can be found out by integrating the power spectral density over the frequency range of interest. It is always easier to calculate the noise power from the frequency domain.

Normally, a fast Fourier transform (FFT) program is run on the computer to check the frequencies contained in a signal sample, irrespective of whether the signal is periodic, aperiodic or random.

12.1.3 Narrowband Signal

When white noise is passed through a narrowband filter, the result is interesting. For the time being, let us consider it being passed through a near ideal low pass filter. The signal description in the frequency domain is as shown in Fig. 12.4(a). By applying inverse Fourier transform, we find that the time domain signal is no longer uncorrelated as the white noise.

More interesting things happen, if this narrowband noise is made to ride a carrier. Let the bandwidth of the noise signal B be small compared to the carrier frequency, f_c that is B/f_c is small. For small but finite B/f_c, narrowband noise looks like a wave with slow and random amplitude and phase variations, Fig. 12.4(b). Such signals definitely interfere with actual message signals.

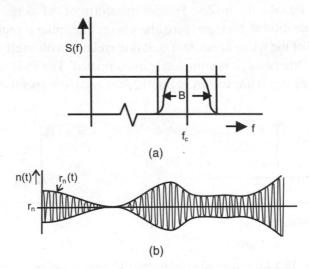

Fig. 12.4 (a) A narrowband signal of bandwidth B riding a carrier f_c and (b) narrow band random noise signal in the time domain

The narrowband noise can be described as a phasor of frequency f_c with random amplitude r_n (Rayleigh distributed) and random phase ϕ_n (uniformly distributed). The important things to note are that the mean value of the envelope, r_n, is not zero as it is never negative and that $\sigma^2 = 2N$, is twice the noise power.

12.1.4 Shannon's Sampling Theorem

Whenever data is to be transmitted digitally, the analog signal is sampled in time and digitized, that is, converted to digital words constituted of bits. However, there is a restriction on the rate of sampling in order that the original analog signal can be recovered from the sampled signal by reconstruction. This is given by Shannon's *sampling theorem*. The theorem is in two parts.

1. If the highest significant frequency contained in a signal is W, or alternatively the signal is band-limited in W, it is completely described by instantaneous sample values uniformly spaced in time, $T_s \leq 1/2W$. Sampling rate of $f_s = 2W$ is called the Nyquist rate.

2. If a signal has been sampled at Nyquist rate or greater ($f_s \geq 2W$) and the sample values are represented by weighted impulses, the signal can be exactly reconstructed from its samples by an ideal low pass filter (LPF) of bandwidth B, where $W \leq B \leq (f_s - W)$.

Practical sampling rate is 5-8 times the highest significant frequency contained in the signal and practical samples are not exactly impulses in time, but pulses of finite duration. It hardly makes any difference as far as signal reconstruction is concerned. The proof of the above theorem is easier in the frequency domain.

Let the band limited continuous-in-time signal has a frequency domain representation as shown in Fig. 12.5(a). If the above signal is "AND"ed with a rectangular pulse train, we get sampled signals of finite duration the frequency domain representation of which is as given in Fig. 12.5(b). Hence by fitting a low pass filter (shown dotted), the continuous- in-time signal can be recovered.

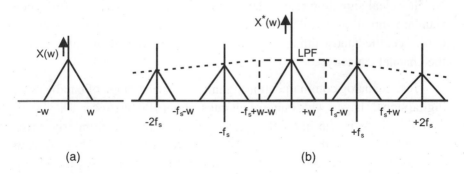

Fig. 12.5 (a) Message Spectrum and (b) Spectrum of the sampled signal

If any data is to be digitized for the purpose of data transmission, Shannon's sampling theorem should always be taken care of. If the signal to be transmitted is digital itself, like status signals, sampling theorem is of no consequence.

12.1.5 Bandwidth, Signal-to-Noise Ratio, Signaling Speed and Error Probability

Two important parameters for analog communication systems are bandwidth (BW) and signal-to-noise (SN) ratio. If in an amplitude modulated (AM) signal the bandwidth is no greater than 2W, where W is the message bandwidth, the post detection signal to noise ratio never rises above 30 to 40 decibel (dB). The decibel unit is defined as $10\log_{10}$ (S/N), where S and N are the signal and noise powers in watts respectively. The same analog signal when frequency modulated (FM), requires a bandwidth of 10 to 15 times message bandwidth. However, there is an attendant improvement in signal-to-noise ratio giving rise to clearer reception in FM broadcast. Thus we see that the bandwidth and signal-to-noise ratios are exchangeable with each other.

If the defining parameters in the analog communication systems are bandwidth and signal to noise ratios, the corresponding defining parameters in digital communication systems are signaling speed given in bit rate (bits/ sec), and the error probability (probability that a '1' gets changed to a '0', or vice versa). The Hartley-Shannon law puts an upper limit to the channel capacity defined by

$$C = B \log_2 (1+ S/N), \tag{1}$$

where B is the channel bandwidth and there is an average signal power constraint S. The noise is additive Gaussian in nature and the noise power is N. This ideal signaling rate is never reached, it is used as a performance standard and all practical channels work below this limit. Here also as we could see, the bandwidth and signal-to-noise ratio is exchangeable within the channel capacity.

Just as analog communication signals have wide range of bandwidths, digital signals have a wide range of bit rates, about 100 bits/sec in a keyboard to 10^9 bits/sec in the internal operation of a computer. Similarly, error probabilities vary in the range of 10^{-3} to 10^{-5} (one in a thousand bits to one in 1, 00,000 bits), 10^{-4} being a representative error probability. We would afterwards see that error probability is related to signal-to-noise ratio, S/N. Similarly, the signaling speed is related to the bandwidth,

$$s_{max} = kB, \ 1 \leq k \leq 2, \tag{2}$$

where k is the pulse shape factor as sharp pulses require larger transmission bandwidth.

However, digital data transmission could incorporate system refinements like data regeneration, error control coding and coherent detection (matching with the possible transmitted waveform) and scores above the analog communication systems. In analog communication systems, system regeneration could not eliminate noise, no error control is possible excepting cross correlation for fishing out small sinusoidal signals (e.g., radar echo embedded in noise), and synchronous detection could be used in some cases to improve signal-to-noise ratio to some extent.

12.2 Different Kinds of Modulation [1,2]

12.2.1 Amplitude Modulation

Input signals as they come out of the transducers, are not suitable for transmission. A carrier wave, which is better suited to transmission medium, is modulated by the signal for transmission. Thus modulation, *which in more general sense includes coding also*, is the alteration of a carrier wave by the message or modulating signal. Modulation is done for many reasons, which include ease of radiation, reduction of noise and interference, for channel assignment including multiplexing and to overcome equipment limitations. Thus for a 100 Hz signal to be radiated by a half wavelength antenna, the antenna required will be a few hundred kilometers across; whereas when the same is radiated in the range of 88-108 MHz, the antenna required will be hardly one and a half metre long. Other points will be clear as we proceed.

All modulation can be classified into three categories: amplitude (or linear) modulation, angle (or exponential) modulation and pulse modulation. The first two are continuous wave (c-w) modulations and the carrier is a sinusoid. For pulse modulation, the carrier is a pulse train. However, most pulse modulation system involve c-w modulation as a final step in transmission.

Linear or amplitude modulation is essentially direct frequency translation of the message spectrum; double sideband (DSB) modulation is precisely that. Minor modifications of the translated spectrum yield amplitude modulation (AM), single side band (SSB) modulation, vestigial sideband (VSB) modulation and compatible single sideband modulation. We start with conventional AM which is also historically the first to emerge, and then move on to others and study their *waveforms, spectra, detection methods, transmitters and receivers and system performance in the face of noise and interference.* We would assume an arbitrary message signal, x (t), which is an ensemble (collection) of all probable messages emanating from a given source. There is a definite upper limit to the frequency content of this signal, call it W and W is called the message bandwidth. This upper limit may be a few hundred Hertz in a telegraph signal to several megahertz in television (video) signals. Greater extremes are found in telemetry and computer systems.

Also, the message power x^2 (t) $\leq$ 1. Sometimes, for the purpose of analysis, we use tone modulation. A single tone refers to a single sinusoid: x (t) = A_m cos w_mt, $A_m \leq 1$, $w_m = 2\pi f_m$ with $f_m \leq$ W. Multiple tones are required for non-linear (exponential) analysis, where the principle of superposition does not hold.

12.2.1.1 Amplitude Modulation (AM)

In AM, the message signal is literally made to ride on the carrier. The modulated signal is

$$x_c \ (t) = A_c \ [1 + mx \ (t)] \ \cos \ w_c t \tag{3}$$

where A_c cos w_ct is the carrier and m is the modulation index. The modulated carrier amplitude

$$A_c \ [1 + mx \ (t)]$$

is a linear function of the message. The envelope of the modulated signal has the same shape as that of x (t) [Fig. 12.6(a) and (b)] provided the following two conditions are satisfied:

$$f_c >> W \text{ and } m << 1. \tag{4}$$

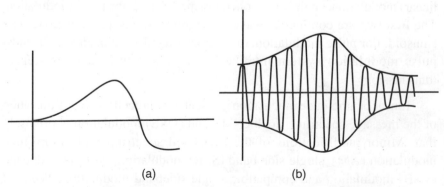

(a)   (b)

Fig. 12.6 (a) Message Signal and (b) modulated AM Signal

The particular waveform gives the ease of envelope detection. There is no phase reversal in the modulated signal. Any phase reversal will distort the envelope which could happen if m>1. The bandwidth of transmission, $B_T = 2W$. The carrier power is $P_c = 1/2 \, A_c^2$ .The carrier does not contain any information and a substantial part of the transmitted power, more than 50%, is wasted in the carrier. However the wasted power contributes to the ease of envelope detection. The peak power requirement of the transmitter is $4A_c^2$, as $x_{c\,max} = 2A_c$.

12.2.1.2 Double Side Band (DSB) Modulation

In the double side band suppressed carrier modulation, the carrier term is dropped. Doing away with the modulation index,

$$x_c (t) = x (t) \, A_c \cos w_c t, \tag{5}$$

The transmitter power, $P_T = 2P_{SB} = 1/2 \times A_c^2$ and

$$X_c (f) = A_c/2 \, [X (f-f_c) + X (f+f_c)] \tag{6}$$

DSB spectra and waveform, modulated and unmodulated, are shown in Fig. 12.7(a) and (b). There is a phase reversal as x (t) crosses through zero and though the envelope is proportional to | x (t) |, it is difficult to differentiate between positive signals and negative signals by looking at the envelope. The transmission bandwidth is still $B_T=2W$ and the peak power requirement is reduced to A_c^2.

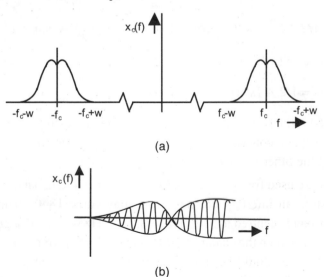

Fig. 12.8 (a) Double side band spectra and (b) double side band waveform

For detection of DSB, we do de-modulation, that is, a downward translation of frequency by multiplying the modulated signal with $A_{LO} \cos w_c t$, where A_{LO} stands for the local oscillator synchronised with the carrier. A low pass filter of cut-off frequency W, will retain

$$x_D(t) = K_D x(t) \tag{7}$$

This is what is known as *synchronous detection*. However, synchronizing the local oscillator to the carrier is not an easy task, particularly when the carrier is not present at all in the signal.

12.2.1.3 Signal Side Band (SSB) Modulation

The upper and lower side bands are symmetric about the carrier frequency and given the amplitude and phase of one, the other can always be constructed. Note that the amplitude has an even symmetry and the phase has an odd symmetry. Transmission bandwidth is cut into half if only one of the sidebands is transmitted. For SSB $B_T = W$ and $P_T = 1/4 \times A_c^2$.

SSB is readily visualized in the frequency domain. The time domain description of the waveform is difficult. With tone modulation,

$$x_c(t) = 1/2 A_m A_c \cos(w_c \pm w_m) t, \tag{8}$$

the +ve or -ve sign depending on whether the USB is retained or the LSB is retained.

$$= 1/2 A_m \cos w_m t A_c \cos w_c t - 1/2 A_m \sin w_m t A_c \sin w_c t \tag{9}$$

Generalising,

$$x_c(t) = 1/2 A_c[x(t) \cos w_c t - \hat{x}(t) \sin w_c t], \tag{10}$$

where $\hat{x}(t)$ is the Hilbert transform of $x(t)$. The interesting thing is that the modulated wave consists of two carriers of the same frequency, one in phase and the other in quadrature.

SSBs are used for voice telephony. Say, for 0-4 kHz bandwidth voice transmission, an intercity coaxial cable may carry 1860 channels with subcarrier bands centred at 312 kHz to 8,234 kHz with a 10% guard band provided in between the channels. SSBs have poor low frequency response as the spectrum should be hollow near the zero frequency to fit in the separating band pass filters.

12.2.1.4 Vestigial Side Band (VSB) Modulation

VSB is obtained by filtering DSB in such a fashion that one of the side bands is passed almost completely, while just a trace, *or vestige*, of the other side band is included. This vestige could be about 30%. They are good for transmission of signals with significant low frequency content. Examples are television (video), facsimile and high speed data signals. They conserve the bandwidth and, at the same time, avoid the poor low frequency characteristics of the single side band signals.

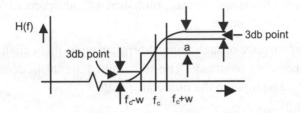

Fig. 12.9 Side band filter for vestigial sideband modulation

The design of the sideband filter is important (Fig. 12.9). If a = 0, we have DSB, while if a = 0.5, we have SSB. Try to visualize how the filter characteristic changes for either case.

12.2.1.5 Compatible Single Sideband Modulation

Instead of filtering DSB, we filter AM. The idea is to combine the ease of detection with bandwidth conservation by suppressed sideband. Adding a carrier term and a modulation index, the resultant modulated wave is

$$x_c(t) = A_c \{ [1+mx(t)] \cos w_c t \pm m q(t) \sin w_c(t) \}, \qquad (11)$$

where q (t) is the quadrature component. The envelope of x_c (t) is found in the usual fashion

$$r(t) = A_c \sqrt{[1+mx(t)]^2 + [mq(t)]^2} \qquad (12)$$

which is a distorted AM envelope. As more and more of one of the sidebands is allowed, envelope detection requires m « 1. Alternatively, if a sideband can be tolerated, but | q (t) | is small compared to [1+mx (t)] most of the time, a large modulation index can be used and a check is retained on the relative carrier power. In fact television video signals are transmitted as VSB plus carrier, the distortion can be sizable without detracting from picture quality and a transmission bandwidth only 30 percent greater than message bandwidth is required.

12.2.1.6 Transmitters: Switching Modulators

The active device, vacuum tube or the switch driven at carrier frequency closes every $1/f_c$ second. The load, called a tank circuit, *tuned to f_c*, rings sinusoidally. The steady state load modulation is then $e_p(t) = E \cos w_c t$. Adding the message, x (t), to the supply voltage, say via a transformer gives,

$$e_p(t) = [E + Nx(t)] \cos w_c t, \tag{13}$$

where N is the transformer turns ratio. If E and N are correctly proportioned the desired modulation has been accomplished without appreciable generation of undesired components.

The performance of such transmitters can be further enhanced through the use of negative feedback. The transmitted signal is demodulated to recover x (t). The input to the modulator is then

$p=[x(t) + \beta x_d(t)]$,
where β is the feedback factor. The closed loop gain is $A_f = A_o / [1+\beta A_o]$ where A_o is the forward loop gain. If $[1+\beta A_o] \gg 1$, $A_f = 1/\beta$ and performance is largely independent of the imperfections in the modulation circuitry (Fig. 12.11).

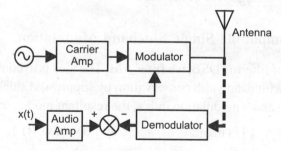

Fig. 12.11 Feedback incorporation in the modulator

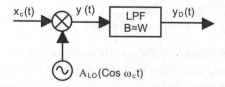

Fig. 12.12 Synchronous detection

The above is an AM modulation. A balanced modulator is used for DSB generation. Single side band or vestigial sideband modulation is produced when the output of the balanced modulator is processed by the sideband filter. In the same way, compatible SSB is produced from AM by sideband filtering.

12.2.1.7 Synchronous Detection

We have already discussed synchronous demodulation or detection, wherein the received modulated signal is multiplied by the local oscillator signal exactly in synchronism with the carrier (Fig. 12.12).

Let

$$x_c(t) = [k_c + k_m x(t) \cos w_c t + k_m q(t)] \tag{14}$$

On multiplication by $A_{LO} \cos w_c t$

$$y(t) = A_{LO}/2 \{[k_c + k_m x(t)] + [k_c + k_m x(t)] \cos 2w_c t$$
$$+ k_m q(t) \sin 2w_c t\} \tag{15}$$

After low pass filtering

$$y_D(t) = k_D[k_c + k_m x(t)] \tag{16}$$

The dc component $k_D k_c$ corresponds to the translated carrier if it is not suppressed in the modulation. The same can be blocked by a capacitor, or a transformer. As we see, synchronous detection or demodulation is a kind of downward frequency translation (as distinguished from upward frequency translation in modulation).

In DSB, the modulated signal spectrum and the noise spectrum are both shifted down to the zero frequency. However, the translated signal sidebands overlap in a coherent fashion, whereas the noise side bands sum incoherently. In the SSB, the translated positive frequency and negative frequency sidebands exactly make up the spectrum. In VSB, provided the sideband filters have characteristics as described earlier, the double sided spectrum overlap to make up the message spectrum (Fig.12.13(a) and (b)).

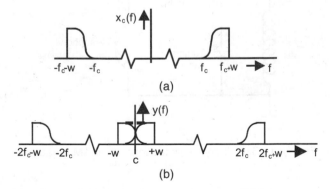

Fig. 12.13 (a) Modulated VSB Spectra and (b) VSB demodulation

The crux of the problem is to synchronize the local oscillator to a sinusoid that may not even be present in the incoming signal (if the carrier is suppressed). To facilitate synchronization, suppressed carrier systems may have a small carrier reinserted in x_c (t) at the transmitter (c.f., compatible single sideband modulation). The pilot carrier is picked off at the receiver by a narrowband filter, amplified and used in place of LO. This is the principle of *homodyne* (as distinguished from heterodyne) *detection*.

Nonetheless, some degree of phase and frequency drift is always present. For frequency drift, a pair of tones is produced in DSB while one tone is preserved in SSB though with a shift which may produce discord in music transmission. For phase drift, DSB vanishes for $\phi = \pi/2$ while slowly varying ϕ gives an apparent fading effect. Phase drift in SSB appears as delay distortion. Subjective listener tests have shown that for voice transmission via SSB, frequency drifts of 10 Hz are tolerable. Summarizing, phase and frequency synchronization requirements are modest for voice transmission by SSB. But in data, facsimile and video signals with suppressed carrier, careful synchronization is a must.

12.2.1.8 Envelope Detection

The simplest crystal detector set for AM broadcast is as shown in Fig 14. The crystal does rectification. The headset does not respond to the high (carrier) frequency and blocks the dc. Practical receivers based on envelope detection works on *super heterodyne principle* as shown in the block diagram in Fig. 12.15.

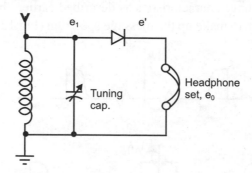

Fig. 12.14 Simple crystal detection set for AM broadcast

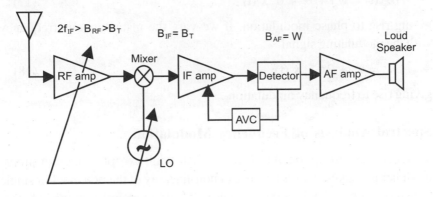

$2f_{IF} > B_{RF} > B_T$ $B_{IF} = B_T$ $B_{AF} = W$ Loud
Speaker

Fig. 12.15 Receiver based on super heterodyne principle

Note that if we are trying to receive a station at $f_c = f_{LO} - f_{IF}$, we also pick up $f_c' = f_{LO} + f_{IF}$, which is called the image frequency. The image frequencies are related by fc' -fc = $2f_{IF}$. The purpose of the radio frequency (RF) amplifier is to reject the image frequency. The intermediate frequency (IF) amplifier, which is fixed tuned, provides most of the gain and selectivity (from adjacent channels). The audio frequency (AF) amplifier which follows the detector provides the power level necessary for the speaker. The principle of mixing is known as heterodyning and because of the two stages of selection, it is a superhet receiver. An automatic volume control (AVC) is incorporated for signal variation.

12.2.2 Angle (Exponential) Modulation

In contrast to linear modulation, exponential modulation is a non-linear process. The modulated spectrum is not related to the message spectrum in a simple way. The transmission bandwidth required for effective angle modulation is much greater than twice the message bandwidth. Compensating for the bandwidth is the increased post-detection signal to noise ratio without increase in transmitter power. This is known as *wide band noise reduction* property.

Time Description of the Angle Modulation

$x_c (t) = A_c \cos \theta_c t.$

Let us explore the possibility of transmission of information x (t) by varying angle θ_c (t) of the carrier. There are two ways of doing the same: phase modulation (PM) and frequency modulation (FM). If angle θ_c (t) is varied linearly with x (t), then

$$\theta_c(t) = w_c t + \theta_o + \theta_d x(t) \tag{17}$$

giving rise to phase modulation. If we vary the instantaneous frequency with the modulating signal

$$w_i = w_c + 2\pi f_d x(t) \tag{18}$$

giving rise to frequency modulation.

Spectral Analysis of Frequency Modulation

An exact description of FM spectra is difficult except for certain simple modulating signals. This is because of non-linearity of the process. For single tone modulation, that is, $x(t) = A_m \cos w_m t$, instantaneous frequency $f_i(t)$ is given by $f_c + f_d A_m \cos w_m t / 2\pi f_m$. The FM modulation index β is defined as $[A_m f_d / f_m]$, where f_m is the frequency of the single tone.

FM Bandwidth

FM requires infinite bandwidth for transmission. But practical FM systems having finite bandwidth do exist and perform quite well. This is because sufficiently far away from the carrier frequency, the spectral components are small and can be discarded. Omitting any portion of the spectrum will cause distortion, but distortion can be minimized by retaining all significant spectral components.

Significant Sideband Lines: $J_n(\beta)$, the coefficient of the spectral lines, falls off rapidly for $J_n / \beta > 1$, particularly if $\beta >> 1$. Assuming that the modulation index β is large, we can say that $| J_n(\beta)|$ is significant only for $|J_n| > \beta = A_m f_d / f_m$. Then all significant sideband lines are contained in the frequency range $[fc \pm A_m f_d]$. If $\beta << 1$, the significant sideband lines are fc $\pm$ fm.

We shall estimate B_T from the worst case tone modulation analysis assuming that any component of $x(t)$ smaller in amplitude and frequency will require a smaller bandwidth than this B_{max}. This is given by Carson's rule of thumb:

$$B_T = 2(D+1)W, \tag{18}$$

where D, the deviation ratio, is defined as equal to f_d / W, or better still by

$$B_T = 2(D+2)W \tag{19}$$

For commercial FM: $f_d = 75$ kHz and W = 15 kHz, so D = 5, which gives $B_T = 210$ kHz. The FM broadcast is in the range of 88-108MHz. The fractional bandwidth is (approximately) still 2×10^{-3}.

FM Generation and Detection Methods

Conceptually, direct FM requires nothing more than a voltage controlled oscillator (VCO) whose frequency has a linear dependence on applied voltage. This is readily implemented in the microwave band ($f_c = 1$ GHz), where devices such as Klystron tube have linear VCO characteristics over a substantial frequency range, typically several megahertz. If a lower carrier frequency is desired, the modulated signal can be down converted by heterodyning with the output of a fixed frequency oscillator, often another Klystron. At lower frequency, phase locked loop (PLL) chip could also be used for frequency modulation.

In an indirect frequency modulation system, the signal is integrated and then phase modulated by a stable crystal oscillator. The narrow band FM is frequency multiplied by several frequency doublers and triplers. Usually, the deviation ratio is kept small to avoid distortion inherent in NBFM. Finally, the spectrum is down converted by mixing with another local oscillator (LO) to bring it in the required frequency band.

FM Detection / Frequency Discriminators

In a balanced discriminator (Fig. 12.16), two resonant circuits are employed -one tuned above f_c and the other below f_c. Thus as f_i changes, the amplitude variations are in opposite directions and taking the difference gives the frequency-to-voltage characteristics, the typical S-curve.

They are readily adapted to microwave band, with resonant cavities serving as tuned circuits and crystal diodes for envelope detectors. The dc component is automatically cancelled; no separate dc block is necessary but the low frequency response is preserved. Alternatively, phase locked loops (PLLs) could be used as frequency discriminators.

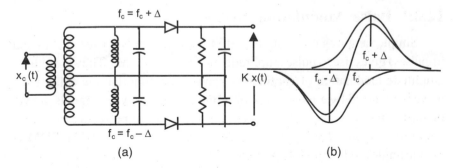

Fig. 12.16 (a) Circuit for balanced discriminator and (b) Typical S-curve

Most FM receivers are of super heterodyne variety. They differ in two respects from AM superhet receivers: i) a limiter discriminator (limiter to clip any change in amplitude variation) replaces the envelope detector and (ii) automatic frequency control (AFC) in place of AVC is provided to correct the frequency drift of LO. Commercial FM radios have tuning ranges of 88-108 MHz, f_{IF} = 10.7 MHz and IF bandwidth of 200-300 kHz. Note that the fractional bandwidth of the IF amplifier is about 2×10^{-2}, the same as for AM super heterodynes where f_{IF} =455 kHz and B_{IF}=10 kHz.

Noise Performance: As far as noise performance of FM is concerned, including the de-emphasis, (S/N)D = 640z. Other factors being equal, a 1-watt FM system could replace a 640-W base band or AM system with no reduction in signal-to-noise ratio. However, we are always considering r_n « A_c. FM also has a threshold effect and it comes in about 13 dB. FM improvement above threshold is quite impressive. Sometimes, threshold extension methods can be used to lower the threshold for strategic communication systems.

FM Applications: FM is used for manifold applications: Commercial FM and television audio, the latter being transmitted on a separate carrier with a deviation of 25 kHz. At microwave frequencies, both noise reduction and constant amplitude properties are advantageous. With threshold extension technique, FM is employed in satellite and space probe communications, where transmitter power is at a premium. The balanced discriminator has excellent low frequency performance and FM can equal DSB or VSB without bothersome synchronization problem.

FM sub-carrier modulation is used in telemetry for less active input signals. For same reasons, high quality magnetic tape recorders are equipped with FM mode recording. Unstable transmission and multi-path propagation often preclude wideband modulation, ionospheric radio being a case in point.

12.2.3 Pulse Modulation System

Simple sampling of the signal with finite width pulses (instead of ideal impulses) gives rise to pulse amplitude modulation (PAM). They are akin to amplitude modulation of continuous wave, excepting that the continuous wave is replaced by a pulse train, or a burst of pulses when signals arrive. Instead if the message signal varies the width of the pulse, we get pulse width modulation (PWM) or pulse duration modulation (PDM). PDM is akin to angle modulation of continuous wave.

Whatever information is contained in pulse duration modulation is also contained in pulse position modulation (PPM), wherein the pulses in PDM or PWM modulation is differentiated and we get two spikes, one at the leading edge of the pulse and the other at the trailing edge of the pulse. If the pulse width is varied in one direction only, the leading edge is fixed and occurs at uniform intervals, while the trailing edge position varies depending on the signal amplitude. PPM (with one side pulses) contains all the information that is necessary for signal recovery.

The above three are analog pulse modulation methods and have hardly ever been used. Whereas PAM is part of the sampling theorem, PWM has been used in controlling the thyristor chopper voltage, PPM has sometimes been used in telemetry because they form low energy pulse bursts with regular pulse shapes undergoing small distortion during travel.

What is more important is the digital pulse modulation system. In pulse code modulation, the sampled analog signal is coded into a fixed number of well formed pulses, that is, converted into a digital word consisting of a number of binary digits (bits) and transmitted as such.

12.2.3.1 PCM Generation System

Pulse code modulation (PCM) being radically different from whatever we have read so far, we start with a description of PCM generation system. PCM generation consists of sampling, quantizing and coding in that order (Fig. 12.17).

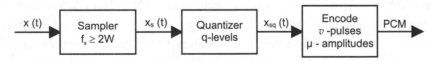

Fig. 12.17 PCM generation system

The sampled and quantized wave $x_{sq}(t)$ is discrete both in time (by virtue of sampling) and amplitude (by virtue of quantization). Let v be the number of pulses (digits) in each code group (digital word), each having one of the μ discrete values. There are μ^v different combinations of pulses with q amplitudes. We require $\mu^v \geq q$ for unique coding or $v = \log_\mu q$. The most commonly used form is $\mu = 2$ (binary) and $q = 2^v$.

Because several coded pulses are required for each message sample, the PCM bandwidth will be much greater than message bandwidth. Quantized samples occur at the rate of $f_s \geq 2W$ samples per second. So there must be vf_s codes sent per second.

12.2.3.2 PCM Receiving System

A PCM receiving system is as shown in Fig. 12.18. The first task of the decoder is recognition of the binary digits. This can be done with a decision circuit whose crossover is set at zero volts. When the waveform is positive, it is taken to be a '1', etc. Clearly, the decision should be made at the centre of each pulse for maximum reliability. Hence, synchronizing signals are usually supplied to trigger the decoder at the optimum times.

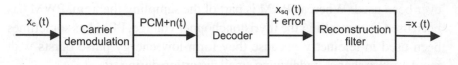

Fig. 12.18 PCM receiving system

As long as the magnitude of the noise is less than pulse amplitude, things proceed quite smoothly. But, occasionally the random noise peaks will cause the decoder to make a wrong decision, resulting in error. The frequency with which such error happens depends on the relative strength of the noise and its statistical properties. The probability of decoding errors is a function of $A/\sqrt{N}$, assuming the noise to be zero mean and $\sigma = \sqrt{N}$. The errors decrease rapidly with increased pre-detection signal-to-noise ratio, $(S/N)_T = A^2/N$.

With $(S/N)_T = 16$, the error probability is less than 10^{-4} or 1 digit in 10,000 will be in error. Therefore provided $(S/N)_T$ exceeds some error threshold, the random noise has virtually no effect on system performance. But if $(S/N)_T$ drops below the threshold, the severity of the error will depend on where the error occurs (MSB or LSB position). As a rule of thumb, mutilation is negligible if error probability is about 10^{-4} or less.

There is off course error detection and error correction codes (with redundancy added to the coded signal) to detect or correct errors in digital data transmission. Binary PCM is less vulnerable to errors than μ-ary PCM ($\mu > 2$), with fixed $(S/N)_T$. The spacing between the levels (relative to the rms noise) decreases with increasing μ and hence the error probability increase.

12.2.3.3 Quantization Error

Despite the fact that decoding errors can be ignored, the final output at the receiver will not be identical to the original message - recall that encoded signal is x_{sq} (t) and not x_s (t). When $x_{sq.}$ (t) is low pass filtered, the reconstructed signal differs from x (t) because the quantized samples are not the exact sample values. Furthermore there is no way of obtaining exact values at the receiver, that the information was discarded at the transmitter in the quantizing process. Therefore, perfect message reconstruction is impossible in pulse code systems, even when random noise has negligible influence. This is a basic limitation of coded systems, just as random noise is a limitation of conventional analog systems.

Noise reduction in PCM is an exponential exchange of BW for S/N ratio. This exchange is more dramatic then that of wideband analog modulation, where $(S/N)_D$ increases linearly, or as square of the BW ratio. Today's compact discs (CDs) use digital mastering for audio and video reproduction. In practice, a *tapered* quantization is accomplished with uniformly spaced level, the message being non-linearly compressed prior to sampling; a complementary expansion process restores the wave shape. The combined process of compression and expansion is called companding.

12.2.3.4 Delta Modulation

In differential pulse code modulation (DPCM), the successive bit strings sent as sample measures are the difference bits between the previous and the succeeding sample representation. This helps in preserving the bandwidth as the bits required per sample becomes less in number. In the extreme case, DPCM evolves into delta modulation (Δ-M).

Delta modulation is a single bit pulse code modulation. With conventional PCM modulation, each sample of the analog signal is converted into a digital word consisting of several (binary) digits. With delta modulation, rather than multiple bit representation of the sample, a single bit is transmitted to indicate whether the current sample is larger or smaller than the previous sample. If the sample is larger a '1' is sent; if it is smaller, a '0' is sent.

A delta modulation transmitter consists of the blocks shown in Fig. 12.19. The input analog signal is compared to the output of the DAC. The DAC equals the magnitude of the previous sample. The up-down counter is incremented or decremented depending upon whether the current sample is larger than or smaller than the previous sample. The output of the comparator (a '1' or a '0') constitutes the delta modulated output.

A delta modulation receiver works much in the same way as the transmitter except that the comparator is no longer needed and the output of the DAC (which is a pulse amplitude modulated wave) is low pass filtered to give the received continuous signal. The delta modulation receiver is shown in Fig. 12.19(b). However there are two problems associated with delta modulation, those of slope overload and granularity.

Slope Overload happens when the analog signal changes at a faster rate than the DAC can keep up with. Granularity is the problem when the analog signal has a very gentle slope and the DAC keeps oscillating around it falling into a steady pattern of increment decrement cycle.

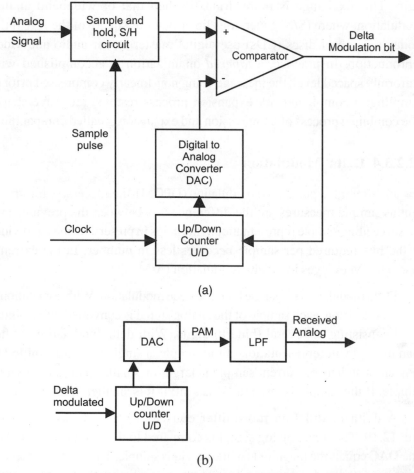

(a)

(b)

Fig. 12.19(a) Delta modulation transmitter and (b) Delta modulation receiver

Adaptive Delta Modulation: In adaptive delta modulation, the step size of the DAC is automatically adjusted depending on the slope of the analog input signal. A sequence of three consecutive 1s or 0s is normally taken as indicative of the DAC loosing track of the input signal and the step size is adjusted by a factor of 1.5 to catch up with the analog signal. Various other algorithms have been used for adaptation. Adaptive delta modulation reduces both distortion due to slope overload (problem of a steep slope of the input analog signal) and granular noise (a problem of gradual slope in the input signal).

12.3 Diverse Topics

12.3.1 Multiplexing

When several messages are transmitted using the same transmission facility - be it a pair of wires, co-axial cable, electromagnetic radiation, optical fibre or satellite channel - the process is called multiplexing. Multiplexing could be for a telephone network, FM stereo, a space probe telemetry system and variety of other purposes. The two basic multiplexing techniques are: frequency division multiplexing (FDM) and time-division multiplexing (TDM). They may be used together but at different hierarchical levels.

12.3.1.1 Frequency Division Multiplexing

In frequency division multiplexing, several signals are sent simultaneously over the same channel, but they remain distinctly separate in the frequency domain. The principle of FDM could be, say, where three input messages modulate the sub carriers f_{c1}, f_{c2} and f_{c3}. If the sub carrier modulation is SSB, the modulated signals remain separate in the frequency domain, though they are hopelessly jumbled up in the time domain. The modulated signals are then summed up to produce the base band signal with the spectrum $x_b(f)$.

The modulated sub carriers are separated by band pass filters in parallel, following which the messages are individually demodulated and detected. The major practical problem of FDM is cross talk, that is, coupling of one message with the other. Intelligible cross talk arises primarily because of nonlinearities in the carrier modulation system. Unintelligible cross talk occurs due to imperfect spectral separation by the filter bank. The modulated message spectra are provided with guard bands in to which the filter roll off characteristics is fitted.

In FM stereophonic broadcasting, left speaker and right speaker signals are first matrixed to produce (L+R) and (L-R) signals. The (L+R) signal is directly inserted into the base band (of 15 kHz bandwidth), while (L-R) DSB modulates a 38 kHz sub carrier derived from a 19 kHz supply. The 19 kHz pilot tone is for receiver synchronization, and the (L+R) signal only is heard in a monophonic receiver. The DSB is employed to preserve fidelity at low frequencies. Off course, the deviation ratio for FM transmission is now much reduced.

12.3.1.2 Time Division Multiplexing (TDM)

TDM is the technique of transmitting several messages on one facility by dividing the time domain into slots. Unlike FDM, the signals remain distinctly separate (interleaved) in time domain whereas they are mixed up in frequency domain. Evidently, FDM is used for analog signals, whereas TDM is used for discrete-in-time or pulsed, signals.

The essentials of TDM are quite simple. Several signals, all band limited in W, are sequentially sampled by a commutator. The switch makes at least one complete revolution in 1/2W sec, extracting sample from each input. The commutator output is a PAM waveform containing the individual message samples interlaced in time. A set of pulses containing one sample from each input is called a frame. Actually, TDM synchronization is not a problem in the sense that if you can identify a frame (usually by a separate frame pulse), the individual message samples can be recovered in the sequence they are obtained.

Compared to FDM, TDM instrumentation is simpler. Further, TDM is invulnerable to the usual sources of FDM inter-channel cross talk because of imperfect channel filtering and cross modulation due to nonlinearities. In fact there is no cross talk if the pulses in TDM are non-overlapping since message separation is obtained by gating in time. Cross talk due to pulse shape imperfections, gradual roll off kind of pulses at the receiving end etc., can be effectively reduced by providing guard times between pulses.

In *statistical* time division multiplexing, the time slots are allotted depending on the requirement and availability. It requires separate destination header to be added to the message.

Similarly, in *spread spectrum communication* the frequency slots in a large band are statistically allotted for *cellular telephony*, originally the spread spectrum technique was meant for military purposes of frequency hopping though.

12.3.2 Digital Data Communication System [3]

So far we have confined ourselves to analog signals only. In PCM, the analog signal is sampled and digitized to form a discrete-in-time system. However, to-day enormous quantities of direct digital data are being generated in government, commerce, science and industries. The data is mostly alpha-numeric {combination of alphabets and numerals} but they are represented by binary code groups, e.g., Hollerith, ASCII 1 or 2, EBCDIC, etc. The need to eliminate manual handling and human error, has led to electronic data processing systems. These developments coupled with decentralization have made the transmission and distribution of digital data on a *network*, a major task of electrical communication.

The design of the practical data system is based on two elementary considerations: (1) the gross source rate in bits per sec and (2) desired transmission reliability, that is, the tolerated error rate. These factors dictate signaling speed and error probability requirement. In fact as mentioned earlier, these two are defining parameters of digital systems, equivalent to bandwidth and signal-to-noise ratio for analog systems.

In the analog transmission, the goal is to reproduce the actual message waveform as closely as possible. In digital transmission, the goal is to deliver the message information as accurately as possible. Since a digital message (data string) consists of a finite number of different symbols (0s and 1s in case of binary coded data), the designer is not bound to any waveform. The receiver has a dictionary.

There are three basic forms of digital modulation systems:

   (i) amplitude shift keying (ASK), or equivalent to AM,

  (ii) frequency shift keying (FSK), or equivalent to FM, and

 (iii) phase shift shift keying (PSK), or equivalent to PM.

All other higher level m-ary transmission methods are variations, or combinations of the above. Minimum shift keying (MSK) is a variation of FSK in which the frequencies are commensurate with the bit rate, making the transitions without sharp discontinuities.

12.3.3 Hybridization, Networking and Multiple Multiplexing

In the next few subsections, we introduce interfacing between analog and digital systems either way, data networking facilities in the public and private domain and higher capability multiplexing methods, called multiple multiplexing.

12.3.3.1 Data Modems

The data modems, also called DCE, a dataset, or data phone or simply a modem, interfaces the digital terminal equipment to an analog communication channel. At the transmit end the modem converts the digital pulses from the serial interface to analog signals, and at the receive end, the modem converts the analog signal to digital pulses. Note that the word modem is a combination of modulator and demodulator. Similarly, codec is a combination of coder and de-coder, transceiver is a combination of transmitter and receiver and can work either in 'transmit' mode or 'receive' mode.

Modems are generally classified either as asynchronous or synchronous and use either FSK, PSK or QAM modulation. With synchronous modems clocking information is essential to be generated and recovered in the receive modem. Asynchronous modems use PSK modulation and are restricted to low-speed applications. Synchronous modems using PSK or QAM modulation are used for medium and high speed applications.

Asynchronous modems are primarily used for low speed dial-up circuits. It could have half-duplex operation with two wire network, or full duplex with four-wire private line circuits. To operate full duplex with a two-wire dial-up circuit, it is necessary to divide the usable bandwidth of a voice band circuit in half creating two equal-capacity data channels. In Bell System 103 modem, low-band channel occupies a pass-band from 300 to 1650 Hz and high-band channel occupies a pass band from 1650 to 3000 Hz. The mark and space frequencies for the low-band channel are 1270 and 1070 Hz, and the same for high-band are 2230 and 2030 Hz

Because of the clock and carrier recovery circuit, a synchronous modem is more complicated and expensive than its asynchronous counterpart. QPSK is used with 2400 bps data modems and 8 PSK is used with 4800 bps modems. They can operate over two-wire dial-up circuits but only in the simplex mode. Off course, half duplex is possible. High speed synchronous modems operating at 9600 bps use 16 QAM modulation. 16 QAM has a bandwidth efficiency of 4 bps/Hz, the baud rate and minimum bandwidth for 9600 bps modems are 2400 baud and 2400 Hz. The synchronous modem is designed for full duplex operation on 4-wire private line circuits. Today the capabilities of the modems are hugely increased and a 56 kbps modem is quite common.

An asynchronous data format is used with asynchronous modems and a synchronous data format with synchronous modems. However, asynchronous data is occasionally used with synchronous modems (isochronous transmission); synchronous data is never used with asynchronous modems.

12.3.3.2 CODEC

A Codec is a large-scale integration (LSI) chip designed for use in the telecommunication industries for private branch exchanges (PBXs), central office switches, digital handsets, voice store and forward systems and digital echo suppressors. Essentially the codec is applicable for any purpose that requires the digitizing of analog systems, such as in a PCM-TDM carrier system.

'Codec' is a generic term that refers to coding and decoding functions performed by a device that converts analog signals to digital codes and digital codes to analog signals. Recently developed Codecs are called 'Combo' chips because they combine codec and filter functions in the same LSI package. The input/output filters perform the following functions: analog sampling, encoding/decoding and digital companding. The 2913/14 combo chip provide the analog-to-digital and digital-to-analog conversions and the transmit and receive filtering necessary to interface a full duplex (four wire) voice telephone circuit to the PCM highway of a time-division multiplexed (TDM) carrier system.

12.3.3.3 Public Data Networks (PDN)

A public data network (PDN) is a switched data communication network similar to public telephone network except that PDN is designed for transferring data only. Public data networks combine the concept of value added networks (VANs) and packet switching networks. A value added network 'adds value' to the service like error control, enhanced connection reliability, dynamic routing, failure protection, logical multiplexing and data format conversions. The organization may have communication lines from carriers like AT&T and add new type of communication services like e-mail facility and develop VANS like GTE, Telnet, etc.

The common switching techniques used with PDN are three in number: circuit switching, message switching and packet switching. Circuit switching is used for making a standard telephone call on the public telephone network. The call is established, information is exchanged in real time, and then the call is disconnected. In the duration of the call, the circuit is not available to others. Message switching is a form of store and forward network. Data, including source and destination identification codes, are transmitted into the network and stored in a switch. The network transfers the data from switch to switch when it is convenient to do so. There is a delay time between message transmission and message reception. A message switch does more than simply transfer the data. It can change the format and bit rate of the data for transmission, and then convert back the data to its original form or an entirely different form at the receiving end.

With packet switching, the data is divided into smaller segments called packets prior to transmission. Because a packet can be held in memory at a switch for a short period or time, packet switching is sometimes called a hold and forward network. With packet switching, a message is divided into packets and each packet can take a different path through the network. Consequently, all packets do not necessarily arrive at the receive end at the same time or in the same order in which they are transmitted. However, because packets are small, the hold time is generally quite short and message transfer is in near real time and blocking cannot occur.

However, packet switching networks require complex and expensive switching mechanism and complicated protocols. A gateway is used to interface two public data networks. In fact, in ISO (International Standards Organisation) protocols there are seven hierarchical layers (each layer adds value to services provided by the set of other layers) to run a distribution network; the layers from bottom up are: Physical layer, Data Link layer, Networks layer, Transport layer, Session layer, Presentation layer and finally, the Application layer for open system interconnection (OSI). If all these layers are addressed, less than 15% of the transmitted message is source information. Today a simpler four layer architecture is in use using TCP/IP protocol.

12.3.3.4 Local Area Network (LAN)

A local area network (LAN) is a data communication network that is designed to provide two-way communications between a large variety of data communication terminals within a small geographical area. LANs are privately owned and operated and are used to interconnect data terminal equipment in the same building or building complex. The topology and physical architecture of a LAN identifies how the stations are interconnected. The most common configurations used are star, bus, ring and mesh topologies. Presently all LANs use co-axial cables as the transmission medium, although it is likely that in near future fibre optic cables will replace the co-axial cables. Sometimes, a separate fibre optic *backbone* with higher traffic volume handling capability is added to the network.

There are two basic transmission formats: *base band* and *broadband*. Base band transmission uses the connecting cable as a single channel device; however time-division multiplexing is possible. Broadband transmission uses the connecting medium as a multi-channel device. Each channel occupies a different frequency band (frequency division multiplexed). The different channels can contain different encoding schemes and operate at different bit rates. A broadband network permits voice, digital data (could be text), and video to be transmitted simultaneously over the same transmission medium. However, broadband instrumentation will be more complex. There are different ways of channel accessing also; it could be carrier sense multiple access with collision detection (CSMA/CD), token passing either for bus topology or ring topology. There is no end to the vista. And today it is a world wide web (www) for multimedia based on hypertext transmission protocol (http) and dependent on access to Internet.

12.3.4 Multiple Multiplexing

CCITT (Comite Consultif Internationale Telephonique et Telegraphique), later changed to International Telecommunication Union - Telecommunication Standards sector (ITU-T), gives the frame alignment for PCM-TDM system as follows. A 125 μs frame is divided into 32 equal time slots. Time slot 0 is used for frame alignment pattern and for an alarm channel. Time slot 17 is used for a common signaling channel. The signaling for all the voice band channels is accomplished on this common signaling channel. There are 30 voice band channels time division multiplexed into each CCITT frame. With the CCITT standard, each time slot bas 8 bits. Total number of bits per frame is 256 bits and the line speed 2.048 Mbps. North American digital hierarchy is different where in T3 line type there are 672 channels (transmitting at 44.736 Mbps) carrying voice band telephones, picture phones and broadcast quality television.

Similarly in FDM the basic voice band circuit is a 3002 channel of 300 to 3000 Hz which can be subdivided into 24 narrower 3001 (telegraph) channels. The same voice band frequency (0-4 kHz) can be used for data transmission. A group is the next higher level in the FDM hierarchy: a basic group consists of 12 voice band channels stacked together (off course frequency division multiplexed). The 12-channel group is standard building block for most broadband communication systems. Five such groups are combined to form a super group. Ten such super groups are combined to form a master group. A radio channel comprised three master groups. The system can be increased from 1800 voice channels to 1860 by adding an additional super group.

With hybrid data it is possible to combine digitally encoded signals with FDM signals and transmit them as one composite base band signal. There are four primary types of hybrid data: data under voice (DUV), data above voice (DAV), data above video (DAVID) and data in voice (DIV).

12.3.5 Transmission Media

The transmission medium could be a wire pair, co-axial cable, fibre link or free space. Accordingly one has wire or wireless communication. More explicitly, the links could be just a pair of wires, power line carrier, co-axial cables for LAN network to city-to-city trunk dialing facilities, fibre optic links, electromagnetic radiation - directed or undirected (through antenna or satellite up link and down link converters). We discuss a few of them below, either because they are unusual, or because they are new and comparatively promising technologies.

12.3.5.1 Power Line Carrier Communication [4]

In electric utilities, the high voltage power lines (forming a grid) could also be used for communication purposes. Though the power at 50/60 Hz and the communication signals at 150 kHz or above travel over the same transmission lines, they are assiduously separated out before they enter a power station or substation. The equipment for separation or for making the communication signal ride the power channel is quite elaborate and is illustrated in Fig. 12.20. Besides communication, power line carrier is also used for phase comparison relaying in the distance relay.

The line trap removes any vestige of the carrier signal going into the station bus. The carrier signal is coupled to the transceiver through a high voltage (insulating) capacitor. Any power line signal coming through this coupling is effectively grounded by the choke coil. The transmission lines carry a combination of power and carrier signal (> 150 kHz). They are well removed in frequency with respect to each other. Thus effective grid communication is established for voice frequency, supervisory control and data acquisition (SCADA) system and power line protection (relaying). Power line carrier communication eliminates the necessity of laying down communication cables for hundreds of kilometres of transmission distance.

Outdoor

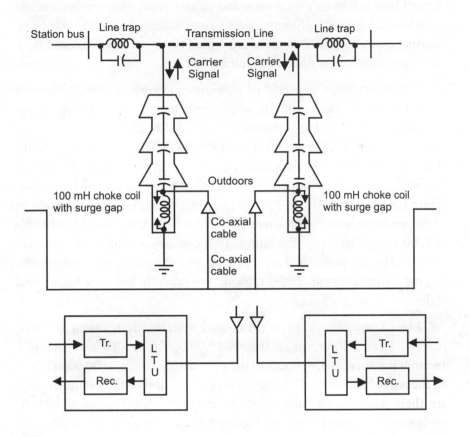

Fig. 12.20 Power line carrier communication system

Indoor

Circuits in similar lines are now being contemplated by Internet service providers (ISPs) for carrying Internet data to households through normal power supply lines.

12.3.5.2 Fibre Optic Links for Communication

The last decade has seen fantastic increase in communication. A phenomenal increase in voice, data and video communication has caused an increasing demand for more economical and larger capacity communication systems. Terrestrial microwave systems have long since reached their capacity and satellite systems can at best be the bridges for long-haul systems. Communication systems that use light as the carrier of information have recently received a great deal of attention. Systems that use glass or plastic fibres to contain a light wave and guide it from source to destination are becoming increasingly important. Communication systems that carry information through a fibre cable are called fibre optic systems and they constitute today's fibre optic communication.

The information capacity of a communication system is directly proportional to its bandwidth: the wider is the band width, the better is the information highway. For comparison purposes, we always talk of relative bandwidth. A VHF radio system operating at 100 MHz, say, has a bandwidth of 20 MHz (Recall commercial FM radio operate in the band 88-108 MHz). Its useful relative bandwidth (expressed as a percentage of mid-frequency, approximately 100 MHz) is 20%. A microwave radio system operating at 6GHz having that much of relative bandwidth will have a useful bandwidth of 1200 MHz. Definitely, the information capacity or number of channels that could be accommodated even with wideband modulation is much more and the frequency multiplexed information capacity becomes higher and higher.

Light frequencies used in fibre optic systems are between 10^{14} and 10^{15}Hz. Even 10 percent of that is 1, 00,000 GHz. To meet today's communication needs, or needs of the foreseeable future, 1, 00,000 GHz is an excessive bandwidth. The other plus points of fibre optic communication are their relative immunity to electromagnetic interference (EMI), or environmental effects (compared to their metallic counterparts). They are lighter, intrinsically safe and cannot be tapped as easily. The only significant disadvantage is higher initial cost of installing a fibre optic system.

The three primary building blocks are the transmitter, the fibre guide and the receiver. The light source is either a light emitting diode (LED) or an injection laser diode (ILD). The amount of light emitted is proportional to current. The source to fibre coupler is a mechanical interface. The optical fibre consists of glass or plastic fibre core, a cladding and a protective jacket. The fibre to light detector is also a mechanical coupler. The light detector is very often either a PIN (positive- intrinsic-negative) diode or an APD (avalanche photo diode). Both the PIN and the APD diode convert light energy to current. Hence a current-to-voltage converter is used. The modulation could be done by analog signal direct also, but then impedance matching and limits of input signal amplitude become matter of concern at either end.

12.3.5.3 Satellite Communication (Transponder)

Essentially a satellite is a radio receiver in the sky. The word 'transponder' relates to transmitter and responder. The system consists of a transponder, a ground station to control its operation and a user network of earth stations that provide the facilities for transmission and reception of communications. Satellite transmissions are categorized as either bus or payload. The bus includes control mechanisms that support the payload operations. The payload is the actual user information that is conveyed through the satellite system.

Satellite System Link Models

Essentially a satellite link model consists of three basic sections, the uplink, the satellite transponder and the downlink.

Uplink: The uplink section of the satellite system is the earth station transmitter. It consists of an IF modulator, an IF to RF microwave up converter, a high power amplifier (HPA) and an output band pass filter. The IF modulator converts the input base band signal to either an FM, a PSK or a QAM modulated intermediate frequency. The up converter (mixer and band pass filter) converts the IF to an appropriate RF carrier frequency. The HPA provides adequate sensitivity and output power to propagate the signal to the satellite transponder. HPA commonly used are Klystron and travelling wave tubes.

Transponder: A typical satellite transponder consists of an input band limiting device (BPF), an input low noise amplifier (LNA), a frequency translator, a low level amplifier and an output band pass filter. This transponder is an RF-to-RF repeater. The output of the LNA (A common device used as an LNA is a tunnel diode) is fed to a frequency translator (a shift oscillator and a BPF) which converts the high-band uplink frequency to the low band downlink frequency. The low level power amplifier, which is commonly a travelling wave tube, amplifies the RF signal for transmission through the down link to the earth station receivers. Each RF satellite channel requires a separate transponder.

Downlink: An earth station receiver includes an input BPF, an LNA, and an RF-to-IF down converter. Again, the BPF limit the input noise power to the LNA. The LNA is a highly sensitive, low noise device such as a tunnel diode amplifier, or a parametric amplifier. The RF-to-IF down converter is a mixer / band pass filter combination which converts the received RF signal to an IF frequency.

Crosslink: Occasionally, in an application it is necessary to communicate between satellites. This is done using satellite cross links. A disadvantage of inter-satellite link (ISL) is that both the transmitter and receiver are space bound. Consequently, both the transmitter output power and the receiver input sensitivity are limited.

Satellites used for communication are geo-stationary satellites. The C-band transponder works in the range of 3.4 to 6.425 GHz and the K-band transponder works in the range of 10.95 to 14.5 GHz.

12.4 References

1.   A Bruce Carlson: Communication Systems, An Introduction to Signals and Noise in Electrical Communication, International Student Edition, McGraw Hill-Kogakusha, Tokyo, 1968.

2.   Wayne Tomasi: Electronic Communication Systems, Fundamentals through Advanced, Prentice Hall, NJ, 1985.

3.   S Haykin, Digital Communications, John Wiley, 1988.

4.   Electrical Transmission and Distribution Reference Book, Westinghouse Electric Corporation, Oxford Book Company, 1950.

   For further reading on data communication, the following two books may be referred:

5.   William Stallings, Data and Computer Communications, 5th edition, Prentice Hall India, 1998.

6.   B A Forouzan, Data Communications and Networking, Tata McGraw Hill, 2000.

Chapter 13

High Order Filter Design

A. Barua

Abstract

Active filters consist only of operational amplifiers, resistors and capacitors. Complex roots are realized by the use of feedback, dispensing of inductors. Many signal conditioning applications require filters of order greater than two. One of the ways to realize a high order filter is to cascade second order filters. If the order of the filter is odd a first order section will be necessary. All modern signal conditioning system include various types of continuous time filters that the designer has to realize in an appropriate technology. In this chapter we present the steps for design of high order active RC filters.

13.1 Introduction

The synthesis of a filter is a three stage problem. The task of a filter designer is to determine (a) the filtering operation required, (b) the filter topology to be implemented and (c) the network component values of the filter circuit.

The filter design starts with the specifications such as passband and stop-band corner frequencies, maximum attenuation tolerated in the passband and minimum attenuation required in the stopband. There are different approaches to realize a given high order transfer function using active RC circuits, each with its own merits and demerits. Most high order filter realizations use first order and second order transfer functions as building blocks. A large number of circuits are available to implement a second order transfer function. These second order circuits show widely varying performance on different performance measures. Hence for any given requirements, one of the circuits will be more suitable than the others.

13.2 Approximation Function

Let us consider the ideal lowpass gain characteristics which is shown in Fig. 13.1. The amplitude of the transfer function is constant from $\omega = 0$ to $\omega = \omega_p$ and zero for all frequencies greater than ω_p. This type of characteristic is called brick-wall type of response. However such a stringent requirement cannot be realized with practical design constraints. The only way we can come out of this brickwall type response is to approximate the transmission characteristic.

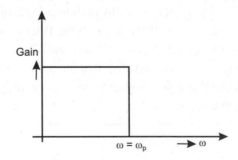

Fig. 13.1 Ideal lowpass gain characteristics

Therefore in real lowpass gain characteristic there is some maximum allowable loss in the passband, some minimum attenuation in the stopband and a finite transition band. Essentially, the approximation problem is that of finding a transfer function whose attenuation (or gain) characteristics match the given requirements. It can first be solved for a lowpass transfer function, and any lowpass transfer can be transferred to highpass, symmetrical bandpass and bandreject filters, by employing various frequency transformation [1, 2, 3].

A filter designer should consider all filtering functions namely Butterworth, Chebyshev, inverse Chebyshev, Elliptic and Bessel types.

13.3 Second Order Filter

Here the transfer function is a second order polynomials in both numerator and denominator. Such a transfer function is called a biquadratic transfer function and the network implementing this transfer function is called "biquad" [4, 5, 6]. A biquadratic transfer function is of the form

$$T(s) = \frac{ms^2 + cs + d}{s^2 + as + b} \tag{1}$$

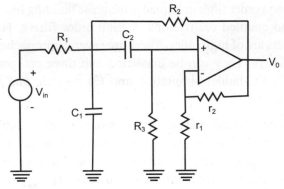

Fig. 13.2 Single amplifier bandpass filter

We can derive any type of frequency response by properly choosing the coefficients m, c, d, a and b in $T(s)$. Various network structures are available for implementation of the biquadratic transfer function given in 1, Such as, Sallen and Key filter, Friend biquad, Steffan allpass, Tow Thomas biquad and Akerberg-Mossberg circuit. Again each circuit has several choices of passive elements leading to nine different second order filter structures. The biquads realized by means of active circuits incorporating op-amps (operational amplifiers) are usually high input impedance and low output impedance blocks. A second order bandpass filter implemented with a single amplifier is shown in Fig. 13.2. The three amplifier feedforward which can make any type of filter response is shown in Fig.13.3.

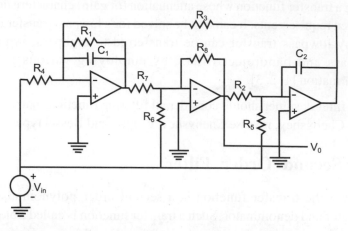

Fig. 13.3 Three amplifire feed forward biquad

13.4 High Order Filter

The second order filter or biquad is the basic building block used in both cascaded and coupled realizations of high order filters. However, most practical filters are of high order. There are several approaches for realizing high order filters. They can be classified into three categories (1) direct synthesis, (2) LC-ladder simulation and (3) cascade and multiple loop feedback [3].

The direct synthesis technique is the method of realizing high order filter using a particular configuration comprised of one or more active elements and appropriately placed RC network, determined by classical driving point immittance synthesis. However filter parameters (pole frequency, pole selectivity etc.) of such a structure is highly sensitive to component variations and hence this approach is rarely used. The LC-ladder simulation approach consists of first synthesizing an LC ladder realization and subsequently simulating this passive network either on operational basis or on a component simulation basis.

In other methods, a high order filter is realized by either cascading two or more active biquadratic sections, or employing negative feedback around groups of biquads in cascade.

The simplest possible realization for an n th order filter will be $\frac{n}{2}$ biquads in cascade, if n is even, or at best $\frac{(n-1)}{2}$ biquads and a first order block in cascade if n is odd. The biquads realized by means of active circuits incorporating op-amps are usually high input impedance and low output impedance blocks. A simple decomposition of the high order transfer function into second order transfer functions is enough.

The considerations for decomposition [7] are:

1. Pole-zero pairing for each biquad:

   Here the goal is to keep the ratio F_{jMax}/F_{jMin} as close to unity as possible, where $F_j(\omega) = \left| T_j(j\omega)/K_j \right|$ is the normalized magnitude squared function of the jth biquad section. For maxima F_{jMax} the entire frequency range $0 \leq \omega \leq \infty$ is to be considered, since no signal should be allowed to overload the filter. On the other hand for the minima F_{jMin} only the filter passband is to be considered, since the filter is supposed to reject the signal out side of the per band and their signal-to-noise ratio is of no concern.

   For convenience a logarithmic measure of flatness can be expressed as

   $$d_j = log \left(\frac{F_{jMax}}{F_{jMin}} \right)$$

   and algorithm has been developed to keep it as close to zero as possible.

2. Distribution of the gain for achieving maximum dynamic range:

   Gain should be distributed among the various sections of the cascade such that all the intermediate transfer function have equal maxima.

3. Ordering of the pole-zero pairs:

   In this part of the solution, the decomposition problem is to find the optimum sequence of cascading filter sections so that the magnitude responses of all the intermediate transfer functions ara as flat as possible. For a Nth order filter there are $(N/2)!$ possible sequences. Complex optimization procedures for solution of the sequencing problem are given in the literature.

   The trial and error method consists of sketching the magnitude response of each individual section. Then, an attempt has to be made to cascade the section in a way that spreads out the maxima and minima over the passband. In addition, one may wish to place a bandpass or lowpass section at the beginning of the cascade. This will prevent internally generated low frequency noise from appearing at the output. Another guideline is to place the sections with the flattest response at the input, followed by sections with lesser flatness, and so on. This means that the biquad Q_p factor goes on increasing from input to output in the cascade.

To implement these rules of thumb in a computer program one should note that the selection of sections for the first and last stages of the cascade is determined by frequency response. For selection of sections for the intermediate stages two guidelines are given:

(a) Spread out maxima over passband.

(b) Increase Q_p factor from input to output.

The application of each guideline separately may result in different sequences; hence both should be combined in a single criterion. For spreading out maximas, select an ith stage biquad such that $\left|\omega_{Max(i)} - \omega_{Max(i-1)}\right|$ is the maximum, where $\omega_{Max(i)}$ denotes the frequency of maxima of the ith stage biquad. For increasing the Q-factjor, select an ith stage biquad such that Q_i is at a minimum, where Q_i denotes the Q-factor of the ith stage biquad. Combining the two criteria, select the ith stage biquad such that the quantity $\dfrac{\left|\omega_{Max\,(i)} - \omega_{Max\,(i-1)}\right|}{Q_i}$ is the maximum. Once a biquad section is assigned to any one stage of the cascade, it is not available for other stages. Hence a set variable which indicates the sections that have already been used, must be maintained.

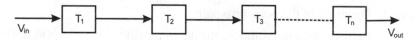

Fig. 13.4 Cascade structure block diagram

However, the sensitivity figures (for component variation) are rather large in cascade design and may not fulfill the condition imposed on the sensitivity figure in many filter designs. A cascade structure is shown in Fig. 13.4. An improvement in the sensitivity figure is obtained in the cascade design, by introducing negative feedback between the stages. In the simplest form of follow-the-leader feedback (FLF), all the biquads are identical and feedback is from output of each stage to the input summing point (Fig. 13.5). Each biquad has its optimum gain to maximize dynamic range of the filter.

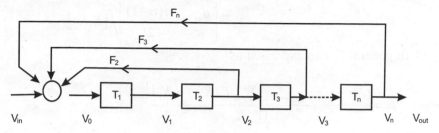

Fig. 13.5 FLF structure block diagram

The feedback path around the first block is suppressed, as it is absorbed in the Block itself. The method is known as Hurtig's Primary Resonator Block (PRB) technique [8] and is ideal for realizing high order all-pole bandpass filter functions with low sensitivity figures.

In continuation of the above all-pole realization, Laker and Ghausi [9] optimized Q_p values of the different sections to arrive at a certain minimum sensitivity measure. The method is complicated and requires computer aided minimization of a performance index function. Laker and Ghausi's method is referred to in the literature as FLF method.

13.5 FLF Structure

The complete set of design equations for allpole bandpass filters using the FLF structure is given below.

Let the LPP be characterized by

$$T(s) = \frac{ma_0}{s^n + a_{n-1}s^{n-1} + \cdots + a_0}$$

where m is the desired centre frequency gain of the bandpass filter.

Calculate the constant k from $k = \frac{a_{n-1}}{n}$

Evaluate the gain constant k to be associated with each of the bandpass sections from

$$K_i = \frac{Max\left[|T(j\Omega)|\left(1+\frac{\Omega^2}{k^2}\right)^{(n-i+1)/2}\right]}{Max\left[|T(j\Omega)|\left(1+\frac{\Omega^2}{k^2}\right)^{(n-i)/2}\right]} \tag{2}$$

where $i = 1,2\ldots, n$.

Evaluate the feedback coefficients F_i from

$$F_2 = \left[\frac{a_{n-2}}{k^2} - \frac{n(n-1)}{2!} \right] \Big/ [K_1 K_2]$$

In general F_i can expressed as

$$F_i = \frac{\left[\frac{a_{n-i}}{k_i} - \frac{n!}{(n-i)!i!} - \frac{1}{(n-i)!} - \sum_{i=2}^{i-1} F_l(K_1 \cdots K_l) \frac{(n-l)!}{(i-l)!} \right]}{\Pi_{i=1}^{i} K_i}$$

The input summing coefficient K is given by

$$K = ma_0 / k^n \sum_{i=1}^{n} K_i$$

If the bandpass filter is required to have a centre-frequency ω_0 and a band width B, that is an equivalent Q-factor $\hat{Q}_p = \omega_0 / B$, then each biquad section has bandpass transfer function with centre frequency ω_0 and Q-factor $Q_p = \hat{Q}_p / k$ and centre frequency gain K_i given in (2).

Tow addressed the question of including the first feedback link in the PRB design. In the extreme case when we use an integrator instead of a first order block in the lowpass prototype, we get a companion form state variable realization. Tow's design is known as generalized follow the leader feedback (GFLF) structure. PRB or FLF could be treated as particular cases of Tow's generalized design. The state variable form or shifted companion form realized by Tow's design [10] showed no sensitivity advantage over the PRB technique.

Approaching the problem at another end, the classical passive double terminated LC ladder filter (Fig. 13.6) has remarkably good sensitivity measure. The low sensitivity performance of LC ladders provided a strong motivation for designing active filters based on simulating LC ladder prototypes. There are two approaches to the problem.

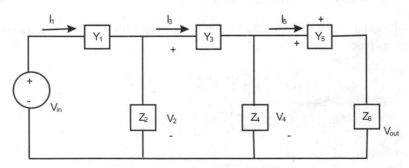

Fig. 13.6 The ladder sturcture

In the component simulation approach inductors are replaced by simulated (active circuit) inductances. Antoniou's generalized immittance converter is one of such inductance simulators. However, for 'floating' inductor simulation certain kinds of transformation, frequency dependent negative resistance (FDNR), will be necessary. Component simulation method is not economic as far as the number of op-amps used are concerned.

Close on heels comes the idea of using operational simulation of LC ladder-block by block, end sections included, rather than component simulation. Equations of the same are given by (3) to (8).

$$I_1 = Y_1(V_{IN} - V_2) \tag{3}$$
$$V_2 = Z_2(I_1 - I_3) \tag{4}$$
$$I_3 = Y_3(V_2 - V_4) \tag{5}$$
$$V_4 = Z_4(I_3 - I_5) \tag{6}$$
$$I_5 = Y_5(V_4 - V_6) \tag{7}$$
$$V_o = V_6 = Z_6 I_5 \tag{8}$$

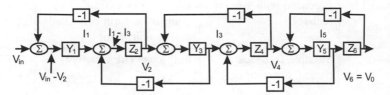

Fig. 13.7 Leap frog realizations of ladder structure

Those may also be represented by the leapfrog realization of ladder structure (Fig. 13.7) in which the output of each block is fed back to the input of the preceding block. As a result, the individual blocks are not isolated from one another and any change in one block affects the voltages and currents in the other blocks. The blocks can be realized by biquads, and the resulting realization is known as Leap Frog (LF) realization [11].

Thus we find that we proceed from the passive LC-ladder structure to a coupled biquad design again with a viewpoint to minimize sensitivity measure (to component variation). LF could be classified as a multiple feedback structure. The whole gamut of high order filter design is illustrated in Table 13.1.

13.6 Selection of Approximation Function

The choice of a particular approximation function is based on the application. There are three main criteria for choosing the approximation function: (1) economy, (2) delay characteristics and (3) characteristics in passband. By "economy" is meant the number of active components required to realize the transfer function. From the economic point of view, the lowest order filter, hence the lowest number of active components is always preferred. Among the approximation functions considere, elliptic gives the lowest order filter. Butterworth gives the highest order. Chevyshev and inverse Chevyshev give the same order, between two extreme limits.

Table 13.1: High order filter design: Spectrum

Building Blocks	Filter structure	Sensitivity to component variation	No. of op-amps used
Biquads	Cascaded	high	economic
-Do-	Cascaded with simple fedback	Lower (PRB)	-Do-
-Do-	Cascaded with simple fedback	Significant improvement (Laker and Ghausi's FLF design large dynamic range)	-Do-
-Do-	Cascaded with general fedback	Lower (Tow's GFLF, companion form or shifted companion form)	-Do-
-Do-	Multiple feedback biquad structure or block substitution method	Still lower (LF structure)	-Do-
-Do-	Component substitution lower FDNR and Gyrator	(active filters using gyrators or FDNR, derived from LC ladder) not economic	
Passive LC (doubly ter -minated)	Classical LC ladder filter	very low	Does not apply

In some specific application one may want to have moderately constant delay over the frequency range of interest. It is not possible to satisfy magnitude as well as delay characteristics simultaneously. Among the approximations Butterworth and Inverse Chevyshev give better delay performance than the other two. However, Butterworth has fast transient response. The user should look at the delay characteristics as well as the magnitude characteristics for all the approximation functions for a given filtering requirement before making a particular choice. Bessel approximation is considered only for satisfying some delay characteristics and not magnitude characteristics.

13.7 Selection of High Order Filter Architecture

The alternatives available in this selection are cascade, FLF, Leapfrog and Component Simulation of LC-ladder. The selection is mainly based on two criteria-sensitivity and economy.

13.7.1 Sensitivity

Synthesis of a network is carried out assuming ideal behaviour of the elements, viz., no deviation from the nominal values of elements or infinite gain bandwidth product of op-amps. In practice, real circuit components will deviate from the nominal design values due to manufacturing tolerances and environmental conditions. A factor of great importance to any filter designer is the effect of nonzero component tolerances and parasitic elements on the performance of a particular filter. Sensitivity is a measure of the change in the overall transfer function with respect to the change in an element in the network. Mathematically, one can relate changes in parameter (Δy) to variations in elements (Δx) in the following manner.

$$\frac{\Delta y}{y} = S \frac{\Delta x}{x}$$

where S is sensitivity of y to variation in x. it is always the objective of a filter designer to keep the sensitivity as low possible.

13.7.2 Economy

When a filter circuit is realized in a specific topology, either with discrete components or in integrated circuit form, the number of op-amps finally decides the economy. No op-amp in general is used in isolation, an op-amp always has some RC circuit associated with it. Therefore, a reduction of the number of op-amps means a low count of components, both passive and active, in the filter circuit. Therefore, the approach that requires the minimum number of op-amps, while not sacrificing performance unduly is considered to be the best.

Ideally, the user wants both the number of components and sensitivity as low as possible. But it is not possible to minimize both the criteria simultaneously, hence some compromise should be made. Number of op-amps used by these structures for realizing the n-th order filter are given in Table 13.2. This table will aid the engineer to make his decision for high order filter selection.

The cascade structure is by far the most economic structure, but its sensitivity performance is the poorest among these structures. Leapfrog and component simulation of LC ladder are the two least sensitive structures but require large number of op-amps, seriously degrading the economy. The FLF structure is the compromise between the two extremes. Its sensitivity performance is better than the cascade.

Table 13.2: Number of op-amps required by different structures

Filter Structure	Frequency Response	Number of op-amps required
Cascade	X	$\left[\dfrac{N+1}{2}\right]$
FLF	BP, BR	$\left[\dfrac{N}{2}\right]+\left[\dfrac{N\!/\!2+1}{2}\right]$
	LP	$N+\left[\dfrac{N+1}{2}\right]$
Leapfrog	LP	$N+\left\lfloor N\!/\!2\right\rfloor$
	BP	$\left[\dfrac{N}{2}\right]+\left[\dfrac{N}{4}\right]$
Component simulation	Highpass	N

As pointed out earlier, we have to compromise our demands on two performance criteria. Therefore user will make its own choice regarding the number of op-amps required in different high order structure for a particular filtering operation, along with its respective sensitive figures, Other factors which affect the selection process are starting point in the design of the structure and frequency response of the filter. The FLF structure design starts with the transfer function of lowpass prototype while the design of leapfrog and component simulation starts with specifications of the filter. Hence, if the users input the transfer function of the filter then FLF, leapfrog and component simulation cannot be choice.

13.8 Biquad Selection

Given a biquadratic transfer function, various biquad circuits are available to implement it. Each circuit has got its own merits and demerits and is suitable for a particular situation.

The important performance measures for these circuits are: sensitivity to component variations, element spread, number of components and orthogonal tuning. While considering sensitivities both active and passive sensitivities are to be considered. At low frequency passive sensitivities dominate, while at high frequency active sensitivities dominate. Let us compare different biquad circuits [12].

13.8.1 Sallen and Key Circuits

It is basically a single amplifier positive feedback circuit, suitable for realization of low pass, handpass and highpass filter. Two types of design are available in this category viz., unity gain amplifier design and Saraga design. Unity gain amplifier is an extreme design. Here all the passige Q_p sensitivities are minimized by employing minimum possible feedback. The design will give minimum possible sensitivity with acceptable element spread for pole selectivity close to unity. Saraga design is another way of synthesizing Sallen and Key filter. It has optimum compromise between low element spread and low sensitivity requirements. Sensitivities and element spread of this design are acceptable upto $Q_p = 5$. The Sallen and Key lowpass circuit is shown in Fig. 13.8.

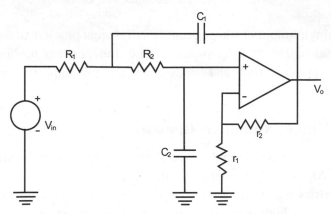

Fig 13.8 Sallen and key lowpass filter

13.8.2 Friend Biquad

It is a multiple negative feedback single amplifier biquad circuit which can realize any filter function except lowpass. Moreover, it has a good compromise between active and passive sensitivities. The circuit is suitable for medium Q_p range ($5 \leq Q_p \leq 20$). The Friend biquad is the most widely used single amplifier filter circuit. An important aspect of the Friend biquad is that the component tolerances are taken into consideration right at the start of synthesis by calculating the optimum element spread for a given component tolerances [13]. However the Friend biquad is not suitable for low Q_p allpass filter. Stefffan's allpass circuit can be used for this purpose [14]. A multiple feedback second order filter circuits is shown in Fig. 13.9.

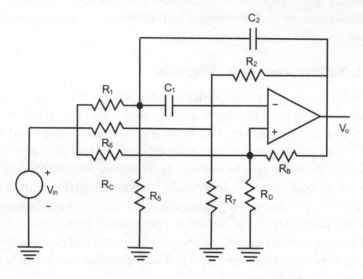

Fig. 13.9 Multiple feedback filter circuit

All single amplifier filter circuits have inherent problem of tuning. For orthogonal tuning of filter parameter i.e. independent tuning of pole frequency, pole selectivity and gain constant, multiple amplifier circuits have been preferred.

13.8.3 Three Amplifier Biquads

Two popular three amplifier biquad circuits are the Tow-Thomas circuit and the Akerberg-Mossberg circuit. Both circuits have the desired characteristics of orthogonal tuning facility. Moreover, these circuits remain stable even for high pole selectivity. These circuits are not economic as they require three op-amps and they dissipate more power as well. Tow Thomas circuit requires tight component tolerances for realizing highpass and notch functions. This stringent requirement can be avoided by using the Akerberg-Mossberg circuit.

In the selection process of biaquad circuit our objective is to keep sensitivity and number of components as low as possible with realizable element spread. If orthogonal tuning is necessary, then there is no choice but to use a three amplifier biquad. However if the operating frequency is above 10KHz, the active sensitivity will play a key role in selecting a particular topology from the entire gamut of biquad circuits. The guidelines in selection of biquad circuit is presented in summarized form in Table 13.3.

Table 13.3: Second Order Filter Selection

Orthogonal tuning required	Very high weightage to economy	Range of Q_p function	Type of	The choice
Yes	X	X	LP, BP	TT
Yes	X	X	AP	TTFF
Yes	X	X	HP, LNP HPN, BR	AM
No	X	Very low	LP, BP, HP	SKF UG
No	X	Low	LP, HP	SKF Saraga
No	X	Low	BP	Friends
No	X	Very low Low	AP	Steffens
No	X	Medium	HP, BP, AP, Notch	Friends
No	X	Medium	LP	LPSAB
No	X	X	Notch	Friends
No	No	High	LP, BP	TT
No	No	High	AP	TTFF
No	No	High	Notch	AM
No	Yes	High	X	Friends

Legend:

X → Don't care.; LP → lowpass.; HP → highpass.; BP → bandpass.; BR → bandreject.; AP → allpass.; TT → Toe-Thomas biquad.; TTFF → Tow-Thomas with feedforward.; AM → Akerberg-Mossberg.; SKF → Sallen anf Key filter.; LPSAB → Lowpass single amplifier filter.

13.9 Conclusion

A systematic approach to design of high order filters has been presented in this article. It is a complete design technique of active filter from the approximation function to a detailed circuit. Three basic topologyies of high order filter have been considered. Auhors have developed several expert system based tools for design and synthesis of high order filter [15, 16, 17]. All the designed circuits have been simulated with PSpice and it has been found that the response exactly matches the input specification.

13.10 References

1. G. Daryanani. *Principles of Active Network Synthesis and Design*. John Wiley, New York, 1975.

2. R. Schaumann, M. S. Ghausi, and K. R. Laker. *Design of Analog Filters: Passive, Active RC and Switched Capacitor*. Prentice Hall Inc., Englewood cliffs, 1990.

3. M. S. Ghausi and K. R. Laker. *Modern filter design: Active RC and switched capacitor*. Prentice Hall Inc., Englewood cliffs, 1981.

4. F. W. Stephenson. RC *Active Filter Design Handbook*. John Wiley, New York, 1985.

5. W. K. Chen. *Passive and Active Filters: Theory and Implementations*. John Wiley, New York, 1986.

6. W. K. Chen, editor. *Circuits and Filters handbook*. CRC Press Inc., Florida, 1995.

7. A. S. Sedra and P. O. Brackett. *Filter Theory and Design: Active and Passive*. Matrix Publishers, Portland, Oregon, 1978.

8. G. Hurtig. Voltage tunable multiple bandpass active filters. In *Proceedings of IEEE International symp. Circuits and Systems*, San Francisco, Calif., 1974.

9. K. R. Laker and M. S. Ghausi. A comparison of active multiloop feedback techniques for realizing high order bandpass filters. *IEEE Trans. Circuits and Systems*, CAS- 21(6): 774-783, November 1974.

10. J. Tow. Design and evaluation of shifted companion form active filters. *Bell system Tech. Journal*, 54:545-568, March 1975.

11. F. E. J. Girling and E. F. Good. Active filters 12: The leap frog or active-ladder synthesis. *Wireless World*, 76:341-345, 1970.

12. Alok Barua and Satyabroto Sinha. *Computer aided analysis, synthesis and expertise of Active filters*. Dhanpat Rai and Sons, Delhi, 1994.

13. P. E. Fleischer. Sensitivity minimization in a single amplifier biquad circuit. *IEEE Trans. Circuits and Systems*, CAS-23(1):45:55, January 1976.

14. J. J. Friend, C. A. Harris, and D. Hilbermann. STAR : An active biquadratic filter section. *IEEE Trans. Circuits and Systems*, CAS-22(2):115-121, February 1975.

15. Alok Barua and Kedar A. Choudhary. EXSHOF: An artificial intelligence approach to high order filter synthesis. *Engineering Application of Artificial Intelligence*, 6(6):533-547, 1993.

16. Alok Barua and Nitin Patel. EXSHOF-II: Active filter design from approximation function to graphic display of the circuit. *Engineering Application of Artificial Intelligence*, 8(6):709-721, 1995.

17. Alok Barua. Obtaining Expert Advice: A knowledge based approach to analog filter design. *IEEE Circuits and Devices*, 15(1):17-25, 1999.

PART D

Image Based Instrumentation

Chapter 14

Boundary Extraction and Contour Propagation of Heart Image Sequences

A. Mishra
P. K. Dutta

Abstract

In this article an automatic detection of the boundary of Left Ventricle (LV) in a sequence of cardiac images has been presented. The LV boundary is quite noisy and general edge-detection techniques fail to produce even a moderately continuous boundary without heavy low-pass filtering which shifts the boundary from its original position. The contour detection algorithm is formulated as a constrained optimization problem based on active contour model. The optimization problem has been solved using Genetic Algorithm (GA). The result obtained by the GA based approach is compared with conventional nonlinear programming methods, e.g., sequential quadratic programming. Validation of the computer-generated boundaries is done after comparing them with manually outlined contours by expert observers. The performance of the algorithm is comparable to inter-observer anomalies.

14.1 General Introduction

Image processing has found a very potential and promising application in biomedical instrumentation and engineering. There are many areas of biomedical processing where image processing can be applied profitably. Some of them are-

- Restoration and Enhancement of X-ray, Ultrasound and MRI images
- Information extraction to aid in diagnostic tools
- 3-D Modelling of inner organs to facilitate surgical and anatomical simulations.

One of the important task for interpretation of biomedical images is segmentation [1][2]. Most of the biomedical images, in particular, ultrasonic images are difficult to segment [3][4][5][6]. Very often one has to apply some prior knowledge of the shape of the object to be segmented. Here two examples of image segmentation will be taken - the first one with echocardiographic ultrasonic image sequences and the other with MRI image sequences of heart. Automatic segmentation reduces painful process of manual extraction of the boundaries. Because manual extraction involves expert medical practitioners, it is almost impossible to carry out this task through out the cardiac cycle manually.

The human heart can be considered as a two-stage pump, physically arranged in parallel where blood passes through the pumps in a sequential manner. The circulation of blood through lungs for oxygenation is called *pulmonary circulation* and the circulatory system that supplies the blood to the rest of the body is called *systemic circulation*. If we consider it from an engineering standpoint the systemic circulation needs high pressure to circulate the blood in the whole body. The pumping cycle is divided into two major phases, systole and diastole. Systole corresponds to the period of contraction of heart muscles specifically the ventricular muscles when the blood flows for pulmonary circulation and into the aorta. Diastole is the period of dilation of the cavities as they fill up. With a time delay the heart repeats this process. The whole process of cardiac motion is governed by an electrical excitation system.

Researchers in the area of cardiac image analysis have long sought to analyze the detail cardiac motion patterns of the cardiovascular walls [1]. Subjective evaluation of the medical images on a large set of clinical data is an important step to establish validity and clinical applicability of the algorithms. Here, an approach to track and quantify the non-rigid and non-uniform motion of the left ventricular wall from 2-D cardiac images has been discussed. The work carried out includes image segmentation using two different imaging modalities (MRI and Echo-cardiographic image sequences).

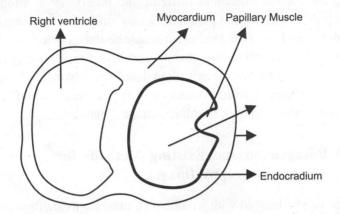

Right ventricle   Myocardium   Papillary Muscle

Endocradium

Fig. 14.1 Axial view of cardiac anatomy showing both the chambers (left and right ventricles).

Even though heart is a 3-D spatially deforming body the results obtained in outlining the boundary, predicting the detail motion vectors from 2-D sequential images [19][20][21] provide rich testing ground for algorithm development and volume estimation. The attention of the computer vision community working in this area primarily focuses on left ventricular motion because of two important reasons i.e.,

(i) It works as the most powerful biological pump for systemic circulation.

(ii) Visual inspection of motion patterns of left ventricles provides most of the information regarding myocardial artifacts.

However, the algorithms may be extended for similar study in other cardiac chambers. Fig. 14.1 shows the middle level cardiac anatomy with a view of the left and right ventricles.

14.1.1 Boundary Detection for Echocardiographic Images

To get quantitative information regarding heart's physical performance a reliable automated and minimally interactive contour extraction method is essential. Tracing epicardial and endocardial boundaries of the left ventricle is essential for quantification of cardiac functions reported in earlier work of *Folland et al.* [1]. The contours on end-diastolic (ED) and end-systolic (ES) positions allow clinically important measures such as ejection fraction, stroke volume and regional wall thickenings. Moreover, tracing the borders of sequential images covering the entire cardiac cycle allows quantitative interpretation of LV contraction dynamics [2]. In this situation, manually outlining the contours becomes difficult and interobserver anomalies in evaluating the segmentation parameters may suffer from subjective bias. This leads to the clinical need for an automatic boundary detection algorithm so that heart can be physically characterized over the entire cycle rather than just at systole and diastole as is often done in case of manual contouring. Stacking 2-D image derived contours with a particular spatial resolution helps 4-D visualization of the cardiac surface dynamics.

14.1.1.1 Background and Existing Methods for Echocardiography Images

Heart can be imaged with a number of imaging modalities associated with different boundary extraction techniques. Several researchers have attempted to identify the LV boundaries on echocardiographic images either automatically or semi-automatically. Many depend on image gray value information to identify the cardiac borders. Due to poor contrast at the image boundaries and low signal to noise ratio (SNR) image gray value information may not be always sufficient to correctly extract it.

In most of the cases a priori information including some kind of user interaction are necessary to get useful results. It requires an initial curve to study the performance of the algorithm and the influence of the user interaction to study the performance of their algorithms in detecting the boundaries.

Work of Chalana *et al.*[6][10] provides an algorithm for detecting the LV endocardial and epicardial boundaries on short axis echocardiographic image sequences that can integrate multiple source of information. The algorithm is basically an extension of active contour model originally proposed by Kass *et al.* [9]. Comparing the computer-generated boundary with that outlined manually by four expert observers, the results are validated on 44 clinical data sets. The performance of the algorithm is evaluated by comparing the area enclosed by the computer-generated boundary and that enclosed by manually drawn contours and the mean boundary distance between them. Both the above performance evaluation parameters are comparable to inter observer anomalies. The correlation coefficient between the areas enclosed by the computer-generated boundary and average manually outlined boundaries was 0.95 for epicardium and 0.91 for endocardium and the algorithm was insensitive to the initial approximation of the boundary in most of the cases [11].

14.1.1.2 Performance Evaluation

Due to lack of a definitive gold standard to evaluate algorithm performance for accurate assessment of boundaries normally the results are compared with manually outlined contours by expert observers. Even though such segmentation suffers from inter-observer bias they are taken as a reference or standard by incorporating some statistical parameterization of data volume with more than one expert observer [6]. To define the evaluation metric most of the researchers have used parameters derived from the boundaries such as area or perimeter of contours for comparison. Metric based on distance between boundaries has been considered for evaluation. To address the problem of inter observer variability Detmer *et al.*[4] have independently evaluated the inter-observer variability on a hand segmentation process.

A well defined metric is an issue, where the only information available is boundaries outlined by multiple expert observers. In this case, the results of the segmentation algorithm can be evaluated against the multiple observers' opinion. Statistical evaluation of the contours has been performed to find whether computer-generated boundaries agrees with the observers' hand outlined boundaries as much as the different observers agree with each other.

14.1.1.3 Image Segmentation in the Field of MRI and other Modalities

Cardiac MRI is a relatively new noninvasive technique developed for heart imaging. Unlike ultrasound or angiography data it doesn't provide continuous images. The data sets are essentially 4-dimensional, consisting of cross-sectional slices varying with time over the cardiac cycle. Clinical cardiac images consist of large number of images, from which quantitative information about the heart function can be obtained by viewing the images in some sequence. To obtain additional quantitative measures for heart's physical performance i.e. to study the detail cavity morphology, the actual motion patterns required to evaluate the degree of infracted muscles, the blood pool border inside the heart chamber must be outlined. A reliable, automated (or minimally interactive) contour extraction method can be used to advantage for this purpose. Then the heart can be physically characterized over the entire cycle, rather than at end systole or diastole, as is often done when the boundaries are outlined manually. An additional benefit is to display and manipulate cardiac surfaces for a dynamic 3-D visualization, which may substantially simplify clinical investigations.

Contour extraction methods for cardiac MRI study have been reported in [13][19]. Staib *et al.* [12] used a probabilistic deformable model parameters using maximum a posteriori estimate.

Significant contributions in the segmentation of MRI images are reported in [27] where contour extraction methods were based on snakes or active contour model [9][13]. Specifically contour propagation methods are considered to make the contours reliable enough despite noise, artifacts and poor temporal resolution. The emphasis was on reliable contour extraction with minimum of user interaction both in case of spin and gradient echo studies. Snake models a deformable contour by processing internal energy in order to impart smoothness to the boundary. When an external energy field acts on this, the contour seeks equilibrium at a local minimum of the energy field by moving and changing the shape. The energy field provides a flexible mechanism to incorporate information about boundary to be extracted, which can come from a priori knowledge, image features etc. Snakes has got a wide spectrum of application in medical imaging and other areas where segmentation algorithms are suggested for a variety of non-rigid bodies using different modalities.

14.1.1.4 Segmentation using Snakes

The problem of digitized intensity image segmentation to extract the traces of the boundary or contours has been addressed with different techniques such as region growing, edge detection and relaxation labeling. These techniques are often found to be numerically expensive and often difficult to implement for non-rigid bodies. Hence a description of the contour based on an active contour model is best suited to simultaneously solve the segmentation and tracking problems popularly known as snake model. Snakes can be represented as energy minimizing splines guided by external constraint forces and image forces such as lines edges and region homogeneities with the image. However, internal spline forces impose smoothness constraints on the modeled contour. The dynamic behavior of the snake is of interest in cardiac images where changes in consecutive frames are not significant.

The digital active contour model [9] parameterizes the contour with respect to its length and the possible configuration of the contour has an energy associated with it. This energy is formulated as a combination of internal and external energies. Features of interest on the images are detected by optimizing the total energy associated with it. Thus the original optimization problem was solved by an iterative gradient descent method. This optimization problem requires an initial approximation to the solution. The iteration converges to a local minimum in case of a dynamic snake where in general the extracted contour depends on the initial approximation. Hence the initial approximation must be in the vicinity of the border to be extracted otherwise the iterations may converge to an undesirable local minimum. At the beginning of the iterations the contour samples are uniformly spaced along it and as the iterations proceed the sampling becomes highly complex.

Convergence can be checked in different ways i.e. iterations may be terminated when the change in energy is below a threshold or when the standard deviation of a consecutive block of iterations is below a threshold value. The former method does not work reliably because the external force (normally the image gradient force) some times becomes too large to make the snake oscillate around the equilibrium point preventing monotonic decrease in energy.

In this study we address the contour optimization functional based on an active contour model originally proposed by Kass *et al.* [9]. A new technique for automatic detection of the boundary of left ventricle (LV) in a sequence of cardiac images has been proposed applying genetic algorithm (GA). Contour extraction is formulated as a constrained optimization problem based on an active contour model. The results obtained, optimizing the contours with standard nonlinear programming techniques are compared with GA. We have validated the final results by comparing the computer-generated boundaries with manually outlined contours by expert observers. The performance analysis of the algorithm finds the outcomes to be comparable to inter-observer anomalies.

The objective function i.e., snake energy is minimized in a constrained feasible solution space using genetic algorithm. GA being computationally expensive in comparison to other non-linear optimization techniques possesses promising results when the initial population of chromosomes (genotype or candidate solutions) is constrained within a search area. Finally the performance of the algorithm is compared with classical iterative solution methods.

14.1.2 Active Contour Model

14.1.2.1 Snake: An Overview

An active contour model refers to a deformable curve $v(s) = (x(s), y(s))$ in a 2-D image parameterized with respect to the normalized contour length s. Snake has been devised to permit non-rigid stretching of a contour. The specialty of active contour model is to allow unequal movement, i.e., displacement, rotation and scaling, of each contour point. Various aspects and interpretations of active contour model in deformable framework is available in the references [9][13][14][15][16][17]. The desired contour is extracted by minimizing the potential energy function E_{snake} in eqn. (1), which permits the snake to have two-dimensional degree of freedom in the x, y plane. Here E_{int}, E_{ext} and E_{image} represents the internal, external and image forces respectively.

$$E_{snake} = \int_0^1 E(v)ds = \int_0^1 [E_{int}(v) + E_{image}(v) + E_{ext}(v)]ds \qquad (1)$$

As a consequence the internal energy of the snake is independent of torsion and the second order partial derivative is sufficient to describe the energy function. The optimum potential energy function is defined as

$$\begin{aligned} E^* = \min_{x,y} \quad & \frac{1}{2}\int_0^1 w_1(s)\left[x_s^2(s) + y_s^2(s)\right]ds \\ & + \frac{1}{2}\int_0^1 w_2(s)\left[x_{ss}^2(s) + y_{ss}^2(s)\right]ds \\ & + \int_0^1 w_3(s)[E_{image}(x(s), y(s))]ds \\ & + \int_0^1 w_4(s)[E_{ext}(x(s), y(s))]ds \end{aligned} \qquad (2)$$

In eqn.(2) w_1 and w_2 are respectively the weighting functions to regulate the stretching and bending effects in the model. A large value of $w_1(s)$ encourages the contour to shrink similarly higher value of $w_2(s)$ discourages sharp bends. E_{image} in the energy equation represents appropriate image features to attract the contour towards high gradient region. Similarly E_{ext} represents some user defined or knowledge-based influences to be incorporated in the model and w_1, w_2, w_3 and w_4 are the weighting functions governing the energy eqn. (1) that minimizes it to a desired shape. The original segmentation problem modeled by Kass *et al*[9]. solves the equation using a gradient descent algorithm, which requires an initial approximation to the solution.

Discretization of eqn. (2) is essential to solve the optimization functional using standard $O(h^2)$ derivative operators.

$$E^* = \min_v \frac{1}{2h} \times \sum_{i=0}^{N-1} w_1(i)|v_i - v_{i-1}|^2 + \frac{1}{2h^3} \times \sum_{i=1}^{N-1} w_2(i)|v_{i-1} - 2v_i + v_{i+1}|^2$$
$$+ h\sum_{i=1}^{N-1} w_3(i)E_{ie}(v_i) \tag{3}$$

The closed contour is uniformly sampled at N number of points and can be assumed to be periodic. If we combine the external forces $E_{ie} = E_{image} + E_{ext}$, E^* will attain the minimum value when $\dfrac{dE^*}{\partial v} = 0$, hence the above equation can be written in a compact form solving the Euler-Lagrange equation.

$$\mathbf{Av} + \mathbf{f} = 0 \tag{4}$$

$$\text{where } \mathbf{f} = \frac{\partial E_{ie}}{\partial v} = \left[\frac{\partial E_{ie}}{\partial x} \times \frac{\partial E_{ie}}{\partial y}\right] = [\mathbf{f}_x, \mathbf{f}_y] \tag{5}$$

A is a sparse banded matrix composed of the coefficient of **v**. Separating the equation into its x and y components we get

$$\mathbf{Ax} + \mathbf{f}_x = 0$$
$$\mathbf{Ay} + \mathbf{f}_y = 0 \tag{6}$$

Here $\mathbf{f}_x$ and $\mathbf{f}_y$ are partial derivatives of the external and image force with respect to **x** and **y** coordinates. The system can be made dynamic by assigning mass density to the snake and allowing the kinetic energy to be dissipated by friction. The dynamic equation of motion can be written (D.Terzopoulos [34]) as

$$(\mathbf{A} + \mu\mathbf{I} + \gamma\mathbf{I})\mathbf{x}_t = 2(\mu + \gamma)\mathbf{x}_{t-1} - \mu\mathbf{x}_{t-2} - \mathbf{f}_x(\mathbf{x}_{t-1},\mathbf{y}_{t-1})$$
$$(\mathbf{A} + \mu\mathbf{I} + \gamma\mathbf{I})\mathbf{y}_t = 2(\mu + \gamma)\mathbf{y}_{t-1} - \mu\mathbf{y}_{t-2} - \mathbf{f}_y(\mathbf{x}_{t-1},\mathbf{y}_{t-1}) \tag{7}$$

where μ is the mass density and γ is the dissipation constants. The above equations are solved using gradient descent algorithm which requires an initial approximation to the contour $(\mathbf{x}_0, \mathbf{y}_0)$. The solution will converge to a local minimum if the initial solution is not defined close to optimum contour. The output oscillates with iterations due to non-uniform sampling and rapid change in external forces at those points. To maintain the sampling uniformity the contour is resampled after each iteration at the expense of computational time.

Observing the variation of the total snake energy normally checks convergence of the iterative procedure. The iterations are terminated when the change in energy is below a certain limit. In case of a dynamic snake model this very often doesn't work reliably. The external force can some times be sufficiently large and erratic to make the snake oscillate around equilibrium point.

14.1.2.2 Model Considerations

The model for contour extraction is quite similar to the original snake model. We optimize the snake energy in a constrained 2-D space using genetic algorithm (GA). Genetic algorithm, being a heuristic search optimization technique is distinguished by its parallel investigation of search space simultaneously manipulating a population of candidate solutions.

An essential preprocessing is required to define a solution space for optimization where the initial contour occurs most likely. A relaxed preprocessing makes the whole process automated, avoiding the user interaction required to go for an initial approximation of the contour (draft contour) in the beginning unlike most of the early methods. Optimization is posed as a constrained problem in this area that incurs a significant improvement in convergence time. The results obtained are consistent to the classical methods both in attaining the global optimum and convergence time as well. The final outcomes are compared with constrained quasi-Newton method, which guarantee super linear convergence by accumulating second order information regarding the KT (Kuhn-Tucker) equations using a updating procedure

14.1.3 Image Preprocessing

Image acquisition using ultra sound transducers is a continuous process. A good temporal resolution can be achieved to extract and propagate the contours in to future frames unlike the MRI where images are confined to limited cardiac phases. The initial solution to the contour detection algorithm is provided by manually outlined contours in most of the cases. However, we have implemented a preprocessing technique to detect an initial boundary followed by an optimization procedure.

A potentially good image is selected in the beginning of systole and is filtered by convolving it with a 3×3 Gaussian low pass filter with variance σ=3. The filtered image is thresholded by the mean value of the blood pull intensity in a small rectangle producing a binary image (Fig 14.2). To eliminate the noise present in the image having equal intensity level as that of the inner cavity we have applied two successive stages of morphological filtration i.e., dilation and erosion Dilation eliminates noise present in the converted binary image applying a 3×3 structuring window twice. The output image $I_{out} = (\ I_{in} \otimes SE_3\ )$ removes the spurious noise in the region. Similarly applying conditional erosion having similar structuring element twice (may be once if the rough boundary of the endocardial wall is found discontinuous), the image restores a smooth profile to the inner boundary. The final output image $\overline{I}_{out} = (\ I_{out} \otimes SE_3\ ) \mid I_{in}$ is shown in Fig. 14.2(e). The rough boundary is coded and the Cartesian co-ordinates are stored in a vector form. These are the initial estimates of the probable contour in the 2-D space. Once the rough contour location is estimated, we define a search area where the boundary occurs most likely.

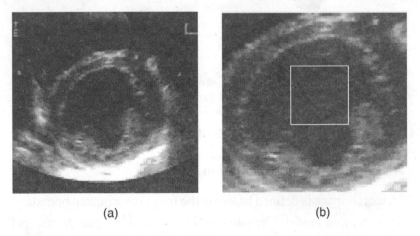

(a)   (b)

Fig. 14.2 (a) Original image. (b) Croped and low pass filtered image applying Gaussian filter with a marked area to calculate mean gray value to threshold the image.

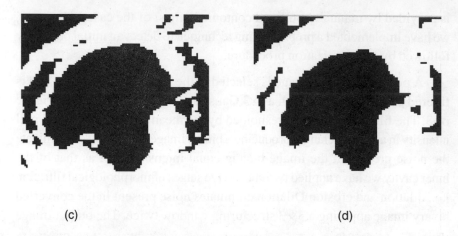

(c)   (d)

Fig. 14.2 (c) Thresholded binary image of the endocardial region.
(d) Binary image after conditional dilation (twice).

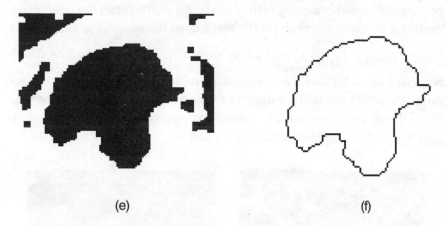

(e)   (f)

Fig. 14.2 (e) Eroded image with similar structuring elements (once).
(f) Rough boundary of the endocardial boarder.

14.1.4 The Optimization Problem

14.1.4.1 Search Area Estimation

A search space is defined based on the roughly estimated boundary for
an active contour model. The initial contour is divided into N number of
samples of equal length and lines approximately perpendicular to the contour
are drawn at each sample location. The dimension of the search width is
quantified by number of pixels (M), which is normally assigned an odd
magnitude keeping sample points at middle (Fig.14.3). The contours joining

possible combination of indexed points on the search grid lines are considered as the candidate solutions to the optimization functional in the model.

14.1.4.2 Discretization

The spatial discretization is performed on the contour at N distinct equidistant points $(v_i, i = 1,2 \cdots N)$. Standard finite difference operators perform discretization of the internal energy.

$$\left[x_s^2(s) + y_s^2(s)\right] \approx (x_i - x_{i-1})^2 + (y_i - y_{i-1})^2 \tag{8}$$

and

$$\left[x_{ss}^2(s) + y_{ss}^2(s)\right] \approx (x_{i-1} - 2x_i + x_{i+1})^2 + (y_{i-1} - 2y_i + y_{i+1})^2 \tag{9}$$

The combined external energy $E_{ie}(v)$ in the model is defined on the basis of image gradient force to attract contour towards the high gradient region. The external energy is defined in the eqn. (10).

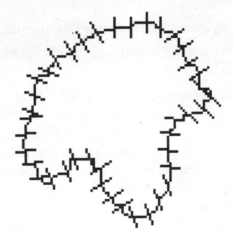

Fig. 14.3 Search area defined by approximate perpendicular lines at sample locations.

$$\int E_{ie}(v)ds = -w_3(s)\int \left\|f(\nabla I)\right\|^2 ds \tag{10}$$

$\nabla I(x, y)$ is the gradient of the image on search grids where the probable contour occurs. The image gradients are computed using Sobel's operator. The normalized gradient g_i is expressed in eqn.(11) where g_{max} is the maximum gradient on the individual search grids.

$$\nabla I = g_i = \frac{g_i}{g_{max}} \tag{11}$$

We incorporate function f as a nonlinear mapping function, which transforms the normalized gradient so that it discourages the model to adhere low gradient region (eqn. 12). The mapping function is tested for different values of a. The importance of the function in optimizing the contour and choice of a and η is discussed in Section 14.6.

$$f(g_i) = \frac{1}{1 + \exp(a \times (\eta - g_i))} \tag{12}$$

Finally the discretized equation for the snake energy can be recasted in this form

$$E_{snake} = \sum_{i=1}^{N} \{w_1[(x_i - x_{i-1})^2 + (y_i - y_{i-1})^2] + w_2 \times$$

$$[(x_{i-1} - 2x_i + x_{i+1})^2 + (y_{i-1} - 2y_i + y_{i+1})^2] - w_3 \times (f(g_i))^2 \} \tag{13}$$

where w_1, w_2 and w_3 are assumed to be constants over the whole boundary. The boundary is detected by minimizing the energy function defined in eqn.(13).

Here, Genetic algorithms has been used to solve the optimization problem. Genetic algorithm starts with a fixed population of candidate solutions and each of the candidates is evaluated with a fitness function that is a measure of the candidate's potential as a solution to the problem. The fitness function maps an individual of a population in to a scalar. Genetic operators like selection, crossover and mutation are implemented to simulate the natural evolution. A population, usually presented by a binary string is modified by the probabilistic application of the genetic operators from one generation to the next. GA has a potential of multidimensional optimization as they work with a population of candidate solutions. Its recent application in computer vision includes edge detection and adoptive image segmentation [25][26].

According to the GA theory [25], possible solution to a global optimization problem is encoded as chromosomes and a population of candidate solutions contains possible solutions to the problem. The size of the population is an important criterion to determine the effectiveness of the solution. The process of selection crossover and mutation improves the fitness of the individual candidate solutions, which are normally binary strings. Offsprings are generated from randomly selected parents carrying genetic information. The fitter individual has better chance of reproduction. However, the solutions with lower fitness are not always rejected from the population set to resist the loss of any otherwise useful genetic materials.

14.1.5 Solution Approach

14.1.5.1 Encoding

The basic task in the optimization problem is to minimize the energy function described in eqn. (13). The energy function is computed using the image sequence derived contours that define the boundary of the left ventricle. Since the original problem is constrained to a predefined solution space the generation of the candidate solution is different from simple genetic algorithm. Initially R number of candidate solutions are generated where the genotype (chromosome representation) is not represented by a binary string as in case of traditional GA, instead we directly use the indexed point sets on grid locations. The input to algorithm is a corresponding point set $P = [Z_r \mid r = 1 \cdots R]$, where the parameter vector Z_r represents corresponding points assigned to a candidate solution or a solution contour. Further we define the parameter vector $Z_r = [z_{ri} \mid i = 1 \cdots N]$ and the individual chromosomes or genes are selected randomly from a minimal subset $[z_{ri}(k) \mid k = 1 \cdots M]$ where M is the number of points on the search grids. The individual genes are indexed by their location on the search grid. The initial population of candidate solutions is randomly generated from the minimal subset restricting the choice within search space. Solution contour C^r can be traced joining the points in Z_{ri} i.e. $C^i \subset \Im^2 \;\; \forall \; r \in [1 \cdots R]$, where $\Im^2$ is a 2-D Euclidean space.

14.1.5.2 Objective Function Evaluation

A given population consists of R individuals and each is characterized by its genotypes with N number of genes, which determines the validity or fitness for survival. In this problem we have directly coded the genotypes as indexed points chosen randomly from a minimal subset. The Cartesian coordinates corresponding to the candidate solutions (x_{ri}, y_{ri}) are the input to the objective function. Objective function in eqn. (13) can be explicitly defined in a matrix form to evaluate the fitness functions.

$$E^r_{snake} = w_1 \times (U \times De^T \times De \times U^T) + w_2 \times (U \times Ds^T \times Ds \times U^T) - w_3 \times (G^T \times G)$$

$$(14)$$

where $U = [x_{r1}, \cdots, x_{rN}, y_{r1}, \cdots, y_{rN}]$

$$De = \begin{bmatrix} D & 0 \\ 0 & D \end{bmatrix}, \quad Ds = \begin{bmatrix} D^2 & 0 \\ 0 & D^2 \end{bmatrix}, \quad G = [f(g_1), \cdots, f(g_N)]^T \text{ and } D \text{ is}$$

defined to be a $N \times N$ first order difference operator a banded matrix

$$D = \begin{bmatrix} 1 & 0 \cdots & \cdots & -1 \\ -1 & 1 \cdots & \cdots & 0 \\ 0 & -1 & 1 & 0 & \vdots \\ \vdots & \ddots & \ddots & & 0 \\ 0 \cdots & & \cdots 0 & -1 & 1 \end{bmatrix}$$

14..1.5.3 Algorithm

Step1: Select the population size (R), crossover rate p_C, the mutation rate p_M.

Step2: Two parents are chosen with probabilities proportional to their relative position in the current population evaluated by their contribution to the mean objective function.

Step3: Different offspring are produced by recombination of two parental genotypes by means of crossover at a given recombination probability p_C, both of these offspring are taken in to further consideration. *Step1* and *2* are repeated until a user defined portion $m \times R$ $(m<1)$ individuals are generated to represent the next generation.

Step4: The rest of the candidates in the next generation are directly selected based on their fitness value. This subset of the population goes through regular selection for mating but is not altered going into the next generation.

Step5: Offspring with a given small probability undergo further modification by means of point mutation working on individual genes by reversing the phase i.e. if the particular gene in the chromosome is indexed by the kth position on the search grid it will be reversed to M-k.

Step6: The process is repeated until the population fitness variance σ is under a threshold value.

Step7: If the maximum number of iterations is reached the best string is retained (elite selection) as the optimal solution.

Retaining a predefined portion of the population in the subsequent generations is called a steady state GA and is particularly suitable for objective functions where number of variables in the solution vector is more.

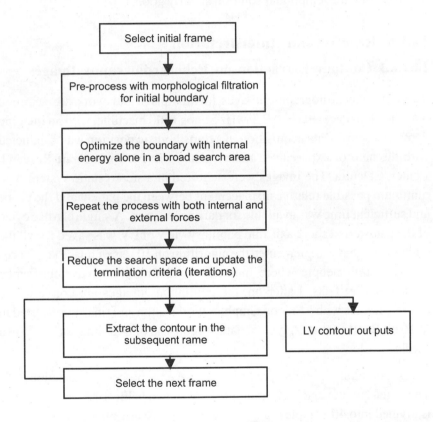

Fig.14.4 Block diagram of a possible method for boundary extraction

14.1.5.4 Contour Initialization and Propagation Strategy

The preprocessed contour after low pass filtering and morphological operations is optimized using genetic algorithm for rough boundary delineation at fixed number of sample points. Initially we optimize the snake with internal energy alone. The weighting function regulating the bending energy (w_2) is made higher so that the initial approximation provides a smooth boundary to the endocardial boarder of the left ventricle. This contour is then projected on the original image and the process is repeated to extract the boundary with both internal and external image forces (Fig. 14.5). The optimized contour in the image frame is considered as a good initial approximation for the next frame in a sequence. However, the constrained region of optimization is decreased once the initial contour is extracted. This saves computational time and avoids a potential risk for the boundary to be trapped at local minima when the temporal resolution is fairly good.

14.1.6 Results and Interpretation

14.1.6.1 Contour Extraction on Echocardiography Images

The echocardiographic images for this particular work are acquired with Hewlett-Packard (HP) 2.5 MHz Sonos 1500 machine, with an imaging depth of 16cms. Data acquisition in a transthorasic position was conducted with the help of experienced radiologists in B. M. Birla Heart Research Center, Calcutta. The images were acquired in a stereo static system with minimum possible relative displacement between the patient and the probe and sufficient time was available for data acquisition. A single cardiac cycle of the parasternal short-axis mid papillary view of LV was saved for all the patients. We have implemented the algorithm on 10 sets of image sequences with different people where majority of them were having confirmed myocardial artifacts. Latter on, the continuous images are converted to discrete frames with a Silicon Graphics workstation and finally converted in to 16 bit BMP format. A typical data set consists of 8-20 image frames in a cardiac cycle.

A potentially good image has undergone preprocessing (Section 2.3). The initial contour and the search space are shown in Fig.14.3 where contour is divided into 40 samples and the search grids are obtained by tracing approximately perpendicular lines to the contour at sample points. The number of pixels is chosen to be 9 on each search line keeping the sample point at the middle, which provides equal freedom to the snake to optimize the energy function by allowing it to crawl on either side of the initial contour. The initial boundary has been outlined in two steps. The first step optimizes the contour with internal energy alone and w_2 is made deliberately high to get a smooth curve (Fig 14.5 (a)).

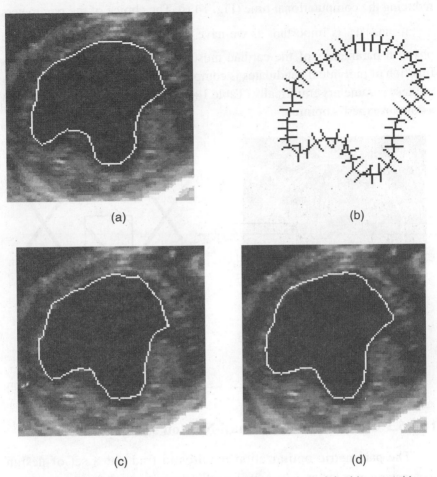

Fig. 14.5 (a)Optimized initial contour projected on the original image taking the internal energy into consideration alone (75 iterations). (b) Resampled contour showing the search area for the next step. (c) Optimized contour with internal and external image forces after 30 iterations. (d) Detected contour after 75 iterations. Images are obtained by linearly interpolating the optimized sample points

In the next step external image force is added to optimize the final contour. Search space dimension is made relatively higher in both cases (9 pixels) so that the initial contour should not be trapped at local minima (Fig.14.5(c,d)). This is done intentionally because the preprocessing does not guarantee the rough contour to occur in close vicinity of the actual contour because morphological operations are performed liberally to get initial boundary. After the contour in the initial frame is estimated the subsequent contours are extracted by reducing the search space. We have achieved consistent boundaries at 5-7 pixels in most of the cases, thereby reducing the computational time (Fig. 14.6). The choice of the parameters w_1, w_2 and w_3 is important as we have sparse knowledge regarding the physical parameters of the cardiac muscles in the left ventricle. Fitness function of individual candidates is computed by selecting constant values of these parameters empirically (Table 14.1) and results are compared with a domain expert's opinion.

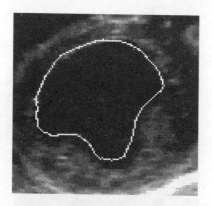

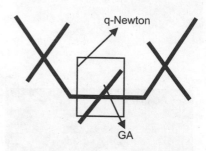

Fig. 14.6 Optimized contour in the next frame with reduced search space (7 pixels) after 40 iterations

Fig. 14.7 Enlarged portion of the rough boundary showing the search space at individual sample locations for quasi-Newton and GA optimization pro cesses

14.1.6.2 Comparison with quasi-Newton's Algorithm

The parametric optimization requires to find out a set of design parameters, $x = \{x_1, x_2,...x_n\}$, that can in some way be defined as optimal. It may be optimization (minimization/maximization) of some system characteristic, which is dependant on x. In this particular case the objective function $f(x)$ to be optimized is subject to constraints (Eqn. 13) in the solution space. In general it may be subjected to constraints in the form of equality constraints $G_i(x) = 0$, $i = 1...m_0$, inequality constraints $G_i(x) \leq 0$ $i = m_0 + 1...m$ and/or parametric bounds $x_l \leq x \leq x_u$ where x_l and x_u are the lower and upper bounds for the design parameters. A general problem is stated below.

$$\arg \min_{x \in \Re^n} f(x) \tag{15}$$

$$\text{subject to } G_i(x) = 0, \qquad (i = 1...m_0)$$

$$G_i(x) \leq 0 \qquad (i = m_0 + 1...m)$$

$$x_l \leq x \leq x_u$$

where x is the vector of design parameters, $(x \in \mathfrak{R}^n)$, $f(x)$ is the objective function that returns a scalar value $f(x) : \mathfrak{R}^n \rightarrow \mathfrak{R}$ and the vector function $G(x)$ returns the value of the equality and inequality constraints evaluated at x i.e., $(G(x) : \mathfrak{R}^n \rightarrow \mathfrak{R}^m)$

An efficient and accurate solution to this type of constrained problem is not only dependent on the size of the problem in terms of the number of constraints and design variables but also on the characteristics of the objective function. When both of them are linear function of the design variable or the objective function is quadratic and is linearly constrained, there are reliable solution procedures to solve this kind of problems. It is more difficult to solve nonlinear problems in which the objective function and constraints may be non-linear function of the design variables. However, we address the above problem to be constrained in the solution space defined on the search grid locations alone. The parametric variables are provided with lower and upper bounds in the absence of equality and inequality constraints but the objective function is highly nonlinear due to the mapping function added as the image force. A solution of the nonlinear programming (NP) problem generally requires an iterative procedure to establish a direction of search at each major iteration.

Table 14.1 Snake parameters, Search space and Termination criteria for contour extraction with a = 0.7

	w_1	w_2	w_3	Search space dimension	Termination criteria (max. no. of iterations)
Initial frame Step 1	0.3	0.8	0.0	9	75
Step 2	0.4	0.6	0.7	9	75
Subsequent frame	0.5	0.6	0.7	7	50

A comparative study has been presented by optimizing the snake with the help of constrained quasi-Newton method, more commonly known as SQP techniques (Sequential Quadratic Programming). At each major iteration an approximation of the Hessian of the Lagrangian function is made using a quasi-Newton updating scheme and then it is used to find search direction. Fig.14.7 shows the region in which the independent variables are constrained to the lower and upper bound as we restrict it at the two extreme points on the search lines. The region of search using GA strictly adheres to the points on the search lines because the genotypes are indexed points on it.

The search for optimal contour can be provided with equal degree of freedom if the genotypes are coded with the co-ordinate values in a similar manner at the expense of computational time. As an example we have optimized the contour in the Fig 14.2.(a) with identical constraints. The snake energy vs. the number of iterations and respective computational times are plotted in Fig. 14.8(a). Even though the optimal energy in both the cases are nearly equal there is a difference in computational time due to the fact that the search region for individual genotypes increases. On the other hand it is difficult to constrain the input as points on the search lines using quasi-Newton method. Fig.14.8(b) shows the variation of energy function vs. number of iterations applying both the methods on the preprocessed initial contour. It is obvious from Fig. 14.8(a) that the optimal or equilibrium position is reached with more number of oscillations when the search space is wider. This is due to the fact that as the search space increases the sampling becomes more nonuniform. The GA converges to a steady state value below 75 iterations in most of the cases and even below 40 iterations when the search space is reduced to 7 pixels. The optimal energy may not necessarily be lower than that obtained by the classical method because of the difference in constraining the variables but produces consistent boundaries comparable to inter-observer anomalies. In case of genetic optimization we have plotted the energy associated with the best candidate after each iteration.

The algorithm is found to be robust to the variation of search space. However, increased search space some times lead to an undesirable shape. The algorithm is fairly insensitive to the initial

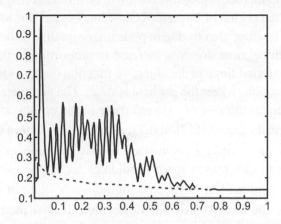

Fig. 14.8 (a) Normalized energy vs. number of iterations to optimize the contour in Fig. 14.5 (b). Both GA and q-N algorithms were subject to equal freedom in the search space. GA converges below 60 iterations while the program is terminated at 75 and 100 iterations for q-Newton and GA method. The objective function corresponds to the energy associated with the best candidate solution

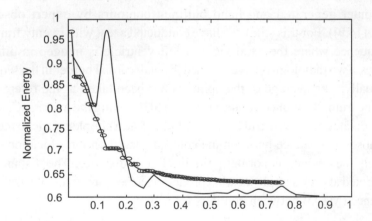

Fig. 14.8 (b) Energy vs.number of iterations with different search space in Fig. 14.5(d) while the search width is reduced to 7. The GA saturates after 40 iterations.

approximation of the curve in the search region when we apply genetic algorithm. Possible combination of contours joining points on the search grid converges to the optimum position provided the search space is defined precisely not to include regions where it will be influenced by external force well out side the region of interest (ROI). It may occur due to the inclusion of epicardial boarder area or due to poor image quality. It is observed that the computational time does not increase in proportion to the search area because the external force in the objective function (eqn. (14)) significantly reduces on locations where the gradient is small. The parameter a (Eqn.(12)) greatly influences the role of external force in the optimization. We have fixed $a=30$ and the threshold η at 0.7 (equation (12)) so that the low image gradient regions ($g_i<0.7$) are subjected to less attention in GA selection process. Due to this reason more candidates are supposed to have poor fitness value and will be subject to mutation. Values of a and η are maintained at the above particular values for our next method to extract contours of MRI images.

14.1.6.3 Performance Evaluation

The aim of performance evaluation of the proposed algorithm is to study the inter observer variability and to quantify the deviation between the computer generated and hand outlined contours by expert observers (Fig. (14.9)). For this study we have conducted a test with twenty frames in a sequence where the axial view of left ventricle is imaged in different phases. Two radiologists have outlined the endocardial border in the sequence manually. The average of the opinions has been taken as the reference or ground truth. Reference contour is obtained by taking the average Cartesian co-ordinates on search grid locations. We define a simple metric to measure the absolute distance between the contours based on their location on the search lines, which is normally the Euclidean distance. The distances are computed at each sample location producing a measure of the mean absolute distance (MAD) between contours (eqn.16).

$$\Re(a,b) = \frac{1}{N} \sum_{i=1}^{N} sqrt\left((x_{ai} - x_{bi})^2 + (y_{ai} - y_{bi})^2\right) \qquad (16)$$

The parameters x_{ai}, y_{ai}, x_{bi} and y_{bi} are the Cartesian co-ordinate of the contour a and b at i^{th} sample location. The above method of performance quantification of the algorithm is influenced by work in [12]. The mean absolute distance between the manually outlined contours and that between the ground truth and computer generated boundaries are computed for the whole set of 20 images. The inter observer mean distance and their standard deviations are shown in Fig. 14.10. Fig 14.11 shows both the.

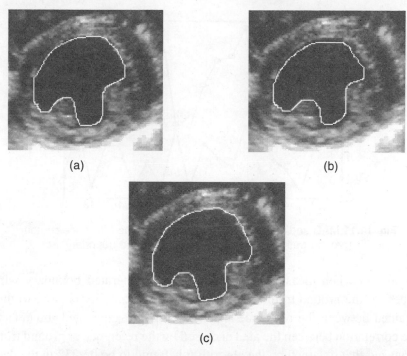

(a)   (b)

(c)

Fig. 14.9 (a,b) Manually outlined contours by expert observers.
(c) Contour detected by GA at the end systolic position

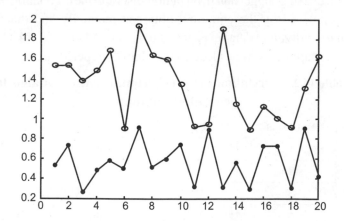

o o o   Mean absolute distance (MAD).
* * *   Standard deviation

Fig. 14.10 MAD and Standard deviation at sample locations between
manual contours drawn by two observers.

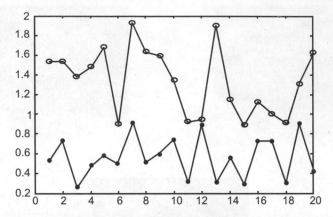

Fig. 14.11 MAD and standard deviation between the reference contour
(ground truth) and the computer generated boundary

parameters when measured for the computer-generated boundary with
respect to the ground truth. The inter observer variability is close to that
obtained between the reference and the computer generated boundary.
The correlation between the area enclosed by the reference or ground truth
curve and that obtained by the algorithm is found to be 0.9232 in the data
set. We have compared the computer-generated boundaries with manually
traced contours in a similar manner for the rest 9 data sets. These boundaries
were outlined taking suggestion from radiologists and the correlation between
the areas enclosed by the contours, obtained manually and that by our
algorithm for individual data set is presented in table 14.2. Fig. 14.11 shows
the observer anomalies to be close to the algorithm performance.

**Table 14.2 Correlation between ground truth and algorithm
generated contour**

Data sets	C.C. 1	C. C. 2
1	0.9321	0.9124
2	0.9311	0.9372
3	0.9317	0.9234
4	0.9327	0.9319
5	0.9404	0.9307
6	0.9314	0.9416
7	0.9507	0.9401
8	0.9218	0.9184
9	0.9293	0.9307

A similar experiment with images having lower temporal resolutions has been conducted with genetic and snake parameters as in our previous analysis. A sequence of images having 12 frames in a cardiac cycle with outlined endocardial boundary is presented in Fig. 14.12. The method could successfully delineate the boundary of the endocardium up to 12 frames in a cardiac cycle for human. However, resolution below 12 frames/cycle could not be tracked successfully even after we increase the search space dimensions. We have shown the mean correlation between the areas enclosed by observer boundary with our approach varying the resolution in table 14.3 for 10 sets of data. The correlation decreases when the numbers of frames are less in a cycle because the contour may trap to an undesirable shape when the actual ROI does not fall within the search space. In this situation increasing the search space alone does not solve the purpose due to the fact that the area may include other regions that influence the image force producing undesirable boundaries.

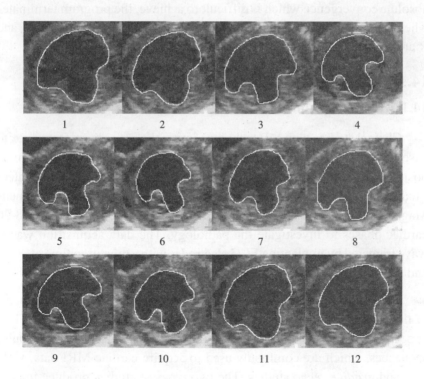

Fig. 14.12 A short axis mid papillary view of the endocardial boundary of 12 sequential frames outlined using genetic algorithm.

Table 14.3 (Average correlation vs no. of frames in a cycle)

Frames/ cycle	8	10	12	14	16	18	20
Correl- ation	0.8525	0.8814	0.9127	0.9211	0.9227	0.9348	0.9412

14.1.6.4 Genetic Parameters

Selection of parameters (population size R, crossover rate P_C and mutation rate P_M) for the genetic algorithm is an important criterion to make the algorithm efficient in handling constrained multivariable problems. As an example a small population may encourage premature convergence where a large population requires more evaluations per generation. A user defined portion m×R (m=0.7) is produced in the current population to represent the next generation in steady state GA. We fix the population fitness function variance σ at 0 and the maximum iteration at 75 and 50 respectively for the initial and subsequent images in this work. A small value of σ=0 represents absolute convergence which is difficult to achieve, the program terminates when the maximum iterations is reached. The typical values of P_C, P_M and R are fixed at 0.5, 0.08 and 40.

14.1.7 Boundary Detection on MRI Images

14.1.7.1 Description of Cardiac MRI Images

Clinical studies with MR technique image the entire heart from apex to base using 5-11 cross-sectional tomographic slices. A view of the axial positions fixed before imaging in the chest region of a normal volunteer (male 29 years) is shown in Fig 14.12. The images are acquired by Siemens Magnetom Impact 1.5-T Scanner. Each spatial slice is imaged at 8-20 cardiac phases to investigate the pathology. The data acquisition was R wave triggered by implanting ECG electrodes with a repetition time of 40ms and an echo time of 7ms with a field of view 350×350mm.

Cardiac MRI studies provide a representation of four main constituents of the chest cardiac region, i.e. muscle, fat, blood, and air (lungs). Their appearance in images depends on the pulse sequence used. The two pulse sequences, which are commonly used to acquire cardiac MRI data, yield *spin* and *gradient echo* studies. The two types of studies produce images with different characteristics and we consider both in this work.

In spin echo images, regions of blood flow appear dark, whereas in gradient echo images they appear bright. In both types of studies, muscle

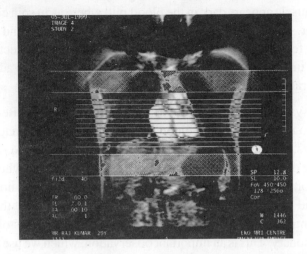

(a)

Fig. 14.13 (a) View of the axial positions fixed before imaging the cardiac region using gradient echo technique. The field-of-view (FOV) is fixed at 450×450mm.

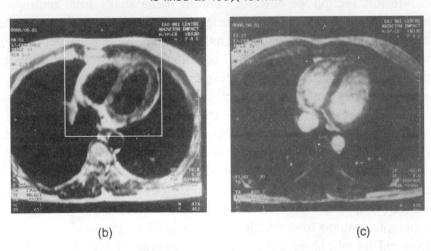

(b)   (c)

Fig. 14.13 Mid axial view of cardiac anatomy (b) spin echo, (c) gradient echo images

has mid gray intensities and air is dark. Fat appears bright relative to muscle inboth types of studies, though this contrast seems more in spin echo studies. The contrast between the blood pull and the cardiac wall plays a vital role in image segmentation. Fig. 14.13 shows two different images acquired using both techniques at mid axial level showing the short axis anatomy of left and right ventricles. In subsequent images just below the base level the papillary muscle appears dark in the blood pull where as in spin echo images it is lighter. Even though MRI images are to the scale and stationary anatomical features are quite less noisy in comparison to the sector scanned ultrasound images the dynamic blood flow in cardiac studies adds noise in few temporal sequences.

After systole, the rush of blood in to the ventricles is often turbulent and leads to signal loss in the blood pull in gradient echo images. In spin echo images, on the other hand bright regions may be found due to slow flow. In the data it is found that the image quality degrades when the heart expands from systole to diastole exhibiting contrast loss and noise. This is due to rapid expansion of heart and out of plane motion at this stage of cycle. Potentially good images are usually obtained at end diastole when the heart is stationary. This is an important observation, which we use throughout for an efficient contour propagation algorithm.

14.1.7.2 Sequencing and User Interaction

In this section we describe contour using snake from multislice multiphase spin and gradient echo images. Unlike the ultrasound images the MRI images are having a limitation on temporal resolution that depends on maximum number of image frames acquired during a cycle and similarly the spatial resolution or inter-slice gap. In case of healthy human the maximum time gap between to successive frames is approximately 60ms. Propagation of spatial slices in the other hand is basically decided from anatomical considerations. The temporal and spatial resolutions are quite enough for endocardial boundary to undergo significant change in shape. In addition to that the image quality is variable over the cardiac cycle. Keeping these two salient features in mind we go for a method, which is similar to our earlier approach used for ultrasound images. Hence it is reasonable to propagate contours from middle slice to other slices in spatial domain and from end diastolic phase to end systolic phase in case of multi-phase images.

To counteract the artifacts due to varying blood flow rate in different phases we initially filter the images with a Gaussian of spread $\sigma = 1.5$. The initial contour or seed is preferably a potentially good image that can be thresholded at the mean value of the blood pull intensity in a small rectangle producing a binary image. Fig. 14.14 shows spin echo images in the middle slice. Subsequently we perform similar morphological filtration operation over the binary image to delineate the rough boundary. The conditional dilation and erosion operation is normally performed convolving with similar structuring 3×3 windows as in our previous work. However, in the present context we suggest two conditional dilation followed by two erosion operations to preserve overall shape and continuity of the rough boundary. In addition to that spurious noise can be eliminated so that the seed should not trapped to undesirable shape.

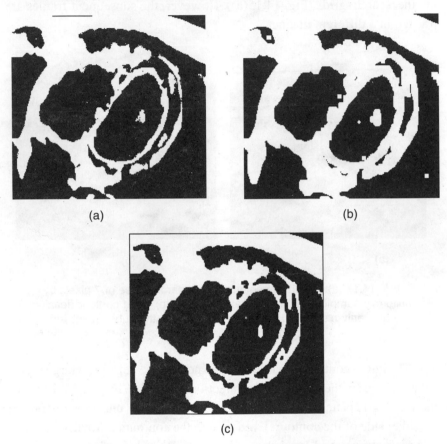

(a)

(b)

(c)

Fig. 14.14(a) Binary image of marked portion in Fig. 14.13 (a), (b) and (c) images after conditional dilation and erosion. Two successive dialation and erosion resulted in to the boundary shwn in the next image being optimized after 40 iterations (Fig. 14.15(a)).

14.1.7.3 Contour Propagation Strategy in a Spatial Sequence

Snakes can reliably extract contours in adjacent images by tracking local energy minima if the interimage sampling is fine enough. Otherwise the snake can get trapped in incorrect local minima and extract the wrong contours. In cardiac studies, the interslice distance can be more than a centimeter, within which cardiac anatomy can change sufficiently to lead the snake to incorrect contours. Therefore we suggest little modification to the original algorithm in our previous work. i.e., by increasing the search space dimension in a predefined fashion. As an example we start with the 5 axial short axis spin echo images with an interslice gap of 5mm. The initial contour is optimized with empirically chosen constants w_1, w_2 and w_3 with an search space dimension of 7 pixels on the search grids (Fig. 14.15(a)). However, the subsequent frames are tracked in a different manner.

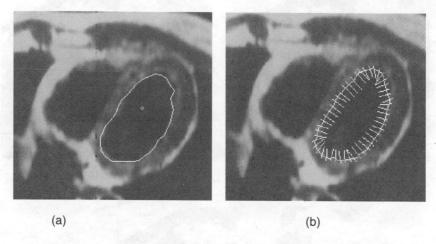

(a)   (b)

Fig. 14.15 (a) optimized contour with a search space of 7 pixels by uniformly sampling the contour at 48 points. (b) initial contour defining an unevenly distributed search space to track the subsequent frame while the heart contracts.

The initial contour is overlaid on the next frame defining a large search solution space in the direction of contraction/expansion. Fig. 14.15 (b) shows the search grid pattern on the contour where the space is unevenly distributed on either side of the contour. Even though the contours normally shrink as we proceed from mid-level slices the small outward search space provides regional flexibility to the subsequent contours because it may fall outside the expected region locally due to spatial rotation. Fig. (14.16) shows three alternate frames tracked in a sequence.

A similar GA based search algorithm for the optimal solution contour is applied to minimize the snake energy. The performance of the process is evaluated by comparing the results with expert opinions. The empirically determined parameter constants for the study are shown in table 14.4. However, the steady state GA parameters i.e. selection, crossover and mutation etc. are taken to be same as in our previous analysis.

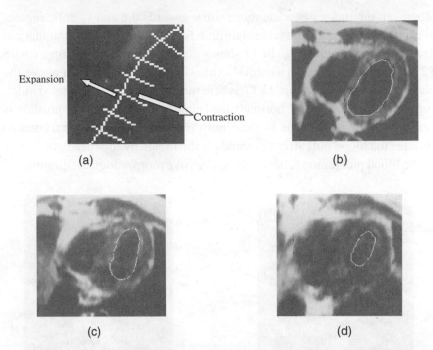

Fig. 14.16 (a) A portion of the boundary showing uneven distribution of the search space i.e., 11 pixels on the contracting region and 3 pixels on expansion side. (b),(c) and (d) are the optimized contours in three alternate spati al sequences

Table 14.4 Snake parameters and search space for MRI images

	Weighting functions			Search space dimension	Maximum number of iterations	Number of samples
	w_1	w_2	w_3			
Initial contour	0.3	0.5	0.8	7	40	48
Subsequent	0.3	0.5	0.8	13	60	48

14.1.7.4 Contour Extraction in a Temporal Sequence

The basic difference between the spin and gradient echo images in outlining the blood pull is the image gray value distribution. The blood appears bright and the endocardial muscle appears dark relative to the blood pull.

However, the images become more noisy towards the end systolic region. We have implemented the same algorithm for delineation of the boundary in gradient echo images. Fig. 14.17 shows a set of multiphase images where 12 images are acquired in a cardiac cycle. The axial level is kept above the base line (3rd axis view, Fig. 14.17, where the papillary muscle is not visible). A potentially good image, normally the frame at the end diastolic position is selected as the initial seed in the algorithm. The converted binary image isolates the blood pull after thresholding the image by the mean gray value of the blood pull region followed by successive morphological filtrations.

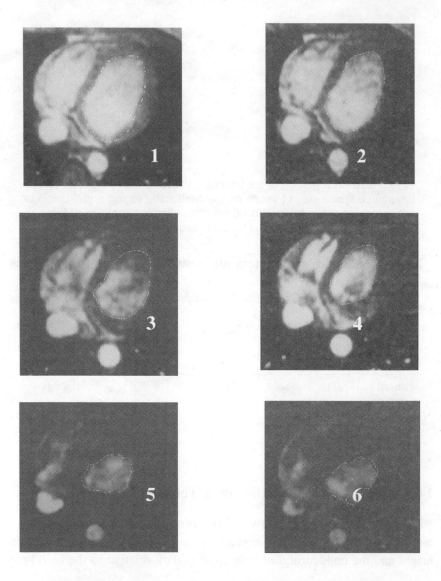

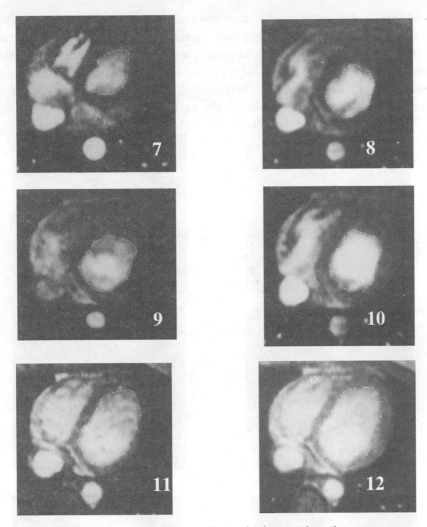

Fig. 14.17 A sequence of 12 gradient echo images in a time sequence from ED to ES. The endocardial boundary is out lined with ED frame 1 as the seed with a search space dimension 7 pixels. The contour propagates to the subsequent frames with an unevenly distributed search space with 15 pixels where 11 pixels are aligned in the direction of contraction/expansion.

The set of images shown in Fig.14.18 are contoured after being acquired with lower temporal resolution i.e. 8 frames in a cardiac cycle. The initial contour in this case is outlined by a similar method. However, we increase the search length in the direction of motion to search for an optimal contour in case the resolution is poor. The endocardial boarder is well outlined even though the LV cycle has four major phases, two of them that involve only shape changes with volume changes. Due to the presence of noise in the end systolic region the empirically selected value of coefficient regulating the bending force is kept deliberately higher so that undesirable increase in image force due to artifact will be suppressed. The values of the coefficients are presented in table 5 to reliably propagate the contours in a multiphase sequence. Fig.14.19 shows the convergence of the normalized snake energy for the initial and second contour in Fig.14.17.

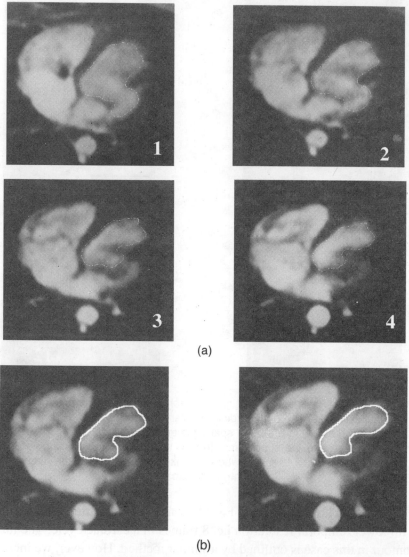

(a)

(b)

Fig. 14.18 (a) Isolated left ventricle in a 8 image data where 4 phases of temporal motion are displayed from ED-ES. The capillary muscle appears as a depression in all the images. (b) Manually outlined contours with mouse where boundary for comparison are coded by the inner pixels.

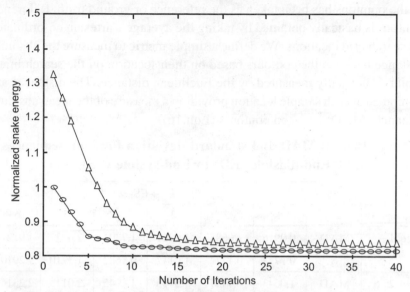

Fig. 14.19 Normalized snake energy vs. number of iterations for the initial and next frame in Fig.17 (Snake energy is normalized with respect to the initial frame). Program terminated after 40 iterations.

Table 14.5 Snake parameters and search space for temporal sequence

	Weighting functions			Search space dimension	Maximum number of iterations	Number of samples
	W_1	W_2	W_3			
Initial contour	0.3	0.7	0.5	7	40	48
Subsequent	0.3	0.7	0.5	15	60	48

14.1.7.5 Performance Evaluation

The algorithm's potential to outline the exact contour of the region of interest is difficult to evaluate due to the non-availability of definite gold standard methods and very often depends on observer's opinion for pathological investigations. Many of researches have followed multiple observer opinion based scoring to evaluate the performance. However, using the performance evaluation procedure in section 14.1.6.3, two radiologists have outlined the endocardial boarder in the sequence manually. The average of the opinions has been taken as the reference or ground truth. Reference contour is basically obtained by taking the average Cartesian co-ordinates on search grid locations. We define a simple metric to measure the absolute distance between the contours based on their location on the search lines, which is normally measured by the Euclidean distance. The distances are computed at each sample location providing a measure of the mean absolute distance (MAD) between contours (Eqn.16).

Table 14.6 (a) MAD and standard deviation for four sequences End-diastole (ED) to End-systole (ES)

Frames→		1	2	3	4	5	6
		\multicolumn{6}{c}{$ED \rightarrow ES$}					
Seq. 1	MAD	0.9916	1.0682	1.0291	1.7301	2.0931	1.9127
	σ	0.7722	0.4772	1.0117	0.5682	1.3710	1.2910
Seq. 2	MAD	1.0643	0.9457	1.5113	1.8945	2.0113	2.2314
	σ	0.4223	0.8141	0.6651	0.4150	1.2021	1.1013
Seq. 3	MAD	1.2261	1.7390	1.9218	2.0908	2.0310	2.1316
	σ	0.2912	0.6317	0.7111	0.9234	1.0033	0.9344
Seq. 4	MAD	1.2727	1.4430	1.7097	1.6552	1.5460	1.9714
	σ	0.3715	0.7551	0.5568	0.3159	0.8410	0.7121

Table 14.6 (b) MAD and standard deviation for four sequences ES to ED

Frames→		1	2	3	4	5	6
		\multicolumn{6}{c}{$ED \rightarrow ES$}					
Seq. 1	MAD	1.8718	1.7593	1.4554	1.3017	1.2701	1.4320
	σ	0.6921	0.7470	0.6122	0.4581	0.6014	0.5152
Seq. 2	MAD	2.0730	1.8300	2.1127	1.5302	1.5309	1.0362
	σ	1.1347	0.9930	0.9081	0.6232	0.5314	0.4914
Seq. 3	MAD	1.6540	2.3201	1.4470	1.5371	1.9290	1.3540
	σ	0.8713	1.1028	0.8490	0.7234	0.6045	0.4007
Seq. 4	MAD	2.0117	1.8932	1.9449	1.3101	1.4182	1.3070
	σ	1.0982	1.2110	0.8623	0.9111	0.6702	0.6122

The MAD between the average observer opinion, the ground truth and that produced by our approach in 4 axial locations covering the LV region with 12-frame sequences is shown in tabular form (Table 14.6).

14.1.8   Conclusion

The boundary detection algorithm presented in Section 2 has been tested for automatic boundary detection of the endocardial boarder of left ventricle in a set of echocardiographic images. Subsequently a slight modification of the snake model parameters, image pre-preprocessing and propagation strategy is suggested for a boundary extraction of multi slice and multi phase MRI images. Even though application of GA in non-linear optimization problems is computationally expensive the results found to be promising when the inputs are constrained, because GA has a potential for multidimensional optimization as they work with a population of solutions rather than a solution in the process to optimize the objective function. The results obtained through GA have been compared with a standard optimization technique i.e., quasi-Newton algorithm. There is an appreciable reduction in computational time in GA when the solution space is constrained to search for the optimal contour on the search grids. However, the solution contour may not always guarantee the global minimum but converges to a near optimal solution, which is comparable to observer opinion. In addition to that contour optimization is insensitive to the initial approximation in the search space region. Because of this reason optimization of the snake energy is followed by a relaxed delineation of the rough boundary in the preprocessing.

The algorithms work quite satisfactorily provided the patient is properly imaged in a stereo static external co-ordinate system i.e., there should not be any relative motion between the transducer and patient. Finer temporal resolution provides a detail motion study and reduces the search space dimension, which ultimately reduces computational time. However, the limitation on number of frames in MRI sequence are well handled defining an unevenly distributed search space. The results using MAD and the standard deviations are compared with the inter observer variability. The algorithm can be used to outline the boundaries of other chambers with minor modifications in the algorithm that basically refers to the pre-processing of the images.

In the last few years, however, there are considerable developments in specialized MRI instruments. Now it is possible to get tagged MRI images with metal markers, most of the cases with pig-heart, to verify the efficiency and accuracy of the reconstruction and tracking algorithms.

14.2 References

1. S.G. Goldberg, "Analysis and interpretation of thickening and thinning phases of left ventricular wall dynamics," *Ultrasound Med. Biol.,* vol.10, pp.797-802,1984.

2. H. E. Melton, S. M. Collins and D. J. Skorton, "Automatic real-time endocardial edge detection in two-dimensional echocardiography," *Ultrason. Imag.,* vol. 5, pp. 300-307, 1983.

3. J. W. Klinger *et al.,* "Segmentation of echocardiographic images using mathematical morphology," *IEEE Trans. Biomed. Eng.,* vol. 35, no. 11, pp. 925-934, Nov. 1988.

4. P. R. Detmer, G. Basein and R. W. Martin, "Matched filter identification of left-ventricular endocardial boarders in transesophageal echocardiograms," *IEEE Trans. Med. Imag.,* vol. 9, pp. 396-404, 1990.

5. P.Lilly, J. Jenkins and P. Bourdillon "Automatic contour definition of left ventriculograms by image evidence and multiple template based models," *IEEE Trans Med Imag.,* vol. 8, no. 2, pp. 173-185, June 1989.

6. V. Chalana, D. T. Linker, D. R. Haynor and Y. Kim, "A multiple active contour model for cardiac boundary detection on echocardiographic sequences," *IEEE Trans. Med. Imag.* vol.15, no. 3 pp. 290-298, Jun 1996.

7. I. L. Herlin and N. Ayache, "Feature extraction and analysis methods for sequence of ultrasound images," *Image Vision Comput.,* vol. 10, pp. 673-682, 1992.

8. J. E. Perez, A. D. Waggoner, B. Barzilai, H. E. Melton, J. G. Miller and B. E. Sobel, "On-line assessment of ventricular function by automatic boundary detection and ultrasonic backscatter imaging," *J. Amer. College Cardiol.,* vol. 19, pp. 313-320, 1992.

9. M. Kass, A. Witkin, and D. Terzopoulos, 'Snakes: Active contour models," *Int. Jr. Comp. Vision,* vol. 1, pp. 321-331, 1988.

10. V. Chalana, T. C. Costa and Y. Kim, "Integrating region growing and edge detection using regularization," in *Proc. SPIE Conf. Med. Image.,* M. H. Loew, Ed., 1995, vol. 2434, pp. 262-271.

11. V. Chalana and Y. Kim, "A methodology for evaluation of boundary detection algorithms on medical images," *IEEE Trans. Med. Image.,* vol. 16, no. 5, October, 1997.

12. L. H. Staib, and J. S. Duncan, "Boundary finding with parametrically deformable models," *IEEE Trans. Pattern. Anals. Machine Intell.,* vol-14, no-11, pp. 1061-1074, Nov. 1992.

13. S. Ranganath, "Contour extraction from cardiac MRI studies using snakes," *IEEE Trans. Med. Imag.* vol. 14. no. 2, pp. 328-338, Jun 1995.

14. I. Kompatsiaris, D. Tzovaras, V. Koutkias and M. G. Strintzis, "Deformable boundary detection of stents in angiographic images," *IEEE Trans. Med. Imag.* vol. 19, no. 6, June 2000.

15. K. F. Lai, "Deformable contours: Modeling, extraction, detection and classification," *Ph. D dissertation*, Univ. Wisconsin, Madison, 1994.

16. F. Leymarie and M. D. Levine, "Tracking deformable objects in the plane using an active contour model," *IEEE Trans. Pattern Anal. Mechine Intell.*, vol. 15, no. 6, pp. 580-591, June 1993.

17. F. Leymarie, "Tracking and describing deformable objects using active contour models," McGill Univ., Rep. TR-CIM-90-9, Feb, 1992.

18. A. Mishra, P. K. Dutta & M. K. Ghosh, " Active contour optimization for boundary detection of non-rigid bodies a comparative study using GA and classical methods," CERA-01, Roorkee, India.

19. A. Mishra, P. K. Dutta, M. K. Ghosh, " Non-rigid cardiac motion quantification from 2D image sequences based on wavelet synthesis", Image and Vision computing, Elsevier sciences, vol. 19, pp 929-939, 2001.

20. C. Slagger, T. Hooghoudt, P. Serruys, J. Schurbiers and J. Reiber, G. Meester, P. Verdoew and R. Hugenholtz, "Quantitative assessment of regional left ventricular motion using endocardial landmarks," *JACC*, vol. 7, no. 2, pp. 317-326, 1986.

21. C. Kumbhametu and D. Goldgof, "Point correspondence recovery in nonrigid motion," *Comput. Vision Pattern Recog.*, pp. 222-227, June 1992.

22 A. Amini and J. S. Duncan, "Bending and stretching models for LV wall motion analysis from curves and surfaces," J. *Image & Vision Comput.*, vol. 10, no. 6, pp. 418-430, 1992.

23. J. C. McEachen, A. Nehorai, and J. S. Duncan, "A recursive filter for temporal analysis of cardiac motion," in *Proc. IEEE Workshop on Biomedical Image Analysis,* pp. 124-123, 1994.

24. L. Axel and L. Dougherty, "MR Imaging of motion with spatial modulation of magnetization," in *Radiol.,* vol. 171. pp. 841-845, 1989.

25. D. E. Goldberg "Genetic Algorithms and evolutionary algorithm come of age*,"* *Comm. ACM*, vol. 37, no. 30,1994.

26. M. J. D. Powell, "Variable Metric Methods for Constrained Optimization," *Mathematical Programming: he state of the Art,* (A. Bachem, M. Grotschel and B. Korte, eds.) Springler Verlag, pp. 288-311, 1983.

27. A. Mishra, P. K. Dutta, M. K. Ghosh, "A GA based approach for boundary detection of left ventricle   with echocardiographic image sequences", Image and Vision computing, Elsevier sciences, Vol 21, No. 11, 2003, pp 967-976.

Chapter 15

Remote Sensing– An Overview

S. Sengupta

Abstract

A brief outline of the basic principles of remote sensing is presented in this article. The principles of space borne sensors, both optical and microwave are also included. Imaging parameters of a few sensors are shown in tables. References include the names of some of the textbooks for further reading.

15.1 Introduction

We perceive our surrounding world with our five senses out of which the senses of touch and taste require contact of our sensing organs with the objects we sense. But the senses of sight and hearing, through which we acquire much knowledge, remotely sense our surrounding without coming in physical contact with it. In another word, we are performing **Remote Sensing** all the time.

In satellite imaging the information regarding the earths features passes through the intervening medium of atmosphere and are gathered by the space borne sensors. The information carrier is the electromagnetic radiation reflected from the earth's features after being illuminated by the sun-light (Fig.15.1). The output of this is an image representing the amount of electromagnetic energy reflected by the earth features in different wavelength bands. A few image types are shown in Figs. 15.2, 15.3 and 15.4.

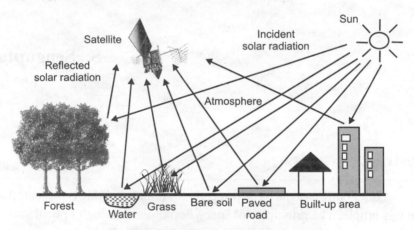

Fig.15.1 Reflection from earth's features

Fig.15.2 PAN Image by IRS

Fig. 15.3 Multi - spectral image by IRS

Fig.15.4 PAN Image by IKONOS

15.2 Electromagnetic Radiation

Electromagnetic waves travels with a velocity c = 2.99792458 x 10^8 m/s in the form of a combined electric and magnetic fields. All electromagnetic waves travel through space with the velocity of light (**c**). An electromagnetic wave is characterized by a **frequency (n)** and a **wavelength** (λ). These two quantities are related to the speed of light by the equation,

$$c = n \; \lambda$$

The frequency (and hence, the wavelength) of an electromagnetic wave depends on its source. In our physical world we encounter a range of frequencies, ranging from the low frequency of the electric waves generated by the power transmission lines to the very high frequency of the gamma rays originating from the atomic nuclei. This wide frequency range of electromagnetic waves constitutes the **Electromagnetic Spectrum**.

The electromagnetic spectrum can be divided into several wavelength (frequency) regions. Out of this entire range only a narrow band from about 400 to 700 nm is visible to the human eyes. The various wavelength regions of the electromagnetic spectrum are:

Radio Waves: 10 cm to 10 km wavelength.

Microwaves: 1 mm to 1 m wavelength. The microwaves are further divided into different frequency (wavelength) bands: (1 GHz = 10^9 Hz)

- o P band: 0.3 - 1 GHz (30 - 100 cm)
- o L band: 1 - 2 GHz (15 - 30 cm)
- o S band: 2 - 4 GHz (7.5 - 15 cm)
- o C band: 4 - 8 GHz (3.8 - 7.5 cm)
- o X band: 8 - 12.5 GHz (2.4 - 3.8 cm)
- o Ku band: 12.5 - 18 GHz (1.7 - 2.4 cm)
- o K band: 18 - 26.5 GHz (1.1 - 1.7 cm)
- o Ka band: 26.5 - 40 GHz (0.75 - 1.1 cm)

Infrared: 0.7 to 300 µm wavelength. This region is further divided into the following bands:

- o Near Infrared (NIR): 0.7 to 1.5 µm.
- o Short Wavelength Infrared (SWIR): 1.5 to 3 µm.
- o Mid Wavelength Infrared (MWIR): 3 to 8 µm.
- o Long Wavelength Infrared (LWIR): 8 to 15 µm.
- o Far Infrared (FIR): longer than 15 µm.

The NIR and SWIR are also known as the **Reflected Infrared**, which are the main infrared component of the solar radiation reflected from the earth's surface. The MWIR and LWIR are the **Thermal Infrared**.

Visible Light: This narrow band of electromagnetic radiation extends from about 400 nm (violet) to about 700 nm (red). The various color components of the visible spectrum fall roughly within the following wavelength regions:

- o Red: 610 - 700 nm
- o Orange: 590 - 610 nm
- o Yellow: 570 - 590 nm
- o Green: 500 - 570 nm
- o Blue: 450 - 500 nm
- o Indigo: 430 - 450 nm
- o Violet: 400 - 430 nm

Ultraviolet: 3 to 400 nm

X-Rays and Gamma Rays: less than 400nm

15.3 Effects of Atmosphere

The earth's surface is covered by a layer of atmosphere consisting of a mixture of gases and suspended solid and liquid particles. The gaseous material extends to several hundred kilometers in altitude, though there is no well-defined boundary for the upper limit of the atmosphere. The first 80 km of the atmosphere contains more than 99% of the total mass of the earth's atmosphere.

The atmosphere is broadly divided into four layers: **troposphere**, **stratosphere**, **mesosphere** and **thermosphere**. The tops of these layers are known as the **tropopause**, **stratopause**, **mesopause** and **thermopause**, respectively.

The electromagnetic radiation are absorbed or scattered by the constituent particles of the atmosphere. Molecular absorption converts the radiation energy into excitation energy of the molecules. The energy of the incident beam is redistributed to all directions by scattering. The overall effect is the removal of energy from the incident radiation.

15.3.1 Atmospheric Transmission Windows

Each type of molecule absorbs different wavelength bands of the electromagnetic spectrum. As a result, only the wavelength regions outside the main absorption bands of the atmospheric gases can be used for remote sensing. These regions are known as the **Atmospheric Transmission Windows (Fig. 15.5).**

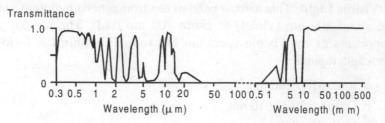

Fig. 15.5 Atmospheric transmission characteristics

The wavelength bands used in remote sensing systems are usually designed to fall within these windows to minimize the atmospheric absorption effects. These windows are found in the visible, near infrared, certain bands in thermal infrared and the microwave regions.

15.3.2 Effects of Atmospheric Absorption on Remote Sensing Images

Atmospheric absorption affects mainly the visible and infrared bands. Optical remote sensing records the solar radiation from the earth's surface. The solar radiance within the absorption bands is reduced due to absorption by the atmospheric gases. The reflected radiance is also attenuated after passing through the atmosphere. This attenuation is wavelength dependent. Hence, atmospheric absorption will alter the apparent **spectral signature** of the target being observed.

15.3.3 Effects of Atmospheric Scattering on Remote Sensing Images

Atmospheric scattering is important only in the visible and near infrared regions. Remotely sensed images are degraded due to scattering of radiation by the constituent gases and aerosols in the atmosphere. Scattering of the solar radiation in the atmosphere produces a **hazy appearance** of the image. This effect is particularly severe in the blue end of the visible spectrum due to the stronger Rayleigh scattering for shorter wavelength radiation.

Furthermore, a target outside the field of view of the sensor may scatter into the field of view of the sensor. This effect is known as the **adjacency effect**. The adjacency effect results in an increase in the apparent brightness of the darker region while the apparent brightness of the brighter region is reduced near the boundary between two regions. Scattering also produces **blurring** of the targets in remotely sensed images due to spreading of the reflected radiation in different directions, resulting in a reduced resolution image.

15.4 Spaceborne Sensors

Spaceborne sensors are mounted on-board a spacecraft (space shuttle or satellite) orbiting the earth. Such sensing from space provides the following advantages:

o Frequent and repeated synoptic view of a large area;

o Quantitative estimation of characteristics of ground features with radiometrically calibrated sensors using computerized processing;

o Lower cost per unit area of coverage than conventional methods.

Though the satellite imagery has a generally lower resolution compared to aerial photography but with the advancement in technology images with ground resolution of one meter are available in public domain.

15.4.1 Optical Sensors

Optical sensors use a CCD based scanning system with an array of detectors. These sensors have a narrow field of view (IFOV) and scan the earth's surface to produce a two-dimensional array of data representing the intensity of light energy reflected from a surface feature. The intensity is expressed as integer number in a gray scale of 8 bit or less. Multispectral scanner (MSS) uses a scanning system to image the earth's surface over a set of different wavelengths. The methods of scanning employed are either along track or across track (Figs.15.6a, 6b). Across track scanners uses a rotating mirror (A) to focus the image on a scanner element (B) and gathers the data as an array of scan lines oriented perpendicular to the direction of motion of the sensor platform. Along track scanners uses a linear array of detectors (push broom array A) placed at the focal plane of the optical system (C). The IRS series satellites (Fig. 15.7) use this technique.

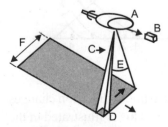

Fig. 15.6a Across track scanning

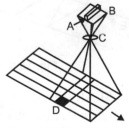

Fig. 15.6b Along track scanning

Fig.15.7 IRS Satellite Camera

15.4.2 Microwave Imaging

The imaging system used in microwave scanning is different from optical imaging (Fig. 15.8). The space borne platform travels forward in the flight direction (A) and the microwave is transmitted obliquely at right angles to the flight direction illuminating a swath (C). The across track dimension is *range* (D) and the along track dimension is *azimuth* (E). In a Side Looking Airborne Radar (SLAR) the image is formed by illuminating the ground with a transmitted pulse and recording the backscattered signal from the ground objects.

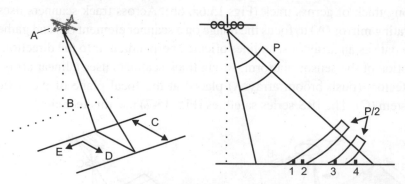

Fig. 15.8 Microwave imaging

In case of SLAR the resolution in range and azimuth direction changes due to the change in inclination of the radar beam. This is illustrated in the Figs. 15.9 and 15.10.

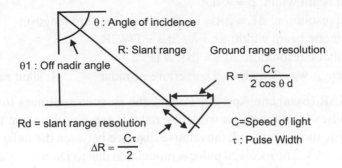

$$\Delta R = \frac{C\tau}{2}$$

R: Slant range

Ground range resolution

$$R = \frac{C\tau}{2\cos\theta\,d}$$

θ : Angle of incidence

θ1 : Off nadir angle

Rd = slant range resolution

C=Speed of light

τ : Pulse Width

Fig. 15.9 Range resolution

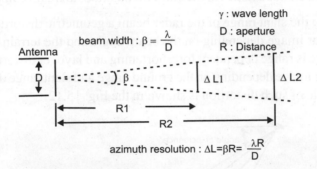

γ : wave length

D : aperture

R : Distance

beam width : $\beta = \frac{\lambda}{D}$

azimuth resolution : $\Delta L = \beta R = \frac{\lambda R}{D}$

Fig. 15.10 Azimuth resolution

The resolution in the azimuth direction depends on the beam width and the distance to the target. It increases with shorter wavelength and bigger antenna size. However, it is difficult to place such a large antenna in the orbiting satellite. This difficulty is overcome by simulating a large aperture with a small antenna through pulse compression and Doppler shift principles. The relation between real aperture and synthetic aperture is shown in the Fig. 15.11.

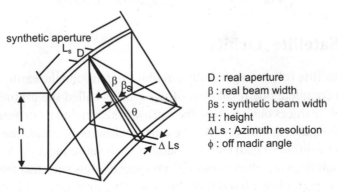

D : real aperture
β : real beam width
βs : synthetic beam width
H : height
ΔLs : Azimuth resolution
φ : off madir angle

Fig. 15.11 Relation between real aperture and SAR [11]

Real beam width: $\beta = \lambda / D$

Real resolution: $\Delta L = \beta R = Ls$ (synthetic aperture length)

Synthetic beam width: $\beta s = \lambda / 2Ls = D / 2R$

Synthetic resolution: $\Delta Ls = \beta sR = D / 2$

Where λ: wavelength   D: aperture of radar   R: slant range

In SAR (Synthetic Aperture Radar), the antenna continues to receive return pulses from the target while the radar is sending the beam to it. As the satellite moves forward, the relative distance between the radar and the target changes. The received pulse is modulated due to Doppler effect. The azimuth resolution is increased by using a matched filter corresponding to the reversed characteristics of the chirp modulation (azimuth compression).

Due to the slant range of the radar beam a geometric distortion occurs on the radar image depending on the beam angle and the terrain relief. As radar image is range dependent foreshortening and layover distortion occur in the radar image depending on the ground range and slant range difference. An example of such distortion is shown in the Fig. 15.12.

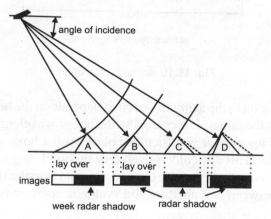

Fig. 15.12 SAR image geometric distortion

15.5 Satellite Orbits

A satellite follows a generally circular orbit around the earth. The time taken to complete one revolution of the orbit is called the orbital **period**. The satellite traces out a path on the earth surface, called its **ground track**, as it moves across the sky. As the earth below is rotating, the satellite traces out a different path on the ground in each subsequent cycle. Remote sensing satellites are often launched into special orbits such that the satellite repeats its path after a fixed time interval. This time interval is called the **repeat cycle** of the satellite.

15.5.1 Geostationary Orbits

A satellite appears to be stationery if its orbit is on the equatorial plane of the earth and it rotates around the earth from west to east with the same period of rotation of the earth (24 hours). Such an orbit is called a **geostationary** orbit. These satellites orbits are located at a high altitude of 36,000 km. The advantage of these orbits is that the satellite appears to be stationary viewing the same area on the earth. Due to its great height the satellite can also cover a large area of the earth. The geostationary orbits are commonly used by **meteorological and communication satellites**.

15.5.2 Sun Synchronous Orbits (Fig 15.13)

Remote sensing satellites circle around the earth at a height of around 800 to 900 kms above earth's surface in a sun-synchronous orbit. A sun synchronous orbit is a near polar orbit whose altitude is such that at given latitude the satellite will always pass over a location at the same local solar time. This enables the satellite to image an area under same solar illumination except for seasonal variation.

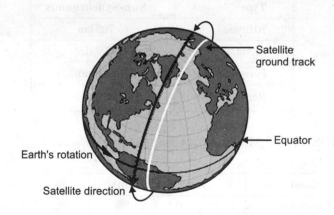

Fig. 15.13 Sun synchronous orbit

15.5.3 Remote Sensing Satellites

Presently a large number of remote sensing satellites launched by different nations are orbiting the earth. The purpose of launching these satellites is also different in various cases and the sensors are designed to suit a specific purpose. Some of these are:

Optical/Infrared Remote Sensing Satellites

- o LANDSAT
- o SPOT
- o MOS
- o ADEOS
- o IRS
- o RESURS

Microwave Remote Sensing Satellites

- o ERS
- o JERS
- o RADARSAT

Meteorological Satellites

15.5.3.1 LANDSAT-4, 5 Orbit parameters

Type	Sun-Synchronous
Altitude	705 km
Inclination	98.2 deg
Period	99 min
Repeat Cycle	16 days

LANDSAT 4, 5 MSS Sensor Characteristics

Band	Wavelength (μm)	Resolution (m)
Green 1	0.5 - 0.6	82
Red 2	0.6 - 0.7	82
Near IR 3	0.7 - 0.8	82
Near IR 4	0.8 - 1.1	82

LANDSAT 4, 5 TM Sensor Characteristics

The TM sensor has a ground resolution of 30 m for the visible, near-IR, and mid-IR wavelengths and a spatial resolution of 120 m for the thermal-IR band.

Band	Wavelength (µm)	Resolution (m)
Blue 1	0.45 - 0.52	30
Green 2	0.52 - 0.60	30
Red 3	0.63 - 0.69	30
Near IR 4	0.76 - 0.90	30
SWIR 5	1.55 - 1.75	30
Thermal IR 6	10.40 - 12.50	120
SWIR 7	2.08 - 2.35	30

15.5.3.2 SPOT (Satellite Pour l'Observation de la Terre), France

The SPOT series of optical remote sensing satellites were placed in the orbit to image the earth for land use, agriculture, forestry, geology, cartography, regional planning and water resources.

SPOT Orbit parameters

Type	Sun-Synchronous
Altitude	832 km
Inclination	98.7 deg
Period	101 min
Repeat Cycle	26 days
Off-Nadir Revisit	1 to 3 days

Sensor Characteristics

	Multispectral Mode (XS)	Panchromatic Mode (P)
Instrument Field of View	4.13 deg	4.13 deg
Ground Sampling Interval (Nadir Viewing)	20 m by 20 m	10 m by 10 m
Pixel per Line	3000	6000
Ground Swath (Nadir Viewing)	60 km	60 km

Mode	Band	Wavelength (µm)	Resolution (m)
Multispectral	XS1	0.50 - 0.59 (Green)	20
Multispectral	XS2	0.61 - 0.68 (Red)	20
multispectral	XS3	0.79 - 0.89 (Near IR)	20
Panchromatic	P	0.51 - 0.73 (Visible)	10

SPOT Twin HRV Imaging System

SPOT satellite has the unique capability to observe an area not only in a direction vertically down (nadir view) but also at an angle ranging from $-27°$ to $+27°$. Such views enable one to have a stereoscopic vision of an area which can be utilized to study topographic variation and generate a digital elevation model of the region. The revisit time for required for stereo vision is in the interval of 1 to 3 days.

15.5.3.3 IRS

The Indian Remote Sensing (IRS) satellite series has features common to both the Landsat MSS/TM sensors and the SPOT HRV sensor. The third satellite in the series, IRS-1C (launched in December, 1995) and IRS-1D have three sensors: a single-channel panchromatic (PAN) high resolution camera, a medium resolution four-channel Linear Imaging Self-scanning Sensor (LISS-III), and a coarse resolution two-channel Wide Field Sensor (WiFS).

IRS Sensors characteristics

Sensor	Wavelength Range (mm)	Spatial Resolution	Swath Width	Revisit Period (at equator)
PAN	0.5 - 0.75	5.8 m	70 km	24 days
LISS-II				
Green	0.52 - 0.59	23 m	142 km	24 days
Red	0.62 - 0.68	23 m	142 km	24 days
Near IR	0.77 - 0.86	23 m	142 km	24 days
Shortwave IR	1.55 - 1.70	70 m	148 km	24 days
WiFS				
Red	0.62 - 0.68	188 m	774 km	5 days
Near IR	0.77 - 0.86	188 m	774 km	5 days

The panchromatic sensor, in addition, to have high spatial resolution, can be tilted to have a stereoscopic vision (tilt angle $+26°$ to $-26°$ across-track), and a revisit periods as long as five days. The WiFS sensor is similar to NOAA AVHRR bands and the spatial resolution and coverage is useful for regional scale vegetation monitoring.

15.5.3.4 ERS European Remote Sensing Satellite, European Space Agency

The ERS-1, 2 satellites use active and passive remote sensing to study global measurements of sea wind and waves, ocean and ice monitoring, coastal studies and land use. ERS-1 was launched in July 1991 and ERS-2 was launched in April 1995. It uses a **synthetic aperture radar (SAR)** instrument to acquire images of ocean, ice and land regardless of cloud and sunlight conditions.

ERS-2 is practically identical to ERS-1, with the addition of the GOME sensor for global ozone monitoring. ERS-2 follows the same ground track as ERS-1, except for a 1-day delay. This provides an opportunity to radar interferometry an area using the **synthetic aperture radar** on the two satellites. As ERS-1 has outlived its planned operational life and was deactivated after the tandem mission, only ERS-2 remains in active operation.

ERS-1, 2 Orbit parameters

Type	Sun-Synchronous
Altitude	782 km
Inclination	98.5 deg
Period	100 min
Repeat Cycle	35 days

The ERS has four sensors - AMI, RA-1 ATSR, GOME

The *AMI* operates at a frequency of 5.3 GHz (**C-band**, **VV-polarised**) combines the functions of a Synthetic Aperture Radar (SAR) and a Wind Scatterometer (WNS) and produces a high resolution images (SAR mode) of the Earth's surface

RA-1 (Radar Altimeter): The Radar Altimeter measures variations in the satellite's height above sea-level.

ATSR (Along Track Scanning Radiometer): This passive instrument monitors the thermal emission of the seas and oceans, from which the global sea surface temperature is derived.

GOME, on ERS-2 (Global Ozone Monitoring Experiment): It is a passive spectrometer for monitoring the ozone content of the atmosphere.

15.5.3.5 JERS - 1

NASDA (Japanese Space Agency) launched JERS-1 in February 1992. This satellite carries an L-band SAR and an optical sensor for generation of global data set in order to survey resources and to establish an integrated Earth observation system.

JERS-1 Orbit parameters

Type	Sun-Synchronous
Altitude	568 km
Inclination	97.7 deg
Period	96 min
Repeat Cycle	44 days

15.5.3.6 RADARSAT, Canada

RADARSAT is a Canadian satellite launched in November 1995, with the launch service provided by NASA, USA. This satellite is operated by the Canadian Space Agency (CSA)/Canadian Center for Remote Sensing (CCRS) and is used for gathering global data on ice conditions, crops, forests, oceans and geology. Using a single frequency (C-Band), the RADARSAT SAR has the unique ability to shape and steer its radar beam over a 500 kilometer range. A variety of beam selections can be made to change image swath from 35 kilometers to 500 kilometers with resolutions from 10 meters to 100 meters respectively. Incidence angles range from less than 20 degrees to more than 50 degrees.

Orbit parameters

Type	Sun-Synchronous
Altitude	798 km
Inclination	98.6 deg
Period	100.7 min
Repeat Cycle	24 days

Sensor characteristics

SAR (Synthetic Aperture Radar): The SAR is able to operate in several beam modes:

o Standard: Seven beam modes with incidence angle ranging from 20 to 49 deg nominal, 100 km swath width and 25 m resolution.

o Wide: Three beam modes with varying incidence angles, 150 km swath width.

o Fine: Five beam modes with 50 km swath width and resolution better than 10 m.

o Scansar: Wide swath width (300 - 500 km) with a coarser resolution of 50 to 100 m.

o Extended mode.

MODE	RESOLUTION (m) Range 1 x azimuth (m)	LOOKS 2	WIDTH (km)	INCIDENCE ANGLE 3 (degrees)
Standard	25 x 28	4	100	20-49
Wide - 1	48-30 x 28	4	165	20 - 31
Wide - 2	32-25 x 28	4	150	31 - 39
Fine resolution	11-9 x 9	1	45	37 - 48
ScanSAR narrow	50 x 50	2 - 4	305	20 - 40
ScanSAR wide	100 x 100	4 - 8	510	20 - 49
Extended (H)	22-19 x 28	4	75	50 - 60
Extended (L)	63-28 x 28	4	170	10 - 23

RADARSAT SAR Instrument Characteristics

Frequency/wavelength	5.3 GHz (C band)/ 5.6 cm
Polarization	Linear HH
Bandwidth	11.6, 17.3 or 30.0 MHz
Peak power	5 kW
Antennae size	15 m x 1.5 m
Incidence angle	Mode dependent
Resolution	Mode dependent

15.5.3.7 Meteorological Satellites

- *NOAA-GOES (Geostationary Operational Environmental Satellite), USA*: A series of geostationary satellites for meteorological application.

- *NOAA-POES (Polar-Orbiting Operational Environmental Satellite), USA*: A series of satellites in sun-synchronous near-polar orbits for continuous meteorological observation, sounding of atmospheric profiles and measurement of energy budgets. The current satellites are named as *NOAA-9, NOAA-10, NOAA-11 and NOAA-12*, with more satellites being planned. Sensors onboard the satellites include the *AVHRR* (Advanced Very High Resolution Radiometer), a passive optical and infrared sensor for global measurement of cloud coverage, sea surface temperature and vegetation.

- *METEOSAT, European Space Agency*: A series of geostationary satellites covering Europe, Africa and the Atlantic ocean for weather forecast and meteorological studies.

- *Fengyun-1, Peoples Republic of China*: A series of meteorological satellites in sun-synchronous near-polar orbits.

- *Fengyun-2, Peoples Republic of China*: Geostationary meteorological satellite.

- *GMS (Geostationary Meteorological Satellite), Japan*: A series of geostationary meteorological satellites, known by the name Himawari.

- *DMSP (Defense Meteorological Satellite Program), USA*: A series of meteorological satellites launched by the US Department of Defense.

- *INSAT (Indian National Satellite System), India:* A series of geostationary satellites for meteorological observation and telecommunication over India and the Indian Ocean.

- *METEOR, FSU/Russia*: Series of meteorological satellites in sun synchronous or non-sun synchronous orbits.

15.6 Digital Image

In digital remote sensing features on the earth's surface are mapped simultaneously by several digital cameras with filters corresponding to the wavelength bandwidth. A digital image in gray scale is a data matrix having integer values representing the average of light energy reflected from a square area on the earth's surface (Fig. 15.14). A multi-spectral scanner aboard a satellite, thus, generates a set of data matrices each representing the image of the same area of the earth in narrow wavelength bands. Overlaying these matrices by mapping each one in a different color plane of the computer graphic creates a pseudo-color (false color) image of the earth's surface (Fig. 15.3).

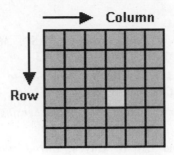

Fig. 15.14 Two dimensional array of pixels

Each element in a digital image is called a pixel which represents a square area on the ground. The spatial resolution of a satellite image is measured by the pixel dimension. The spectral resolution of an imaging system is measured by the wavelength bandwidth and number of such bands (number of cameras - each recording one band). The radiometric resolution of the imaging system is a measure of the number of bits used to represent the gray scale intensity values of the reflected light.

Images from different satellites can also be combined to generate a multilayer image. For example in a three band pseudo color combination one band can be taken from IRS, one from the SAR and another representing the digital elevation data of an area.

15.7 Conclusions

The science of remote sensing covers a wide spectrum and this article gives a view through a small window. The materials of this article have been gathered from many sources, some which are from articles and notes obtained through internet. A detail list of references are omitted to keep this article short. Readers who are interested to learn more about remote sensing may refer to these books for further information. The diagrams and a few images are taken from personal collection of information from various sources available in internet domain.

15.8 References

1. American Society of Photogrammetry (ASP), *Manual of Remote Sensing*, 2nd ed,. ASP, Falls Church, VA 1983.

2. Barett,E.C., and L.F. Curtis, *Introduction to Environmental Remote Sensing* , 3rd ed., Chapman & Hall, New York, 1992.

3. Campbell, J.B., *Introduction to Remote Sensing*, Guilford, New York, 1987.

4. Cracnell, A.P, and L.W.B. Hayes, *Introduction to Remote Sensing*, Taylor & Francis, Washington D.C., 1991.

5. Elachi, C., *Introduction to Physics and Techniques of Remote Sensing*, Wiley, New York, 1987.

6. Lillesand, T.M., and R.W. Kiefer, *Remote Sensing and Image Interpretation*, 3rd ed., John Wiley & Sons, Inc., New York, 1994

7. Lo, C.P., *Applied Remote Sensing*, Wiley, New York, 1986.

8. Schanda,E., *Physical Fundamentals of Remote Sensing*, Springer - Verlag, New York, 1986.

9. Slater, P.N., *Remote Sensing: Optics and Optical Systems*, Addison - Wisely, Reading, MA, 1980.

10. Swain,P.H., and S.M. Davis, (eds.), *Remote Sensing: The Quantitative Approach*, Macgraw Hill, New York, 1978.

Fundamentals of Remote Sensing, JARS.

PART E

Intelligent and Virtual Instrumentation

| Chapter 16 |

Sensor Fusion & Estimation
in Instrumentation

S. Bhattacharya
R. Perla
S. Mukhopadhyay

Abstract

Signals from several sensors, measuring different process parameters, can be used to compute an estimate of a measurement. The method of combining several individual sensors is known as "sensor fusion" and can be used for purposes such as, improvement of accuracy of measurement or estimating a process parameter in case of failure of the corresponding sensor. Two methods of estimation have been discussed in this chapter: One using ANN for a complex industrial process; the other using Kalman filtering in presence of noise. The efficiency of the methods has been elaborated using two illustrated examples.

16.1 Introduction

There are some situations where several sensors are to be employed for the measurement of a given quantity. This may be for reasons of direct measurement being infeasible, or to improve accuracy, reliability and speed of measurement. The individual sensors provide their own individual measurements which are then combined or "fused" in a suitable manner to produce the desired measurement which is to be used further for other purposes - typically, control and command generation. Such a technique of combining a number of individual physical sensor measurements to compute a "virtual measurement" of a system related quantity, is called "sensor fusion" or "data fusion". The process of "fusion" uses estimation techniques, which essentially are signal processing methods that compute signals and/ or parameters related to a system, based on measurements of one or more signals, and system as well as signal models.

Situations in which sensor fusion is used can mainly be categorised into the following cases. Among these cases, we provide two case studies in this article. Other cases may be found elsewhere in the book, as mentioned below.

A. If the quantity of interest cannot be measured directly given the requirement of speed and constraints of the equipment set up, the measurand is expressed as a function of several directly measurable quantities, from which the value of the measurand is estimated. Typical examples of these would be on-line measurement of tool wear in machining (for details refer to Chapter 17 on Monitoring of Manufacturing Systems), on-line measurement of product quality in a chemical process, the measurement of the volume of a steel ingot for a given length or volumetric flow measurement for a single phase, say gas, in a two-phase flow situation such as gas/liquid (for details refer to Chapter 23 on Measurement of Two Phase Flow Parameters).

B. For mission critical applications, multiple sensors are often used in redundant configuration to achieve fault tolerance. Typical examples include triple sensors voting schemes in process application such as steam boilers, for power plants. In other cases where detection of sensor faults is needed to make the system failure tolerant, a redundant measurement of a given sensor is created using model based estimation techniques based on sensors other than the given one. Here, we provide a case study of sensor failure tolerant scheme for a chemical process using Artificial Neural Networks.

C. **For reasons of improved accuracy and reliability, several measurements of the same variable are made using a number of different sensors, often using different measurement principles.** These measurements are then combined by suitable signal processing techniques to create a more accurate estimate of the quantity of interest. Typical examples include target-tracking applications for aerospace systems. Under this category, we provide a case study of a Kalman filter based estimator that employs measurements from a number of tracking sensors.

D. **For making measurements of attributes in spatially distributed systems, multiple sensors are needed.** Sometime one needs measurement of a quantity that is required to be constructed from several local measurements spread over an extent of space. Typical examples include hot spot measurement in equipment such as reactors, boiler tubes or power transformer windings, temperature profile measurements.

Below we present two case studies that demonstrate how a "virtual" measurement of a process variable can be made using estimation techniques from measurement of some physical process variables. The case studies presented below have been developed in the course of executing student as well as sponsored research projects in the Department of Electrical Engineering, IIT Kharagpur.

16.2 Multiple Sensor Fusion using Analytical Redundancy for Sensor Fault Tolerance

In complex processes, such as chemical processes, power plants, automated vehicles, and aircraft a large number of sensors are normally used for monitoring and control. Changes in sensor values directly affect the outputs of the controllers or the decisions of the operators. Failure in generating appropriate control actions in presence of invalid sensor values may lead to total system shutdown or hazards creating significant economic losses and sometimes even endangering both system and human safety. Hence, the safety, reliability, and performance of the system are largely dependent on the validity and accuracy of the sensors that are used. To improve the operational reliability it is necessary to validate the measured sensor data, isolate any failed sensor, and recover the failed measurement.

Sensor fault detection in dynamic processes has been achieved using various methods including neural networks [1-12].

The basic philosophy used to achieve sensor validation is that of analytical redundancy. For determining whether the sensor being validated is generating meaningful data, the sensor data must be compared against measurements from other sensors, to fit known dynamic relationship a priori as the process model. A way of doing this is to estimate a 'redundant' or additional copy of the sensor data using other sensors of the process and the model. As long as the sensor is valid, the original and the redundant copies are supposed to match reasonably closely. Thus, measurements from different process sensors are fused into a redundant version of the sensor measurement using the process model. However, for many processes, it is indeed difficult to build a process model from process physical chemistry. In such cases a general model structure such as that of an ANN has to be tuned based on process data using a data fitting technique such as back propagation learning algorithm. Presented below is a case study for such an application of sensor fusion to a well known chemical process.

16.2.1 Case Study: Sensor Validation using Artificial Neural Networks

In this case study, a methodology is presented for intelligent sensor validation and sensor fault detection for the Multi Component Distillation Column (MCDC), which is used very commonly in chemical plants, petroleum refineries, and also for the final stages of purification, where products are most valuable and quality specifications are most rigid. Distillation column consists of number of trays, reboiler, condenser and an accumulator. The trays are used for intimate mixing of two phases, namely, vapor phase and liquid phase. The manipulated variables are the reboiler heat input, feed rate, and reflux rate. Different temperature sensors are located on different trays for controlling and monitoring of the MCDC. The measured variables are the bottom tray temperature (T_1), temperature of 10th tray (T_{10}), and top tray temperature (T_{15}). For a relative constant pressure distillation column, the top and bottom product purities are related to the bottom and top tray temperatures respectively. The tray temperatures are controlled using cascade control strategy. Additional temperature sensor at the 10th tray is used for monitoring purpose. The detailed description of the process and open loop simulation of MCDC is given in Luyben[13], Ramirez [14].

Fault detection is carried out by comparing the residual, e.i. difference between the measured and estimated values, to a specified threshold and then isolation methods are applied. A well-known approach for sensor fault detection and isolation (FDI) in linear systems is based on Dedicated Observer scheme [1] or Generalized Observer scheme proposed by Frank [12]. The same kind of approach is extended for complex non-linear systems using Neural Generalized Observer Scheme (NGOS), proposed by Marcu, Mirea and Frank [23]. The NGOS consists of as many numbers of observers as the outputs to be supervised. Each observer is driven by all the process inputs and all outputs except the sensor to be supervised. The basic architecture for single sensor fault tolerant control in case of three sensors is shown in Fig. 16.1.

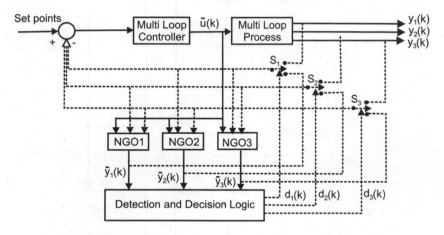

Fig. 16.1 Neural Generalized Observer Scheme for three sensors

Here, local set points include all the set points in the process. The multiloop controller also includes all control loops. A Neural Generalized Observer (NGO) estimates each sensor value. The NGO is driven by all process inputs and the outputs, except the sensor to be supervised. The measured variables from the process are fed to the bank of ANN-based estimators. Each of the ANN estimator estimates the sensor value from the remaining sensors and inputs. All these estimated values are then sent to the decision block. When a single sensor develops a fault, since the corresponding estimator takes inputs from the other sensors which are normal, the ANN is able to estimate correctly the normal values of the process variable for the failed sensor. Since this is likely to be significantly different from the actual output of the failed sensor, a fault can be detected. Upon detection, firstly the feedback of the process variable corresponding to the faulty sensor is switched over from sensor measurements to the estimates of the corresponding ANN estimator. The switches S_1, S_2 and S_3 are controlled by the decision signals $d_1(k)$, $d_2(k)$, and $d_3(k)$ generated by the decision block. Secondly a similar switchover is made for the other ANN estimators that use the process variable corresponding to the faulty sensor as an input. A more generalized case is shown in Fig. 16.2, in which a set of switches are controlled by the dashed arrows originating from the decision block of the corresponding estimator.

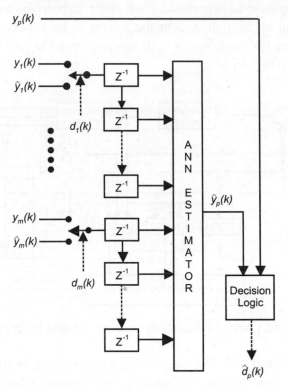

Fig. 16.2 ANN Estimator in SFTC architecture

Training Data Selection

In any dynamical system, the outputs are dependant on the past values of the inputs and the outputs are well correlated. Hence, the discrete-time model to be estimated by the neural network, is described by the difference equation:

$$\hat{y}_i(k) = \hat{f}(\overline{y}_1, \overline{y}_2 \ldots, \overline{y}_{i-1}, \overline{y}_{i+1}, \overline{y}_{i+2}, \ldots \overline{y}_p) + g(\overline{u}_1, \overline{u}_2 \ldots, \overline{u}_m)$$

where,

$$\overline{y}_j = [y_j(k), y_j(k-1) \ldots, y_j(k-n+1)],$$

$$\overline{u}_i = [u(k-k_d), u(k-k_d-1).....,u(k-k_d-n+1)]$$

Here $\hat{f}$ represents the nonlinear correlation function of a sensor with the present and past values of other sensors, $\hat{g}$ represents the nonlinear function between the inputs and the sensor value to be estimated, $\overline{u}_i$ is the input vector consisting of past n values considering its delay samples, and $\overline{y}$ is the output vector consisting of the past n values, $\hat{y}_i$ is the estimate of the ith sensor using past n values of the input and other output signals except the ith output. Using this mathematical approximation model between a sensor value and input vectors, the neural network architecture is selected. In order to train the neural network, the training data is taken such that it includes all kinds of possible changes in the operating range of each sensor. For this purpose, the operating range of each input is divided into three sub ranges. The set points are chosen from each sub range and in different sequences so that the operating range is covered well. The validation data for the ANN is also generated using the same method as for training, but with different sequence of set points. The generated training and validation sets, in the open loop process, are sampled with a suitable sampling time.

ANN Model Selection and Training

Feed-forward multi-layer perceptron ANN is selected as the estimator and the nonlinear activation function for the network is taken as a sigmoid function. As this sigmoid function lies between 0 and 1, the sampled data is normalized such that the normalized values lie between 0 and 1. The normalized data is used for training and validating the neural network. Each network is trained with the past three values of all manipulated variables and the past three values of all the sensors except the sensor to be trained. Training is performed using the standard back propagation algorithm [24] and the network is trained in batch-mode. The performance of the network is estimated using prediction from a different test set using Root Mean Square Error (RMSE) as the stopping criteria. Three individual networks are trained, each representing a single sensor to be monitored from other sensors and inputs.

16.2.2 Estimator Performance in FDI

After the proper training of each individual network, the Neural Generalized Observers are put in the closed loop plant operation. The performance of the NGOS is demonstrated using the following case studies.

Case- a

In this case the estimator performance in the closed loop is investigated. Different set point changes are applied to bottom tray temperature with an interval of 3 hours and the plant is operated for 15hrs. The data for estimation is selected at each time instant before the measurement is fed back to the controller. The estimator performance in the closed loop plant operation is shown in Fig.16.3. This figure shows that the estimator response is very close to the measured value and can be used for sensor fault tolerant control.

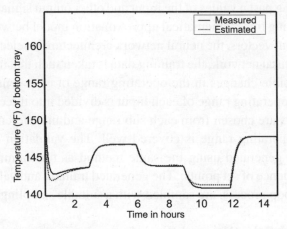

Fig. 16.3 Bottom tray temperature estimator performance without fault

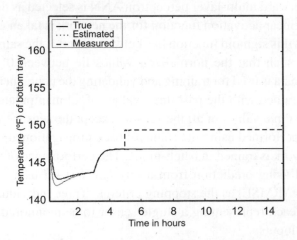

Fig. 16.4 Bottom tray temperature estimator performance with fault

Case- b

In this case the estimator is tested with a step bias fault in bottom tray temperature sensor. A step fault of 10% of normalized value is applied to bottom tray temperature sensor at 5 hours and the plant is operated for 15 hours in closed loop. The performance of the estimator in case of this single sensor fault is shown in Fig. 16.4. The estimated value tracks well the true sensor value and deviates largely from the faulty measured value. The difference between the measured and the estimated values of each sensor is calculated and suitable thresholds are chosen for each estimator. As the bottom tray temperature shows more error than the other two estimator errors and crosses the threshold, it can be concluded that the bottom tray temperature sensor is faulty.

Case- c

The estimator robustness, with disturbance after fault occurrence, is tested. The disturbance is applied by changing the feed rate. As in the previous cases a step bias fault of 10% on the normalized value of the bottom tray temperature sensor is applied at 5 hours and then disturbance is applied in this faulty condition. The feed rate is changed by 20% of its normalized value and applied at 6 hours, 9 hours and 12 hours. The estimator response is shown in Fig. 16.5. The error between the estimated values and the faulty measurement value is very large, the true value and estimated values are coinciding.

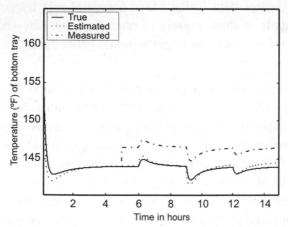

Fig. 16.5 Bottom tray temperature estimator performance with fault and set-point change

Observations

This case study has demonstrated an effective sensor fault tolerant control strategy for a realistic simulation of distillation process. The fault detection is based on comparison of the measured sensor data with the value estimated by an ANN based estimator. The scheme is shown to be effective in single sensor failures.

16.3 Sensor Data Fusion to Improve Measurement Accuracy in Presence of Noise

Often, two or more sensors are employed to measure the same set of measurands. These sensors may be utilising different physical phenomena for sensing. It is therefore likely that, the signal related to the measurand is going to be correlated. But the errors in the signals are not likely to be so, since, they are generated through distinct physical phenomena. One may therefore obtain an estimate of the measurand, which is better than any of the individual measurements, by rejecting the uncorrelated parts of the measurements. For example, in the case of tracking an airborne target, a dynamic target is tracked by N disparate sensors such as various types of radar, optical sensors, IR sensors as well as inertial navigation systems, each with different measurement dynamics and noise characteristics. The problem is to obtain the estimates of the kinematic state of the target based on the observations of the sensors. But it is not possible to deduce a comprehensive picture about the target state from the evidences of individual sensors alone, due to the inherent limitations of their technical features characterizing spatial and temporal coverage, noise characteristics etc. In this context sensor data fusion plays a crucial role by combining the measurements from these sensors to obtain a joint state-vector estimate representing the motion of the target, which is better than the measurements of individual sensors.

There are various data fusion approaches to resolve this problem, of which Kalman filter based technique is one of the most significant. For an introductory exposition to Kalman filters, among many excellent references, the reader is referred to [20]. Among different algorithms for Kalman Filter-based sensor fusion, two commonly employed techniques are (i) State-Vector Fusion and (ii) Measurement Fusion [21]. The state-vector fusion method uses covariance of the filtered output of individual noisy sensor data to obtain an improved joint state estimate. Whereas, the measurement fusion method directly fuses the sensor measurements to obtain a weighted or combined measurement and then uses a single Kalman Filter to obtain final state estimate based on the fused measurement. The two approaches are depicted pictorially in Figs. 16.6(a) and 16.6(b). Both the approaches have their own merits and demerits. Although the measurement fusion method provides better overall estimation performance, state-vector fusion has lower computational cost and has the advantage of parallel implementation and fault tolerance. Judicious trade-off between computational complexities, computational time and numerical accuracy is to be made for selection of algorithm for practical application. A brief outline of these sensor fusion techniques in the context of target racking is discussed in the following section.

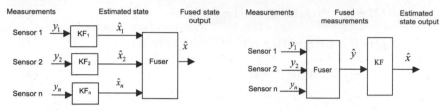

Fig 16.6(a) State-vector Fusion   **Fig 16.6**(b) Measurement Fusion

State Space Model of Target and Measurement Dynamics

Consider the kinematic model of a target, which is tracked by N sensors. The target dynamics and the sensors are modeled by the following discrete-time state-space model:

$$x(t) = A(t)x(t - 1) + B(t)u(t) + G(t)v(t) \tag{1}$$

$$y_j(t) = C_j(t)x(t) + w_j(t), \quad j = 1, 2 ..... N \tag{2}$$

where, t represents the discrete-time index, $x(t)$ is the state-vector, $u(t)$ the input vector, $y_j(t)$ measurement vector, $v(t)$ and $w_j(t)$ zero-mean white Gaussian noise with covariance matrices $Q(t)$ and $Rw_j(t)$, respectively. $A(t)$ is the system matrix, $B(t)$ the input matrix, $G(t)$ system noise dynamics, and $C_j(t)$ the measurement matrix for j- th sensor, each with appropriate dimension. It is assumed that the measurement noise sequences of different sensors are independent. Since, the presently discussed sensor fusion techniques rely on the estimates from Kalman filters, a very brief description of the Kalman Filter is enumerated here.

16.3.1 The Kalman Filter

It is a recursive linear estimator that provides an unbiased and optimal estimates (in the sense of minimum estimate covariance) of the state of a dynamic system as well as a measure of accuracy of its estimation in terms of state error covariances, based on the assumed model of the system and measurement dynamics. It utilises the second order statistics of the system and measurement noise characteristics. The two fundamental steps of Kalman filtering are (i) time update or prediction, and (ii) measurement update or correction.

In time update, the estimated state and state error covariances at $(t-1)$-th instant are projected ahead for t-th instant using the knowledge of system dynamics and deterministic inputs to the system as per the following equations:

$$\hat{x}(t \mid t-1) = A(t)\hat{x}(t-1 \mid t-1) + B(t)u(t) \tag{3}$$

$$\hat{P}(t \mid t-1) = A(t)\hat{P}(t-1 \mid t-1)A^T(t) + G(t)Q(t)G^T(t) \tag{4}$$

$$\hat{y}(t \mid t-1) = C(t)\hat{x}(t \mid t-1) \tag{5}$$

where,

$\hat{x}(t-1 \mid t-1) =$ estimated state at $(t-1)$-th instant using measurements from 0 to $(t-1)$-th instant,

$\hat{P}(t-1 \mid t-1) =$ estimated state error covariance at $(t-1)$-th instant using the measurement from 0 to $(t-1)$-th instant,

$\hat{x}(t \mid t-1) =$ predicted state at t-th instant using measurements from 0 to $(t-1)$-th instant,

$\hat{P}(t \mid t-1) =$ predicted state error covariance at t-th instant using measurements from 0 to $(t-1)$-th instant,

$\hat{y}(t \mid t-1) =$ predicted measurement at t-th instant using measurements from 0 to $(t-1)$-th instant.

The predicted state and state error covariances at t-th instant are corrected using the measurement at t-th instant and estimated state and state error covariances are as follows:

$$K(t) = \hat{P}(t \mid t-1)C^T(t)\left[C(t)\hat{P}(t \mid t-1)C^T(t) + R(t)\right]^{-1} \tag{6}$$

$$\hat{x}(t \mid t) = \hat{x}(t \mid t-1) + K(t)\left[y(t) - \hat{y}(t \mid t-1)\right] \tag{7}$$

$$\hat{P}(t \mid t) = \left[I - K(t)C(t)\right]\hat{P}(t \mid t-1) \tag{8}$$

Here,

$K(t) =$ Kalman gain at t-th instant,

$y(t) =$ Measurement at t-th instant,

$I \quad =$ Identity matrix with appropriate dimension.

A simple illustration can show the efficacy of Kalman filter for estimating the kinematic state (specially the higher order ones, i.e. velocity, acceleration etc.) of a target for the situation when a single sensor tracking the target produces noisy measurements of position only. The most intuitive way to estimate the velocity is by numerical differencing of the available position measurement. However, this leads to a very undesirable situation as evident from Figure 16.7. The figure shows velocity estimates based on noisy position measurement by numerical difference as well as by Kalman filter. The noise in measurement gets magnified while numerical differencing and thus results in extremely noisy velocity estimates, which does not reflect the actual state of the target. However, Kalman filter produces practicable solution in this regard, by utilizing the dynamics of the target and also through measurement and noise statistics of the sensor.

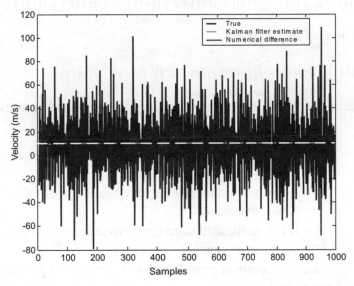

Fig. 16.7 Velocity estimation from noisy position measurement

Further, the formulation of Kalman filter is extremely useful for use in practical application, as the Kalman gain sequence is automatically adjusted (as per eqn. (6)) based on assumed target dynamics and noise characteristics. Again, the state error covariances produced by Kalman filter not only provides a measure of degree of accuracy of estimates, but also opens up the opportunity for other Kalman filter based algorithm to exploit this information. For example, the state-vector fusion algorithm uses the weighted combination of state error covariances of individual sensor based estimates to provide a combined improved state estimate. Further, the recursive nature of the filter formulation avoids the requirement of storing the past measurements, and thus becomes useful for different measurement fusion algorithms, which have to handle large amount of observations.

State-vector Fusion

In this method, individual sensor data is filtered using a Kalman Filter (KF) that utilizes measurement noise characteristics of respective sensors. The state and state-error-covariance estimate of Kalman filter (KF) for each of the sensors are then used to obtain the fused state according to the following equations [22]. The following expressions are shown for two sensors.

Fused state:

$$\hat{x}(t \mid t) = \hat{x}_1(t \mid t) + \hat{P}_1(t \mid t) \left[\hat{P}_1(t \mid t) + \hat{P}_2^T(t \mid t) \right]^{-1} \left[\hat{x}_2(t \mid t) - \hat{x}_1(t \mid t) \right] \qquad (9)$$

Covariance of fused state:

$$\hat{P}_F(t \mid t) = \hat{P}_1(t \mid t) - \hat{P}_1(t \mid t) \left[\hat{P}_1(t \mid t) + \hat{P}_2^T(t \mid t) \right]^{-1} \hat{P}_1^T(t \mid t) \qquad (10)$$

where $\hat{x}_1(t \mid t) =$ KF estimates of target state based on sensor 1 at t-th instant

        $\hat{x}_2(t \mid t) =$ KF estimates of target state based on sensor 2 at t-th instant

        $\hat{P}_1(t \mid t) =$ estimated state error covariance of sensor 1 at t-th instant

        $\hat{P}_2(t \mid t) =$ estimated state error covariance of sensor 2 at t-th instant

In this case, the common process noise affecting the target dynamics corresponding to each of the sensors is assumed to be negligible.

Measurement Fusion

Measurement fusion methods directly fuse the sensor measurements to obtain a weighted or combined measurement and then use a single Kalman filter to obtain the final state estimate based upon the fused observation. Currently there exist two commonly used measurement fusion methods using the Kalman filter. The first (referred to here as Method I) simply merges the multi sensor data through the observation vector of the Kalman filter, whereas the second (referred to here as Method II) combines the multi-sensor data based on a minimum-mean-square-error criterion.

Measurement Fusion Method I

The measurement fusion Method I integrates the sensor measurement information by augmenting the observation vector as follows [25, 26]:

$$y^{(I)}(t) = [y_1(t) \cdots y_N(t)]^T \tag{11}$$

$$C^{(I)}(t) = [C_1(t) \cdots C_N(t)]^T \tag{12}$$

$$R^{(I)}(t) = diag[R_1(t) \cdots R_N(t)] \tag{13}$$

This augmented measurement vector is then used by standard Kalman filter (Equations (3)-(8)) for estimating the target state.

Measurement Fusion Method II

The measurement fusion Method II obtains the fused measurement information by weighted observation as [19]:

$$y^{(II)}(t) = \left[\sum_{j=1}^{N} R_j^{-1}(t)\right]^{-1} \sum_{j=1}^{N} R_j^{-1}(t) y_j(t) \tag{14}$$

$$C^{(II)}(t) = \left[\sum_{j=1}^{N} R_j^{-1}(t)\right]^{-1} \sum_{j=1}^{N} R_j^{-1}(t) C_j(t) \tag{15}$$

$$R^{(II)}(t) = \left[\sum_{j=1}^{N} R_j^{-1}(t)\right]^{-1} \tag{16}$$

This weighted combination of measurement vector is used in standard Kalman filter equations for obtaining state estimates.

Information Filter form of the Kalman Filter

If we define an information state vector:

$$\hat{z}(t_1 \mid t_2) \equiv \hat{P}^{-1}(t_1 \mid t_2)\hat{x}(t_1 \mid t_2),$$

where $\hat{P}^{-1}(t_1 \mid t_2)$ is called the information matrix, then the standard Kalman filter can be transformed into the following information form.

Time update:

$$\hat{z}(t \mid t-1) = \hat{P}^{-1}(t \mid t-1)A(t)\hat{P}(t-1 \mid t-1)\hat{z}(t-1 \mid t-1) + \hat{P}^{-1}(t \mid t-1)B(t)u(t) \tag{17}$$

$$\hat{P}(t \mid t-1) = A(t)\hat{P}(t-1 \mid t-1)A^T(t) + G(t)Q(t)G^T(t) \tag{18}$$

Measurement update:

$$\hat{z}(t\,|\,t) = \hat{z}(t\,|\,t-1) + C^T(t)R^{-1}(t)y(t) \tag{19}$$

$$\hat{P}^{-1}(t\,|\,t) = \hat{P}^{-1}(t\,|\,t-1) + C^T(t)R^{-1}(t)C(t) \tag{20}$$

The information form of the Kalman filter, or the information filter, is functionally equivalent to the Kalman filter, but have some computational advantages, especially in multi-sensor data fusion where the innovations covariance matrix $\left[C(t)\hat{P}(t\,|\,t-1)C^T(t) + R(t)\right]$ is usually of high dimension and non-diagonal. The major difference between the standard Kalman filter and the Information filter lies in their measurement update phase. The measurement update of the information filter is simpler as the gain matrix $K(t)$ in the standard Kalman filter is more complex than the term $C^T(t)R^{-1}(t)$ in the information filter, especially in multi sensor data fusion where the inversion $\left[C(t)\hat{P}(t\,|\,t-1)C^T(t) + R(t)\right]^{-1}$ in $K(t)$ may become computationally prohibitive, when the number of sensors is large.

16.3.2 Simulation Results

For experimental comparison of the state-vector fusion method and the two measurement fusion methods, simulation results are obtained in this section using the following target and sensor models:

$$x(t) = \begin{bmatrix} 1 & T \\ 0 & 1 \end{bmatrix} x(t-1) + \begin{bmatrix} T^2/2 \\ T \end{bmatrix} v(t) \tag{21}$$

$$y_j(t) = C_j x(t) + w_j(t) \tag{22}$$

where the sampling time $T = 0.1$ sec, $v(t) = N(0, Q)$ and $w_j(t) = N(0, R_j)$ for a multivariate normal distribution of mean μ and variance Σ. The simulation studies here are focused on the comparison of the fused state estimate covariance matrices of state-vector fusion and the two measurement fusion methods for a scenario where the target described by equation (21) is being tracked by two sensors (i.e. $j = 1, 2$). The model described by (21)-(22) satisfy the conditions under which the fused state estimate covariance $\hat{P}(t\,|\,t)$ will converge to a steady-state value [27], which is denoted here by

$$\begin{bmatrix} \hat{P}_{11} & \hat{P}_{12} \\ \hat{P}_{12} & \hat{P}_{22} \end{bmatrix}.$$

The results are shown in the form of plots of $\hat{P}_{11}$, $\hat{P}_{12}$ and $\hat{P}_{22}$ versus process noise covariance (Q) for Q varying from 0.0001 to 100 m^2/sec^4. The following cases are studied.

Case I

Two sensors are assumed to have identical measurement matrices, which are set as $C_1 = C_2 = [1 \ 0]$, with same sensor noise variances $R_1 = R_2 = 4m^2$. The results for state-vector fusion and two measurement fusion algorithms are shown in Figs. 16.8(a)-3(c).

Case II

In this case also the measurement matrices of the two sensors are identical, with $C_1 = C_2 = [1 \ 0]$, but sensor noise variances are different, e.g., $R_1 = 4m^2$ and $R_2 = 25m^2$. The results are shown in Figs. 16.9(a)-(c).

Case III

Here, the two sensors are having different measurement matrices set as $C_1 = [1 \ 0]$ and $C_1 = [1 \ 0.5]$. The sensor noise variances are same for both the sensors $R_1 = R_2 = 4m^2$. The results are plotted in Figs. 16.10(a)-(c).

Case IV

In this case, the two sensors have different measurement matrices as $C_1 = [1 \ 0]$ and $C_2 = [1 \ 0.5]$, but have same sensor noise variances, with $R_1 = 4m^2$ and $R_2 = 25m^2$. The results are shown in Figs. 16.11(a)-(c).

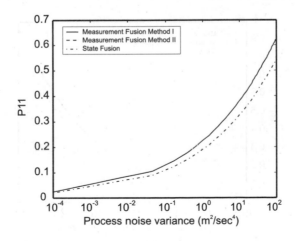

(a)

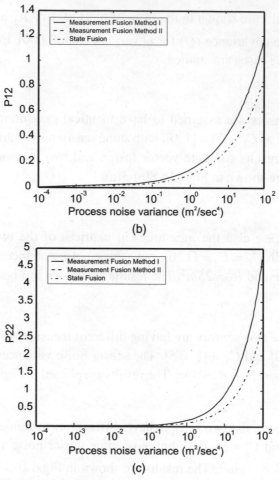

Fig. 16.8(a)-(c) Results for Case I with $C_1 = C_2$ and $R_1 = R_2 = 4m^2$

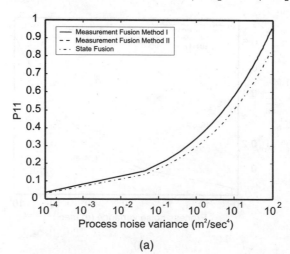

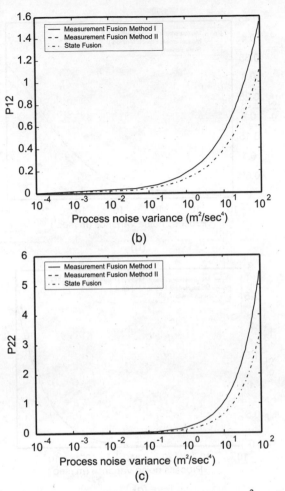

Fig. 16.9(a)-(c) Results for Case II with $C_1 = C_2$, $R_1 = 4m^2$ and $R_2 = 25m^2$

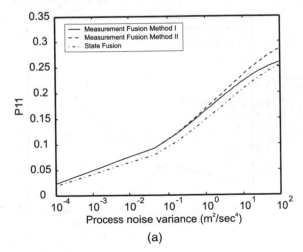

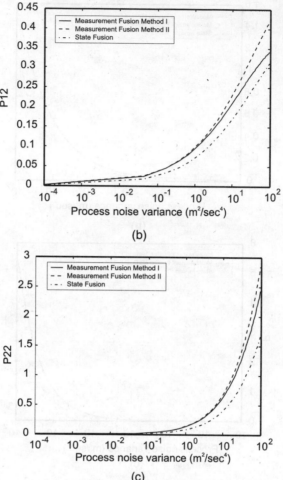

Fig. 16.10(a)-(c) Results for Case III with $C_1 = [1\ \ 0]$, $C_2 = [1\ \ 0.5]$ and $R_1 = R_2 = 4m^2$

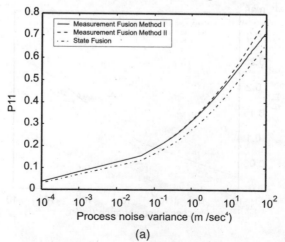

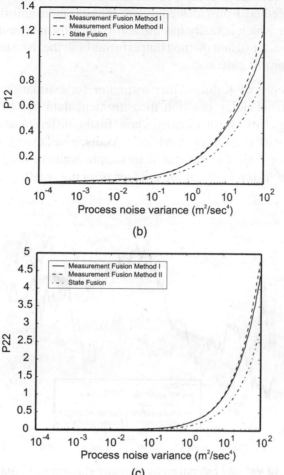

Fig. 16.11(a)-(c) Results for Case IV with $C_1 = [1\ \ 0]$, $C_2 = [1\ \ 0.5]$ and $R_1 = R_2 = 4m^2$ and $R_2 = 25m^2$

From the results it is apparent that the performance of the two measurement fusion algorithms are identical when the sensors have identical measurement matrices, even if their measurement noise variances are different (Case I and II and Fig. 16.8 and 16.9). It can be shown that for identical measurement matrices, the two measurement fusion algorithms are equivalent [21]. However, the state-vector fusion method performs better in both the cases as compared to measurement fusion methods. Moreover, the computational requirement of state-vector fusion and measurement fusion Method II is less than that of measurement fusion Method I.

But when the measurement matrices are different for different sensors (Case III and IV and Figs. 16.10 and 16.11) the performance of the measurement fusion Method I is superior in terms of reduced state error covariance estimates as compared to measurement fusion Method II. This is because Method I provides complete measurement information to the Kalman filter which actually has the ability of information integration by itself. State-vector fusion method outperforms both the measurement fusion techniques for this case also.

The essence of Kalman filter estimator for estimating higher order kinematics from noisy position measurement data has been shown in Fig. 16.7. Fig. 16.12 further clarifies how fusing different sensors data can further improve the estimate of velocity. As discussed above, the efficiency of state-vector fusion as compared to simple Kalman filter and the two measurement fusion methods are evident from the results.

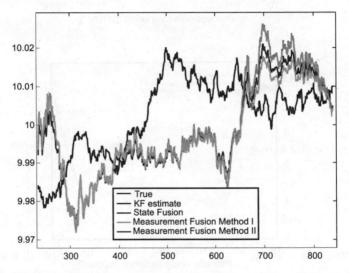

Fig. 16.12 Velocity estimation by standard Kalman filter, State-vector Fusion, Measurement Fusion Method I and Measurement Fusion Method II

In addition, by comparatively examining the formulations of the two measurement fusion methods, we note that Method I is more flexible and will become more efficient in the sense of computational cost, when the number of sensors increases and the measurement matrices of sensors are different, and especially when the measurement matrices and noise characteristics are time-varying. Also, by using the information filter, the inverse of the high-dimensional innovations covariance matrix in Method I can be avoided. On the other hand, computation resource requirement of state-vector fusion is less compared to the other two algorithms and it shows its capability to provide improved estimates than the measurement fusion methods for all the studied cases. The above results are useful in the design of practical multi-sensor data fusion systems for navigation and target tracking.

16.4 Conclusion

There are several contexts of measurements where the measurement of a quantity is enabled, enhanced or eased using measurements of two or more other appropriate quantities. In some cases it enables measurement of some quantities, which are difficult to measure given the situation. In other cases, it enables measurement of a quantity even when the sensor for the quantity fails and therefore renders the system tolerant to sensor failures. In still other cases it achieves improved measurement accuracy in presence of significant measurement noise. Thus, using such methods of estimation and sensor fusion one can effectively construct "instruments" which are superior in measurement compared to the physical instruments that are available.

16.5 References

1. Clark, R.N" Instrument fault detection ", *IEEE trans. Aerospace and Electronic Systems*, 14,3, pp 456-465,1978.
2. Ray, A. and Desai, M., (1984)" A Calibration and Estimation Filter for Multiply Redundant Measurement Systems", *ASME Journal of Dynamic Systems, Measurement, and Control*, vol. 106, pp. 149-156,.
3. Glockler, O., Upadhyaya, Morgenstern, B. R., V. M. and Olvera, J., 1989, "Generalized Consistency Checking of Multivariable Redundant Measurements and Common-Mode Failure Detection", *7th Power Plant Dynamics, control and Testing Symposium Proceedings*, Knoxville, TN, p. 82.01-82.22, vol. 2.
4. Turkcan, E., (1991)"Sensor Failure Detection in Dynamical Systems by Kalman Filtering Methodology", Proceedings of *Dynamics and Control in Nuclear Power Stations*, London, 22-24 October.
5. Guo, T. H. and Nurre, J. (1991) Sensor Failure Detection and Recovery by Neural Networks, *Int. Joint Conference on Neural Networks*, Seattle, USA, p. 221-6, vol. 1.
6. Eryurek, E. and Turkcan, E., (1992), "Neural Networks for Sensor Validation and Plantwide Monitoring", *Nuclear Europe Worldscan*, vol. 12, no. 1-2, pp. 72-74.
7. Venkatasubramanian V. and Chan K. (1989) A neural network methodology for process fault diagnosis. American Institute of Chemical Engineers Journal,35(12), 1993-2005.
8. VenkataSubramanian, V Vaidyanathan., R. and. Yamamoto Y, Process fault detection and diagnosis using neural networks- I. Steady-state processes. Computer chem. Engg. Vol. 14, No. 7, pp. 699-712,1990.
9. Jones W.P. and Hoskins J.C. (1987) Back-propagation, a generalised delta learning rule. BYTE Mag. Oct, 155-162.
10. Hoskins J.C. and Himmelblau, D.M (1988) Artificial neural network models

of knowledge representation in chemical engineering. Computer chem. Engng. Vol 12. No. 9/10, pp. 881-890.

11. Himmelblau D.M. (1978) Fault Detection and Diagnosis in Chemical and Petrochemical Processes. Elsevier, Amsterdam.

12. Rich S.H. and Venkatasubramanian V., (1987) Model-based reasoning in diagnostic expert system for chemical process plants. Comput. Chem. Engng 11, 111-122.

13. William L. Luyben," *Process Modeling Simulation and Control for Chemical Engineers*, McGraw Hill Publishing Company, second edition, pp 132-141, and 1990.

14. Ramirez, W. F." *Computational Methods for Process Simulation*", Butterworth Heinemann, pp. 235-246, 1997.

15. Chang, K. C., Tian, Z., and Saha, R. K. (1998) Performance evaluation of track fusion with information filter. In Proceedings of International Conference on Multisource-Multisensor Information Fusion, 648-655.

16. Chang, K. C., Saha, R. K., and Bar-Shalom, Y.(1997) On optimal track-to-track fusion. IEEE Transaction on Aerospace and Electronic Systems, 34, 4, 1271-1276.

17. Saha, R. K., Chang, K. C. (1998) An efficient algorithm for multi-sensor track fusion. IEEE Transaction on Aerospace and Electronic Systems, 34, 1, 200-210.

18. Saha, R. K.,(1996) Track-to-track fusion with dissimilar sensors. IEEE Transaction on Aerospace and Electronic Systems, 32, 3 , 1021-1029.

19. Roecker, J.A., and McGillem,C.D.(1988). Comparison of two-sensor tracking methods based on state vector fusion and measurement fusion. IEEE Transaction on Aerospace and Electronic Systems, 24, 4, 447-449.

20. Bar-Shalom, Y., and Li. X. R. (1995). Multitarget-Multisensor Tracking: principles and Techniques. Storrs, CT: YBS Publishing,

21. Gan, Q.; Harris, C. J.; '*Comparison of Two Measurement Fusion Methods for Kalman-Filter-based Multi-sensor Data Fusion*', IEEE Transaction On Aerospace and Electronic Systems, Vol. 37, No. 1, January 2001; pp 273-280.

22. Saha, R K; '*Effect of Common Process Noise on two-track Fusion*'; Journal of Guidance, Control & Dynamics; Vol. 19, No. 4, July-August 1996; pp 829-835.

23. Marcu T, Mirea, L and Frank, P M" Neural observer schemes for robust detection and isolation of process faults", *UKACC international Conference on CONTROL'98, Swansea, UK*, pp. 958-963, Sept. 1998.

24. Haykin, Simon"*Neural Networks*", Pearson Education Asia, second edition, 2001

25. Manyika, J., Durrant-Whyte, H. (1994). Data Fusion and sensor Management: a Decentralized Information-Theoretic Approach. New York: Ellis Horwood, 1994.

26. Doyle, R. S., Harris, C. J. (1996). Multi-sensor data fusion for helicopter guidance using neuro-fuzzy estimation algorithms. The Royal Aeronautical Society Journal (June/July 1996), 241-251.

27. Bar-Shalom, Y., Fortman, T. E. (1988). Tracking and Data Association. New York: Academic Press, 1988.

| Chapter 17 |

On-line Monitoring of Manufacturing Processes

A Case Study of Tool Condition Monitoring in Face Milling

S. Mukhopadhyay
A. Patra
A. B. Chattopadhyay
P. Bhattacharyya
D. Sengupta

Abstract

This chapter describes an experimental research work that illustrates how immeasurable process parameters related to a manufacturing process can be estimated accurately and on-line, using real-time measurements from a number of sensors and an estimation algorithm. Steps such as, data acquisition to signal processing, feature extraction and finally the estimator have been explained in the context of the face milling process carried out using a CNC machine.

17.1 Introduction

The global manufacturing industry is striving to enhance productivity and improve product quality through excellence in operation. To support this, significant investments have been made in upgrading instrumentation, data acquisition and computing infrastructure. The expectation is that, with more process and product data readily available, useful information and better process knowledge can be gained in a timely fashion. Such information and knowledge can be used for various operations such as:

- Equipment level signal processing and control
- System level supervision including optimal operational command generation, process monitoring and fault diagnostics
- Data Acquisition and Management functions for tracking quality and progress of production including real-time data records and event logs as well as data archival
- Management functions including resource scheduling, quality management, specialized decision support
- Communication via computer networks to other software

Manufacturing operations, such as turning, milling, drilling, etc. usually have definite quality requirements, but maintaining that quality is sometimes difficult because the cutting tools themselves wear and deform in minute increments as they remove metal from the work piece. At some point, the tool ceases to produce satisfactory results. In a large-scale manufacturing setting, there may be hundreds of machines operating simultaneously. When an individual tool degrades beyond tolerance thresholds, or even breaks entirely, often no one realizes it until parts start appearing with poor quality. At that time the line must be stopped, the broken or worn tool replaced, and the ruined products scrapped. This is expensive. The general solution at present is to replace tool bits at conservatively estimated preset intervals. However, metal cutting is a complicated process, and many factors affect the growth of tool wear rate. In order to avoid as many tool failures and downtime as possible, the tools usually get replaced long before the end of their useful lives. Even with these expensive precautions, there is no way to predict a sudden tool breakage. Direct wear measuring involves taking the bit out and directly (optically) measuring the wear with a microscope. It is therefore important to find a reliable way of monitoring tool wear for reduction of machine downtime to improve productivity, as well as for improving the quality of machining. This requires accurate on-line estimation of tool wear using measurement of process signals, like cutting forces and vibration signals in three major directions, spindle vibration, sound pressure level, acoustic emission, spindle current and power signals etc. Industrial acceptability of a method of On-Line Tool Condition Monitoring (OLTCM) crucially depends on the cost and maintainability of sensors as well as the associated operational problems, such as mounting, environmental protection, noise immunity etc.

Cutting forces in the tangential and longitudinal feed directions are among the most effective signals for estimation of tool wear although cutting forces are measured with dynamometers, which are relatively expensive and restrain high productivity machining. Spindle motor current and the power signals are much more favorable from these considerations. Several authors have reported results of tool condition monitoring in a variety of manufacturing applications such turning, milling and drilling using a number of process measurements, such as force, current, vibration and acoustic emission. More details on these techniques can be found in the references [1-10]. Many of the present day methods rely on cutting force measurement using dynamometers and nonlinear neural networks for estimation of tool wear. In this chapter we describe a technique of real-time tool condition monitoring without the use of costly cutting force based method that uses novel signal processing methods such as,

1. Estimation of carrier frequency, demodulation and segmentation
2. Use of simple time domain features for real time applicability
3. Feature space filtering for elimination of random high frequency variations of features due to vibrations or local variations of material properties
4. Use of a simple linear estimation model without sacrificing accuracy and reliability of nonlinear models such as using artificial neural networks

17.2 Experimental Set up and Data Processing

17.2.1 Experimental Set up

The architecture of a typical condition monitoring system is shown in Fig. 17.1. To estimate cutting tool wear condition from signals tapped from machining environment, signals that are most affected by tool condition should be natural selection. On the basis of the previous works in TCM and expert knowledge, following sensors (and corresponding peripherals) and signals were selected.

- 3-axes dynamometer for cutting force signals (Fx, Fy, Fz)
- Hall effect current probe for spindle current (C)
- Voltage sensor for spindle voltage (V)

Signals from these sensors are digitized and acquired in a PC. Raw signals undergo several stages of signal processing to reduce noise, segment the machining time signals and extract several simple but real-time computable features. Actual (main cutting edge average) tool-wear values are manually measured (by an optical microscope) and augmented with the feature vector to estimate the parameters of the off-line predictor.

Once the predictor parameters are estimated, they may be used for online tool-wear prediction. The extracted features are then directly fed to the on-line estimator and the tool-wear estimation is obtained in real-time (less than 1 second). For acceptability in the industry a graphical user interface (GUI) is developed so that general machining personnel can use it to acquire data, extract features, train the system and monitor the predicted tool wear value, while the machining operation is continuing. Further one can change the machining parameters (manually or programmatically) for optimal productivity.

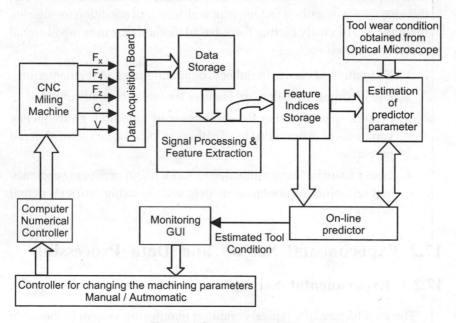

Fig. 17.1 System architecture

A series of experiments both at the laboratory and industrial set up were undertaken, as reported in [12]. Two datasets acquired from the tool life experiments have been used for development of the estimator. The first dataset pertains to experiments conducted on a conventional face-milling machine in the laboratory and the second dataset is from experiments on an industrial CNC Plano-miller. The work piece material for all such experiments was C-60 steel. The machining, on the conventional machine, was conducted at 140 m/min cutting speed with linear feed of 0.22 mm/tooth and a depth of cut of 1.5 mm. The same values for the industrial machine were 212 m/min, 0.16 mm/tooth and 2.4 mm. Experiments were conducted upto a maximum wear value of 631/370 microns. The experiments were carried out using a single cutting tool. However the method could easily be extended to multi-tool cutters also, by incorporating an indexing mechanism.

17.2.2 Signal Processing

In this section the nature of typical process signals are presented. It is seen that the signals have two distinctly different characteristics, one during the period the tool is actually removing metal and the other when the tool is not engaged with the work-piece. Note that this effect is pronounced since only one insert was used in the tool holder. In the actual case where many inserts are used, a non-cutting period may not exist, since one insert may engage before the previous one disengages.

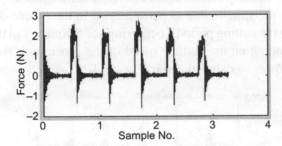

Fig 17.2(a) Force in the longitudinal Direction (Scale X - 1:10⁴; Scale Y - 1:500)

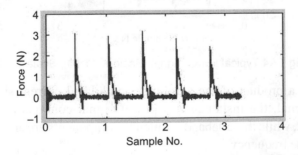

Fig 17.2(b) Force in the transverse direction (Scale X - 1:10⁴; Scale Y - 1:500)

The cutting forces in the two directions, as shown in Figs. 17.2(a) and (b), are roughly periodic with a fundamental period corresponding to the rotational speed of the spindle. Ideally, the signal for the non-cutting region should be zero, however, high frequency noise is present in the signal due to various factors such as vibration, sensor noise etc. The cutting lobe only is relevant for the purpose of estimation. However, as the cutting forces increase with wear, increased vibration results, especially when the tool disengages from the job. Signals therefore need to be considered including the tool disengagement time. The nature of spindle motor current is shown in Fig. 17.3.

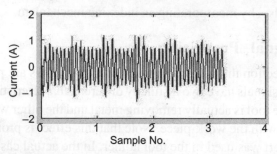

Fig 17.3 Spindle Motor Current signal (Scale X - 1:10⁴; Scale Y - 1:10)

The line frequency current signal is amplitude modulated by the rotational frequency of the spindle due to periodic rise in the torque demand in the motor during the cutting periods, occurring at a frequency in the range of 2-10 Hz, depending on the cutting speed of the process. A similar trend is seen in the power signal depicted in Fig. 17.4.

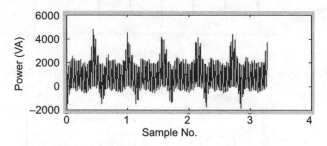

Fig 17.4 Typical power signal (Scale X - 1:10⁴; Scale Y - 1:1)

Being a product of the amplitude modulated current signal and the voltage signal, the instantaneous power signal contains a baseband dc component, while the sideband frequency components are around twice the nominal line frequency.

Before feature extraction, it is necessary to segment all signals to remove much of the non-cutting period containing no information relevant to tool wear. Further, the current signal with suppressed carrier amplitude modulation is to be demodulated to recover the baseband signal using an estimate of the line frequency, which may vary around its nominal value. For the power signal, the base band components can be isolated with simple low-pass filtering.

17.2.2.1 Demodulation of the Current Signal

The first step in the demodulation is to estimate the line frequency and phase of the signal. This is achieved by modeling the signal as $y_i = a_1$ $sin(2\pi ft) + y_i = a_2 \, cos(2\pi ft)$ and minimizing the error sum of squares

$$e^2 = \sum_i (y_i - a_1 sin\,(2\pi ft_i) - a_2 cos\,(2\pi ft_i))^2$$

with respect to a_1, a_2 and f. The detailed steps for estimation are as follows:

(a) Assume an initial value for f (48 Hz). Estimate a_1 and a_2 by the least square fit method for the assumed f.

(b) Calculate error as $e^2 = \sum_i (y_i - a_1 sin\,(2\pi ft_i) - a_2 cos\,(2\pi ft_i))^2$

(c) Increment f by 1.

(d) Repeat steps (a) - (c) up to some final value of f (52 Hz).

(e) Choose f as the one giving minimum e^2.

(f) Choose increment as one tenth the old increment. New start value for frequency $f = f - 10$ x increment and final value as $f = f + 10$ x increment.

(g) Repeat the search till the increment in f is 0.0001.

The signal is then demodulated using the above estimated carrier frequency. This is shown in Figs. 17.5(a) and (b).

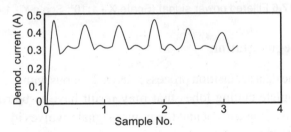

Fig 17.5(a) Demodulated current signal (Scale X - 1:10⁴; Scale Y - 1:10)

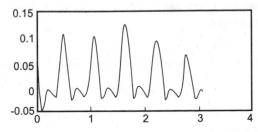

Fig 17.5(b) Demodulated signal after removal of initial distortion and drift
(Scale X - 1:10⁴; Scale Y - 1:10)

Note that RMS values of the original current signal in the different segments have been used for tool condition monitoring. The demodulated waveform has only been utilized to segment the original current signal based on the zero crossings of its slope.

17.2.2.2 Filtering of the Power Signal

The power signal consists of a double carrier frequency band, together with a baseband component. The dc in the baseband corresponds to idle running of the machine, and is again eliminated by a least square fitted line. The other baseband components correspond to the cutting operation and have been filtered out using a fourth order Butterworth filter with cutoff 17.5 Hz. Some initial sample points are again ignored before segmentation. The resulting signal is shown in Fig. 17.6. Note that, the filtered signals show remarkable resemblance with the force.

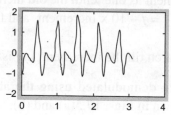

Fig 17.6 Filtered power signal (Scale X - 1:10⁴; Scale Y - 1:4000)

17.2.2.3 Segmentation

Since the data acquisition process is asynchronous with spindle rotation, some incomplete cutting lobes that may result have to be removed. The segmentation of the current and the power signal involves identification of time points that approximately indicate the beginning and end of a given lobe. The detailed steps of the algorithm are presented below:

- Calculate the no. of sample points corresponding to one rotation of the cutter (sp):

$$Sp = F_s \times \frac{\pi \times D}{V_c}$$ where F_s = Sampling frequency, V_c = cutting speed and D = cutter diameter

- Select S_m as the maximum filtered signal value in 2x Sp points from the start

- Select the threshold as $Th = 0.7 \times S_m$

- Let S_i denote the filtered signal value at the i-th local optimum. Then index of the signal at the beginning of any cutting lobe is i-1, where $S_i > Th$, $S_{i-1} < Th$. Similarly, the index for end of the same cutting lobe is i+1 $\forall i \geq 2$ and $\forall i \leq N$ -1, where N= total no. of optima points in the filtered signal

For the industrial data, significant high frequency noise may be introduced due to the Power Electronic drives. Since presence of such noise may change the estimated value of carrier frequency, the raw current signal is to be filtered first with a low-pass filter to remove the high frequency components before frequency computation.

17.2.3 Features and Their Trends

17.2.3.1 Feature Computation

Here, in view of suitability for real time applications, time domain features have been considered for estimation. Since, the energy required for cutting increases as the tool gets increasingly blunt due to wear, for current signals, features such as average mean squared value, square of the average standard deviation and square of the average largest singular values of the segmented signal can be used. For the power signal, RMS value, average standard deviation and average of largest singular values over the number of cutting lobes are the corresponding features. Here results pertaining to RMS values only are shown.

17.2.3.2 Feature Trends

In this section the trends of the above features are shown for various datasets. Fig. 17.7 depicts the trend of tool-wear with time. In Figures 17.8-10 the trends of the average of mean square feature with time for the different methods of segmentation, namely, based on force, power and current signals are shown.

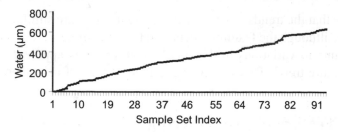

Fig 17.7 Trend of tool wear with time

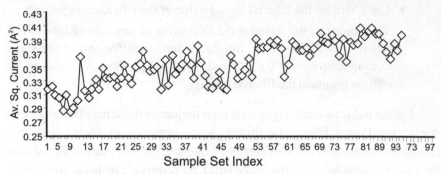

Fig 17.8 Trend of average mean of squares for current signal
(Scale Y-1:100)

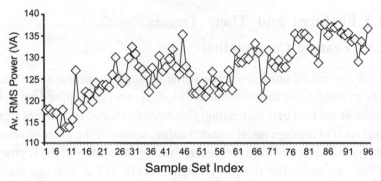

Fig 17.9 Trend of average root mean square value for power signal (Scale Y-1:40)

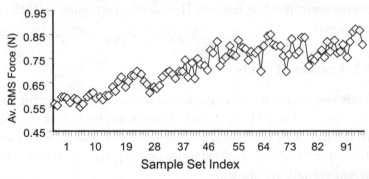

Fig 17.10 Trend of average mean of squares for force signal ((F$_x$) Scale Y- 1:10)

Note that the trends in voltage and current signal are similar to that of the force, although the features calculated from the force signals, however, are immune to variations in the supply voltage, as is apparent from the above feature trend. This is an important advantage of using force-based methods.

17.2.4 Feature Space Filtering

Tool wear is essentially a monotonic phenomenon. The high frequency variations in the feature values have therefore been removed. In view of the non-uniformity in data indices, exponential weighting has been employed for smoothing purpose. The principle is as given mathematically below:

$$Y_i = \frac{\sum_{j=1}^{i} w^{i-j} \times x_{i-j}}{\sum_{j=1}^{i} w^{i-j}}$$

Where, w is a predefined weighting factor (0.99 in the present case), x_i is the i-th sample value of the feature and i is the corresponding wear value which serves as the index of the feature values. In real time operation, data will be processed on some regular time intervals, which could then be used as indices for smoothing. The trends of filtered feature values are shown in Figures 17.11(a) to (c).

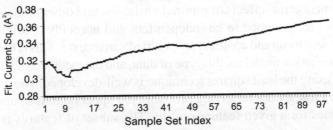

Fig 17.11(a) Filtered Current Squared (Scale Y-1:100)

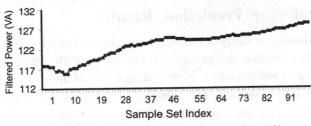

Fig 17.11(b) Filtered Power (Scale Y-1:40)

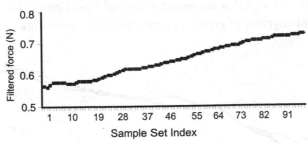

Fig 17.11(c) Filtered Force (Scale Y-1:500)

The important point to note here is that at the initial sample points the filtering effect is insignificant in the absence of sufficient number of previous points and hence the filtered feature is not smooth in the initial part. However, this does not affect the performance of the estimator as the tool is fresh in this area of operation and the resulting wear values need not be very accurate.

17.3 Tool-wear Prediction

17.3.1 Prediction Model

The smoothed features are used as predictors in a multiple linear regression model of the form

$$y_i = \hat{\beta}_0 + (\hat{\beta}_1 \times x_{i1} + \cdots + \hat{\beta}_n \times x_{in}) + e_i$$

where y_i is the i-th wear value, x_{ij} is the jth feature value, $\hat{\beta}_0, \hat{\beta}_1$.. are the unspecified model coefficients, and $e_1, \ldots e_n$ are the model errors representing measurement error, effects of omitted variables, and other modeling error. The errors are assumed to be independent and normally distributed with common zero mean and common unspecified variance σ^2. This is the simplest known prediction model for this type of data, and the methodology for model building using the least squares technique is well-developed. The model also provides a framework for giving probabilistic upper bounds of unobserved wear values for a given feature profile. The subset of features is selected using the correlation criterion.

17.3.2 Tool-wear Prediction Results

The estimation is fairly accurate as is apparent from Figures 17.12(a) to (d) which compare the estimated values with the measured ones corresponding to various cases. Note that because of an incidence of drop in the power supply voltage, towards the later part of the first data set a sudden dip in the estimated values resulted after the initial monotonic trend. From the plots, it is apparent that estimation based on the combination of current and power signal is comparable to that based on the force signal. Further, the histograms of errors indicate that large errors are few.

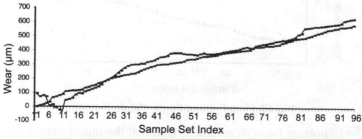

Fig. 17.12(a) Results based on power signal

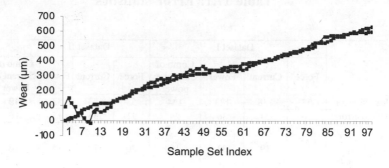

Fig. 17.12(b) Results based on current signal:

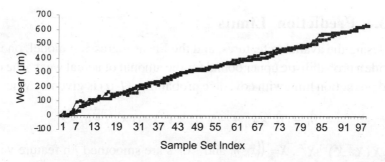

Fig. 17.12(c) Results based on a combination of current and power signals

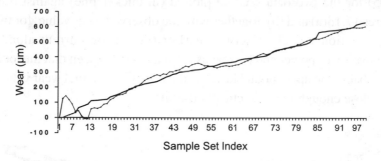

Fig. 17.12(d) Results based on force signal

17.3.3 Error Statistics

In Table 17.1, the maximum absolute error, the average absolute error for the two data sets as also the same errors ignoring some (10 for the Ist dataset and 3 for the IInd dataset) initial data points are given. The conclusion for the regression statistics holds in this case also. The performance is the best for the force based method with that of combination of current and power based method competing evenly almost on statistics.

Table 17.1: Error statistics

	Dataset I				Dataset II			
	Force	Current	Power	Comb of Current & power	Force	Current	Power	Comb of Current & Power
Max abs error	57	138	123	134	53	73	53	49
Av abs error	16	24	36	24	21	35	17	16
Av abs error leaving some sample points	15	19	32	19	20	30	15	16

17.3.4 Prediction Limits

Using the available features, and the linear regression model, one can provide a probabilistic upper bound for the amount of actual wear. The one-sided prediction limit with coverage probability (1-α) is given by where

$$y_i = \hat{\beta}_0 + (\hat{\beta}_1 \times x_{i1} + \cdots + \hat{\beta}_n \times x_{in}) + \sigma \times \sqrt{1 + h_i} \times t_{\alpha, n-2}$$

$h_i = X(X^T X)^{-1} X^T$, $X = ((x_{ij}))$ and x_{ij}'s are smoothed jth feature values for the ith sample point. The loci of the probabilistic upper bound (for α = 0.05) for the two data sets are plotted (in thicker line) against time in Figures 17.13(a) and (b), together with the observed wear value, for two of the prediction models described earlier (i.e., using current alone and combination of power and current as estimators). It is seen that the locus of the probabilistic upper bound lies generally above the actual wear value, but stays close enough to be practically useful.

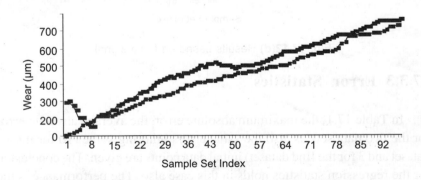

Fig 17.13(a) Prediction limit for combination of current and power

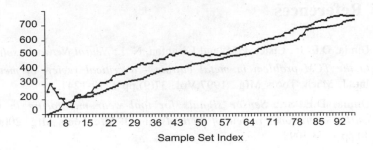

Fig 17.13(b) Prediction limit for current

17.4 Conclusions

From the results presented in this chapter, it is clear that it is possible to "measure" tool wear effectively using real-time process signals that can be physically measured easily, such as, current or power. It is also seen that in general, estimates are improved if they are based on a number of sensors rather than a single one. This is borne out from the results derived from a combination of current and power signals. The linear regression model used here is simpler although nonlinear Neural Network based methods can also be used [12]. For industrial implementations several operational factors such as sensor maintenance and useful life, cost, degree of intrusion into manufacturing operations etc., need to be considered in addition to accuracy of the results. For example, it is demonstrated here that a force-based methodology, although it provides the best results, may not be favoured, since it is costlier and difficult to install and maintain in industrial environments. Both current and power based methods, being noninvasive and cheaper are favourable from these considerations and provide accuracy comparable to that of the force based method. In this work a single model is assumed to hold for a new tool/work-piece combination under similar cutting conditions. However, fresh off-line estimation of predictor parameters may be necessary for a new installation and also for significantly different cutting conditions.

Acknowledgements

The authors would like to thank the Department of Information Technology, Govt. of India for providing financial support to set up the experimental facility at Indian Institute of Technology, Kharagpur, India. They would also like to acknowledge the contributions of the project team members Prof. S. Paul, Prof. A. R. Mohanty, Prof. A. K. Chattopadhyay, Mr. N. Ghosh and Mr. Y. B. Ravi towards setting up of the facility and data collection.

17.5 References

1. Dimla, D.E. Jr., Lister, P.M. and Leighton, N. J.: *Neural Networks solutions to the TCM problem in metal cutting - a critical review of methods,* Int. J. Mach. Tools Mfg., 1997,Vol. 37(9),pp 1219-1241.

2. Dimla, D.E. Sr.: *Sensor signals for tool wear monitoring in metal cutting operations- review of methods* ,Int. J. M/c Tools Mfg. 2000,Vol 40,pp 1073-1098.

3. Belazinski, Marek, Czogala, Ernest, Jamielnaik, Krzysztof and Leski, Jacek : *Tool condition monitoring using artificial intelligence,* Engg. Application of Artificial Intelligence.

4. Szecsi, Tamas.: *A DC motor based cutting tool condition monitoring system,* J. of Material processing Technology,1999,92-93,pp 350-354.

5. Szecsi, Tamas.: *Automatic Cutting-Tool Condition monitoring on CNC lathes,* J. of Material processing Technology, 1998,77,pp 64-69.

6. Du, R.: *Signal understanding and tool condition monitoring, Engg. Application of artificial intelligence,* 1999, Vol 12,pp 585-597.

7. Prickett, P.W. and Johns, C.: *An overview of approaches to end-milling tool monitoring,* Int J. M/c Tools Mfg., 1999,Vol 212, pp 105-122.

8. Chen, Shang-Liang, and Jen, Y.W.: Data fusion neural network for tool condition monitoring in CNC milling machining, Int. J. M/c Tools Mfg. 2000 ,40, pp 381-400.

9. Dimla, D.E. Sr., and Lister, P.M. : *On line metal cutting tool condition monitoring I: Tool state classification using multi-layer perceptron neural networks,* Int. J. M/c Tools Mfg.,2000,Vol 40 pp 739-768.

10. Dimla, D.E. Sr., and Lister, P.M. : *On line metal cutting tool condition monitoring II: Force and Vibration analysis,* Int. J. M/c Tools Mfg.,2000,Vol 40 pp 769-781.

11. Dutta, R.K., Paul, S., Chattopadhyay, A.B.: *Applicability of the modified back propagation algorithm in tool condition monitoring for faster convergence,* J. Mat. Proc. Tech., 2000,Vol 98, pp 299-309.

12. Ghosh, N.: *Estimation of Tool Wear Using Sensor Fusion for Intelligent Machining Applications,* M.S.Thesis, Department of Electrical Engg., Indian Institute of Technology , Kharagpur, 2003

Chapter 18

Real Time Power System Frequency Estimation Techniques

A. Routrary

A. K. Pradhan

Abstract

The chapter summarizes a number of frequency estimation techniques using voltage and current signals of the power system. The exact frequency information is necessary for power system relaying, stability monitoring, synchronization, load shedding and over all condition monitoring. The larger the system, the smaller is the deviation in the frequency for changes in the system operation. Therefore, along with the speed, accuracy of the estimation algorithm is very important. Several methods have been proposed in the literature for power frequency measurement. Zero Crossing Detector (ZCD), Discrete Fourier Transform (DFT), Least Squares (LS), Least Mean Square (LMS) based methods and variants of Kalman Filtering techniques are some of the important methods which are discussed in this chapter. Simulation results at different levels of Signal to Noise Ratio (SNR) have been carried out for each of these methods. As an example case the LMS based algorithm has been implemented on a TI6711 series processor for tracking the frequency of a given signal.

18.1 Introduction

Frequency is an important operating parameter of a power system. It is always essential to maintain the system frequency very close to its nominal value (50Hz or 60Hz). The frequency starts to decrease if the total generation is less than the sum of the loads and the losses and vice versa. The fluctuations in the frequency can lead to several undesirable phenomena in a power system. For example, it can maneuver extreme vibrations due to mechanical resonance in the steam turbine blades or motor drives. The decrease in system frequency also reduces the reactive power supplied by the capacitors and transmission lines which upsets the voltage profile of the system.

It is not always possible to maintain a perfect generation vs load balance although active control systems (governors) attempt to do this by continuously adjusting the generator power input. Fig.18.1 shows the nature of frequency deviations against generation response for two extreme cases. Slow and precise adjustment of the governor mechanism makes the frequency come back slowly to the nominal value. In Fig.18.1b the governor response is fast but imprecise leading to steady state frequency drift.

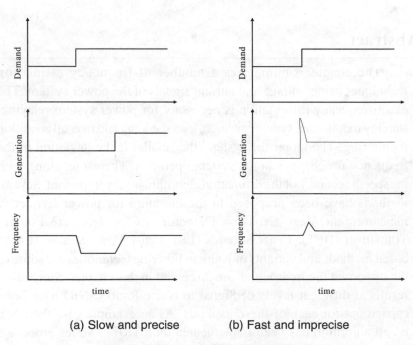

(a) Slow and precise   (b) Fast and imprecise

Fig.18.1 The response of the Governor

Small mismatches between generation and load result in small frequency deviations, which can be tolerated as it does not degrade reliability. But large changes can damage equipment, degrade load performance, and interfere with the system protection schemes, which may ultimately lead to system collapse (Fig.18.2). The effect of a given change in frequency depends on the system capacity. For instance a drift of 0.01 Hz in frequency may not be a serious phenomenon in smaller power systems but such changes might lead to undesirable consequences for larger systems. For monitoring, control and protection of power systems, it is important to carry out on-line estimation of supply frequency and its variations. Because of the stiff load characteristics in modern power systems high precision measurements are required.

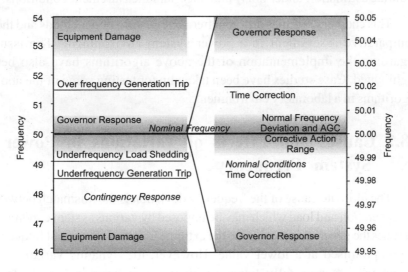

Fig.18.2 Effects of Frequency Variation

For sinusoidal and stationary waveforms zero-crossing detectors are the natural choice. With the proliferation of power electronically switching loads the power system voltage and the current waveforms no longer remain sinusoidal. Further the modern protection philosophy aims at faster diagnosis and isolation of faults to cater reliable supply to increasingly sensitive loads. Waiting for a cycle to obtain the measurement may be too slow for these systems. Therefore, in the last 20 years there has been a lot of research for quick and accurate estimation of power system frequency.

The methods for power frequency estimation can be broadly classified under the following headings [1, 2]

1. Zero Crossing Detection
2. Discrete Fourier Transform (DFT)
3. Newton's Method
4. Recursive Least Square (RLS)
5. Kalman Filtering
6. Other Iterative Methods

The conventional methods such as a zero crossing detector assumes that the power system voltage waveform is purely sinusoidal and therefore the time between two zero crossings can be used to calculate the system frequency. Such a technique is more sensitive to harmonic and noise. However the iterative methods such as Kalman filtering can give a very accurate estimation under noisy and uncertain measurement conditions.

This chapter describes different frequency estimation methods and their comparative assessment in a power system environment. The issues regarding the implementation of the above algorithms have also been highlighted. Case studies have been taken up for testing some of the above algorithms in a laboratory environment.

18.2 Causes and Effects of Variations in Power System Frequency

The ultimate cause of the frequency deviation is the mismatch between the generation and load which may be imposed by various system conditions. At times the generation is driven to the extreme limit and the overall frequency is compromised at a lower value. However, the dynamic variations in frequency are generally taken seriously as it heralds a cascade of unacceptable events in the system. This dynamic variation of the mismatch which leads to the frequency deviations can be attributed to the following reasons [3]

1. Load switching
2. Generator shut down
3. Line Outages
4. Inter-area oscillations
5. Controller Actions.

The distribution of a typical power system frequency around the nominal value has been shown in Fig.18.3 [4].

A change in frequency can affect the system depending on its stiffness. Even a small change in frequency can be indicative of some major disturbances in the large systems. At one hand the change in frequency from the nominal value may be the result of one or many of the above mentioned reasons. On the other hand a frequency change may lead to further deterioration of the system in a cumulative manner. The following paragraphs discuss some of the major events caused by frequency deviation.

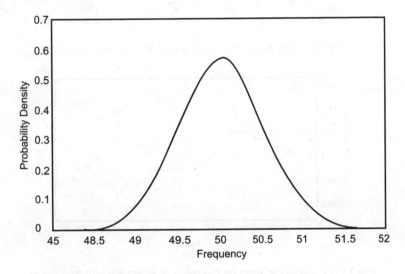

Fig.18.3 Typical Distribution of Frequency in a 50Hz power system
Case I Power System Protection

Case I **Power System Protection**

The power system protection systems use the fundamental components of the voltage and current signals for detection, diagnosis and prognosis of any abnormal conditions. For instance occurrence of a fault can be detected by the magnitude of the fundamental component of the relevant current or voltage waveforms. Distance protection schemes continuously monitor the voltage and current phasors to estimate the impedance. The indication of fault in a specified zone is obtained from the magnitude and orientation of the impedance in the R-X (Resistance-Reactance) plane. If the frequency variations are not taken into account the phasor and impedance estimation algorithms are substantially affected.

Fig.18.4 shows the single line diagram of a 3 phase, 50Hz radial power system. This case-study demonstrates the effect of the frequency drift on the calculation of the impedance for a distance protection scheme. A

3-phase fault has been created at a point which is 100 kms away from the relay. Fig.18.5 and Fig.18.6 show the apparent impedance calculated by the relay algorithm by dividing the fundamental components of the voltage and current estimated from one-cycle Discrete Fourier Transform (DFT). In both the cases the frequency assumed by the relay algorithm is 50 Hz.

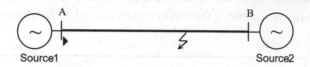

Fig.18.4 The single line diagram of the power system

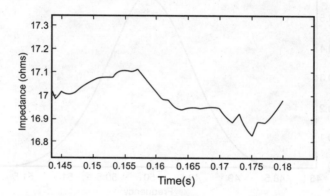

Fig.18.5 Impedance seen by the relay at 50Hz system frequency

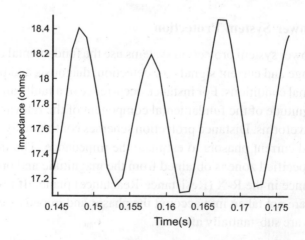

Fig.18.6 Impedance seen by the relay at 52Hz system frequency
(Here the relay algorithm assumes 50Hz for phasor estimation.)

Fig.18.5 shows the impedance variations during the fault when the system frequency is 50Hz. Whereas Fig.6 shows the impedance variations when the system frequency is 52 Hz. In Fig.6 relatively larger modulations of the seen impedance are observed. In this case DFT estimations are modulated because of the mismatch in the system frequency (52 Hz) and the one assumed by the relay algorithm (50 Hz). The apparent impedance seen by the relay is inaccurate in the latter case. Thus the drift in frequency may lead to erroneous trip decisions and imprecise location of the fault. Therefore, for accurate estimation of the impedance the frequency used by the algorithm should be updated in accordance with the system frequency.

Case II Drive System Operation

The speed of AC machines is proportional to the supply frequency. The three phase motors maintain a constant torque and speed as long as the supply voltage magnitude and frequency remain constant. Torque and speed fluctuations are the immediate consequences of the variations in system frequency. These dynamic changes in the motor-torque are the direct input to the mechanical subsystems coupled to it. The response of these mechanical subsystems depends on several factors such as the natural frequencies, rigidity and inertia etc. These torque oscillations might induce unacceptable vibrations leading to system failure. Fig.18.7 shows the photograph of a very long conveyor belt driven by a pair of synchronous motors. The changes in supply frequency beyond a certain limit can not be tolerated in such type of loads as it may lead to violent oscillations along the conveyor system.

Fig.18.7 A 14.6-km long conveyor belt driven by Synchronous Motors for transporting the ore from the mines at NALCO M & R complex (India)

  This case study demonstrates the effect of the frequency variations on a typical drive system. Fig.18.8 shows a typical Induction Motor Drive system fed directly from the supply. The drive system has been simulated with the supply frequency change as

$$f_{supply}(t) = \left[50 + 1.5\sin(\frac{2\pi}{3}t) \right] Hz$$

  Such type of frequency variations results from over all dynamic mismatches in the load and generation in a typical power system.

  Fig.18.9 shows the speed and torque fluctuations due to the above frequency oscillation. The torque fluctuations percolate through the mechanical load causing vibrations and resonance in some cases. These frequency variations need to be detected to take preventive actions such as shutting down of the drive system.

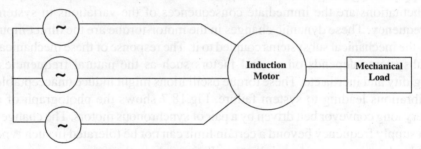

Fig.18.8 A typical Induction Motor Drive System

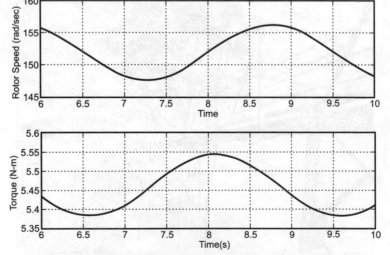

Fig.18.9 Torque and Speed Fluctuations due to Oscillations in System Frequency

Systems Using Real-Time Frequency Information

Variation in the supply frequency is a result of several events in a power system. This also leads to other consequences as mentioned earlier. Therefore, the real time estimation of the system frequency is essential for power system monitoring, protection and control. The following issues address the utility of frequency information.

Monitoring

The present and past changes in the frequency observed over a stretch of time can be informative about the system state and its evolutions. Decisions on Generation scheduling, load management and dynamic reactive power control, can be more effective with this frequency variation charts.

Protection

Frequency relays are widely used for load-shedding and generator-protection. Rapid changes in system frequency demand emergency control strategies like islanding operations, generator shutdown, large load shedding and activating FACTS devices (Flexible AC Transmission System) etc. Besides the adaptive protection schemes need continuous frequency information for accurate decisions.

Control

The drift in frequency can be used by the governors, power system stabilizers and reactive power controllers for improving the dynamic stability. The frequency information is also effectively utilized by the power electronic-controllers used in HVDC (High Voltage DC Transmission), FACTS devices and Reactive Power Compensators. Accurate frequency information is necessary to prevent unsymmetrical operations of these converters.

18.3 Frequency Measurement Methods

The traditional method for power system frequency measurement is based on zero-crossing detectors. Faster and accurate estimation of the frequency is necessary because of the evolution of new protection schemes. The frequency information is also necessary for accurate estimation of voltage and current phasors. This facilitates in effective monitoring, control and protection as described earlier. The following paragraphs shall discuss some of the potential frequency estimation methods based on Zero Crossing Detection (ZCD), Discrete Fourier Transform (DFT), Newton's Method, Recursive Least Square (RLS) and Kalman Filtering. Generally in a power system the voltage signals remain more stationary and clean as compared to the current signals. Therefore, in all the following methods the sampled values of the voltage waveform have been considered for finding the frequency.

18.3.1 Zero Crossing Detection Methods

This is a popular method for power frequency measurement. While using zero crossing methods one determines the time between the two zero crossings of the signal to determine frequency. So in this method it becomes important to determine the accurate zero crossing instants. Fig.18.10 shows a conventional zero crossing detection circuit. The output of this detector is fed to a counter to measure the time between zero-crossings, which can be used to determine the frequency.

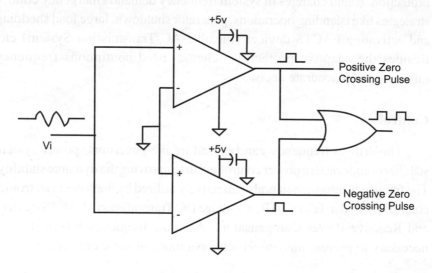

Fig.18.10 A conventional zero-crossing detector

However, in practice the 50 Hz signal is seldom free from distortions. Hence, the exact location of the zero crossing is affected. Further inaccuracies can be introduced by the measurement and sensing circuits. A concise comparative summary of existing solutions to noise reduction in zero crossing detector was published by Weidenburg *et al.* [5]. Low pass filtering, Predictive Filtering and online waveform reconstruction are some of the method suggested to overcome problems with zero-crossing detectors. Vainio *et al.* [6] have proposed a multistage filter for eliminating noise and harmonics before the zero crossing detection circuit. A median filtering precedes a predictive filtering block operating at different sampling rates. The median filtering eliminates the notch type of disturbances whereas the predictive filtering improves the delay. Thus the waveform is smoothened and interpolated for improving accuracy of the zero crossing detection. In a subsequent article an improvised version of the method has been reported [7]. The scheme as shown in Fig.18.11 has two filtering blocks. The first block is a three-point median filter which improves the disturbing impulses. It is a non-linear filter as it sorts the samples inside the window and chooses the middle value i.e. the median as the output. This method completely removes isolated impulses while introducing one sample delay. The restoration and delay compensation is carried out by the following block i.e. the predictive filter. This filter is a two-step ahead predictor one to compensate for the delay and the other to allow interpolation rather than extrapolation in the final processing stage. The time resolution in this case depends on the output sampling rate. The instantaneous frequency is estimated by counting the number of cycles of the clock between the zero-crossings of the filtered sinusoids.

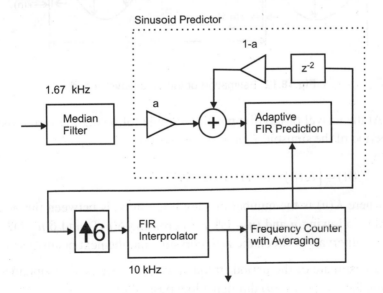

Fig.18.11 Block diagram for the multi-stage digital filter based zero-crossing detector

Friedman [8] has proposed a novel zero-crossing algorithm for estimating the frequency of a single sinusoid in white noise. The frequency is estimated by estimating the time period between the negative to positive going zero-crossings (Fig.18.12).

For a particular case of single sinusoid, its digital form can be expressed as

$$y_n = X \sin 2\pi \frac{nT_s}{T_{sin}} \tag{1}$$

where T_{sin} is the time period of the sinusoid

             T_S it the sampling time, X is the amplitude

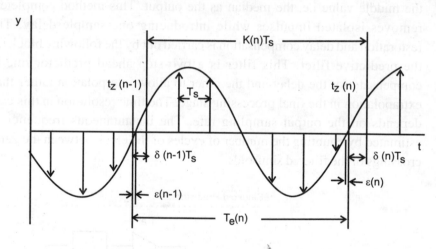

Fig.18.12 Estimation of the sine-wave period

At the arrival of the first positive sample following the zero crossing the period of the sinusoid $T_e(n)$ is computed as:

$$T_e(n) = \left[K(n) - \delta(n) + \delta(n-1) \right] T_s \tag{2}$$

where $K(n)$ is the number of sampling intervals between the positve samples following n and $(n - 1)^{th}$ zero crossing $\delta(n)T_S$ and $\delta(n-1)T_S$ are the time intervals between the zero corssing and the next positive samples

An estimate of the period (frequency) of the sinusoid is obtained by passing the samples $T_e(n)$ through a low pass filter.

If $\varepsilon(n)$ is the error made in the computation of the nth zero crossing

$t_z(n))$ i.e.

$$t_z(n) = n \cdot T_{sin} \qquad (3)$$

Then the estimation of the sinusoid period T_e is equal to

$$T_e(n) = t_z(n) - t_z(n-1) = T_{sin} + \varepsilon(n) - \varepsilon(n-1) \qquad (4)$$

or in z-domain

$$T_e(z) = T_{sin} + \varepsilon(z) \cdot (1 - z^{-1}) \qquad (5)$$

and in frequency domain

$$\left| T_e(f) \right|^2 = T^2 \delta(f) + \left| \varepsilon(f) \right|^2 \cdot 4 \sin^2 \left(\frac{\pi f}{f_{sin}} \right) \qquad (6)$$

The spectrum $\left| T_e(f) \right|$ consists of a dc component and the spectrum of the error signal shaped by

$$H(f) = 4 \sin^2 \left(\frac{\pi f}{f_{sin}} \right) \qquad (7)$$

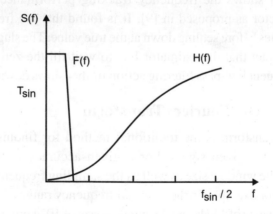

Fig.18.13 The filter transfer function and the spectrum of the output of the ZCD[8]

where $S(f)$ = amplitude axis
         $H(f)$ = spectrum of the error term
         $F(f)$ = desirable filter transfer function
         T_{sin} = frequency of interest

From the frequency response of T_e (Fig.18.13) it is apparent that the frequency can be computed to an arbitrary degree of accuracy by using a low pass filter of appropriate order.

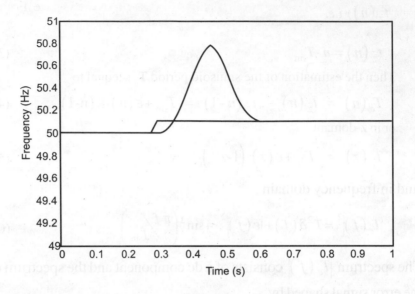

Fig. 18.14 Frequency Tracking by ZCD
Reference Frequency is changed from 50 Hz to 50.1 Hz

Fig.18.14 shows the frequency tracking performance of the zero-crossing detector as proposed in [9]. It is found that the frequency takes around 17 cycles before settling down at the true value. The sluggish response is due to the fact that the estimator has to wait till the zero-crossings to occur. Subsequent low-pass filtering action further slows down the response.

18.3.2 Discrete Fourier Transform

Fourier transform is the traditional method for finding the spectral components of a given signal. For a given accuracy in the frequency information, the window size as well as the sampling frequency needs to be determined. For example, in the nominal frequency range of 45-55 Hz, at a sampling frequency of 1 kHz and the window size of 1024 provide a frequency resolution of 0.9766 Hz. As discussed earlier for large power systems the required resolution in frequency should be higher. On the other hand, for an accuracy of 0.001 Hz at a sampling frequency of 1 kHz the required window size will be 10^6. Handling this data on-line is not possible with the present standards of computing power. Therefore, alternative methods have been developed for accurate estimation of the fundamental frequency with reduced window size and computational complexity. The following paragraphs shall discuss the application of Recursive Discrete Fourier Transform for on-line calculation of the fundamental frequency.

Discrete Fourier Transform (DFT) with an initial assumption of the nominal frequency has been used in [10] to estimate the frequency deviation. In [11] a modified form has been presented to estimate the frequency which also incorporates harmonics in the model. The derivation of the DFT equations can be given as follows:

Consider a sinusoidal input signal of frequency ω given by

$$x(t) = X \cos(\omega t + \phi) \tag{8}$$

Let the signal is sampled at N times per cycle of 50 Hz waveform to produce the sample set $\{x_k\}$

where,

$$x_k = X \cos(\frac{2\pi}{N} k + \phi) \tag{9}$$

The DFT of $\{x_k\}$ contains a fundamental frequency component given by

$$\bar{X}_1 = \frac{2}{N} \sum_{k=0}^{N-1} x_k e^{-j\frac{2\pi}{N}k} = X_c - jX_s \tag{10}$$

where

$$X_C = X \cos \phi$$

$$X_C = X \cos \phi$$

when the frequency of the signal is 50 HZ (denoted by f_0)

The recursive form of the above equation becomes,

$$X_{r+1}(f_0) = \left[X_r(f_0) + \frac{2}{N}\{v(r+N) - v(r)\} \right] \cdot e^{j\frac{2\pi}{N}} \tag{11}$$

When, the frequency of the signal deviates by Δf, slow oscillations in the estimated amplitude is observed with the use of recursive equation (11). This is shown in Fig.18.15 for a frequency of 50.1 Hz.

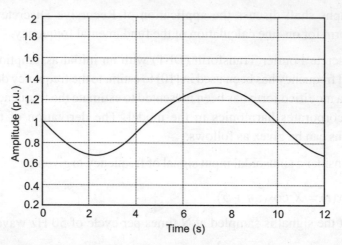

Fig.18.15 The variation in estimated amplitude at a frequency of 50.1 Hz

Assuming significant amplitude of the fundamental and negligible harmonics and inter-harmonics, the analytical expression for a frequency deviation of Δf the amplitude is given as:

$$X_r(f_0) = \frac{2}{N}\sum_{k=0}^{N-1} v(r+k)\cdot e^{-jk\frac{2\pi}{N}}$$

$$= \frac{2}{N}\sum_{k=0}^{N-1}\left\{X\cos\left((r+k)\frac{2\pi}{N}\left\{1+\frac{\Delta f}{50}\right\}+\phi\right)\right\}\cdot e^{-jk\frac{2\pi}{N}}$$

$$X_{r+1}(f_0) = \frac{2}{N}\sum_{k=0}^{N-1} v(r+k+1)\cdot e^{-jk\frac{2\pi}{N}}$$

$$= \frac{2}{N}\sum_{k=0}^{N-1}\left\{X\cos\left((r+k+1)\frac{2\pi}{N}\left\{1+\frac{\Delta f}{50}\right\}+\phi\right)\right\}\cdot e^{-jk\frac{2\pi}{N}}$$

$$X_r(f_0) = \frac{Xe^{j\phi}}{N}\cdot\frac{\sin\pi\frac{\Delta f}{50}}{\sin\frac{\pi}{N}\frac{\Delta f}{50}}\cdot e^{j\frac{\pi}{50N}\{\Delta f(2r+N-1)+100r\}}$$

$$+\frac{Xe^{-j\phi}}{N}\cdot\frac{\sin\pi\frac{\Delta f}{50}}{\sin\frac{2\pi}{N}\left(1+\frac{\Delta f}{100}\right)}\cdot e^{-j\frac{\pi}{50N}\{\Delta f(2r+N-1)+100(r+N-1)\}} \tag{12}$$

For small frequency deviations the second term can be neglected.

Now,

$$X_r(f_0) = \frac{Xe^{j\phi}}{N} \cdot \frac{\sin \pi \dfrac{\Delta f}{50}}{\sin \dfrac{\pi}{N} \dfrac{\Delta f}{50}} \cdot e^{j\frac{\pi}{50N}\{\Delta f(2r+N-1)+100r\}} \tag{13}$$

The phasor thus obtained recursively undergoes two modifications: one is the magnitude factor and the other is the phase factor. The magnitude factor is independent of 'r' and is small for minor changes in frequency. The frequency deviation can be obtained either from the magnitude factor or phase factor. However, the phase factor is more sensitive to frequency deviation and provides a direct measure of Δf.

Denoting the phase factor by $e^{j\Psi}$, the recursive equation for phase is obtained as:

$$e^{j\Psi_r} = e^{j\frac{\pi}{50N}\{\Delta f(2r+N-1)+100r+\phi\}} \tag{14}$$

$$\Psi_r = \frac{\pi}{50N}\{\Delta f(2r+N-1)+100r+\phi\} \tag{15}$$

$$\Psi_r = \Psi_{r-1} + \frac{2\pi}{N} \cdot \left(1 + \frac{\Delta f}{50}\right) \tag{16}$$

therefore the frequency can be computed from

$$f = 50 + \Delta f = \left(50 + \frac{1}{2\pi} \cdot \frac{d\Psi}{dt}\right) \text{Hz}$$

$$= \left(50 + \frac{1}{2\pi} \cdot mean\left(\frac{\Delta\Psi}{T_s}\right)\right) \text{Hz} \tag{17}$$

The mean is taken over a data window of 2 cycles

Because of the derivative component the above equation may not give the correct estimate in presence of noise and uncertainties. Therefore, the actual computation is performed by averaging the derivatives as:

$$\Psi_e = f_e = \frac{1}{2N} \sum_{k=-N}^{N} \frac{\Psi_{k+1} - \Psi_k}{T_s} \tag{18}$$

where

f_e is the estimated frequency

Ψ_e average phase change

T_s is the sampling frequency

The window for computing f_e is $2N + 1$ sample size (approximately 2 cycles).

Fig. 18.16 and 18.17 show the step response of the estimator when the frequency deviates by 0.1 Hz. The estimator output oscillates around the nominal frequency with a small error. It has been found that the effect of the second term in (13) can not be neglected for frequency deviations beyond a limit which depends on the window size N. Fig.18.16 and Fig.18.17 show the frequency estimation for one-cycle and five-cycle DFT. The sampling time has been maintained at 1 kHz. The accuracy has been improved in the later case.

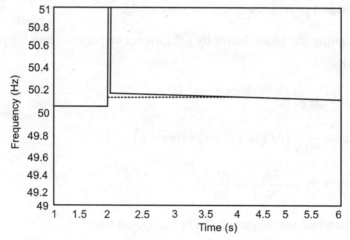

Fig.18.16 Frequency Tracking by Recursive DFT (The overshoot goes up to 55.932) Hz. Reference Frequency is changed from 50 to 50.1. The DFT window size (N) is 20

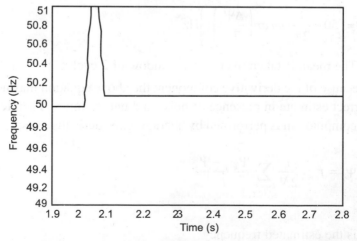

Fig.18.17 Frequency Tracking by Recursive DFT (The overshoot goes up to 51.08) Hz.Reference Frequency is changed from 50 to 50.1 the DFT window size (N) is 100

18.3.3 Least Square Methods

Literature reports a number of frequency estimation methods based on Least Square Error minimization. Sachdev et. al [12] reported a least error square method for estimating the frequency deviations of the voltage signal. The first three terms of the Taylor series expansion have been used to linearize the signal model around the nominal frequency. Dash et. al. [13] have proposed a Least-Mean-Square (LMS) formulation of the frequency estimation problem. They have used a connectionist-architecture in the form of an Adaptive Linear Combiner (Adaline) to implement an improvised LMS algorithm for real time tracking of the nominal frequency. The following paragraphs shall discuss each of these methods in a nutshell.

18.3.3.1 Least Error Square [12]

The power system voltage signal can be modeled as

$$v(t) = X \sin(\omega t + \phi) = X \cos\phi \sin\omega t + X \sin\phi \cos\omega t \qquad (19)$$

Using Taylor series expansion up to three terms in the neighborhood of the nominal frequency,

$$v(t) = a_1 x_1 + a_2 x_2 + a_3 x_3 + a_4 x_4 + a_5 x_5 + a_6 x_6 \qquad (20)$$

where,

$$x_1 = X \cos\phi, \quad x_2 = \Delta f \cos\phi$$

$$x_3 = X \sin\phi \quad x_4 = \Delta f \sin\phi$$

$$x_5 = (\Delta f)^2 X \cos\phi \quad x_6 = (\Delta f)^2 X \sin\phi$$

and

$$a_1 = \sin(2\pi f_0 t) \qquad a_2 = 2\pi t \cdot \cos(2\pi f_0 t)$$

$$a_3 = \cos(2\pi f_0 t) \qquad a_4 = -2\pi t \cdot \sin(2\pi f_0 t)$$

$$a_5 = -2(\pi t)^2 \sin(2\pi f_0 t) \qquad a_6 = -2(\pi t)^2 \cos(2\pi f_0 t)$$

frequency deviation,

$$\Delta f = f - f_0$$

where

f_0 is the nominal frequency

f is the actutal frequency

Taking 'm' observations at time instants $(t_1, t_2 ..., t_m)$ the equations can be written in the matrix form as:

$$\underset{m\times6}{\mathbf{A}} \; \underset{6\times1}{\mathbf{x}} = \underset{m\times1}{\mathbf{V}} \qquad (21)$$

where

$$\mathbf{A} = \begin{bmatrix} a_1(t_1) & a_2(t_1) & a_3(t_1) & a_4(t_1) & a_5(t_1) & a_6(t_1) \\ a_1(t_2) & a_2(t_2) & a_3(t_2) & a_4(t_2) & a_5(t_2) & a_6(t_2) \\ \vdots & \vdots & \vdots & \vdots & \vdots & \vdots \\ a_1(t_m) & a_2(t_m) & a_3(t_m) & a_4(t_m) & a_5(t_m) & a_6(t_m) \end{bmatrix} \quad (22)$$

$$\mathbf{x} = \begin{bmatrix} x_1 & x_2 & x_3 & x_4 & x_5 & x_6 \end{bmatrix}^T \quad (23)$$

$$\mathbf{V} = \begin{bmatrix} v(t_1) & v(t_2) & \cdots & v(t_m) \end{bmatrix}^T \quad (24)$$

This is a standard least square problem and hence the unknown $\mathbf{x}$ can be found out by taking the pseudo inverse of $\mathbf{A}$.

$$\mathbf{x} = \left(\mathbf{A}^T \mathbf{A} \right)^{-1} \cdot \mathbf{A}^T \cdot \mathbf{V} \quad (25)$$

From the elements of x the frequency deviation can be computed as:

$$(f - f_0)^2 = \frac{x_2^2 + x_4^2}{x_1^2 + x_3^2} \quad (26)$$

Fig.18.18 shows the frequency tracking performance of the LES method. The signal frequency has been abruptly changed from 50.1 to 50.2 Hz. It has been found that the filter output jumps up to a higher value before settling at the new value within 2 cycles. The additive noise has been kept at very low level. (High SNR)

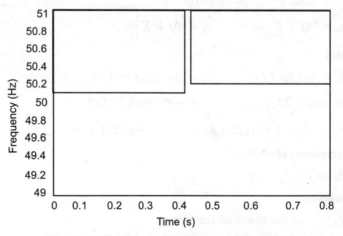

Fig.18.18 Frequency Tracking by LES method

18.3.3.2 Least Mean Square Method

Recursive methods in the form of Recursive Least Squares (RLS) and Least Mean Squares (LMS) can be developed from the basic equations to predict frequency with lesser computational burden. The following paragraphs present an LMS approach implemented in the form a connectionist architecture known as Adaptive Linear Combiner or Adaline [14]. The recursive least square methods can be further optimized by using Kalman filtering approach discussed in the subsequent sections.

A power system voltage or current signal can be represented as:

$$v(t) = X \cos(\omega t + \phi) + \varepsilon_t \tag{27}$$

or in the discrete form

$$v_k = X \cos(k\omega T_s + \phi) + \varepsilon_k = \hat{v}_k + \varepsilon_k \tag{28}$$

where, $\hat{v}_k = X \cos(k\omega T_s + \phi)$

    k is the sampling instant

    T_s is the sampling time

    ε_κ is the additive noise (assumed to be Gaussian)

Three consecutive samples of the signal can be expresses in the form:

$$\hat{v}_k = 2\cos \omega T_s \cdot \hat{v}_{k-1} - \hat{v}_{k-2} \tag{29}$$

This can be represented in the network form as shown in Fig.19.

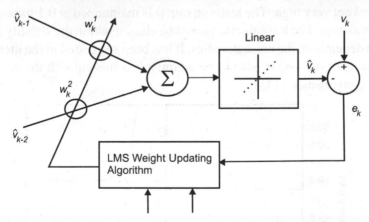

Fig.18.19 The Adaline based Frequency Estimator using LMS algorithm

The LMS weight updating algorithm can be written as,

$$\mathbf{w_{k+1}} = \mathbf{w_k} + \mu \cdot e_k \cdot \mathbf{v_k} \tag{30}$$

where,

$$\mathbf{v_k} = \begin{bmatrix} v_{k-1} \\ v_{k-2} \end{bmatrix} \qquad \mathbf{w_k} = \begin{bmatrix} w_k^1 \\ w_k^2 \end{bmatrix} \tag{31}$$

$e_k = v_k - \hat{v}_k = $ error

$\mu = $ the learning rate

After convergence assuming the noise in the signal to be zero

$$w_k^1 = \cos \omega T_s \quad \text{and} \quad w_k^2 = -1 \tag{32}$$

However with the presence of noise they will deviate from these values.

In presence of noise and harmonics the above algorithm converges slowly. Modifications have been proposed [13] for better convergence and noise reduction in the estimated weights. The weight updating equation becomes,

$$\mathbf{w}_{k+1} = \begin{cases} \mathbf{w}_k + \mu \cdot e_k \cdot \dfrac{\mathbf{v}_k}{\mathbf{v}_k^T \cdot \mathbf{v}_k} & \text{when } \mathbf{v}_k^T \cdot \mathbf{v}_k \neq 0 \\[3mm] \mathbf{w}_k & \text{when } \mathbf{v}_k^T \cdot \mathbf{v}_k = 0 \end{cases} \tag{33}$$

where $\mathbf{v}_k^T = $ transpose of $\mathbf{v}_k$.

Fig.18.20 shows the simulation results of frequency tracking. The nominal frequency is changed from 50.0 to 50.2 Hz at 1 second. The estimated frequency is obtained from (32). In this case the signal to noise ratio is kept very high. The learning rate μ is maintained at 0.1 throughout the simulation. The learning rate should be chosen within the stability limits which depends on the noise statistics. It has been suggested in the literature that the learning rate μ should be adaptively adjusted with the error for faster convergence [14].

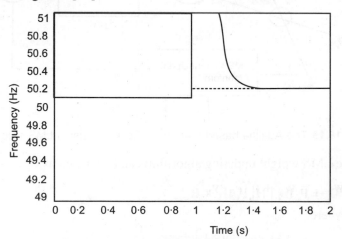

Fig. 18.20 Frequency tracking by LMS method

18.3.3 Kalman Filter based Methods

The determination of power system frequency can be parametrically modeled as a state estimation problem. The Kalman filtering based methods are optimal estimators well-suited for systems described by state variables [15]. Both linear and non-linear Kalman filter formulations have been proposed in the literature [16].

18.3.4.1 Linear Kalman Filter

In the implementation of a linear Kalman filter, the signal is represented in a state variable form as

$$x_{k+1} = \varphi_k x_k + \zeta_k \tag{34}$$

where

x_k is the process state vector,

φ_k is the state transition matrix, and

ζ_k is the uncorrelated sequence with known covariance structure

The measurement of the process is of the form

$$z_k = H_k x_k + \eta_k \tag{35}$$

where

z_k is the measurement vector at step k,

H_k is the ideal connection between the measurement and state vector, and

η_k is the measurement error assumed to be uncorrelated with known covariance structure.

It is also assumed that the system noise and measurement noise are uncorrelated.

With an *a priori* estimate $\bar{x}_k$ and its error covariance P_k the recursive Kalman filter equations are provided below.

(1) Compute Kalman Filter gain

$$K_k = \bar{P}_k H_k^T \left[H_k \bar{P}_k H_k^T + R_k \right]^{-1} \tag{36}$$

where,

R_k = measurement error covariance

(2) Update estim. ate with the measurement $\mathbf{z_k}$

$$\hat{\mathbf{x}}_\mathbf{k} = \hat{\bar{\mathbf{x}}}_\mathbf{k} + \mathbf{K_k}\left[\mathbf{z_k} - \mathbf{H_k}\hat{\bar{\mathbf{x}}}_\mathbf{k}\right] \tag{37}$$

(3) Compute error covariance $\mathbf{P_k}$

$$\mathbf{P_k} = \left[\mathbf{I} - \mathbf{K_k}\mathbf{H_k}\right]\bar{\mathbf{P}}_\mathbf{k} \tag{38}$$

(4) Project ahead

$$\hat{\bar{\mathbf{x}}}_{\mathbf{k+1}} = \boldsymbol{\varphi_k}\hat{\mathbf{x}}_\mathbf{k} \tag{39}$$

$$\bar{\mathbf{P}}_{\mathbf{k+1}} = \boldsymbol{\varphi_k}\mathbf{P_k}\boldsymbol{\varphi_k}^T + \mathbf{Q_k} \tag{40}$$

$\mathbf{Q_k}$ is the process noise covariance
The power system signal can be represented as

$$v_k = X\cos(k\omega T_s + \phi) + \varepsilon_k = \hat{v}_k + \varepsilon_k \tag{41}$$

$$\hat{v}_k = X \cdot \cos\left[k\left(\omega_0 + \Delta\omega\right)T_s + \phi\right]$$

$$= X\cos\left(\phi + k\Delta\omega T_s\right)\cdot\cos k\omega_0 T_s - X\sin\left(\phi + k\Delta\omega T_s\right)\cdot\sin k\omega_0 T_s \tag{42}$$

where,

$\omega_0 =$ is the nominal frequency

$\Delta\omega =$ is the deviation

$$\mathbf{H_k} = \left[\cos k\omega_0 T \quad \sin k\omega_0 T_s\right] \tag{43}$$

$$\mathbf{x_k} = \begin{bmatrix} X\cos\phi_k' \\ X\sin\phi_k' \end{bmatrix} \tag{44}$$

where,

$$\phi_k' = \phi + k\left(\Delta\omega\right)T_s \tag{45}$$

and

$$\mathbf{z_k} = v_k \tag{46}$$

Fig.18.21 shows the frequency tracking performance of the above formulation. The signal to noise ratio has been maintained at a very high value. The frequency has been changed to 50.2 Hz from the nominal value of 50 Hz at 0.5 seconds. The settling time of the filter is approximately 0.05 seconds (2.5 cycles). The covariance matrix is initialized to identity matrix to track the changes.

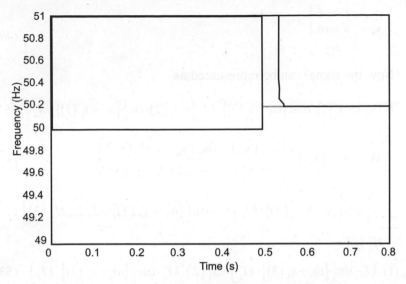

Fig.18.21 Frequency estimation by Linear Kalman Filter

18.3.4.2 Extended Kalman Filter

In the previous section the frequency is estimated as the deviation from the nominal value which resulted in a parametric model with a time varying measurement matrix H_k. There can also be an alternative model which results in a non-linear state space formulation. The Kalman filter in this case can be approximated by the first order Taylor series expansion. This is known as Extended Kalman Filter (EKF) in signal processing literature [17]. This is a modified version of the linear Kalman filter to be applied on systems with non-linear process and measurement equations. In each step of estimation, these non-linear equations are linearized at the latest estimate to form linear process and measurement equation. The formulation of EKF algorithm is as follows.

Consider the signal with a frequency deivation af $\Delta\omega$ from the nominal frequency ω_0.

$$v_k = X\cos(k(\omega_0 + \Delta\omega)T_s + \phi) + \varepsilon_k = \hat{v}_k + \varepsilon_k \tag{47}$$

where,

$$\hat{v}_k = X\cos(k(\omega_0 + \Delta\omega)T_s + \phi) \tag{48}$$

Let the different states be

$$\mathbf{x_k} = \begin{bmatrix} X\cos\phi \\ X\sin\phi \\ \Delta\omega \end{bmatrix} \tag{49}$$

Now the signal can be represented as

$$\hat{v}_k = h_k(\mathbf{x_k}) = x_k(1)\cdot\cos\left[\left\{\omega_0 + x_k(3)\right\}\cdot kT_s\right] - x_k(2)\cdot\sin\left[\left\{\omega_0 + x_k(3)\right\}\cdot kT_s\right] \tag{51}$$

$$\mathbf{H}_k = \nabla h_k(\mathbf{x_k}) = \left[\frac{\partial h_k(\mathbf{x_k})}{\partial x_k(1)} \quad \frac{\partial h_k(\mathbf{x_k})}{\partial x_k(2)} \quad \frac{\partial h_k(\mathbf{x_k})}{\partial x_k(3)}\right] \tag{52}$$

$$= \left[\cos\left[\left\{\omega_0 + x_k(3)\right\}\cdot kT_s\right] \quad -\sin\left[\left\{\omega_0 + x_k(3)\right\}\cdot kT_s\right] \quad H_k(3)\right]$$

$$H_k(3) =$$

$$-x_k(1)\cdot kT_s\cdot\sin\left[\left\{\omega_0 + x_k(3)\right\}\cdot kT_s\right] - x_k(2)\cdot kT_s\cdot\cos\left[\left\{\omega_0 + x_k(3)\right\}\cdot kT_s\right] \tag{53}$$

The recursive filter equations are given as

(1) Compute Kalman Filter gain

$$\mathbf{K_k} = \bar{\mathbf{P}}_k\mathbf{H}_k^T\left[\mathbf{H}_k\bar{\mathbf{P}}_k\mathbf{H}_k^T + \mathbf{R_k}\right]^{-1} \tag{54}$$

(2) Update Estimate with the measurement z_k

$$\mathbf{z_k} = v_k \tag{55}$$

$$\hat{\mathbf{x}}_k = \bar{\hat{\mathbf{x}}}_k + \mathbf{K_k}\left[\mathbf{z_k} - h_k\left(\bar{\hat{\mathbf{x}}}_k\right)\right] \tag{56}$$

(3) Compute Error Covariance $\mathbf{P_k}$

$$\mathbf{P_k} = \left[\mathbf{I} - \mathbf{K_k}\mathbf{H_k}\right]\bar{\mathbf{P}}_k \tag{57}$$

(4) Project ahead

$$\bar{\hat{\mathbf{x}}}_{k+1} = \varphi_k\hat{\mathbf{x}}_k \tag{58}$$

$$\bar{\mathbf{P}}_{k+1} = \varphi_k\mathbf{P_k}\varphi_k^T + \mathbf{Q_k} \tag{59}$$

The results pertaining to the above EKF formulation is shown in Fig.22. A step change in the frequency (from 50Hz to 50.2 Hz) is initiated at 0.21 seconds. The Signal to Noise Ratio (SNR) is kept at higher value. It can be seen that the filter is able to track the new frequency approximately within 7-cylces.

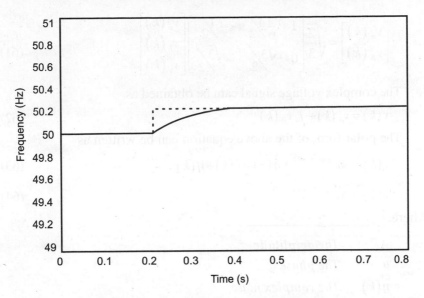

Fig.22 Frequency tracking by EKF

18.3.4.3 Extended Complex Kalman Filter

In power systems the availability of a balanced three phase signal can be used to our advantage by generating a pair of orthogonal signals. The noise and other uncertainties in the measurements can be de-correlated to some extent with the three-phase to two-phase transformation [18]

The frequency estimation problem can be formulated in a complex domain and the formulation leads to an Extended Complex Kalman Filter (ECKF) [19]. The equations of the recursive estimator are given as:

The three phase voltage signals are given as

$$v_a(k) = V_m \cos(k\omega T_s + \phi) + \varepsilon_a(k)$$
$$v_b(k) = V_m \cos(k\omega T_s + \phi - \frac{2\pi}{3}) + \varepsilon_b(k) \qquad (60)$$
$$v_c(k) = V_m \cos(k\omega T_s + \phi + \frac{2\pi}{3}) + \varepsilon_c(k)$$

where,

$\varepsilon_a(k), \varepsilon_b(k), \varepsilon_c(k)$ are the noise components

The α, β – Transformation of the above signals are

$$\begin{bmatrix} v_\alpha(k) \\ v_\beta(k) \end{bmatrix} = \sqrt{\frac{2}{3}} \begin{bmatrix} 1 & -\frac{1}{2} & -\frac{1}{2} \\ 0 & \frac{\sqrt{3}}{2} & -\frac{\sqrt{3}}{2} \end{bmatrix} \begin{bmatrix} v_a(k) \\ v_b(k) \\ v_c(k) \end{bmatrix} \tag{61}$$

The complex voltage signal cam be obtained as,

$$v(k) = v_\alpha(k) + j \cdot v_\beta(k) \tag{62}$$

The polar form of the above equation can be written as

$$v(k) = A e^{j(\omega kT_s + \theta)} + \eta(k) = \hat{v}(k) + \eta(k) \tag{63}$$

$$\hat{v}(k) = A e^{j(\omega kT_s + \theta)} \tag{64}$$

where,

A *The amplitude*

θ *The phase*

$\eta(k)$ *The complex noise*

The state space model of the above signal for Kalman filter formulation is given as

$$\mathbf{x_{k+1}} = \begin{bmatrix} x_k(1) \\ x_k(1) \cdot x_k(2) \end{bmatrix} = \mathbf{F}(\mathbf{x_k}) \tag{65}$$

$$where \quad \mathbf{x_k} = \begin{bmatrix} e^{j\omega T_s} \\ A e^{j(k\omega T_s + \theta)} \end{bmatrix} \tag{66}$$

$$\hat{v}(k) = A e^{j(k\omega T_s + \theta)} = \mathbf{H} \mathbf{x}_k \tag{67}$$

$$where \quad \mathbf{H} = \begin{bmatrix} 0 & 1 \end{bmatrix} \tag{68}$$

$$\mathbf{F_1}\big|_k = \nabla \mathbf{F}\big|_{\mathbf{x}_k} = \begin{bmatrix} 1 & 0 \\ \hat{x}_{k|_k}(2) & \hat{x}_{k|_k}(1) \end{bmatrix} \tag{69}$$

where

$\hat{\mathbf{x}}_{k|_k}$ is the estimated states at k^{th} instant.

$$\hat{x}(k|_k) = \hat{x}(k|_{k-1}) + K(k)\{v(k) - H\hat{x}(k|_{k-1})\} \tag{70}$$

$$\hat{x}(k+1|_k) = F\left(\hat{x}(k|_k)\right) \tag{71}$$

$$K(k) = \hat{P}(k|_{k-1}) H^{*T} \left[H\hat{P}(k|_{k-1}) H^{*T} + R \right]^{-1} \tag{72}$$

$$\hat{P}(k|_{k}) = \hat{P}(k|_{k-1}) - K(k)H\hat{P}(k|_{k-1}) \tag{73}$$

$$\hat{P}(k+1|_{k}) = F_{1}(k)\hat{P}(k|_{k})F_{1}^{*T}(k) \tag{74}$$

The performance of the ECKF while tracking an abrupt change in frequency from 50Hz to 50.2 Hz has been shown in Fig.18.23. The SNR is kept at higher value. The estimated frequency jumps to a higher value and takes about 4 cycles before settling at the new value.

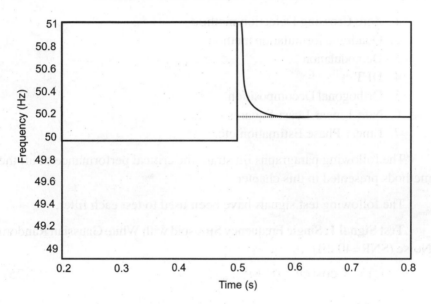

Fig.18.23 Frequency tracking by ECKF

18.4 Comparative Assessment

Frequency estimation is an age-old research area in communication systems. The traditional methods mainly focus on estimation of frequency bands. Considerable research and development have been reported on spectral estimation for addressing this issue. However, in power-frequency estimation the focus is to estimate a single frequency with high degree of accuracy and speed imposed by various control and decision making algorithms. Noise and harmonics substantially affect the conventional algorithms demanding newer methods which should be stable, robust and easy to implement.

For a comparative assessment the following qualities of the algorithms should be analyzed.

a. Accuracy   b. Speed   c. Stability
d. Robustness   e. Algorithm Complexity   g. Implementation Issues

An extensive comparative study of EKF based estimators with some of commonly employed algorithms such as Adaptive Notch Filter (ANF) and Multi Frequency Tracker(MFT) has been carried out in [16]. It has been found with lower SNR the ANF and MFT exhibit poor performance.

Thomas and Woolfson [20] have given a very concise discussion on some of the power-frequency estimation methods such as

1. Zero Crossing Detection method
2. Quadratic formulation method
3. Demodulation
4. DFT
5. Orthogonal Decomposition
6. Non-Linear Least Squares
7. Linear Phase Estimation etc.

The following paragraphs illustrate the critical performance of all the methods presented in this chapter.

The following test signals have been used to test each filter.

Test Signal 1: Single Frequency Sinusoid with White Gaussian Random Noise (SNR=40 dB)

$$v(t) = X\cos(\omega t + \phi) + \varepsilon_t \tag{75}$$

where,

$X = 1.0$

$\omega = 2\pi f$

$\phi = \dfrac{\pi}{7}$

$\varepsilon_t = X \cdot 10^{-\left(\frac{SNR}{20}\right)} \cdot v_t$

v_t = Zero mean Gaussian random noise with unity variance

Test Signal 2: Fundamental plus harmonics and noise

$$v(t) = X_1\cos(\omega t + \phi_1) + X_3\cos(3\omega t + \phi_3) + X_5\cos(5\omega t + \phi_5) + \varepsilon_t \tag{76}$$

where,

X_1 = Fundamental amplitude = 1.0
X_3 = Third Hamonic camplitude = 0.1
X_5 = Fifth Harmonic amplitude = 0.05
ω = 2πf = fundamental angluar frequency

$$\varphi_1, \varphi_2, \varphi_5, = \text{ Corresponding Phases } = \frac{\pi}{7}, \frac{\pi}{9}, \frac{2\pi}{11}$$

$$\varepsilon_t = X_1 \times 10^{-\left(\frac{40}{20}\right)} \times v_t$$

$\varepsilon_t = $ Zero mean Gaussian random noise with unity variance

The three test signals are shown in Fig.18.24 through Fig.18.26 respectively.

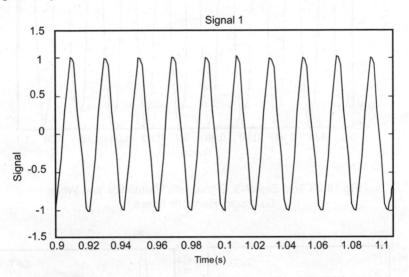

Fig.18.24 Test Signal 1: Single Frequency Sinusoid with White Gaussian Random Noise (SNR=40 dB)

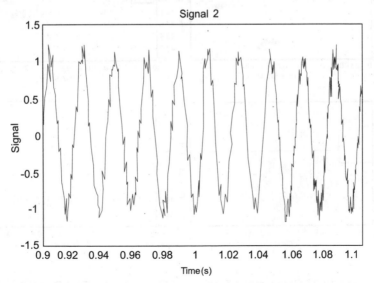

Fig.18.25 Test Signal 2: Single Frequency Sinusoid with White Gaussian Random Noise (SNR=20 dB)

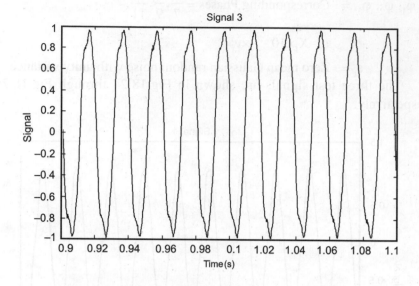

Fig.18.26 Test Signal 3: Signal with Harmonics and White
Gaussian Random Noise

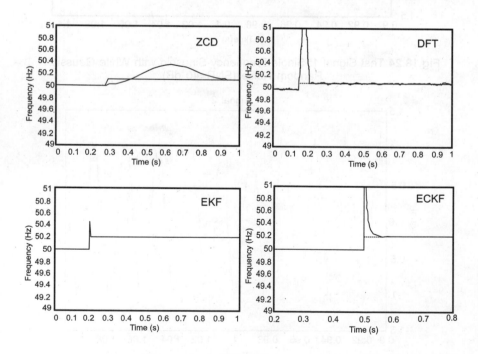

Fig.18.27 Comparative Tracking of Frequency for Test Signal 1

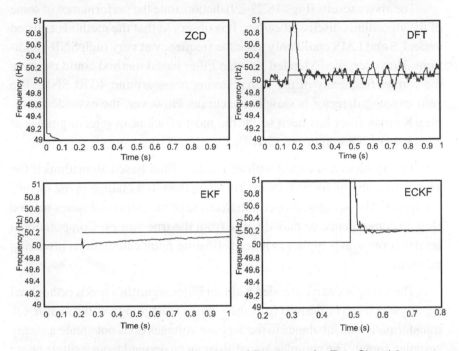

Fig.18.28 Comparative Tracking of Frequency for Test Signal 2

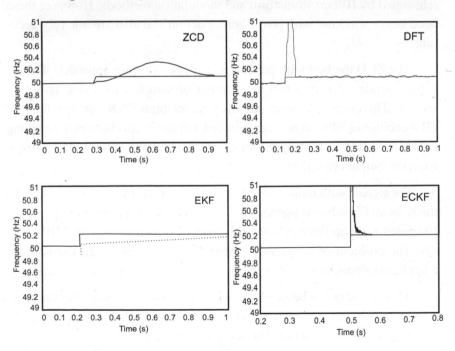

Fig.18.29 Comparative Tracking of Frequency for Test Signal 3

The above results (Figs.18.27-29) demonstrate the performance of some of the algorithms discussed earlier. It is observed that the methods enlisted under LS and LMS could only track the frequency at very high SNR conditions. However, the Extended Kalman Filter based method could estimate the correct frequency of the signal having noise around 40dB SNR. The zero crossing detector is slow but accurate. However, the extended complex Kalman filter has been so far the most efficient in rejecting noise at even lower SNR conditions (20 dB).

The problem associated with all Kalman filter based algorithms is the proper resetting of covariance matrix when there is a change in the incoming signal. Also without proper initialization of the estimated states may be slow in convergence or may diverge from the true values. Computational burden is obviously higher as it needs floating point calculations involving matrices.

The Complex and Extended Kalman Filter algorithm needs orthogonal signals which can be derived from the 3-phase voltage signals through $\alpha - \beta$ transformations. Unbalance in the 3-phase voltages does not create a steady complex signal. The complex signal also can be derived from a single phase real signal by Hilbert -transform and modulation methods. However these need larger windows with post filtering action and also are not very accurate.

The ZCD method with pre-filtering [6] is a simpler approach. The error is introduced here when the number of samples for cycle is not an integer. Therefore, for better accuracy under high SNR (greater than 60 dB) conditions LMS can be recommended. Under abrupt changes the learning coefficient used in this algorithm can be reset using some index derived from the output error [21].

For signals with moderate to higher noise levels (SNR less than 60 dB) the Kalman filter-based methods have to be used. As it is seen the Complex Extended Kalman filter exhibits reasonable performance at an SNR of 20 dB. The problem of covariance resetting is common for all the Kalman filter based algorithms.

This problem has been addressed by various authors in the literature. A hysteresis based method has been suggested in [16] for re-initializing the covariance matrix from the absolute value of the error.

18.5 Real Time Implementation

All the above discussed algorithms need to be tested on real-time for their speed and accuracy before recommended for practice. It is seen that some of these methods like Complex Kalman filter based method perform extremely well under low SNR conditions. However the computational burden may be high for available processors. Other algorithms with lower computational burden such as Zero-crossing detectors are slow and inaccurate for low SNR.

As an example case for real-time testing the Adaline (LMS) based frequency estimator has been chosen. The real-time implementation has been carried out using a floating point Digital Signal Processor TMS320C6711. Such a processor is based on the high-performance, advanced very-long-instruction-word (VLIW) architecture developed by Texas Instruments. It has a complete set of development tools which includes: a C compiler, an assembly optimizer to simplify programming and scheduling, and a windows debugger interface for visibility of source code execution. The overall specifications are given in Table 18.1.

Table 18.1 Specification of DSP Kit [22]

PARAMETER NAME	TMS320C6711-150
Cycle Time (ns)	6.7
Data/Program Memory (bits)	32Kbits L1D Data Cache; 32Kbits L1P Program Cache; 512Kbits L2 Cache
DMA	16 (EDMA)
External Memory Interface	32-bit
Host Port / Exp. Bus / PCI	HPI 16-bit
Timers	32-bit
Core Supply (Volts)	1.8
IO Supply (Volts)	3.3

The schematic diagram of the DSP kit based on TMS320C6711 is shown in Fig.18.30 [21].

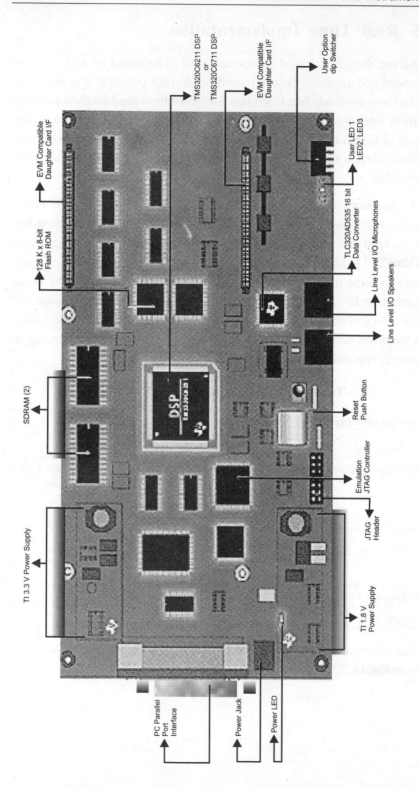

Fig. 18.30 Schematic of the TMS320C6711 DSP Kit

Fig.18.31 shows the experimental setup. The signal is fed from a signal generator into the analog port of the DSP kit. The filtered signal is generated through the D-A channel and recorded on a Digital Storage CRO. The tracking performance of the adaline is shown in Fig.18.32. The inbuilt programmable anti-aliasing filter prevents frequencies higher than the Shannon's frequency at the input of the AD converter. The algorithm is written in C language. The cross-compilation and code transfer is taken care by the Code-Composer Studio and Code-View debugger [23].

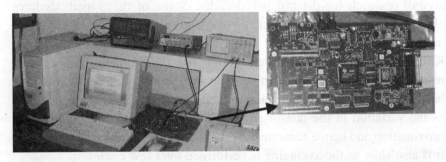

Fig.18.31 The Experimental Setup

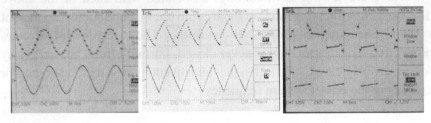

   (i) Sinusoidal   (ii) Triangular   (iii) Square Wave
       Fig.18.32 Signal Tracking (Experimental Results)

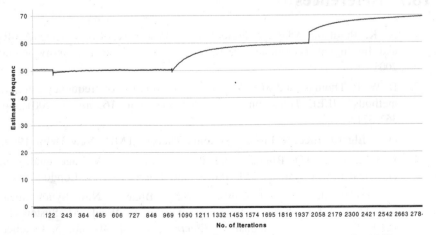

Fig.18.33 Performance for step change in input frequency by LMS
(sampling time=1ms)

Fig.18.33 shows the performance of the LMS based tracker when there are two changes in the frequency in the signal generator. The adaline could track these changes effectively within five cycles of the fundamental time period.

18.6 Conclusion

Accurate estimation of power system frequency has been an active area of research for the last two decades. Some of these methods have already found its way into operation in adaptive relaying. The LMS based method could do it effectively under high SNR situations. The extended Kalman filtering is found to be superior. But the problem of covariance resetting is its major shortcoming. The DFT based method is attractive because of its natural filtering action. The frequency deviation is reflected as the variation in the phasor-angle. The calculated frequency is an approximation and hence accurate only for small deviations in the frequency. It is also slow as the averaging is performed over few cycles.

There are several short-comings associated with each of the above discussed algorithms e.g. none of them could track a continuous variation of system frequency under low SNR conditions. The research on real-time-power frequency estimation is never closed. Further efforts are underway for improvising the existing techniques and inventing new methodologies for estimating the power system frequency.

18.7 References

1. A. K. Pradhan, Adaptive Protection of Power Networks using Artificial Intelligent Techniques, Ph.D. Thesis, Sambalpur University, India, 2001.

2. D. W. P. Thomas and M.S. Woolfson, "Evaluation of frequency tracking methods", IEEE Trans. on Power Delivery, vol. 16, no. 3, 2001, pp. 367-371.

3. O. I. Elgerd, Electric Energy Systems Theory, TMH, New Delhi 1999.

4. J. Schlabbach, D. Blume, and T. Stephanblome, Voltage quality in electrical power systems, IEE Power and Energy Series, London, 2001.

5. Richard Weidenburg, F.P Dawson, and R Bonert, "New Synchronization Method for Thyristor Power Converters to Weak AC-Systems" , *IEEE Transactions on Industrial Electronics*, vol. 40, no. 5. October, 1993, pp. 505-511.

6. O. Vainio and S. J. Ovaska, "Noise reduction in zero crossing detection by predictive digital filtering," *IEEE Trans. Ind. Electron.*, vol. 42, pp. 58-62, Feb. 1995.

7. ____ , "Digital filtering for robust 50/60 Hz zero crossing detectors," *IEEE Trans. Instrum. Meas.*, vol. 45, pp. 426-430, Apr. 1996.

8. Friedman V., "A zero crossing algorithm for the estimation of the frequency of a single sinusoid in white noise", IEEE Trans. on Signal Processing, Vol.42, No.6, June 1994, pp.1565-1569.

9. Olli Vainio, Seppo J. Ovaska, and Matti Pöllä, "Adaptive Filtering Using Multiplicative General Parameters for Zero-Crossing Detection," *IEEE Transactions on Industrial Electronics*, vpl. 50, no. 6. December, 2003, pp. 1340-1342.

10. A.G. Phadke, J.S. Thorpe and M.G. Adamiak, "A new measurement technique for tracking voltage phasors, local system frequency and rate of change of frequency" IEEE Trans. On PAS, vol..102, No.5, 1983, pp. 1025-1038.

11. J. Z.Yang and C. W.Liu, "A precise calculation of power system frequency" IEEE Trans. On PD, vol..16, No.3, 2001, pp. 361-365.

12. M.S. Sachdev, M.M. Giray, "A least error squares technique for determining power system frequency", IEEE Trans. on PAS, vol.104, No.2, page 437-443, 1985.

13. P.K.Dash, D.P. Swain, A. Routray, A.C. Liew, "An adaptive neural network approach for the estimation of power system frequency", Electric Power Systems Research, vol.41, page 203-210, 1997.

14. B. Widrow, S.D. Stearns, "Adaptive Signal Processing", Pearson Education Asia, 2001.

15. A.A. Girgis, T.L.D. Hwang, "Optimal estimation of voltage phasors and frequency deviation using linear and non-linear Kalman filtering: Theory and limitations", IEEE Trans. on PAS, pp. 2943-2951, Vol.

16. A. Routray, A. K. Pradhan, and K. P. Rao, "A novel Kalman filter for frequency estimation of distorted signals in power systems," IEEE Trans.Instrum. Meas., vol. 51, pp. 469-479, June 2002.

17. C. K. Chui and G. Chen, Kalman Filtering with Real-Time Applications,2nd ed. New York: Springer-Verlag, 1989.

18. P. K. Dash, A. K. Pradhan, and G. Panda, "Frequency estimation of distorted power systems signals using extended complex Kalman filter," IEEE Trans. Power Delivery, vol. 14, pp. 761-766, 1999.

19. K. Nishiyama, "A nonlinear filter for estimating a sinusoidal signal and its parameters in white noise: On the case of a single sinusoid," IEEE Trans. Signal Processing, vol. 45, pp. 970-981, 1997.

20. D. W. P. Thomas, M. S. Woolfson, "Evaluation of Frequency Tracking Methods" IEEE Trans. Power Delivery, vol. 16, No.3 pp. 367-371, July 2001.

21.   P. K. Dash, B. R. Mishra, R. K. Jena, and A. C. Liew, "Estimation of power system frequency estimation using adaptive notch filters," *Proc. EMPD'98*, IEEE Cat. No. 98Ex137, pp. 143-148.

22.   TMS320C6000 McBSP: AC'97 codec interface (TIC320AD90)', SPRA528, Application report, CCS user manual.

23.   "TMS320C6000 McBSP initialization', SPRA488, Application report, CCS user manual.

PART - F

Applications

Chapter 19

Automation in Irrigation

A. Joshi

K. N. Tiwari

Abstract

The chapter starts with the introduction of automation in irrigation system; its need and advantages. This section also describes, in short, common methods of irrigation scheduling. The automated irrigation system on the basis of type of control and its scope have been discussed. Various components of an automated irrigation system are described in the following section with emphasis on soil moisture sensors and sensing techniques.

19.1 Introduction

The majority of world's food supply comes from irrigated agriculture. The productivity, profitability and environmental sustainability of irrigated systems are tied with careful management of water. Deciding the timings and the amount of water to meet the plants need is called irrigation scheduling. It is a critical management input for crop production in arid and semi-arid regions. Due to limited knowledge available on irrigation water supply and plant water interaction it leads to over or under irrigation. Under-irrigation leads to a loss in crop quality, yield, market grade and price. On the other hand, over-irrigation leads to a loss in water, electricity for pumping and leaching of nutrients. It also increases crop nutrient needs, fertilizers cost, and nitrogen loss to groundwater. Under-irrigation and over-irrigation can occur during the same season in a given field. Irrigation scheduling is very important during the critical growth stages to ensure an adequate supply of soil moisture thereby minimizing plant water stress.

Usually, an irrigator uses one of several scheduling criteria: intuition and experience, calendar days since the last rainfall or irrigation, plant-based indicator, crop evapotranspiration and soil water measurement.

Intuition and experience is the most widely used method of irrigation scheduling in India. Based on his prior experience, a crop grower decides about the timing and quantity of water to be supplied to the traditionally grown crops. Though this method is widely used, it fails to realise the true potential of a crop both in terms of quality and quantity of produce.

A similar method of irrigation scheduling is based on the calendar days since the last rainfall occurred or irrigation applied is used for deciding the timings of the next irrigation. The amount of water to be irrigated is decided by experience or guess work. This method has the same drawbacks as with the intuition and experience method.

A plant-based indicator approach takes in to account the plant water status itself for scheduling irrigation. This may be considered as the ideal criterion as the plant water integrates the soil, water and climatic factors. It is however not in common practice to measure the plant water status or potential as a standard and cost effective technique.

Estimation of soil water or crop evapotranspiration from climatic parameters provides objective criteria for irrigation management. However, a crop grower has limited knowledge of the dynamics of water evaporation and crop water demand during growing season. Programs that estimate evapotranspiration require large data on climatic parameters, which are seldom available and not easily understood by common crop growers.

Determining soil moisture content through various methods or sensors is the most commonly used method of knowing the soil moisture status, which can then be used to schedule irrigation. Soil water measurement, as compared to other methods of determining soil moisture content, is relatively simple and can be used to schedule irrigation. However, in manual operation, it is very difficult to control irrigation at the predetermined value of soil moisture content. This results either in over or under-irrigation, both of which adversely affect the crop yield. Water requirement of crops varies with change in the climatic conditions, of stage its growth and soil characteristics. It is difficult to decide upon the duration/amount of water application daily and the prevalent manually operated irrigation system is not effective. Hence automated irrigation systems are employed to solve problems associated with manually operated irrigation systems. This also realises the true potential of the irrigation system. An ideal automated irrigation system starts watering just at the predetermined level of moisture content and stops the watering after attaining field capacity, which also accounts for the effective rainfall for scheduling irrigation.

Advantages of automated irrigation system

- Saves water.
- Starts and stops watering just at the right time.
- Takes in to account the effective rainfall for scheduling irrigation.
- Saves labour cost involved with manual operation.
- No need to visit farm at odd times.
- Adequate quantity of water and nutrients supply result in quality produce and high yield.
- No leaching of minerals and nitrogen vital for plants healthy growth.
- Eliminates the long-term ill effects of over irrigation causing development of salinity.

There are different types of automated irrigation systems in use which can be classified on the basis of the type of control unit or their scope of control as shown in the Figs. 19.1 and 19.2. On the basis of the type of control unit, an automated irrigation system can be classified into

  i. Sequential and
 ii. Non-sequential type.

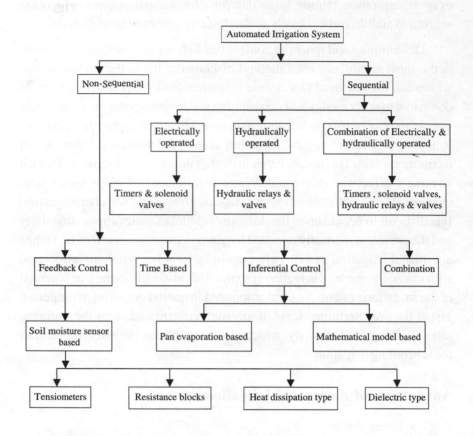

Fig.19.1 Classification of automated irrigation system based
on type of control unit

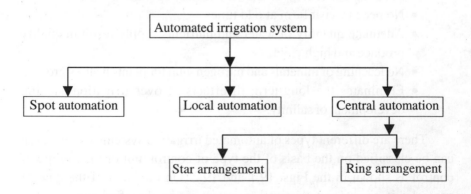

Fig. 19.2 Classification of automated irrigation system
based on scope of control

19.2 Sequential Systems for Automation of Irrigation

In sequential system, the field in divided in to different sub-units, which are irrigated one after the other in a particular sequence. Whereas, in the non-sequential systems the sub-units are irrigated randomly based on the plant water needs, which are operated electrically with or without programming and with possibility of utilizing information collected from the field (feedback) for remote control. A sequential system can be classified further into

　　i. Hydraulically operated
　　ii. Electrically operated and
　　iii. Combination of both hydraulically or electrically operated.

Sequential hydraulically operated automated irrigation for nursery irrigation (Volume based)

The sequential system is particularly suited for irrigation at low discharge rates through small-diameter tubing. At each connection to the main line (except at the end last connection), a unit consisting of a water metering valve and a hydraulic valve is located. At the last connection there is only a metering valve. At the beginning of the irrigation cycle, the metering valves are set for the required volume of water to be supplied to the field. The amount of water is set based on the type of the crop, soil and climatic conditions. Water under pressure passes through the first metering valve to the hydraulic valve on the main line stopping the flow of water further in the main line. After the predetermined quantity of water has passed through the first head, the metering valve closes. The pressure at the head of the hydraulic valve is released, and the water pressure of the main line opens the next valve. Now the irrigation is started in the next group of laterals. Such type of automation is suitable for irrigating green house nurseries.

Sequential hydraulically operated automated irrigation (Volume based)

This type of system is suitable for low to high discharge rates and pipe diameters from 10 mm to 50 mm. The system operates on a main line and automatically opens the valve on the laterals or on a lateral and sequentially open the valve on the dripper heads. The system consists of hydraulic automatic metering valves, with or without an arrangement for cumulative readings of the quantity of water discharged. Each metering valve can operate as an independent unit or can also be used to activate a number of supplementary hydraulic valves. The connection between the various metering valves and the hydraulic valves are achieved by flexible hydraulic tubing of 6-12 mm in diameter. The tubing is buried in the soil along the main line or along the lateral. At the start of irrigation cycle, all the metering valves are pre-set to the quantities of water they are to deliver. When the main valve is opened the water flow reaches the first metering valve. The water pressure is transmitted from the first metering valve through the 12.5 mm hydraulic tubing while the second metering valve is kept closed. When the desired quantity of water has been delivered to the first sub-unit, the first metering valve closes. The pressure on the piston of the second metering valve is released and the water pressure on the main supply line opens the next valve. The procedure is repeated until the lateral is opened and the entire irrigation cycle has been completed. Such type of hydraulically operated automation system is suitable for irrigating orchards and field crops.

Sequential electrically operated system (Time based)

Sequential electrically operated systems are those which operate in sequence the remotely placed solenoid valves with electric current through cables to control irrigation. In these systems the quantity of water delivered to the different plots is regulated by timer clock, which is programmed to start and stop at desired time by the user. If irrigation is not required every day, systems are also available with calendar programming. This type of controller are usually designed with calendar programs so that the watering cycle can be automatically started on the desired day of the week. Most controllers have fourteen day calendar program some are limited to seven days only. This type of system is developed mainly for the irrigation of domestic gardens and for sprinkler irrigation but, in principle, it can be used for any kind of permanent irrigation. Regular solenoid valves are used mainly for low pressure discharge.

Sequential hydraulically–electrically operated automated irrigation system (Volume-time based)

This type of system is used for high pressure discharge. Solenoid valves serve only as control to activate hydraulic valves and the over all operation is hydraulic-electric. Rest of its operation is similar to sequential electrically operated systems.

19.3 Non-sequential Automated Irrigation Systems

Electrically controlled non-sequential systems are automated to a greater extent compared to sequential systems. These systems control electric or hydraulic valves, which operate independently of each other in terms of the quantity of water to be applied or frequency of irrigation. Each unit can discharge a different quantity of water and open at a different time in response to a predetermined program or as per the soil water content. The control panel contains electrical circuits to operate the pump or main valve according to a pre-set schedule, and to measure and supply water to meet crop water needs. Such systems are usually remote-controlled, and are designed to receive feedback of data from the field to regulate irrigation automatically. The adjustments for changes in the pressure and discharge rate of the supply line can also be made in this type of irrigation system.

These systems are further classified as:

(1) Feedback control

(2) Inferential control

(3) Combination of above two methods with time based control

(i) Feedback control system

The main parts of a feedback control system are:

- Soil moisture sensors
- Electronic control unit (comparators/microprocessor/computer)
- Solenoid valves

A farm is divided in to a number of sub-units. Each sub-unit has one soil moisture sensor. Sensors are connected through proper interfacing circuit to the electronic control unit, which could be a microprocessor or a computer with a program to take irrigation scheduling decisions or a comparator circuit. Based on the soil moisture status (as read by soil moisture sensor) of a sub-unit, decision to start and stop the irrigation is taken by the controller. The electronic controller through proper interfacing circuitry and relay switches actuates and de-actuates solenoid valves to start and stop the irrigation respectively.

(ii) Inferential control system

The main parts of a inferential control system are:

- Sensors for measuring various cimatological parameters such as solar radiation, maximum and minimum temperature, wind speed, relative humidity, pan evaporation etc.
- Mathematical model for estimating soil moisture status.
- Solenoid valves.

This system makes use of various evapotranspiraton models to estimate evapotranspiration using different climatic parameters. The estimated evapotranspiration is then used for irrigation scheduling. The solenoid valves are then operated using interfacing circuit and relay switches to put 'on' or to put 'off' the irrigation system.

(iii) Combination of above two methods with time based control

This method uses soil moisture sensors or evapotranspiration models to estimate soil moisture status of the field to start irrigation and timers to stop irrigation after preset value of time for each sub-unit.

Automation of irrigation system based on the scope of control

Automation system can also be classified based on the scope of control as given below

(i) Spot automation

Spot automation refers to an automatic device directly installed on the valve, operating only this valve but with no connection to other valves or systems.

(ii) Local automation

The valves for starting and stopping the irrigation in an subunit are connected to a single controller.

(iii) Central automation

Many local automation units are connected to and controlled by the main central unit.

In central automation networks star or ring configuration is used. These are briefly presented below.

Star Configuration

In this configuration each local unit, is directly connected to the central unit. The cable is of the two-wire type enabling the central unit to send signals to operate solenoid valves. If feedback information is required three-wire cable is chosen.

Ring Network

This arrangement has all units connected in a ring by one cable to the central unit. Cable may be multi wire type in which each local unit is connected by two or three wires to the central unit. Another setting is based on two-wire cable. Two cables are connected to the each one of the local unit. The computer scans continuously the local units by high frequency pulses, identifies each unit, supplies it with relevant information and picks feedback information. This system is cheaper in network cost but requires additional digital circuitry.

19.4 Components of an Automated Irrigation System

Various components of an automated irrigation system are as follows:

I. sensors
II. communication lines
III. controllers and
IV. actuators.

Brief description of these components are stated below.

I. Sensors

There are different types of sensors for measuring/estimating soil moisture content. These sensors utilize different principles such as soil moisture tension, conductivity, soil's complex dielectric, optical properties etc. to sense soil moisture.

Tensiometers

Tensiometer is a device to measure soil water tension, which is used to determine soil water availability for proper scheduling of irrigation. Tensiometer is modified by incorporating transducers to read change in soil moisture tension in terms of change in voltage which can then be connected to the irrigation controller.

Resistance Blocks

This type of soil moisture sensor work on the principle that resistance offered by soil mass placed between two electrodes changes as the soil moisture content changes. But the presence of salt or salinity in irrigation water or soil affects the reading. Resistance blocks could be made up of gypsum or granular matrix.

Gypsum Block

It consists of two electrodes embedded in a block of gypsum. Gypsum neutralizes the affect of salt. Gypsum blocks are easy to use and economical but the inherent disadvantage with these sensors is that they dissolve with water, which changes their calibration curve with the time.

Granular Matrix

Granular Matrix Sensors (GMS) reduce the problems inherent in gypsum blocks (i.e., loss of contact with the soil by dissolving, and inconsistent pore size distribution) by use of a granular matrix mostly supported in a metal or plastic screen (Larson, 1993; Hawkins, 1985). The electrodes inside the GMS are imbedded in the granular fill material above the gypsum wafer. The gypsum wafer slowly dissolves the effect of salinity of the soil solution on electrical resistance between the electrodes. Particle size of the granular fill material and its compression determines the pore size distribution in GMS and their response characteristics (Larson, 1993). Because GMS require little maintenance during the growing season, they are more suited for sensing soil water potential component to controlling irrigation systems automatically. They have advantages of low unit cost and simple installation procedures, similar to those used for tensiometers.

A soil moisture sensor is designed and developed at IIT, Kharagpur which works on the principle that the impedance offered by a soil mass placed between two electrodes changes as the water content of the soil mass changes (Joshi, 2004). Designed granular matrix type soil moisture sensor uses two rectangular corrosion resistant stainless steel plates as the sensing element embedded between the granular matrix of homogeneous sand of particular size and shape (Fig. 19.3). The area of electrode, their spacing, frequency of applied voltage and the particle size of sand have been selected so as to give the maximum sensitivity and to represent the soil surrounding the sensor. There are two replaceable gypsum wafers to buffer against the effect of the salinity contained either in soil or water. Body of the sensor and the support for the stainless steel plate is made up of PVC. The perforated wire mesh cover can be unscrewed to replace the gypsum wafer placed behind the nylon cloth packing.

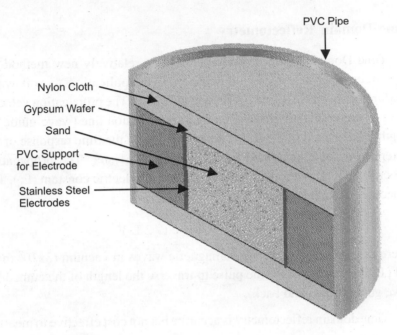

PVC Pipe

Nylon Cloth

Gypsum Wafer

Sand

PVC Support
for Electrode

Stainless Steel
Electrodes

Fig. 19.3 Cut section of the soil moisture sensor

Heat Dissipation

Phene et al. (1971) developed a matric-potential sensor which operates on the principle of the heat dissipation rate in the porous block. A block of ceramic embedded in soil has a heater and a thermo resistive element. The thermal dissipation as sensed by thermo resistive element is proportional to the moisture content of the block hence of the surrounding soil. It is sensitive over a wide range of the matric potential and has an electrical output that can be used for controlling irrigation system automatically. They further improved the design and construction of the original sensor by use of a commercial heater, and the selection of a ceramic material with bubbling pressure of 0.3 bar to improve the sensitivity of the sensor between 0 and 2 bar soil matric potential. The stability of the calibration curves was greatly improved by the use of the ceramic material instead of the gypsum for the porous block.

Dielectric

There are number of soil moisture sensors which make use of the difference in dielectric constant of soil and water to measure the soil moisture content. The dielectric constant of soil is 4 and that of water is 80.

Time Domain Reflectometry

Time Domain Reflectometry (TDR) is a relatively new method for measurement of soil water content. Its first application to soil water measurements was reported by Topp et al.(1980). The propagation velocity (v) of an electromagnetic wave along a transmission line (wave-guide) of length L embedded in the soil is determined from the time response of the system to a pulse generated by the TDR cable tester. The propagation velocity (v=2L/t) is a function of the soil bulk dielectric constant (E_b). The dielectric constant is given by

$$E_b = (c / v)^2 = (c t / 2 L)^2$$

where c is the velocity of electromagnetic waves in vacuum (3×10^8 m/s), and t is the travel time for the pulse to traverse the length of the embedded wave guide (down and back).

Time domain reflectometry is accurate but not cost effective to measure soil water content.

Capacitance (Frequency Domain Reflectometry - FDR) Sensors

Alharthi and Lange (1987) used the dielectric measurement of the moist soils to determine water saturation and established relationship between water content θ and measured dielectric constant E_c. In particular, for sandy soil at 23^0C, the fractional water content is given by,

$$\theta = 0.128(E_c)^{1/2} - 0.204.$$

Velocity Differentiation Domain (VDD)

An electromagnetic wave front is propagated through conductors located in the soil. Increase in transition time between entry and exit points indicate how much water surrounds the conductors. Because of the unique sensing ability of VDD, mist, frost or even a single drop of water can be detected.

Optical Sensors

Under optical sensors category, following types of sensors are developed:

(i)   Polarized light sensors
(ii) Fiber optic sensors
(iii) Near infra-red sensors

Polarized Light

It is based on the principle that the presence of moisture at a reflection surface tends to cause polarization in the reflected beam.

Fiber Optic Sensor

Light attenuation in the fiber varies with the amount of soil water in contact with the fiber because of its effect on refractive index and thus on the critical angle of internal reflection.

Near infrared sensor

Near infrared reflectance technique may be used for monitoring soil moisture content (Stafford, 1988). Such methods depend on molecular absorption by water in the surface layer at distinct wavelength and are therefore not applicable where the moisture distribution is non-homogeneous.

Sensors for climatological parameters

Sensors for measuring various cimatological parameters such as solar radiation, maximum and minimum temperature, wind speed, relative humidity, pan evaporation etc. are interfaced to the computer to estimate evapotranspiration of the cropped field and for irrigation scheduling.

Placement of soil moisture sensor

Soil moisture measured by any method shows the variability in moisture content at different locations in the field. Consequently, a sampling of soil moisture at one location represents the condition only at that particular point. Determination of the average moisture content of the field necessitates a number of measurements at different locations.

II. Communication Lines

Communication lines are used for connecting soil moisture sensors and irrigation valves with irrigation controller. They can be divided into two types.

– Hydraulic and

– Electrical

Electrical communication lines could be of analog, digital or microwave type. Electrical lines are used for receiving signals from soil moisture sensors. If a tensiometer is used as soil moisture sensor, the soil moisture tension is first converted to electrical voltage using a transducer. The electrical communication lines should be protected from interference.

III. Controllers

Irrigation controllers can be of three different types.

– Microcontrollers/ Microprocessors/ Computers

– Timers

– Volume control valves

Now-a-days microcontroller based irrigation controllers are gaining popularity because they are easily programmable.

IV. Actuators

They are of two types:

– Electrically driven and

– Hydraulic

Actuators which make use of both hydraulic and electrical energy are also common in use for irrigation purpose.

Soil moisture measurement and its importance in scheduling irrigation is well recognized. However, the knowledge available on the design and development of soil moisture measuring devices and interfacing these sensors to personal computers/ micro-processors/ microcontroller for automation of irrigation scheduling is limited. Most of such work has been done in developed countries. They use complex irrigation scheduling techniques to automate irrigation system. These systems are commercial, very expensive and their complete design details are not available. The manufacturers make use of commercially available circuitry for interfacing soil moisture sensors and irrigation control valves with the computer which increases cost of the system. These high cost sensors and irrigation systems are not suitable for marginal and small size farm applications.

An appropriate low cost technology for sensing and automating irrigation has considerable potential for development in Indian context.

19.5 References

1. Alharthi A. and Lange J. (1987). Tensiometer irrigation control valve. *Water Resources Research*.23(4):591-595.

2. Hawkins, A. J. (1985). Electrical sensor for sensing moisture in soils. *U.S. Patent 5,179,347*. Date issued: 12 January.

3. Joshi, A. (2004). Design and development of soil moisture sensor and computer controlled automated irrigation system. Ph.D. thesis, Department of Agricultural & Food Engineering, Indian Institute of Technology, Kharagpur.

4. Larson, G. F. (1993). Electrical sensor for measuring moisture in landscape and agricultural soils. *U.S. Patent 4,531,087*. Date issued: 23 July.

5. Phene, C., G. J. Hoffman and S.L. Rawlins. (1971). Measuring soil metric potential in situ by sensing heat dissipation within a porous body, I, Theory and sensor construction, *Soil Science Society of America Proceedings*, (35): 27-33.

6. Topp,G.C., Davis, J. L., and Annan, A.P.(1980). Electromagnetic determination of soil water content: Measurement in coaxial transmission lines. *Water Resource Research*, (16):574-582.

7. Stafford, J.V. 1988 Remote, non-contact and in situ measurement of soil moisture content: a review. *Journal Agricultural Engineering Research*, (41):151-172.

Chapter 20

VLSI in Medical Instrumentation

B. Das

S. Banerjee

Abstract

Advances in the art and science of medical diagnosis (both invasive and non-invasive) and therapy can be attributed largely to the advances in technology in general and VLSI semiconductor electronics in particular. Siliconization – being a potential solution for realization of real time processors, digital signal / image processing algorithms can be implemented on silicon using VLSI technology. Embedded systems may contain a mixture of signal processing, communication, and control algorithms implemented by a variety of technologies such as digital hardware, software, and analog circuits, which is a good platform to realize high quality biomedical instrumentation. This chapter presents a brief account of the-state-of-the-art of medical electronics.

20.1 Introduction

Design of low cost, low power medical system with intelligent adaptation for automated therapeutic purpose is one of the major challenging researches of modern days. In the recent years researches have led to innovations that revolutionized the concepts of medical instrumentation for diagnosis, surgery and therapeutic studies. Diagnostic instruments may be categorized as (a) invasive and (b) non-invasive. The invasive method, an *in-situ* measuring mechanism, though it provides convenience and accuracy of measurement, often interferes with physiological problems and inflicts pain to the patient. Moreover, often being a destructive process such methods of measurement may endanger patient's condition in critical cases, cause cross-contamination of diseases through the instrument. Being an ex-situ phenomenon, the non-invasive methods avoid these. In non-invasive instruments, the subject is normally exposed to some form of transmitted signal, much as magnetic field in case of MRI, ultrasound waves in Doppler Ultrasonography, and Near Infrared (NIR) for carcinoma detection in cancer. The output signal from the concerned region in the human body is processed and displayed. Imaging systems like CT, Doppler Ultrasonography, Elastography, Optical Coherence Tomography, Magnetic Resonance Imaging (MRI) etc. have gone a long way to provide high resolution images and signal for diagnostic, therapeutic studies.

Modern surgery techniques can become bloodless because of the advent of a wide range micro-sensors and micro-actuators. Medical electronics system designers are targeting flexible prototyping systems onto which they can map large designs. Typically these designs may include an analog part, a digital part and software running on a microprocessor or a micro controller. Researchers are trying to utilize the advantages of both alternatives i.e. easy reconfigurability of the digital solutions and the compactness typical of the analog implementation thus allowing for a high degree of flexibility [1,2,3]. Digital signal / image processing algorithms can be implemented on real time processors realized in silicon using VLSI technology. Higher integration and increasing miniaturization has led to a shift from using distributed hardware components towards heterogeneous silicon-on-chip (SOC) designs.

The block level representation of the medical system in its entirety is shown in Fig. 20.1. The sensor (electromechanical interface), (block 1), essentially consists of transducers and other devices that transmit a signal to the human body and collect the output signal (reflected/refracted/backscattered) from the concerned region. The received signal from the subject is processed in different form for display and diagnostic purpose. Since digital signal processing the is obvious choice due to better accuracy and precision compared to analog, the analog signal is digitized using A/D converter for the signal-processing unit (block 4). The output signal needs to be displayed for diagnosis by the medical practitioners. The output of the signal-processing block is often transmitted over the network after being processed and compressed.

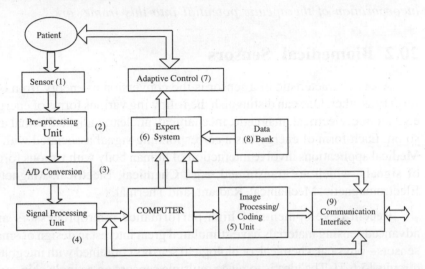

Fig. 20.1 Block level diagram for a complete biomedical system.

The signal may be displayed in the form of spectral representations or digital images as required for diagnostic and therapeutic purposes of the medical practitioners. However, an expert system (block 6) can work as a supplementary device in diagnostic assessment [4]. This system provides quantitative measurement of different parameters from the signal/image. The adaptive feedback from the expert system is often used for therapeutic purposes. Thus, the system not only diagnoses the faulty section but also assists the medical expert to nudge them back to working rhythm. For example, in blood glucose monitoring, the dose of insulin can be monitored automatically based on the glucose level detected by the expert system.

The data from both the DSP block and the expert system can be transmitted over the network to spread critical medical expertise across a desired region around the globe. With data available electronically, analyzing the physiological data, like electro-cardiograms and blood pressure readings, as well as coordinating care with the surgeons, respiratory therapists and others involved in the patient's treatment can be done remotely with telecommunications. Though the requirement of high-bandwidth restricted the application of telemedicine, new hopes are being engendered by the confluence of low-bandwidth telemedicine, improved telecommunication infrastructure and a highly networked medical community.

In a broad spectrum, the research on biomedical instrumentation all over the world is motivated towards design of *low cost, low power medical system with the facility of intelligent adaptability for diagnosis, and extension for telemedicine. One of the future challenges lies in the incorporation of therapeutic potential into this framework.*

20.2 Biomedical Sensors

A key characteristic of a sensor is the conversion of energy from one form to another. One can distinguish the following various forms of energy, e.g., atomic, electrical, magnetic, mechanical, nuclear, radiant, thermal and so on. Each form of energy has a corresponding signal associated with it. Medical applications involve interaction of human body with various forms of signals, which are transmitted, e.g., Chemical, Electrical, Magnetic, Electromagnetic, Mechanical, Radiant, and Thermal.

Recent developments of high-performance microprocessors and advanced sensing materials have stimulated great interest in design of smart sensors – physical, chemical, or biological sensors combined with integrated circuits [5,6,7]. The idea is to place multiple sensors on a single chip, with the integrated circuitry of the chip controlling all these sensors. Since the context should be clear, we also use the term sensor when referring to a smart sensor, i.e., combination of the sensor and the integrated circuit.

Smart sensors are being considered for several medical applications such as a glucose level monitor or retinal prosthesis. These devices require the capability to communicate with an external computer system (base station) via a wireless interface. The limited power and computational capabilities of smart sensor based biological implants present research challenges in several aspects of wireless networking. This is due to the need for having a bio compatible, fault-tolerant, energy-efficient, and scalable design. Further, embedding these sensors in humans add additional requirements. For example, the wireless networking solutions should be ultra-safe and reliable, work trouble-free in different geographical locations (although implants are typically not expected to move; they should not restrict the movements of their human hosts), and require minimal maintenance. This necessitates application-specific solutions, which are vastly different from traditional solutions.

These smart sensors can be relatively inexpensive to build, allowing for the large-scale deployment of networks of smart sensors. Technical advances are expected to improve the capabilities and performance of these devices. Many emerging sensing systems require high performance sensor features that are available at a system level as shown in Fig. 20.2.

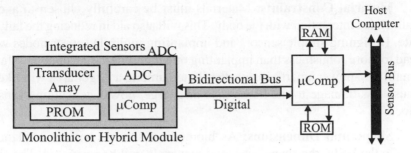

Fig. 20.2 A schematic of a smart sensor

20.2.1 Special Features of Sensors

Biomedical sensors demand few criteria to be fulfilled which are described below.

Robustness and Fault Tolerance: Biomedical sensors are expected to last a long time, as it is not desirable to surgically adjust sensors every week, month, or even year. Ideally, a sensor would last forever, although a more realistic lifetime would be in the range of decades. Because of this, the sensor network should be extremely robust and fault tolerant. In particular, the failure of one node should not cause the entire network to cease operation. The best method of achieving this, is making a distributed network, one where each sensor node can operate autonomously from its neighbors, though still cooperate when necessary. If a sensor does stop working, then the sensors in the surrounding area should still function as normal.

Low Power: Power of the transmitted signal by the sensor must be carefully controlled to avoid damage to the patient. For each application, the power levels for the sensors are defined based on a number of factors – power handling capacity of the organ to be imaged, SNR of the output signal, penetration depth, etc. However, in general one important factor in medical instruments in particular is the requirement of low power.

Wireless sensor networks have power restrictions, whether they are biomedical or otherwise. This is due not only to the small physical size of the sensor, but also because of the absence of wires. An integrated power supply, such as a battery solves the problem. But it supplies power for only a limited time, however, even with conservationist techniques. If the node is implanted in the body, it is not practical to replace the battery as often as would be required. Passive power sources, such as solar and vibration, provide insufficient power for continuous operation. This leads to an external, wireless source of power being the only feasible solution.

Material Constraints: Materials must be carefully chosen to avoid unintended interaction with the body. This will also aid in reducing the failure rate. Designing simple sensors and implanting multiple sensor nodes will lead to more robustness than implanting one monolithic sensor node. Many small, low-power smart sensors distributed among a larger area will likely cause less damage to the tissue than one large, power-consuming sensor block.

Shape and Dimensions: As biomedical sensors will be implanted within the body, the shape, size, and materials will be restricted. The size and shape are determined in part by the application. For example, a smart sensor designed to support the retina prosthesis must be small enough to fit within an eye. Assuming multiple chips, each with many sensors, will be used, each covering a small portion of the retina, each chip must be small enough to accommodate this. A gastrointestinal monitor, on the other hand, needs only be swallowed, so the size of a pill would be sufficient. The sensors may have to curve around the retina or accommodate a bone or cartilage as well, having an impact on the sensor design and material use.

Continuous Operation: Wireless sensor networks have the potential to enhance our lives in many ways, and researchers are thinking of new applications daily. Most of these sensor networks, however, are designed to operate on limited battery power. This might be because the sensors are for military reconnaissance with no hope of retrieval or because of cost constraints. Because of these limitations, much research effort is currently being directed toward improving the battery life of sensors in a network, for the main purpose of prolonging the usefulness of the network. This usually translates into a need for low power requirements, for once the battery has been depleted, the sensor node is defunct.

Diagnostics are another important feature that these systems will need to incorporate. Since these sensors are going to be implanted in and monitor the human body, there are likely to be varying responses from different people. Diagnostics will be vital for testing prototypes of these devices as well. Without them, there would be no clear feedback on what is actually occurring inside the body.

20.3 ADC/DAC for Medical Instrumentation

In biomedical signal processing and telemedicine applications, the data converters are critical building blocks limiting the accuracy and speed of the overall system. Applications mostly require high speed and high resolution. Fuelled by aggressive device scaling in modern integrated circuit technology, practically attainable operating speeds of this converter have increased by almost two orders of magnitude in the last 15 years. For the most part, the trend is explained by the increasing demand for portability as well as recent efforts in system-on-chip (SOC) integration. In SOC implementations, data converters are embedded on the same chip with powerful fine-line digital-signal processing, resulting in a limited budget for their total heat and power dissipation.

To meet the issue of very stringent speed, noise and linearity requirements, small die area and overall power dissipation, a variety of techniques have been developed. Performance of different DAC and ADC architectures suitable for biomedical instrumentation are furnished in Table 20.1 and Table 20.2 respectively.

20.4 Biomedical Signal Processing/Image Processing

Design of a biomedical system, described in Fig. 20.1, involves both signal and image processing units. These sections are followed by communication interface for transmission of the data over net.

Most of the real-world-data examples would not have been practical in real time prior to the availability of digital signal processing in VLSI [2]. Some of the applications where VLSI had made a lot of changes can be listed as:

Table 20.1 Performances of different DAC architectures

No.	Reference	Architecture	No. of Bits	Supply Voltage (V)	Process	Signal Freq.	Sampling Freq. (MSPS)
1	[20]	Hybrid current steering and R-2R ladder. Thermometer coded MSBs	14	5	BiCMOS	2.03	100
2	[21]	Hybrid. Segmented R-2R ladder.	12	5.2	Bipolar	10	72
3	[22]	Current-steering. Interpolation ratio of 4.	14	5	CMOS	5.01	32
4	[23]	Current-steering. Segmented.	10	3.3	CMOS	0.3	10
5	[24]	Current-steering. Binary weighted.	10	5	CMOS	4.43	40
6	[25]	Current-steering. Segmented.	10	3.3	CMOS	20	500
7	[26]	Hybrid current steering. Thermometer coded.	10	5	CMOS	3.9	125
8	[27]	Current-steering. Thermometer coded.	10	5	BiCMOS	10	100
9	[28]	Current-steering. Segmented.	10	1.5	CMOS	3	10
10	[29]	Current-steering. Segmented.	10	3.0	CMOS	490	1000

Table 20.2 Performances of different ADC architecture

SYSTEM ARCHI	TECH NOLOGY	RESO- LUTION	SPEED	SUPPLY VOLT.	DNL	INL	CORE AREA	POWER
SIGMA DELTA[30]	.35 um SP5M CMOS	16 bits	64 MSPS	1.8 V			2.6 mm^2	230 mW
FLASH[31]	0.25-um CMOS	6 bits	1.3 GSPS	1.8/2.5 V	±.8 LSB	±.42 LSB	.13 mm^2	600 mW
SAR[32]	0.35-um CMOS SPQM	6 bits	250 MSPS	0.8-3.3 V	±1.0 LSB	±.65 LSB	.11 mm^2	30 mW
PARALLEL PIPELINE [33]	0.5-um CMOS	10 bits	200 MSPS	3 V	±.8 LSB	±.9 LSB	7.4 mm^2	280 mW
SERIAL PIPELINE [34]	.35 um DP3M	14 bits	75 MSPS	3 V	±.6 LSB	±2 LSB	7.8 mm^2	340 mW
PIPELINED FOLDING [35]	0.5 um CMOS	8 bits	100 MSPS	5 V	±.4 LSB	±1.3 LSB	1.68 mm^2	165 mW
FOLDING & INTERP OLATING [36]	.35 um DP4M CMOS	7 bits	300 MSPS	3.3 V	<±.6 LSB	<±1 LSB	1.2 mm^2	200 mW

*Signal processing:*Digital filters: fixed and adaptive spectrum analyzer, Signal generation, Phase-locked loops, high-speed control.

Speech processing: Analysis, synthesis, enhancement and noise cancellation

Image processing: Image enhancement and computer vision.

*Reconstruction:*Digital x-ray, CAT, PET, MRI tomography, mCT

Integrative application: Artificial organs: nose, eye, ear, printed text-to-voice converter for the blind.

Robotics: Intelligent pilot's assistant, unmanned mobile watchdog, continuous holster monitoring, etc.

Number-crunching: Array processor, Floating – point accelerator, Vector and matrix processor, Transcedental functions: iterative solutions architecture

Communications: High-speed modems, Adaptive equalizers, Data encryption and scrambling, Linear phase filtering, echo canceller, Spread-spectrum communication etc.

Limited Computation: One of the major factors in signal/image processing is the computational complexity. The computation is directly limited due to the limited amount of power, hardware and the real-time requirement.

Communication is very expensive in terms of power, as relatively more power is used to communicate than to compute. Real-time processing demands faster algorithms and lower computational overhead. Also as the computational overhead increases, the hardware requirement goes up, eventually, demanding more space and power and also at situations reducing speed of operation. There are several possibilities to overcome the issues that arise because of this constraint.

Table 20.3 Lists of computation models used in image and video processing

Subsystem	Model of Computation
Audio Processing	1-D dataflow
Digital image processing	2-D dataflow
Image/Video resampling	m-D multirate dataflow
User Interface	synchronous/ reactive
Communication protocols	finite-state machine
Digital control	dataflow
Image understanding	knowledge-based control
Scalable description	process networks

Data Compression: Data compression mechanism has gained significant interest due to bandwidth limitations. This mechanism allows the same amount of data to be transmitted in fewer bits. Different types of algorithms are available for compressing data, each with advantages that are appropriate for different applications. As energy efficiency is the key, the less data to communicate, the less power the sensor will consume. In addition to signal compression, image compression might also be needed for the retina prosthesis. In this instance, the images coming from the external camera will likely be compressed before being transmitted into the internal sensors, reducing the amount of information that needs to be transmitted through the limited bandwidth. One method for compressing the image uses a form of segmentation. A small area of the image is examined and highlights are noted. This might be a point, line, or edge, depending on the method being employed. Point, line, and edge detection are all very computationally intense, however. An external processor will need to assist this process.

20.4.1 Signal Processing

One of the main motivations of signal processing in general is representing a signal in different domains. The idea is to study the signal characteristics with time and space. This is mostly determined by the power and frequency of the signal at the given time/space. In this regard, the orthogonal transforms play an important role. The Discrete Fourier Transform (DFT) enables frequency domain analysis of a time domain signal. Discrete Hartley Transform (DHT) is similar to the DFT only with the difference that it deals only with real computation. Discrete Cosine Transform (DCT) and Discrete Sine Transform (DST) are used in image and speech processing. These transforms are used in a wide range of biomedical signal and image processing, for various imaging techniques and spectral analysis.

Multiscalable transforms like wavelet have proved to be effective as far as signal/image compression is concerned. The new state-of-the-art compression/coding has migrated from the FT/CT/ST family to the wavelet transform.

20.4.1.1 VLSI Design: A unified array architecture to realize DFT, DHT, DCT and DST

The need for real-time on-chip processing has prompted the VLSI system to step in. Since inception, VLSI design and architectures have gone a long way to come up with the highly sophisticated signal processing units.

Design of a state-of-the-art architecture for DFT/DHT/DCT/DST [8,9] is presented in this sub-section.

A number of architectures are proposed for the realization of these transforms. However, a unified architecture, which can compute all these transforms, can serve the purpose of a general DSP chip, and so a unified architecture has been adopted to obtain all the transforms in a single chip. The basic structure of all the transforms, DFT, DCT, DHT and DST, are almost equivalent and this property has been exploited in the design of the unified architecture.

All the transforms are computationally intensive and involve complicated data manipulations. The data queuing makes the process still more complicated. Each of the transforms has a unique data arrangement format. To address these problems, an array architecture has been adopted for the implementation of the DXT (Fig 20.3)- a term coined to accommodate all the transforms, the DFT, the DHT, the DCT and the DST. Moreover, to get rid of the massive hardware overhead required for its asynchronous counterpart, a synchronous design has been used.

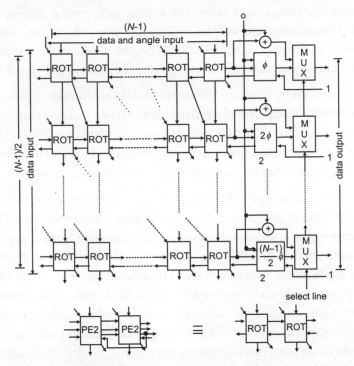

Fig. 20.3 Array architecture for the DXT

The use of a CORDIC processor to achieve a multiplier-less structure is a significant feature of the proposed design, which reduces the hardware overhead drastically. The setup time for the design is reduced by the introduction of the CORDIC element. Low power and high speed can still be achieved by reducing the transistor count and efficient design. TGL instead of the standard CMOS logic can reduce the power and the transistor count of the design to achieve the same. This low-power, high-frequency DFT/DHT/DCT/DST chip is very effective in most signal-processing applications, where it can be used as a generalized chip.

Keeping in view the low-power budget and high-speed requirement, massive parallelism and pipelining of the architecture has been adopted for the design. Parallel architecture provides a high-speed design. At the same time the power budget is reduced owing to the introduction of parallelism. A pipelined structure at the same time provides a high throughput.

20.4.2 Video Processing

For an 800 × 600 image (the computer screen resolution) with 24-bit colour coding, the total image size becomes 1406.25KBytes. Doppler ultrasonography system, for example, generates 150 frames per second. To transmit the image data alone for videoconferencing applications the bandwidth requirement will be » 206MB! Thus, the video compression becomes an obvious choice. The last few decades saw a lot of researches in the field of video image processing and compression in varied forms.

Video coding plays an important role in multimedia systems for bi-directional interactive applications like video-telephony and videoconferencing. The main objective of video coding is reduction of the massive redundancy in spatio-temporal domain, which will have application in telemedicine and videoconferencing for biomedical cases.

A very low bitrate video compression system should smartly combine spatial, temporal and spectral redundancy elimination techniques. The choice of compression also depends on the application environment, e.g., whether the application is broadcast or an interactive environment. Furthermore, many applications require enhanced functionalities and flexibilities along with high compression efficiency. For example, in order to facilitate content-based media processing, retrieval and indexing, and to support user interactions, object based video coding is desired; in order to effectively deliver video over heterogeneous networks (viz. Internet) and wireless channels, error resilience and bit rate scalability are important; and for coded video bitstream usable by different types of digital devices regardless of their computational, display and memory capabilities, resolution/temporal scalability is required. These properties along with high compression of the video signals are accomplished by combining the spatial and temporal compression to form a complete system. The most widely used form of utilizing the interframe dependencies for video coding are (a) motion compensated predictive coding (MCP), (b) object/knowledge based coding strategies, which use elementary motion models and (b) 3-dimensional (3D) sub-band coding (SBC).

20.4.2.1 VLSI Design for Image processing

VLSI based designs are better choice for image/signal processing for real time, high reliability, low cost and low power applications [10-16]. Many architectures for the 2-D DWT have been proposed to meet the real-time, high-efficiency and low-cost requirements. Research works have already been reported for hardware software co-design of 3D-SBC. In 3D SBC was implemented in parallel by dividing the 3D data into parallelopiped on CRAY T3D computers. Partitions were made so that no communication is required in the temporal direction. Kutil and Uhl [14] in proposed a hardware-software co-design for 3D wavelet on shared memory MIMD computers in which the data were decomposed along the time-axis in parallelepiped.

To address some of the problems associated with previous modes of 3D DWT and its architecture, a running 3D wavelet transform [16] can be adopted. The running 3D DWT uses dynamic updating of the transform coefficients in the temporal direction (or z-direction). Both high memory requirement and computational overhead are reduced in this algorithm. Since the dimension of data handled is much less in this algorithm, the implementation complexity also reduces. In parallel computing environment, this algorithm will provide much faster operation and lower memory usage compared to the conventional 3D-SBC.

Figure 20.4 shows steps for realization of the 3D DWT. In an Mth-order transform along temporal direction, M-frames are decomposed each time. The high wait time required for accumulation of group of M frames poses the biggest problem in 3D sub-band coding.

The proposed **running 3D wavelet transform** algorithm uses dynamic updating of the transform coefficients in the temporal direction. The proposed dynamic updating technique for realization of the running 3D DWT is shown in Fig.20.5.

A mixed architecture of running 3D DWT is realized for the temporal and spatial domain decomposition - using (a) QMF lattice filter and (b) a data folding architecture. These architectures have been used in sequence so as to optimize the clock utilization and reduction of timing budget. The data folding architecture takes two samples as input at each clock while the QMF lattice filter takes one sample in one cycle. The QMF structure uses delay elements to compute DWT, and the data folding architecture uses a storage element to store the pre-computed data for use in the later stage. In case of multi-dimensional DWT, the use of data folding architecture is found to be more economic at some stages compared to the QMF lattice structure both in terms of time and peripheral hardware budget like control circuitry and effective memory requirement for the total design.

For an input signal, $x(n)$, the output coefficients after the analysis filter followed by decimation by 2 are given in the z-domain as

$$V_k(z) = \frac{1}{2}\sum_{l=0}^{1} H_k(z^{\frac{1}{2}}W^l)X(z^{\frac{1}{2}}W^l) \tag{1}$$

where

$$W^l = e^{-j\frac{2\pi}{N}} \qquad (2)$$

In a moving image sequence, let the pixel value at the $(m,n)^{th}$ location be given by $f_i(m,n)$, where f_i is the ith frame. The coefficients are given by

$$C_r(m,n) = h(0)f_i(m,n) + h(1)f_{i-1}(m,n) + h(2)f_{i-2}(m,n) + h(3)f_{i-3}(m,n)$$
$$C_{r+1}(m,n) = h(0)f_{i+2}(m,n) + h(1)f_{i+1}(m,n) + h(2)f_i(m,n) + h(3)f_{i-1}(m,n) \qquad (3)$$

where h(n) are the filter coefficients. The property of interleaved data pattern, for formation of the transform coefficients, is utilized for the running DWT. The $(r+1)^{th}$ coefficient for the same location (m,n) can be computed right after the frame f_{i+2} is obtained. Thus, instead of waiting for generation of a GOF the coefficients can be updated after every two frames are obtained (vide Fig.20.5). Figure 20.6 shows the architecture for 3D DWT.

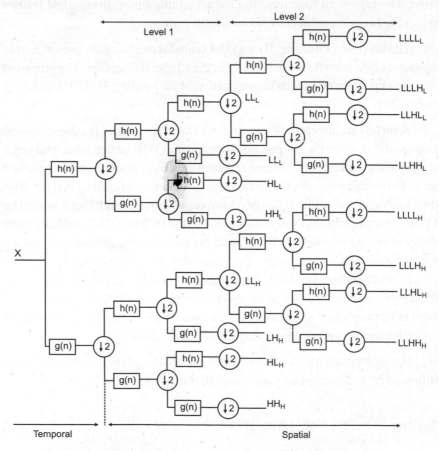

Fig. 20.4 Steps for 3 dimensional wavelet transform

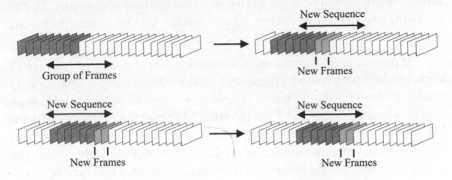

Fig. 20.5 Dynamic image frame updating is done for running 3D DWT. The coefficients are updated at every 2 new frames

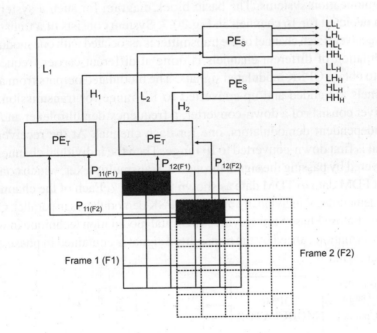

Fig. 20.6 Scheme for 3D wavelet Architecture

20.5 Communications Interface in Medical Instrumentation

For networking the medical community to share the expertise across a region or globe, there is a demand for transmitting data / image/ signal. This engendered technological solutions like telemedicine. The requirement of high bandwidth restricts the application, so need for better telecommunication infrastructure should be made available to the medical community. In this section, we briefly mention some modes of the state-of-the-art communications.

In present telecommunication networks, two standard techniques of digital multiplexing coexist: Frequency Division Multiplexing (FDM) for long distance interconnections, which assigns a specified frequency band to each signal to be transmitted and Time Division Multiplexing (TDM) usually for the shorter or local transmission trunks and switching centers, which interleaves in time, the samples of all the signals appropriately. Today the conversion between the two signal formats is commonly achieved by back-to-back connection of two FDM and TDM terminals.

Satellite communication system is one of the most advanced communication systems. The basic block diagram for such a system has been depicted for 16 channels in Fig.20.7. System consists of a transmitter and receiver. Each channel of the transmitter is associated with one modulator. Modulation of different channels is done at different carrier frequencies (IF) to obtain QPSK modulated signals. The modulated outputs from all the channels are added and up-converted to RF range for transmission. The receiver consists of a down-converter, a frequency demultiplexer and a set of independent demodulators, one for each channel. At the receiver, the signal is first down-converted to IF range. Then the individual channels are recovered by passing the signal through a Transmultiplexer, which converts input FDM data to TDM data as shown in Fig 20.7. Each of the channels is now demodulated individually by using QPSK demodulators in parallel. QPSK or Quadrature Phase Shift Keying is a digital modulation technique in which the information carried by the transmitted signal is contained in phase of the carrier signal.

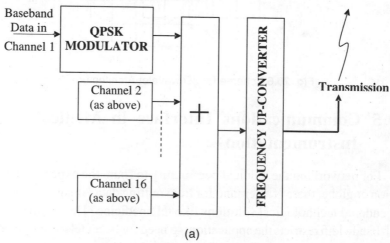

(a)

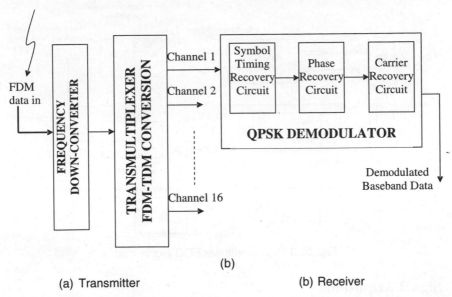

(b)

(a) Transmitter   (b) Receiver

Fig. 20.7 Multi Channel Demodulator Demultiplexer for Satellite Communication

20.6 Application of VLSI Based Biomedical Instrumentation

In this section, we describe a few modern biomedical instrumentation systems along with on the role of VLSI in this area. Four representative medical instrumentation systems are described, e.g., (a) electrocardiograph, (b) Doppler Ultrasonography system, (c) artificial retina and (d) non-invasive blood glucose analyzer.

20.6.1 Electrocardiograph

Electrocardiographs are used for routine checkups and emergency diagnoses, to monitor the effects of medication and surgical procedures, and to help determine the type and extent of cardiac deterioration in patients with heart disease. They may be found in hospitals, clinics, nursing homes, doctor's offices, and emergency medical vehicles.

Typical microcomputer-based cardiographs

Regardless of their differences, all cardiographs have the same basic function.

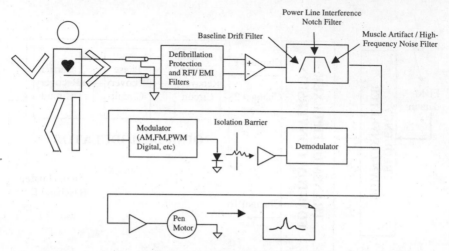

Fig. 20.8 Representative ECG signal path

Signal Acquisition

Figure 20.8 depicts a typical signal path for a single electrode. Buffering, differencing, and filtering functions are performed generally by FET operational amplifiers (Op-Amps). The most critical parameter for active devices in the signal path is noise, which should be under 5 mV P-P over the entire ECG bandwidth. Some form of input protection must be provided for the amplifiers, to prevent damage from static discharge or occasional defibrillation.

After filtering and amplification, the signal is modulated into some form suitable for transmission across the patient isolation barrier. This may involve amplitude, frequency, pulse width or pulse code modulation, or pure digital communications. The measurement unit may be controlled by discrete logic or a microcomputer, locally or from across the isolation barrier.

Main control /Processing/Analysis

Memory requirements for non-analyzing machines are modest (a few Kbytes) unless they include storage and communication facilities. A typical 10-sec 12-lead ECG strip consumes 40K bytes in raw form and 2-10K after data compression. Analyzing cardiographs requires much greater storage for their program code and for the 10 sec of ECG data that must be temporarily stored for analysis. Although electrocardiographs are not critical care devices and are not subject to the stringent reliability requirements imposed on other types of medical equipment, care must be taken to ensure that the machine operates properly at all times. Power-up self-tests, checksum, and background memory tests are often used to reduce the chances for undetected errors.

Faster processors mean faster analysis (which is fast enough already) and spare processor time after the real-time chores are handled. Increasing memory density will provide sufficient room for temporary mass storage of ECG records without disks or tape.

20.6.2 Low Cost Doppler Ultrasonography System

Doppler ultrasound system is used for non-invasive examination of blood flow through the major arteries and veins with ultrasound (high-frequency) waves that echoes off the body. Being one of the most convenient and harmless methods of study, this instrument is widely used [17].

Calculation of the Doppler shift is based on two alterations in the transmitted and received signals. The first Doppler shift occurs as the transmitted signal takes the scatterer and the second when the ultrasound leaves the same. Let f_0 be the incident frequency at an angle θ, and v be the velocity of blood cells. With approximation that the velocity of sound is c, v the Doppler shift Δf can be expressed as:

$$\Delta f = \frac{2 f_0 v \cos \theta}{c}$$

For incident frequency $f_0 = 10$ MHz, and the blood velocity 1m/sec, $\Delta f \approx 9 KHz$, assuming $c = 157 \times 10^3 cm/\sec$ and $\theta = 45^o$.

System Design

Let the shift in frequency due to scattering of frequency f_0 by the blood cells is given by Δf. The back-scattered signal contains frequency component ($f_0 \pm \Delta f$). The sign of the shift is determined by the direction of blood flow. In continuous wave, the shift in frequency is due to the contribution of velocity components of all the red blood cells illuminated by the transmitted signal. This results in a band of shifted frequency components. The signal is detected in the time domain and orthogonal transform is applied to generate the frequency domain signal. The spectrogram of the signal is displayed and studied for diagnostic purposes. Since the frequency shift is directly proportional to the blood velocity, any abnormal effects of the blood flow will be reflected on the spectrogram. The total scheme for the low cost Doppler Ultrasonography system designed at IIT, Kharagpur is given in Fig. 20.9.

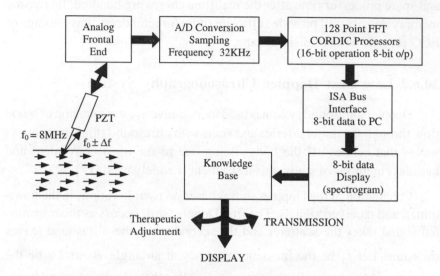

Fig. 20.9 Block Diagram for the Total Doppler Ultrasonography System Designed

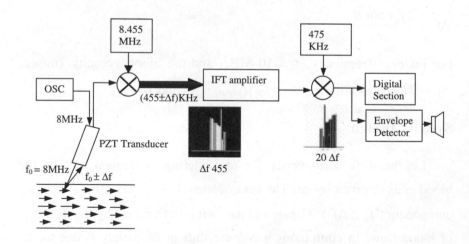

Fig. 20.10 Circuit used for the CW Doppler Ultrasonography Shift Detection

Analog part (Detection of frequency shift)

A PZT transducer is used to transmit an 8MHz signal. The backscattered signal is received, amplified and mixed with a frequency

8:455MHz. Thus, the new band of frequency is (455KHz ± Δ*f*). Using a super heterodyne structure this signal is amplified and AM detection is done for Double Side-Band (DSB), since the frequency shift can simultaneously occur in both directions. Fig. 20.10 shows the block diagram for the analog system. The output of the analog stage is given as input to the next section, which commences with an ADC for analog to digital conversion of the signal.

Digital section (FFT and interface with PC): The digital part of the circuit computes FFT for spectrum analysis of the recorded frequency shift. The ADC produces an 8-bit output, which is fed to the Xilinx FPGA XC4025 chip that computes FFT. Output of 128-point FFT is interfaced with the PC through PCI bus interface. In the subsequent section, the FFT computation using FPGA chip and the PCI interface is described.

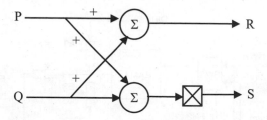

Fig. 20.11 Butterfly structure for 2-point FFT

Figure 20.11 gives the block level diagram for computing a 2-point FFT. A CORDIC (an acronym for Co-Ordinate Rotation Digital IC) based FFT processor is designed which is implemented on Field Programmable Gate Arrays (FPGA). The main advantages of using FPGA are the short design cycle and the scope of re-programmability for improvement in the design, without any additional cost. The choice of CORDIC algorithm for implementing the elementary operation (i.e. the butterfly) saves a lot of hardware without sacrificing the speed.

The Discrete Fourier Transform (DFT) of N complex samples f (n), n = 0, 1, 2,..., N-1 is defined as:

$$F(k) = \sum_{n=0}^{N-1} f(n) W_k^n \quad \text{where k = 0, 1, ..., N-1} \tag{4}$$

where $W = \exp\left(\dfrac{-2j\pi}{N}\right)$

FFT is an efficient method of computing DFT of N number of discrete samples in $O(N\log_2 N)$ time. The FFT algorithm splits the input dataset into even and odd-numbered points. The final form of FFT equation can be shown to be expressed in the form:

$$R_{re} = P_{re} + Q_{re}$$
$$R_{im} = P_{im} + Q_{im} \qquad\qquad (5)$$
$$S_{re} = (P_{re} - Q_{re})\cos(n\theta) + (P_{im} - Q_{im})\sin(n\theta)$$
$$S_{im} = -(P_{re} - Q_{re})\sin(n\theta) + (P_{im} - Q_{im})\cos(n\theta)$$

The last two in the set of Eqn. (5) essentially represent a plane rotation operation, which can be efficiently computed by applying the CORDIC algorithm.

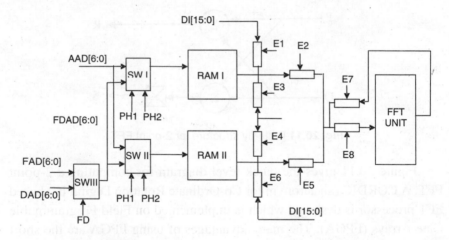

Fig. 20.12 2 RAM structure for accessing data and computing FFT

A two-RAM structure is used for implementation of the CORDIC based 128-point FFT processor (Fig. 20.12). These RAMs are in complementary phase for data reading and display. The lines are time multiplexed with equal time allocated each for computing FFT and display operation through PCI bus interface with computer. Every $4{:}096\text{ms}\left(\dfrac{215}{8MHz}\right)$, each phase signal switches between low and high. Fig. 20.13 shows the circuit layout of the CW Doppler ultrasonography system designed in the Department of Electronics & Electrical Communication Engineering, Indian Institute of Technology at Kharagpur, India.

Knowledge Base: The spectrogram obtained from the Doppler ultrasonography system is analyzed for automatic medical diagnosis. As a first step of extracting arterial abnormalities, some of the salient features of the spectrogram, which describes the spectrogram both qualitatively and quantitatively, are to be extracted. These features include both regional and boundary properties. A region of interest (ROI) is selected from the spectrogram and eight features, viz., (a) acceleration slope, (b) deceleration slope, (c) systolic to diastolic ratio, (d) period, (d) systolic window, (e) spectral broadening, (f) coefficient of variance, and (g) curvature at the systole are derived from the spectrogram enclosed within the ROI. These features are fed to an adaptive back propagation neural network (BPNN), trained in supervised mode. This network is presently trained for detection of five different flow conditions, viz. (a) Normal, (b) Ischemic, (c) Distal stenotic, (d) Proximal stenotic and (e) Flow through vasodilated artery Figure 20.14 (a) and (b) show two cases where proximal stenotic flow pattern and ischemic flow respectively has been detected from the spectrograms using trained BPNN Fig 20.15 shows the block-diagram for knowledge-based system.

Fig. 20.13 Doppler ultrasonography system designed in-house

Flow condition of a number of arteries along an arterial tree is fed to a Bayesian probabilistic framework for final inference in terms of measure of belief (MB) and measure of disbelief (MD). Fig. 20.16 shows the spectrograms for nine different regions for brain. In order to infer any status in transcranial Doppler, these nine regions need to be inspected.

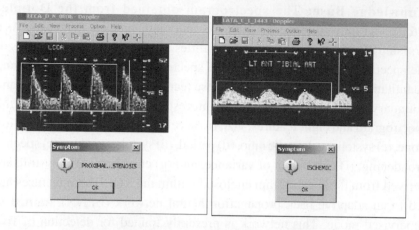

Fig. 20.14 (a) Proximal flow (b) Ischemic flow
detected in a spectrogram using BPNN

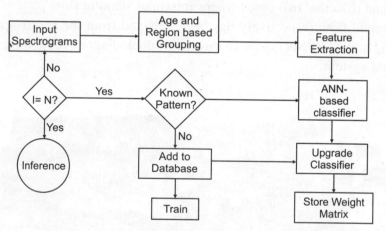

Fig. 20.15 Block diagram for knowledge-base scheme

20.6.3 Artificial Retina

The design of an artificial eye, which is an array of smart sensors, each with 100 m-sensors placed upon the retina is described here [17]. These sensors produce electrical signals which are converted by the underlying tissue into a chemical response, mimicking the normal operating behavior of the retina from light stimulation. The chemical response is digital (binary), essentially producing chemical serial communication. As shown in Fig. 20.17, the front side of the retina is in contact with the microsensor array. Transmission into the eye works as follows. The back side of the retina is stimulated electrically (via an artificial retina prosthesis) by the sensors on the smart sensor chip. These electrical signals are converted into chemical signals by the ganglia and other underlying tissue structures and the response is carried via the optic nerve to the brain. Signal transmission from the smart sensors implanted in the eye works in a similar manner, only in the reverse direction. The resulting neurological signals from the ganglia are picked up by the microsensors and the signals and relative intensity can be transmitted out of the smart sensor. Eventually, the sensor array will be used for both reception and transmission in a feedback system and chronically implanted within the eye. An implantable version of the current ex-vivo microsensor array, along with its location within the eye, is shown in Fig. 20.17. The microbumps rest on the surface of the retina rather than embedding themselves into the retina. A similar design is being used for a cortical implant, although the spacing between microbumps is larger to match the increased spacing between ganglia in the visual cortex.

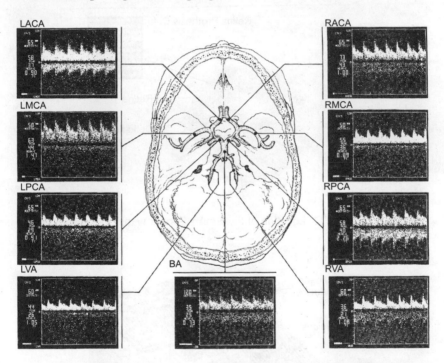

Fig. 20.16 Spectrogram of different regions of the arterial tree in transcranial Doppler

Fig. 20.18 illustrates the design of an artificial retina. It is basically an array of smart sensor chips, each with 100 microsensors, have been built for an ex-vivo testing system of a retina. Fig. 20.18 depicts an illustration of the design. The smart sensor has two components: an integrated circuit and an array of sensors. The integrated circuit is a multiplexing chip, operating at 40KHz, with on-chip switches and pads to support a 10×10 grid of connections. The circuit has both transmitting and receiving capabilities. Each connection has an aluminum probe surface where the micromachined sensor is bonded. This is accomplished by using a technique called backside bonding, which places an adhesive on the chip and allows the sensors to be bonded to the chip, with each sensor located on a probe surface. Before the bonding is done, the entire IC, except the probe areas, is coated with biologically inert substance. The sensors form a 10 X 10 electrode array; each sensor is a microbump that will rest on the retina. The distance between adjacent microbumps is approximately 70 microns. These microbumps start with a rectangular shape and near the end taper to a point. This point is placed on the retina tissue, allowing contact between the smart sensor array and the retina.

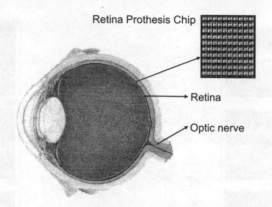

Fig. 20.17 Location of the Smart Sensor within the Eye

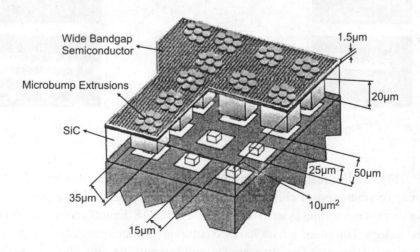

Fig. 20.18 Illustration of the Microsensor Array

Although the microsensor array and associated electronics have been developed, the challenge at this point is the wireless networking of these microsensors with an external processing unit in order to process the complex signals to be transmitted to the array.

The reasons for transmitting formation out of the retina are perhaps less obvious. The purpose of this reverse flow of data is to determine the mapping between the input image and the resulting neurological signals, which enable us to see that image. On-going diagnostic and maintenance operations will also require transmission of data from the sensor array to an external host computer. These requirements are in addition to the normal functioning of the device, which uses wireless communication from a camera embedded in a pair of eyeglasses into the smart sensor array.

Fig. 20.19 depicts the processing steps from external image reception to transmission to the retina prosthesis [18]. A camera mounted on an eyeglass frame could direct its output to a real-time DSP for data reduction and processing (e.g., Sobel edge detection). The camera would be combined with a laser pointer for automatic focusing. The DSP then encodes the resultant image into a compact format for wireless transmission into (or adjacent to) the eye for subsequent decoding by the implanted chips. The setup in Fig. 20.19 shows a wireless transceiver that is inside the body, but not within the retina.

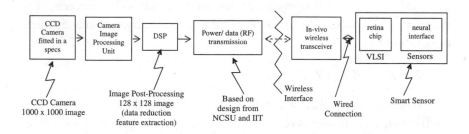

Fig. 20.19 Processing Steps for the Retina Prosthesis Project

20.6.4 Non-invasive Blood Glucose Analyzer

Diabetes mellitus is a complicated, serious and one of the most prevalent non-communicable diseases in the world. It not only affects patients' internal organs, circulating system and eyesight, but also his entire life. Successful management of diabetes involves knowledge of the current blood glucose level to allow the diabetic to compensate by right diet and medical treatment.

Currently glucose measurements are done by pricking a finger and extracting a drop of blood whose chemical composition is then analysed by either chemically sensitive test strip or by glucose oxidation reaction method in conjunction with an optical meter that gives a numerical glucose reading. This requires a daily routine of multiple finger stabs, which is painful.

At present, a widely used method of self-monitoring of blood glucose (SMBG) involves determination of blood glucose concentration with devices utilizing chemical analysis of blood samples taken by puncturing a finger . Even though SMBG has revolutionized the management of diabetes, discomfort and inconvenience are barriers for effective compliance and, therefore, optimum management. These drawbacks may limit the number of blood glucose measurements performed by patients with diabetes and, therefore, result in improper management of the disease. Therefore, there is a great demand for a noninvasive method for frequent or continuous blood glucose monitoring, which would considerably improve the quality of life for diabetic patients, improve compliance for glucose monitoring, and reduce complications and mortality associated with the disease [19].

The measurement is based on the pulsed laser photo-acoustic (PA) spectroscopy in which a pulsed laser energy of wave length 1000 – 1800 nm is used to excite the tissue (Fig. 20.20). Subsequent optical absorption causes microscopic localized heating. The increase in temperature causes rapid thermal expansion, which generates an ultrasound pressure wave detectable by a PZT transducer located at the skin surface. The magnitude of a pulsed PA signal, P, is related to the absorption coefficient of solution by :

$$P = K(\mu_a \beta \sqrt{v})/C_p$$

Where μ_a is the optical absorption coefficient, β is the thermal expansion coefficient, v is the sound velocity, C_p is the specific heat of the solution, and K is a proportionality constant that is related to the bulk modulus of the medium. The C_p of a solution decreases, whereas the acoustic velocity increases with increasing glucose concentration. At a glucose absorption wavelength, change in the PA signal are the result of changes in μ_a, v and C_p. The multiplicative effect increases the PA signal as a function of concentration. The speed of sound and the C_p values change as the total solute concentration changes. When excited by the laser source the absorption is caused by the overtones of O-H and C-H bond vibrations of glucose and other analytes; this absorption is subsequently converted into an acoustic pulse.

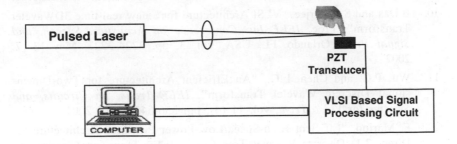

Fig. 20.20 Block Diagram of Non-invasive blood glucose analyzer

This acoustic pulse is detected by the PZT transducer and is converted into corresponding change in voltage. The change in voltage is anlaysed by signal processing circuit to give a measure of glucose level in the blood. This data may be used for automatic insulin therapy using microprocessor/ microcontroller based system. An effort will also be made to replace PZT transducer by Si/ Porous-Si based MEMS.

20.7 References

1. C.Raja Rao, S.K.Guha, "Principles of Medical Electronics and Biomedical Instrumentation", Universities Press (India) Limited, 2000.

2. Willis J. Tompkins (Editor), "Biomedical Digital Signal Processing", Prentice Hall 1995.

3. John G. Webster (Editor), "Medical Instrumentation – Application and Design", John Wiley & Sons. Inc. 1998.

4. Donna L. Hudson, Maurice E. Cohen, "Neural Networks and Artificial Intelligence for Biomedical Engineering", IEEE Press Series in Biomedical Engineering Metin Akay, Series Editor, 2000 by the Institute of Electrical and Electronics Engineers, Inc.

5. Loren Schwiebert, Sandeep K.S.Gupta, Jennifer Weinmann, "Research Challenges in Wireless Networks of Biomedical Sensors", ACM ISBN 1-58113-422-3/01/07, 2001.

6. S.M.Sze, "Semiconductor Sensors", John Wiley & Sons, Inc., 1994.

7. Norman G. Einspruch, Robert D. Gold, "VLSI Electronics Microstructure Science", VLSI in Medicine, vol. 17.

8. B. Das and Swapna Banerjee, "A Unified CORDIC-based chip to realize   DFT/DHT/DCT/DST", IEE Proc.-Computers and Digital Techniques, vol. 149, No. 4, pp. 121-127, July 2002.

9.   http://portal.acm.org/citation.cfm?id=381692&coll=portal&dl =ACM&ret=1.

10. B.Das and S.Banerjee, "VLSI Architecture for a new real-time 3DWavelet Transform", Proc. *IEEE Int. Conference on Acoustics, Speech, and Signal Proc.*, Orlando, FL, USA, vol. 3, pp. 3224-3227, May 13-17, 2002.

11. Wu, P.C. and Chen, L.G., "An Efficient Architecture for Two-Dimensional Discrete Wavelet Transform", *IEEE Trans. on Circuits, and Systems for Video Tech.*, vol. 11 (4), pp. 536-545, April 2001.

12. F. Marino, "Efficient High-Speed/Low-Power Pipelined Architecture for Direct 2-D Discrete Wavelet Transform", *IEEE Trans. on Circuits and Systems-II: Analog and Digital Signal Processing*, vol. 47 (12), pp. 1476-1491, December 2000.

13. H. Nicolas, M. Schutz, and F. Jordan, "Parallel implementation of a 3d-subband decomposition algorithm for digital image sequence compression on the CRAY T3D", *Proceedings of Cray Users's Group Conference*, Denver, CO, USA, pp. 97-102, March 1995.

14. R.Kutil, and A.Utl, "Hardware and software aspects for 3-d wavelet decomposition on shared memory MIMD computers", in ACPC, pp. 347-356, 1999.

15. B. Das and S. Banerjee, "A Memory Efficient 3-D DWT Architecture", *Proc. IEEE Int. Conf. on VLSI and Embedded Systems Design*, New Delhi, India, pp. 208-213, January 4-8, 2003.

16. B. Das and S. Banerjee, "Data Folded Architecture for Running 3D DWT Using 4-tap Daubechies Filter", IEE Proc.-Computers and Digital Techniques, 2004.

17. A. Jensen, "Linear description of ultrasound imaging systems", Notes for the International summer school on Advanced Ultrasound Imaging, Technical University of Denmark, July 5-9, 1999.

18. M. O'donnel, "Applications of VLSI circuits to medical imaging", Proc. IEEE, Vol. – 76, pp. 1160 – 1114, Sept. 1988.

19. G.B.Christison and H.A.Mackenzie, "Laser photoacoustic determination of physiological glucose concentrations in human whole blood", Medical & Biological Engineering. & Computing, Vol. 31, pp. 284-290, May 1993.

20. B.J. Tesch and J.C. Garcia, "A Low Glitch 14-b 100 MHz D/A Converter", IEEE J. of Solid-State Circuits, vol. 32, no. 9, pp. 1465-1469, Sept. 1997.

21. C.G. Martinez and S. Simpkins, "A Monolithic 12-Bit Multiplying DAC for NTSC and HDTV Applications", Proc. IEEE Bipolar Circuits and Technology Meeting, pp. 52-55, 1989.

22. AD9774, Data sheet, Analog Devices Inc., 2000.

23. N. Tan, E. Cijvat, and H. Tenhunen, "Design and Implementation of High-Performance CMOS D/A Converter", Proc. IEEE Intern. Symp. on Circuits and Systems, ISCAS'97, Hong Kong, vol. 1, pp. 421-424, May 1997.

24. C.A.A. Bastiaansen, D.W.J. Groeneveld, H.J. Schouwenaars, and H.A.H. Termeer, "A10-b 40-MHz 0.8-mm CMOS Current-Output D/A Converter", IEEE Journal of Solid-State Circuits, vol. 26, no. 7, pp. 917-921, July 1991.

25. C.H. Lin and K. Bult, "A 10-b, 500-MSample/s CMOS DAC in 0.6 mm2", IEEE J. of Solid-State Circuits, vol. 33, no. 12, Dec. 1998, pp. 1948-1958.

26. S.Y. Chin and C.-Y. Wu, "A 10-b 125-MHz CMOS Digital-to-Analog Converter (DAC) with Threshold-Voltage Compensated Current Sources", IEEE J. of Solid-State Circuits, vol. 29, no. 11, pp. 1374-1380, Nov. 1994.

27. I.H.H. Jorgensen and S.A. Tunheim, "A 10-bit 100MSamples/s BiCMOS D/A Converter", Analog Integrated Circuits and Signal Processing, vol. 12, pp. 15-28, 1997.

28. N. Tan, "A 1.5-V 3-mW 10-bit 50 MS/s CMOS DAC with Low Distortion and Low Intermodulation in Standard Digital CMOS Process", Proc. IEEE Custom Integrated Circuits Conf., CICC'97, Santa Clara, CA, USA, pp. 599-602, May 1997.

29. A.Van den Bosch, A.F. Borremans, M.S.J.Steyaert "A 10-bit 1-GSample/ s Nyquist Current –Steering CMOS D/A Converter", IEEE J. Solid State Circuit, vol. 36, pp 315-323 Mar. 2001.

30. Sandeep K. Gupta, and Victor Fong, "A 64-MHz clock-rate SDADC with 88-db SNDR and –105-dB IM3 Distortion at a 1.5 –MHz signal Frequency", IEEE journal of solid-state circuits, vol. 37, no. 12, December 2002.

31. Uyttenhove Koen and S.J. Michie, "A 1.8-V 6bit 1.3-GHz Flash ADC in 0.25mm CMOS", IEEE journal of solid-state circuits, vol. 38, no. 7, July 2003.

32. Chi-sheng Lin and Bin-Da Liu, "A new Successive Approximation Architecture for Low-Power Low-Cost CMOS A/D Converter", IEEE journal of solid-state circuits, vol. 38, no. 1, December 2003.

33. Mikko Waltari, and Kari A.I. Halonen, "A 10bit 200MS/s CMOS Parallel Pipelined A/D Converter", IEEE journal of solid-state circuits, vol. 36, no. 7, December 2001.

34. W.Yang, D.Kelly, L. Mehr, M.T.Sayuk, and L.Singer, "A 3-V 340-mW 14-b 75-MSample/s CMOS ADC with 85-dB SFDR at Nyquist input", ISSCC Digest of Technical Paper, Feb. 2001, pp. 134-135.

35. Myung-Jun Choe, Bang-Sup Song, K. Bacrania, "An 8-b 100-MSample/ s CMOS pipeline folding ADC", IEEE Journal of Solid-State Circuits, vol. 36, issue 2, pp. 184-194, February 2001.

36. Yunchu Li, and Edgar Sanchez-Sinencio, "A wide input bandwidth 7-bit 300 MSample/s Folding and Current-mode Interpolating ADC", IEEE Journal of solid-state circuits, vol. 38, no. 8, August 2003.

| Chapter 21 |

Robotic Instrumentation

M.K. Ghosh

Abstract

Modern robots are reprogrammable multifunctional manipulators. They employ technologies from multidisciplinary areas such as mechanical, electrical, electronics, computer, instrumentation, engineering. The present article gives a brief review of the concepts of industrial robots, their functional aspects and, in particular the instrumentation technologies involved. The accuracy of performance of a robot, which is the main reason of its use in many cases such as part placement in electronics PCBs, is very much dependent on the sensors used. Robots employ various sensors and instrumentation, classical as well as specialized. The article mainly addresses the latter type of sensors, many of which are innovative and often robot specific.

21.1 Introduction

Robotic Preliminaries

The Czech word 'robota' means 'forced labourer' or 'slave labourer'. The term was first used by Czech playwright Karel Capek in 1921 to picture 'robot' as a machine that resembled people but worked twice as hard. Issac Asimov subsequently coined the word 'robotics' in 1942.

The predecessors of robots [1] are programmable machines of the eighteenth century like mechanical loom, programmable lathe, etc. It was followed by invention of a rotary crane equipped with a motorized gripper, a jointed mechanical arm used for paint spraying and a teleoperator. George Devol, 'father of the robot' developed a magnetic process controller to be used as a general purpose play back device for controlling machines. The first numerically controlled machine tool was developed in 1952. The 'robot age' began in 1954 with the development of the first manipulator with a play back memory by Devol. This device was capable of performing a controlled function from one point to another (i.e. point - to - point motion).In the society and the world of fantasy the term is used in different ways. Even in the scientific world there are lots of variations in use of the terminology. A complete definition is as follows:

A manipulator is a machine, the mechanism of which usually consists of a series of segments jointed or sliding relative to one another for the purpose of grasping and moving objects usually in several degrees of freedom. A robot is a reprogrammable, multifunctional manipulator designed to move material, parts, tools, or specialized devices through variable programmed motions for the performance of a variety of tasks. The industrial robot of today does not look the least bit human. Again bio robots are tiny capsules (micro robots) capable of performing bio medical operations like removal of blockages in artery.

21.2 Classification of Robots

Knowledge about different classes of robots and their functions is necessary to understand the role played by sensors and instrumentation systems in robotic applications. Classification of industrial robots is based on

(a) Type of control i.e. path or motion control

Point to point system is used in machine tool, loading/unloading, assembly, etc.

Controlled path (computed trajectory) system as in seam welding and line tracking.

Continuous path system requires start, finish and path definition. It has a sophisticated method of control.

(b) Capability or intelligence level

Sequence controlled robots are of either fixed sequence or variable sequence systems used in routine handling tasks.

Play back robots are 'taught' to carry out a series of movements by recording a series of points or steps e.g. machine loading/ unloading, welding, spray painting.

Controlled path / computed trajectory robots are programmed to follow a certain path between particular points.

Adaptive robots are provided with sensory feedback to adapt to environmental changes.

Intelligent robots determine their own actions through AI (Artificial Intelligence). They are adaptive -, and learning systems controlled using feedback from sensors.

(c) Mechanical configurations.

The mechanical configuration of a robot depends on the coordinate systems employed for movement of its arm. It fixes the work envelope (i.e. volume of space created within the virtual surfaces swept by the robot arm of maximum and minimum reach) of the robot. Table 21.1 and Figures 21.1, 21.2, 21.3, 21.4, 21.5, 21.6, 21.7 summarize the geometric configurations and their features.

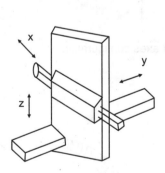

Fig. 21.1 Cartesian Configuration

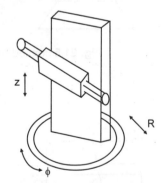

Fig. 21.2 Cylindrical Configuration

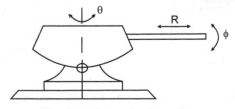

Fig. 21.3 Polar Configuration

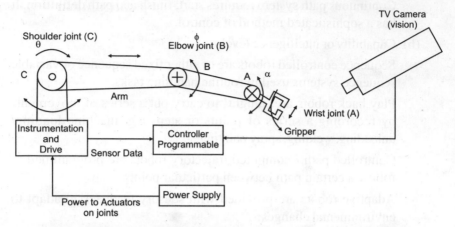

Fig. 21.4 Jointed-arm horizontal axes configuration(along with typical functional components of a robot)

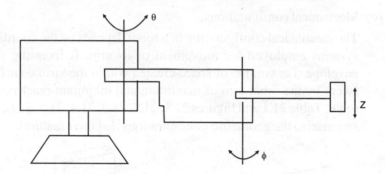

Fig. 21.5 Jointed-arm vertical axes configuration

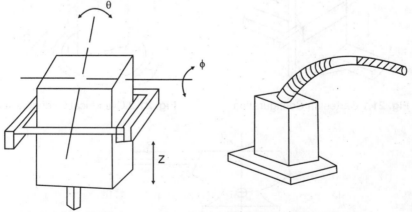

Fig. 21.6 Pendulum area configuration   **Fig. 21.7** Multiple-joint arm configuration

Table 21.1 Classification by Mechanical Configuration

Sl. No.	Geometric configuration		Features	
	Nomenclature	Movement in directions of	Work envelope	General
1.	Cartesian Fig 1	x, y, z	Rectangular	
2.	Cylindrical Fig 2	R, z, θ	Cylindrical	Good work area to floor area ratio, robust
3.	Polar Fig 3	R, θ, ϕ	Spherical	Complex control, less flexible, robust
4.	Jointed – arm horizontal axes Fig 4	θ, ϕ, a	Spherical	Complex control, small base area to work volume
5.	Jointed – arm vertical axes Fig 5	-	Cylindrical	- do -
6.	Pendulum area Fig 6	-	Partial Spherical	Low inertia, high speed, high acceleration
7.	Multiple – joint arm Fig 7	-	Spherical	Excellent flexibility

x - base travel, y,R - reach, z - elevation, θ - base rotation, ϕ - elevation angle, α - reach angle

(d) Mobility

Fixed robots are either mounted on or supported from a fixed (pedestal) base. They are also mounted on a track (for welding, painting) or on an overhead gantry.

Mobile robots are fitted with low cost computers and their movements are on wheels, tracks, legs (two, four, six or more).

(e) Complexity

Complexity of robots depends upon goals and requirements, viz

(i) Type of control

Simple or complex (along two or more axes) force, high positional accuracy (tolerance better than 0.25 mm), precision positioning (< 0.025 mm), sensor directed control (vision, touch, force sensing).

(ii) Application environment

Freedom of movement, working conditions (hot, wet, dusty, hazardous).

(iii) Interaction with other equipment

Synchronization with conveyor (in assembly line), with machine cycle, with another robot arm.

(iv) Range of motion

Arm motion, movement of whole robot.

(v) Speed range

Speed of arms, wrists, grippers - linear or rotary motion

Slow speed	: <300 mm/s; <60°/s
Medium speed	: 300-1500 mm/s; 60°-180°/s
High speed	: >1500 mm/s; >360°/s

(vi) Sensors

Proximity, touch, simple vision, complex vision.

21.3 Basic Functional Components of a Robot

The mechanical, electrical and computational structure of robots may vary considerably. But most have the following basic functional components in common (Fig. 21.4)

(i) *Mechanical Unit*

Manipulator or arm, links and joints, wrists or gripper, end effectors, pedestal base or movement support. Joints contain actuator and sensors. End effector (vacuum cup, electromagnet, clamp, hook, hand etc) costs 20% of the price of a robot. Drives used for movement are electric (dc, step, brush less motors), fluid power (hydraulic, pneumatic). Drive mechanisms are linear (rack and pinion, lead screws, ball screws, etc), rotary (gear trains, timing belt drives, harmonic drives, etc.)

(ii) *Robot Controllers*

The brain of a robot is its controller. It prepares and executes a manipulation and/or locomotion task by coordination of kinematic structure, dynamic control architecture, sensor environment, safety capabilities, programming methodology, interfaces to the environment, and data management and presentation.

A simple and less expensive controller performs open loop control e.g. step sequencer, electronic sequencer and pneumatic logic system. More versatile robots are microcomputer controlled using closed loop control.

The majority of today's controllers are designed for position and trajectory control, and occasionally required to support the interaction with external sensors such as cameras for adaptive control. Different functional units are based on:

Kinematic considerations [2] Involve considerations of motion without reference to mass or force. These involve trajectory planning, trajectory interpolation, coordinate (forward and backward) transformation of all links and joints of a robot arm.

Dynamic considerations involve driving individual links along a desired path (point-to-point, coordinated path or continuous path) despite external forces and torques. Dynamic aspect mainly addresses the lag in response time of the robotic function and the forces and moments necessary to perform a specific duty. The Robots used for movement of the heavy loads require special design consideration in hardware and software.

End effector control combines all kinematic and dynamic considerations of a robot control.

Sensor control is the essence of a closed loop and real time control in a robot.

Drives: electrical, hydraulic and pneumatic.

Interpretation of the program: Kinematic and dynamic mathematical modelling, trajectory equations, computational calculations necessary for executing the coordinate transformations, accuracy and response time constitute the basis for programming the coordinated control of a robot. The controller should be able to interpret the programmes and execute them by sending commands to individual drives with or without interruption by the operator, if needed.

(i) Sensors

Sensors in robots are mainly used for control. Sensing in robotics can involve low level sensors such as shaft angle encoders as well as high level ones like CCD cameras from which complex information has to be extracted using computing methods. Some of the sensors and sensing techniques will be discussed more elaborately in the next section.

(ii) Power supply

It provides the necessary energy to the manipulator's actuators. It can take the form of a power amplifier (servomotor actuated system) or it can be a remote compressor (for pneumatic or hydraulic devices). Power supply to the actuators is controlled by the command signal sent by the robotic controller.

21.4 Robotic Sensing [3]

Robotic sensors constitute a different class of sensors by themselves. They are characterized by miniaturization, high degree of precision and accuracy, and fast response. Though mostly electronic, some of the sensors are mere improved versions of the classical ones. Robotic sensors can be classified as follows [4]:

(a) Proprioceptors or Interoceptors

These sensors are used to measure robot's internal parameters (like location of joints, end effector). They perform functions analogous to 'Kinesthesis' in humans (awareness of the position of one's own limbs). They measure both kinematic and dynamic parameters of the robot.

(b) Exteroceptors

These sensors are used for the measurement of its environmental (external from the robot point of view) parameters. Functionally they are analogous to five sensory organs in humans, viz. vision, touch (tactile), smell (olfaction), taste and hearing (audio). Majority of exteroceptors in robots are based on vision (imaging) and hearing (audio). Some proximity sensors are similar to touch (tactile or contact) sensors. However, olfactory and taste sensors are not common.

(c) Vestibular

This class of sensors is futuristic. The sense of equilibrium and the ability to sense internally the effects of gravity and acceleration on whole body makes a human being, a perfect unstable body from point of view of mechanics, perform stable locomotion. Future robots working in space or under sea may need vestibular senses like humans.

21.5 Robotic Measurement and Instrumentation

This includes sensors to measure kinematic parameters (joint position, velocity, acceleration), dynamic parameters (force/torque, inertia), also temperature, humidity, slip, pressure, etc.

(a) Internal Sensors

(i) Joint Position

- Limit switch/micro switch
  Plunger, lever, roller lever are used to check limits of travel of limbs.

- Potentiometer
  Classical rotary potentiometers are used for not so accurate joint position sensing. They are noise prone. They are mounted directly or linked through a 'timing belt' arrangement. Rack and pinion are used for linear to rotary conversion. For linear accurate positioning LVDT is used.

- Synchro resolver, RVDT (Rotary Varying Differential Transformer) are used for angular position sensing. Inductosyn is used for accurate ($\pm$ 1 arc second) sensing.

- Encoders
  Incremental and absolute optical shaft encoders are used for position sensing. They can also be used for speed sensing and sensing of direction of rotations. Although they provide quantizing information they can be directly interfaced with digital controllers.

- Optical fibre sensors
  Angular position sensing using optical fibre is convenient from point of view of miniaturization and accuracy.

(ii) Speed Sensors

Analogue speed sensing is done by DC and AC tachogenerators. They are essentially permanent Magnet DC or AC machines whose output voltage depends on the rotational speed.

- Magnetic pick up
  The sensor in its simplest form (Fig. 21.8) is a metallic wheel with ferrous teeth on it. The wheel is mounted on a shaft at the joint axis. A permanent magnet serving as the core of an inductive coil is placed near the wheel teeth. The pulse frequency generated is proportional to the joint velocity. The sensor gives digital output. It can operate in severe environments and provide good accuracy for high pulse frequency ($\sim$ KHz), but it is

unsuitable for near zero speeds. Speed resolution increases with number of teeth on the wheel.

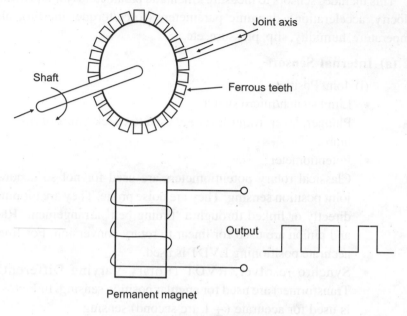

Fig.21.8 Magnetic Pick Up

- Optical tachometer (chapter 4)

  For noncontact speed sensing an optical tachometer can be used. However, there is a need for guarding the signal from noise pick up due to ambient light. Rotary optical shaft encoders can be used in a similar fashion.

(iii)   Acceleration

Miniaturized classical sensors like strain gauges, piezoelectric crystals, capacitive and inductive transducers are widely used for measurement of acceleration in robotics. More recent ones are based on MEMS (chapter 9).

- Micromechanical accelerometer

  IBM, San Jose [1] has developed a smart acceleration sensor using silicon technology. It is a cantilever (0.15 mm long) with gold deposition at the end to increase its sensitivity and inertia, which is used as the pickup. Capacitor position sensor interfaced with a microprocessor measures the acceleration along with necessary (linearization and correction) signal processing.

(iv)   Force/Torque

For robots with 'pay loads', measurement of force is very

important. The locations where force/torque measurements are essential, are

(1) Arm joints - shoulder joint and elbow joint
(2) Wrist joint
(3) Gripper - force exerted by load
(4) Pedestal - total force acting on it that decides mechanical stability of the robot

- Strain gauges

  For joint shaft torque measurement specially profiled strain gauges and strain gauge rosettes are used.

- DC motor armature current

  Wherever the joint is driven by a dc motor, its armature current (motor runs as separately excited armature controlled) is a measure of the joint shaft torque.

- Fluidic drive back pressure

  The back pressure (hydraulic / pneumatic) of a fluidic drive serves as a measure of joint shaft torque.

- Forces on wrist joint

  Wrist joints require measurement of three dimensional lateral forces (F_x, F_y, F_z) and rotational torques (T_x, T_y, T_z).

- Forces on gripper

  Piezo crystals, strain gauge transducers are used.

- Force on pedestal base

  Metal plates and strain gauges are used to measure vertical and horizontal forces.

(b) External sensors

Commonly a robot interacts with its environment through sensing of position (absolute or relative position of the job/obstacle) and force needed to perform the task assigned. Hence measurement of the two main parameters is the main objective. However, depending on the objective and requirements of sensing they vary widely resulting in development of very innovative sensors. Three main categories of external sensors classified on the basis of detection range are:

- Contact (tactile) sensors
- Close range sensors and
- Long range 'far away' sensors

(i) Contact (tactile) sensors.

In robotic applications where detection process permits contact sensing, the sensors used are tactile or contact sensors which measure a multitude of parameters of the touched object surface. Wrist sensors, gripper sensors, end effectors are sensitive, small, compact and not too heavy. Two basic sensors in this group are:

  (a) Displacement sensitive sensors
  (b) Force/torque sensitive sensors

(a) Displacement sensitive sensors

  • Make/break contact sensors, used for two level action or as safety devices, can be plunger type, microswitch type (which function when contact with a surface is made).
  • Displacement sensors
    Classical capacitive sensors, optoelectronics based sensors are used. More sophisticated sensors are based on optical imaging and multisensory imaging. Fig 21.9 represents a binary tactile imaging technique using an array of pressure switches. The sensor array is fitted on the gripper. The image obtained is used to determine the contact surface area and the force on the gripper surface.

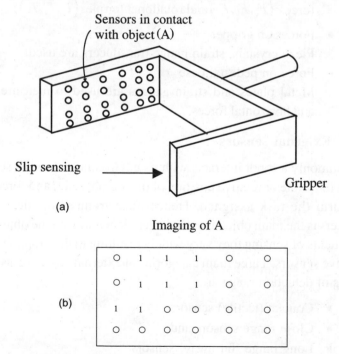

Fig. 21.9 Tactile sensing. (a) Sensor array (b) Tactile imaging

- Slip sensing

  For unknown objects, to adjust the optimum force between the gripper and the object contact surfaces, slip sensors are used. One such sensor is a piezoelectric crystal based vibration sensor. The vibration is picked up using a sapphire needle fitted onto the piezoelectric crystal. The sensor provides indication of slip. (Fig. 21.10)

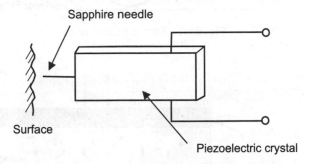

Sapphire needle

Surface

Piezoelectric crystal

Fig. 21.10 Slip sensing

- Proximity sensing (Contact type)

  A novel tactile sensor for proximity sensing [3] uses a whisker to sense presence or absence of an object for collision avoidance and location of the 'job'. The whisker is extended from the gripper finger (Fig.21.11) and supported by air pressure. Smart silicon sensors are fitted on the inner end of the whisker. Any deflection of the whisker due to contact with another object exerts a pressure on the silicon pressure sensor that transmits a signal to the robotic controller

(b)   Tactile sensors for force/torque measurement

Contact force is measured using strain gauge, piezocrystals and piezo resistive semiconductor sensors. Other novel sensors are

- Change of resistance with pressure (presistor).
- Pressure sensitive paint which is a piezoresistive semiconductor mixed with organic material. Its sensitivity is very high e.g. $7M\,\Omega$/gm pressure change.
- Conducting sensing elements

  These sensors change their resistive properties under pressure. The sensing element, sandwiched between two metal electrodes (Fig.21.12), consists of conductive rubber (silicon rubber mixed with metallic compounds) or conducting foam.

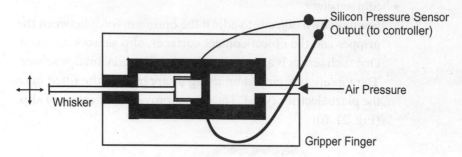

Fig. 21.11 Tactile proximity sensor

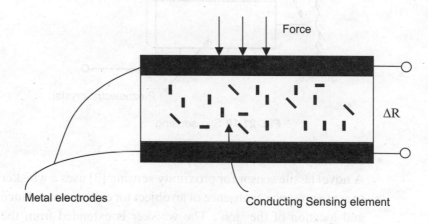

Fig. 21.12 Conducting sensor

(ii) Close range sensors or proximity sensors (non-contact type)

Proximity sensors are also used for close range robotic operations i.e., for collision avoidance, object location and obstruction detection without touching them. Classical sensors used are:

Inductive sensors: Based on change in inductance due to ferromagnetic object movement,

Hall effect sensor: Based on distortion of magnetic field by moving ferromagnetic object.

Magnetic pickup: Based on change in effective permeability due to metallic object approaching.

Capacitive pick up (change in capacitance due to proximity of any solid or liquid).

Ultrasonic pick up (short pulses ~40 KHz reflected from the nearing object and its time of flight is a measure of nearness of the object).

- Optical proximity sensor

Malfunction of the robot during robot programming by the operator is of grave concern. Proximity sensors which are fast, accurate and can sense in all directions are in great demand for safety of personnels and operators. An optical safety system has been developed [5] which consists of optical sources and detectors (Fig.21.13) arranged in a fashion so as to detect the minimum distance between the robot arm and any part of the operator's body that may fall in the plane or direction of movement of the arm in case of malfunction due to operator's mistake during programming. The sensor uses Artificial Neural Networks (ANN) for computation of the minimum distance in 3-D and is programmed to send signal for switching off the power to the arm movement in case the distance equals a threshold.

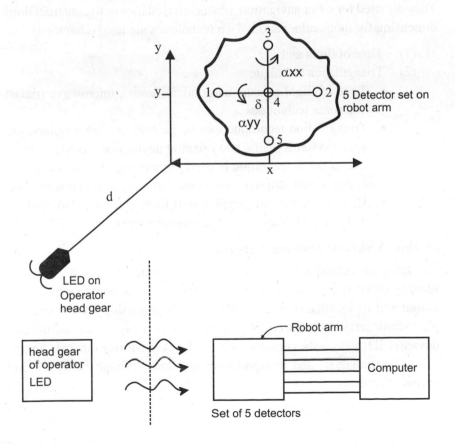

Fig. 21.13 Optical proximity sensor for robot safety

- Voice sensing

  In some CNC machines voice sensing is used for operation in accordance with the voice profile of the programmer stored in its memory.

- Imaging techniques

  Acoustic imaging or ultrasonic imaging is not very effective due to noise corruption. Optical imaging is expensive and for proximity sensing may not be very effective.

(iii) Long Rang sensing

Sensing from a distance has been highly sophisticated for robotic applications. Two categories of sensors are available:

(a) Range sensing, and

(b) Vision or optical imaging

(a) Range sensors

Range sensors measure the distance to object in their operation area. They are used for robot navigation, obstacle avoidance or to construct third dimension for monocular vision. Two techniques are used (chapter 4)

(1) Time of flight and

(2) Triangulation technique

- Time of flight sensors use optical (laser), microwave (radar) and sonar techniques.

- Triangulation technique can be passive or stereo / binocular vision systems (using two imaging devices) and active. Active triangulation technique is used for moving objects. In optical systems light stripe detectors measure lateral movement also.

- Interferometric and Doppler shift techniques are also used for distance and relative motion measurements.

(b) Vision or imaging technique

Imaging techniques are widely used in robotic sensing. It is used to identify objects, understand the object details like shadows, cracks, size, colour and its location or motion. Optical imaging using TV camera and photodiode array or charge coupled device (CCD) array or charge-injection device (CID) array is the common practice for acquisition of images. Image processing, analysis and interpretation are done through hardware and software techniques.

21.6 Sensor Fusion in Robotic Environment

A complete perception of the environment requires coordination between different sensory organs in humans. Modern robots are being built up in the same line with sensor fusion technique. Signals from different sensors are processed simultaneously to extract relevant information. The sensors may be of the same nature or altogether different. In addition to multitude of smart and intelligent sensors used, high level signal processing and subsequent decision support provided by AI, ANN, GA make robotic instrumentation and control an index of advancement in measurement and instrumentation.

21.7 References

1. Khafter, R. D., Chimelewski, T.A. and Negin, M. Robotic Engineering -An Integrated Approach, PHI, New Delhi, 1994.

2. Lee, C.S.G. ' Robot Arm Kinematics, Dynamics and Control', Computer IEEE, Dec 1982, Vol 15, No. 12.

3. Mair, G.M. 'Industrial Robotics' Prentice Hall, NY 1988.

4. Ghosh, K., O'shea, J. and Ghosh, M.K. Some Innovative Approaches to Non-Contact Detection of Human Presence in Automated Work Areas. Proceedings on 'Productivity in a World Without Borders'. Editions of DFL' Ecole Polytechniques, Montreal, Quebec, October 1995, pp 745-754.

5. Murthy, A.V.S.N. et al. A Novel Robot Safety Sensor Using Neural Net. Proceedings of Seminar on Pattern Recognition, Artificial Intelligence, Neural Networks, Dehradun, March 12, 1993.

| Chapter 22 |

Bioprocess Instrumentation

S. Sinha

P. S. Pal

Abstract

The design of a new bio-reactor, its modeling, oxygen transfer characteristics and the related instrumentation are presented in this chapter. The instrumentation includes the all important dissolved oxygen (DO) sensor and soft-sensing or stream inferencing of the variables in a bio-reaction process. Comprehensive platform for computer control of the bio-reaction process is also presented. This is by no means a complete treatise of bio-process instrumentation but hopes to present certain exponents of it at advanced level.

22.1 Introduction

Of late, there has been a growing interest in biotechnology among the researchers. A multi-disciplinary approach is needed to meet the requirement of biotechnology industries. Bioprocesses such as production of fermented foods with the help of micro-organisms, has been known in ancient civilization. Bioprocesses have been developed for large number of commercial products like industrial alcohol, organic solvent, baker's yeast, etc. and special products like antibiotics, therapeutic proteins and vaccines.

Bioprocess operations make use of microbial, animal and plant cells and components of cells such as enzymes, to manufacture new products and to destroy harmful wastes. These processes require effective control techniques due to increased demand on productivity, product quality and environmental responsibility. It is typically so in the case where the biomaterials used in the process are costly and required stringent control over product formation as in animal cell culture.

Bioreactor is where the bioprocess operation takes place and its design and controlled operation are very important for several aspects of production like purity, quantity, efficiency, safety, etc. Basically, there are two kinds of bioprocesses; one *aerobic* and the other *anaerobic*, depending on whether oxygen is required or not required to carry out the bioprocess operation.

There are several types of bioreactors designed and used in the laboratory as well as in large scale industrial applications. Some of these are continuous stirred tank, bubble column, airlift, see-saw and packed bed reactors. As their names indicate, they are meant for aerobic bio-reaction processes. See-saw bioreactor has been developed at Indian Institute of Technology Kharagpur [1, 2].

Bio-reaction is a slow process and takes days or at least hours, to complete. Nevertheless, there is some heat generation (or absorption) and change in pH value of the bioreactor fluid over the time takes place. Temperature and pH values of the bioreactor fluid are considered as *environmental variables* and need to be controlled within a tight band for the micro-organisms to survive. Besides chances of contamination, organic in nature, are there and require to be monitored [3].

Dissolved oxygen content of the bioreactor fluid is probably the most important single process variable for maximizing the product yield. The fan speed in the continuous stirred tank reactor or the see-sawing rate in the see-saw bioreactor acts as the corresponding control variable. Thus, a bio-reaction process requires continuous measurement of the dissolved oxygen content, temperature and pH value of the bioreactor fluid. There are other variables which can not be as easily measured. If a mathematical model of the bio-reaction process could be developed, then one could go for observer based 'soft sensing'.

22.2 Bioprocess Modeling

Most of the bioprocesses are considered as *autocatalytic* reaction. In such process, the catalyst is a product of the reaction. In many of the bioprocesses, enzymes, enzyme complexes or cells work as catalyst. The bioprocess in general is an operation to grow the cell mass up by taking the proper nutrient (substrate) like carbon-di-oxide complex, phosphor, etc. and oxygen. The proper environmental conditions (like temperature and pH) make the organism or plant and animal cell to undergo metabolism, which gives the desired product. The performance of the bioprocess is characterized by process parameters such as the growth rate of cell, product yield from substrate and product yield from oxygen. The reaction thermodynamics is less important as the bioprocesses are *irreversible.*

As the bioprocess involves the living cells or microorganism, the actual mathematical model is very much complex if we consider the growth rate of various cell colonies (depending on their age and genetic condition) differently. A much simpler way to model the dynamics for cell mass, substrate and product is to consider a *unstructured* and *unsegregated* model. The model used here, has been developed assuming *homogeneous* condition of the liquid medium. Temperature and other environmental conditions like pH are considered same throughout the reactor.

The growth rate shows the overall rate of metabolism inside the reactor. Several phases in the cell growth are observed in a batch culture, a typical growth curve considering first-order autocatalytic reaction is as shown in Fig. 22.1.

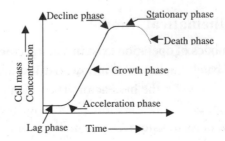

Fig. 22.1 A typical cell growth curve in the bioreactor

22.2.1 Single and Double Limiting Substrate

During the growth and decline phases, the specific growth rate of cell depends on the concentration of substrate in the medium. This limiting effect of substrate concentration on cell growth is modeled by the popular *Monod* equation [4]. Equation (1) shows the cell growth model with single limiting substrate.

$$\mu = \frac{\mu_{max}\, S}{K_s + S} \tag{1}$$

where μ =   Specific growth rate
μ_{max} =   Maximum specific growth rate
S =   Substrate concentration
K_s =   Substrate saturation constant

The *Monod* equation is by far the most frequently used expression for the growth kinetics in bioprocesses. In case of "see-saw" reactor, this model becomes more appropriate if oxygen source is considered as another limiting substrate along with the nutrient concentration [1]. The mathematical model for cell growth used in this work is based on double limiting substrate condition. This is a general condition to explain growth phenomenon in aerobic type bioreactor. Expression shown in equation (2) now stands modified as,

$$\mu = \frac{\mu_{max}\, S\, O_c}{(K_s + S)(K_c + O_c)}, \tag{2}$$

where μ =   Specific growth rate
μ_{max} =   Maximum specific growth rate
S =   Substrate concentration
K_s =   Substrate saturation constant
O_c =   Oxygen concentration
K_c =   Oxygen saturation constant

22.2.2 Mathematical Model

Several modes of operation exist in a bioreactor; those are continuous, batch and fed batch. Inclusion of inflow, outflow and recycle flow defines the operating modes for the bioreactor. The schematic of the bioreactor is as shown in Fig. 22.2. There are three inputs, inflow (F_i), outflow (F_o) and rate of oxygen mass transfer ($K_l a$) in addition to recycle flow F_r.

In batch process mode, inflow F_i, outflow F_o and recycle flow, F_r, are considered as zero. In continuous mode the inflow, F_i, is continuously adding substrate in the reactor while outflow, F_0, is continuously removing product from the reactor. In recycle flow, F_r, some portion of the removed product is fed back to the reactor.

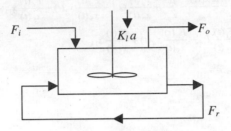

Fig 22.2 Schematic diagram of bioreactor in continuous mode with recycle flow.

The states considered in the mathematical model are

$x_1(t) =$ Cell mass concentration, in gm/lit.

$x_2(t) =$ Substrate concentration, in gm/lit.

$x_3(t) =$ Dissolved oxygen concentration, in gm/lit.

$x_4(t) =$ Product concentration, in gm/lit.

$x_5(t) =$ Operating volume, in m^3

The generalized unstructured, unsegregated and homogeneous model using the mass balance equation in liquid medium, is given by the following basic equation

$$\frac{d\mathbf{x}(t)}{dt} = q_i x_2(t) + q_r x_4(t) - q_o x_4(t) + g[\mathbf{x}(t)], \tag{3}$$

where

$\mathbf{x}(t) =$ Bioprocess variables like cell mass concentration, substrate concentration, oxygen concentration, product concentration, etc.

$q_i(t), q_o(t), q_r(t) =$ Normalized flow rates in the bioreactor like inflow, outflow, recycle flow (normalized w.r.t. the operating volume of the bioreactor).

$g[\mathbf{x}(t)] =$ Material accumulation and consumption rate that describe biological activities.

The detailed mathematical model with double limiting substrate is as below [1]:

$$\frac{dx_1(t)}{dt} = \mu - K_d x_1(t) + \frac{F_i(t)}{x_5(t)} X_{in} - \frac{F_o(t)}{x_5(t)} x_1(t) + \frac{F_r(t)}{x_5(t)} x_{11}$$

$$\frac{dx_2(t)}{dt} = -\frac{\mu}{Y_s} - m_s x_1(t) - \frac{\alpha x_1(t)}{Y_{ps}} - \frac{\beta\mu}{Y_{ps}} + \frac{F_i(t)}{x_5(t)} S_{in} - \frac{F_o(t)}{x_5(t)} x_2(t) + \frac{F_r(t)}{x_5(t)} x_{12}$$

$$\frac{dx_3(t)}{dt} = -\frac{\mu}{Y_o} - m_o x_1(t) - \frac{\alpha x_1(t)}{Y_{po}} - \frac{\beta\mu}{Y_{po}} + \frac{F_i(t)}{x_5(t)} O_{in} - \frac{F_o(t)}{x_5(t)} x_3(t) + \frac{F_r(t)}{x_5(t)} x_{13} + K_l a(O_2^* - x_3(t)) \qquad (4)$$

$$\frac{dx_4(t)}{dt} = \alpha x_1 + \beta \mu + \frac{F_i(t)}{x_5(t)} P_{in} - \frac{F_o(t)}{x_5(t)} x_4(t) + \frac{F_r(t)}{x_5(t)} x_{14}$$

$$\frac{dx_5(t)}{dt} = F_i(t) - F_o(t),$$

where

$$\mu = \frac{\mu_m x_1(t) x_2(t) x_3(t)}{(K_s + x_2(t))(K_c + x_3(t))}$$

F_i = Liquid feed - in rate, F_o = Withdrawal rate, F_r = Feed back flow,

K_l = Oxygen mass transfer coefficient,

a = Surface area through which mass transfer of oxygen takes place,

$X_{in}, S_{in}, O_{in}, P_{in}$ = Influent cellmass, substrate, oxygen and product concentrations respectively,

$x_{11}, x_{12}, x_{13}, x_{14}$ = Recycled cellmass, substrate, oxygen and product concentrations respectively,

α & β are constants,

K_s, K_c = Saturation constants,

K_d = Biomass decay rate,

m_s, m_o = Maintenance coefficient w.r.t. carbon and oxygen source for cellmass,

Y_s, Y_o = Yield coefficient w.r.t carbon and oxygen source for cellmass growth,

Y_{ps}, Y_{po} = Yield coefficient w.r.t carbon and oxygen source for product formation,

O_2^* = Saturation value of oxygen in liquid medium of interest.

22.2.3 Simulation Experiment based on the Mathematical Model

22.2.3.1 Batch Process Simulation

The simulation experiment was for the production of Single Cell Protein (SCP) in batch mode. Oxygen mass transfer co-efficient, $K_l a$, was equal to 0.0128. The other data for the fermentation process were:

Biomass decay rate, $K_d = 0.001$, Saturation constants w.r.t. substrate and oxygen, $K_s = 0.31$ and $K_c = 0.01$ respectively, Yield co-efficient of cell mass and product w.r.t. to substrate and oxygen, $Y_s = 0.6$, $Y_o = 0.7$, $Y_{ps} = 0.62$, $Y_{po} = 0.6$, Maintenance co-efficient for substrate and oxygen, $m_s = 0.01$ and $m_o = 0.00001$. The responses are shown Fig 22.3(a)-(c) plotted against sample No. with a sampling time of 6 mins. The starting values are as follows:

Cell (kg/m³)	Food (kg/m³)	Oxygen (kg/m³)	$K_l a$	O* (kg/m³)	Time
0.001	0.43	0.006	0.0128	0.01	40Hr

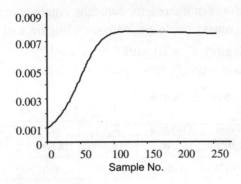

Fig. 22.3(a) Cell Mass Concentration, gm/l

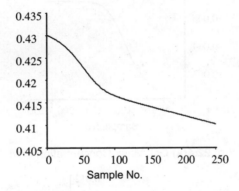

Fig. 22.3(b) Substrate Concentration gm/l

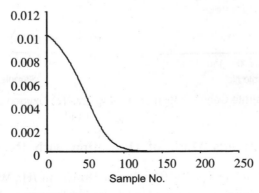

Fig. 22.3(c) Oxygen Concentration gm/l

22.2.3.2 Continuous Mode of Operation without Recycle Flow

In continuous mode the inflow, F_i, is continuously adding the substrate in the reactor while outflow, F_o, is continuously removing product from the reactor. The feed back flow F_r is considered as zero. Liquid feed-in rate, F_i = 0.005 l/min=0.3l/hr and withdrawal rate, F_o = 0.005 l/min=0.3l/hr, keeping the working volume of the reactor constant. Again responses are plotted in Figs. 4(a)-(c) against sample No. with sampling time of 6 mins.

X_{in} =   0.0 gm/l, S_{in} = 10 gm/l,
O_{in} =   0.008 gm/l, P_{in} = 0.0 gm/l.

Initial values are as below:

Cell (kg/m^3)	Food (kg/m^3)	Oxygen (kg/m^3)	K$_l$a	O* (kg/m^3)	Time	X$_5$ (m^3)
0.0001	0.43	0.01	0.0128	0.01	40Hr	20/l

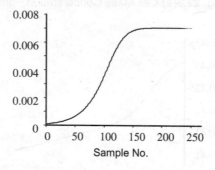

Fig. 22.4(a) Cell Mass Concentration gm/l

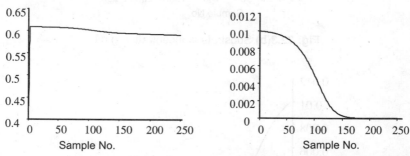

Fig. 22.4(b) Substrate Concentration gm/l   **Fig. 22.4(c)** Oxygen Concentration in gm/l

22.2.3.3 Continuous Mode of Operation with Recycle Flow

Liquid feed-in rate, F_i = 3x10^{-6} m^3/min=18x10^{-5} m^3/Hr, Withdrawal rate, F_o = 5x10^{-6} m^3/min=3x10^{-4} m^3/Hr, Recycles flow rate F_r=2x10^{-6} m^3/

min=12×10^{-5} m³/Hr, X_{in} = 0.0 kg/m³, S_{in} =10 kg/m³, O_{in} = 0.008 kg/m³, P_{in} = 0.0 kg/m³. Initial values are as below:

Cell (kg/m³)	Food (kg/m³)	Oxygen (kg/m³)	$K_l a$	O* (kg/m³)	Time step	X_5 (m³)
0.0001	0.43	0.01	0.0128	0.01	0.1 Hr.	10^{-3}

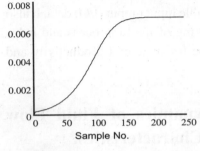

Fig. 22.5(a) Cell Mass Concentration, m³kg/

Fig. 22.5(b) Substrate Concentraintion, kg/m³

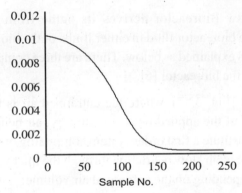

Fig. 22.5(c) Dissolved Oxygen Concentration, kg/m³

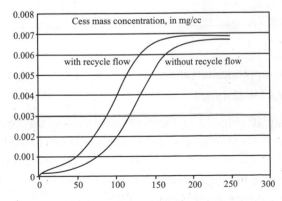

Fig. 22.6 Comparison of Variation in Cell Mass Concentration with and without Recycle Flow

Fig. 22.5(a), (b), (c) are for recycled flow. Fig. 22.6 gives the comparison between continuous mode of operation with and without recycling. In case of recycle flow, a further recycle term, F_r, is added to the reaction equations. *Recycle helps increase the growth rate* [5]. It is difficult to 'repeat' a bioprocess exactly. For many reasons, the underlying mathematical model is also underdetermined, nonlinear and time varying parameter in nature. The simulation experiments save on valuable time and materials when the mathematical model is validated by available input output (I/O) data. It also gives a better understanding of the working of the bioprocess and helps develop an optimized operation schedule for increased productivity and quality [2].

22.3 See-Saw Bioreactor, Dynamics of Fluid Flow and Oxygen Transfer Characteristics.

22.3.1 Dynamics of Fluid Flow in See-Saw Bioreactor

The See-Saw Bioreactor derives its name from the see-sawing movement of the bioreactor fluid in either limbs of the bioreactor. The see-saw movement is explained as below. There are three components affecting the fluid flow in the bioreactor [6].

1. 'A' part (Fig. 22.7), where the entrapped air is first compressed because of the applied large pressure, p_a, air being compressible. This constitute a first order system comprising of a flow resistance (because of air flow restriction through the inlet valve) and capacitance depending on the entrapped air volume.

   The transfer function of this part of the system is given by

   $$\frac{P_0(s)}{P_A(s)} = \frac{1}{RCs+1} = \frac{K_1}{s\tau_1+1} \tag{5}$$

   Using appropriate values, $RC = \tau_1 = 2.1\,\mathrm{s}$, $K_1 = 1$ for the see-saw bioreactor designed.

2. In part 'B' (Fig. 22.7), the gradually building up pressure (through the above first order system) is communicated to the liquid column. The liquid is incompressible. There is a 'gravity spring' force trying to restore the liquid column to x = 0 position for any movement away from the equilibrium position shown.

While experimenting with water, the 'gravity spring' force for a water column of one meter height is hardly recognizable compared to the large forcing function of 4 atmosphere (56 psi). Like as in a flux meter, there is little restoring force present if at all.

The second important force is the drag force between the moving fluid and the limbs and connecting pipe interface respectively. As the liquid displacement is uniform throughout and there is no hold up, the fluid velocity (not the flow velocity) in m/s is different in the limbs and the connecting pipe. The flow regime changes from 'laminar' in the limbs to 'turbulent' in the connecting pipe.

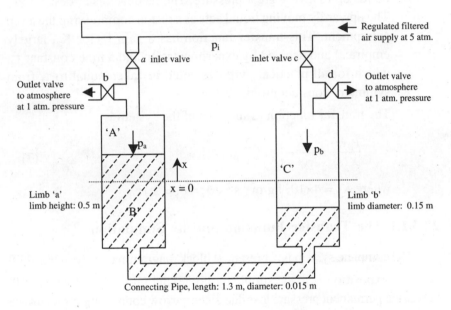

Fig 22.7 The See-saw Bioreactor

The resulting system is a second order system with little or no restoring force as stated earlier but damping generated by the viscous drag force. It can be reduced to a first order system and an integrator in cascade, as explained below.

$$\frac{X(s)}{P(s)} = \frac{K\varpi_n^2}{s^2 + 2\xi_i \varpi_n s + \varpi_n^2}$$

i is different for laminar and turbulent flow

Neglecting 'spring force' the above equation becomes

$$\frac{X(s)}{P(s)} = \frac{K_2}{s(1+s\tau_2)},\qquad(6)$$

where $K_2 = 5\times10^{-8}$ kg^{-1} m^2 s^2 and $\tau_2 = 0.363$s.

3. As soon as the liquid column starts moving, the entrapped air in part 'C'(Fig. 22.7) of the limb b which was held at ambient pressure gets compressed before making exit through the restriction of the now opened outlet valve. The entrapped compressed air volume builds up a back pressure (like back emf of a dc motor) and acts in the feedback loop opposing the applied pressure. This never builds up to a very great pressure. The feedback is at best 'mild'. The upwardly moving liquid acts as a piston compressing the air in entrapment. This transfer function has a gain factor K_3, largely empirical and decided by experimentation and a time constant τ_3, which for all practical purposes could be taken equal to τ_1 from symmetry considerations.

The transfer function of this part of the system is

$$\frac{P_B(s)}{X(s)} = \frac{K_3}{\tau_3 s +1},\qquad(7)$$

where $K_3 = 4\times10^2$ kg m^{-2} s^{-2}, $\hat{o}_3 = \hat{o}_1 = 2.1$s.

22.3.1.1 The Transfer Function and its Simulation

The complete system is represented block diagrammatically in Fig. 22.8

The exposition is made simple by not taking into account the two T-bends, the permanent pressure loss due to the narrow connecting pipe, surface tension effects at either ends, normal distributed force exerted by the tube in the liquid and change in viscosity with temperature, as also by taking isothermal expansion of the gas and glossing over the nonlinearities. The simulation results are given in Figs. 22.9 through 22.12.

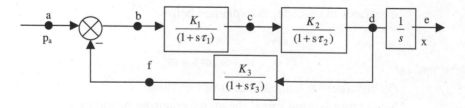

Fig. 22.8 The simplified schematic block diagram for fluid movement with applied pulsating pressure in the bio-reactor.

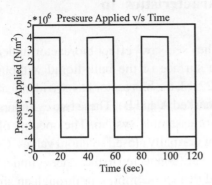

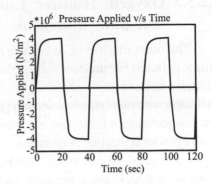

Fig. 22.9 Pulsating Pressure applied to the liquid column at point c of Fig. 22.8

Fig 22.10 Air Pressure developed on the liquid column at point b of Fig. 22.8

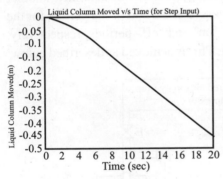

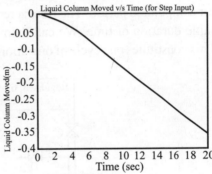

Fig. 22.11(a) Liquid Column movement at point e for a step input with the feedback loop open

Fig. 22.11(b) Liquid Column movement at point e for a step input with the feedback loop closed

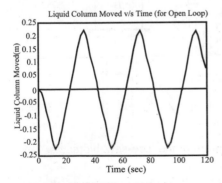

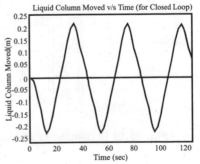

Fig. 22.12(a) Liquid column movement at point e for a pulsating pressure input with the feedback loop open

Fig.22.12(b) Liquid column movement at point e for a pulsating pressure input with the feedback loop closed

22.3.2 Oxygen Transfer Characteristics in See-Saw Bioreactor

The mass transfer of oxygen in the 'see-saw' effect bioreactor under study is based on the renewal of the surface of the bulk liquid without inducing turbulence. As shown in Fig. 22.13, the bioreactor has two identical cylinders with uniform cross section (marked A and B). These two cylinders are connected by a flow pipe and sensor assembly system. The opening of each cylinder is through two numbers of normally closed solenoid valves. C, D are connected to one cylinder and E, F are connected to the other cylinder. Solenoid valves C and E are connected to an air compressor through an air filter and pressure regulator device. D and F are vented to the atmosphere (through air filter).

The aim of this assembly is to transfer liquid from one column to another for a stipulated period. The flow is reversed at the end of the period, for the same duration of time. We call them 'on' and 'off' periods, respectively. This constitutes one cycle of operation. This is achieved as described below:

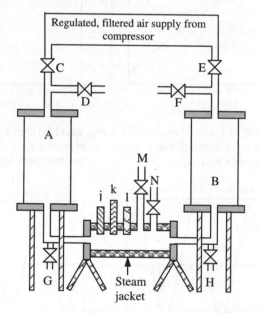

Fig. 22.13 Schematic block diagram of the 'see-saw' bioreactor

1. The on/off-time is controlled by Visual Basic program, the output of the program is transmitted to the solenoid valves through the data acquisition card installed in the computer.

2. During 'on' time, solenoid valves C and F are open. During this period, valves E and D remain closed. During 'off' period, E and D are open and valves C and F are closed.

3. During 'on' time, air from compressor pushes the liquid in column A towards B and the entrapped air in B finds its way out through valve F. As a result, the liquid level in A falls and the liquid level in column B rises equally, the cylinders being similar. This is continued for a predetermined period.

4. During 'off' period, solenoid valves C and F are de-energized and valves E and D are energized. That is, air at high pressure from the compressor enters column B through valve. The liquid in B is pushed downwards. At the same time the liquid column in A rises and the entrapped air in A is vented through valve D. This is maintained for the same predefined time period.

5. The see-sawing of the liquid column is obtained in the experimental bioreactor by alternating through steps 3 and 4.

The mass transfer of gaseous oxygen to the liquid medium takes place through the dynamically falling liquid film on the surface of the vessel wall and through the flat top surface of the liquid in each cylinder.

The oxygen accumulated in the falling film during the recedence of the liquid level in one arm is deposited as dissolved oxygen in the bulk liquid. The see-sawing of the system or the oscillation is visible and is mild for all practical purposes. As shown in Fig. 22.13, the sensors attached to the bioreactor system are a temperature sensor, j(PT-100-type), a pH sensor, k, and a dissolved oxygen sensor, l. G and N are the ports where pH-balancing peristaltic pumps have been connected. M and H are the ports where feed-in and feed-out pumps have been connected. The detailed instrumentation is shown in Fig. 22.14 given over leaf.

22.3.2.1 Modeling of Oxygen Transfer Co-efficient in 'See-Saw' Bioreactor

Theoretical study of the oxygen transfer rate can be divided into two parts:

1. Oxygen transfer in the falling film.
2. Oxygen transfer across the top of the liquid surface in each limb.

These are added up to give the total oxygen transfer in the equipment.

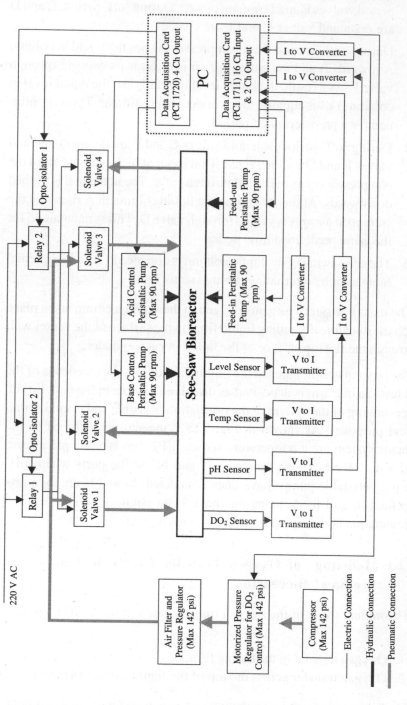

Fig. 22.14 Integrated Instrumentation and Computer Control of the See-Saw Bioreactor in Block Diagrammatic Form

22.3.2.2 Oxygen Transfer through the Falling Film

Mass transfer of oxygen in one limb of the bioreactor is found by integrating it over the total height. The expression takes the following form [2]:

$$G_w = \frac{8}{3} \sqrt{2} \, \pi R \, (C_{ai} - C_{a0}) \sqrt{\frac{D_{ab}}{\pi t_c}} u_s t_c^2 \tag{8}$$

22.3.2.3 Oxygen Transfer through the Flat Top Surface

The mass transfer through the flat surface of one limb of the bioreactor in one cycle of operation is given by the following [2]:

$$G_s = 2 \sqrt{2} \, \pi R^2 \, (C_{ai} - C_{a0}) t_c \sqrt{\frac{4D_{ab}}{\pi t_c}} \tag{9}$$

22.3.2.4 Total Oxygen Transfer in the Bioreactor

The total oxygen transfer (kg/cycle) to the equipment over a cycle of operation is calculated as below,

$$G_t = 2(G_w + G_s)$$

$$= \frac{16}{3} \sqrt{2} \, \pi R \, (C_{ai} - C_{a0}) \sqrt{\frac{D_{ab}}{\pi t_c}} u_s t_c^2$$

$$+ 4 \sqrt{2} \, \pi R^2 \, (C_{ai} - C_{a0}) t_c \sqrt{\frac{4D_{ab}}{\pi t_c}} \tag{10}$$

Amount of oxygen deposited per unit volume per cycle in the bioreactor is given by Gt/v where v is the working volume of the bioreactor. Average transfer rate (in kg/sec) is given by:

$$G_{ta} = \frac{G_t}{2t_c} = \frac{8}{3} \sqrt{2} \, \pi R \, (C_{ai} - C_{a0}) \sqrt{\frac{D_{ab}}{\pi t_c}} u_s t_c$$

$$+ 2 \sqrt{2} \, \pi R^2 \, (C_{ai} - C_{a0}) \sqrt{\frac{4D_{ab}}{\pi t_c}} \tag{11}$$

22.3.2.5 Suggested Modification in the Expression for K_l

From eqn. 11 mass transfer co-efficient of O_2, K_l (in m/s) is given by the following expression

$$K_l = \frac{1}{(R + u_s t_c)} \sqrt{\frac{D_{ab}}{\pi t_c}} \left(\frac{4\sqrt{2}}{3} u_s t_c + \sqrt{2}R \right) \tag{12}$$

Again, $u_s \propto \dfrac{1}{t_c}$, so $u_s t_c$ is constant, say k_1.

$$K_{LT} = \frac{1}{(R + k_1)} \sqrt{\frac{D_{ab} u_s}{\pi k_1}} \left[\frac{4\sqrt{2}}{3} k_1 + \sqrt{2}\, R \right] = k_3 D_{ab}^{0.5} \tag{13}$$

K_{LT} is proportional to $D_{ab}^{0.5}$ for different solutes under the same circumstances. This indicated dependence on D_{ab} is typical of short exposure times, where the depth penetration is small relative to the depth of the absorbing pool. Let us assume that part of the flow is turbulent in violation of assumption. Experience shows a range of exponents on D_{ab} from nearly zero to 0.8 or 0.9 [6].

Then

$$K_{LA} = k_3 D_{ab}^{0.5\alpha}, \tag{14}$$

where α is the factor to be determined experimentally.

Laboratory experiments carried out before [2] gave the O_2 absorption characteristics as depicted in Figs. 22.15 to 22.17. The gradually drooping characteristics of the experimental O_2 absorption is because the $(C_{ai} - C_{a0})$ terms becomes smaller and smaller with each cycle as there is no living organism inside the bioreactor to consume oxygen.

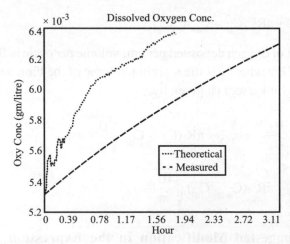

Fig. 22.15 Theoretical and actual dissolved oxygen profile for time period of oscillation = 20sec

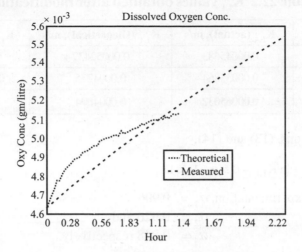

Fig. 22.16 Theoretical and actual dissolved oxygen profiles for time period of oscillation = 25sec

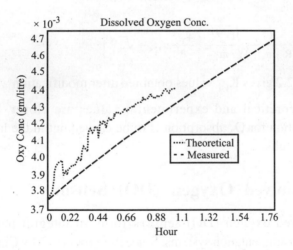

Fig. 22.17 Theoretical and actual dissolved oxygen profiles for time period of oscillation = 35sec

Table 22.1 K_l values obtained from experimentation and from theoretical calculations.

Sl. No.	K_{LA}(actual), m/s	K_{LT}(theor), m/s	K_{LA}/K_{LT}
1	0.00045683	0.00017409	2.6241
2	0.00033864	0.00015788	2.1449
3	0.00065032	0.00013592	4.6486

Table 22.2 K_{LT} values obtained after modification

Sl. No.	K_{LA} (actual), m/s	K_{LT} (theoretical), m/s	K_{LA}/K_{LT}
1	0.00045683	0.00052437	0.96
2	0.00033864	0.0004775	0.712
3	0.00065032	0.0004094	1.58

From eqns. (13) and (14),

$$\ln K_{LA} - \ln K_{LT} = 0.5(\alpha - 1)\ln D_{ab}$$

From experimentation, $\alpha_1 = 0.906$,

$$\alpha_2 = 0.926 \text{ and}$$
$$\alpha_3 = 0.848 \text{ respectively.}$$
$$\alpha_{mean} = 0.893.$$

Including turbulence, modified K_l becomes:

$$K_l = \frac{1}{(R + u_s t_c)}\sqrt{\frac{D_{ab}^{0.893}}{\pi t_c}}\left[\frac{4\sqrt{2}}{3}u_s t_c + \sqrt{2}\,R\right] \tag{15}$$

Table 22.2 gives K_{LT} values obtained after modification

The theoretical and experimental values are closer, but further experimentation for O_2 absorption is to be carried out in the laboratory to check the results.

22.4 Dissolved Oxygen (DO) Sensor

Dissolved oxygen (DO) sensors form an integral part of many bioprocess instrumentation systems. Since its introduction by Clarke in 1956, the membrane-covered dissolved oxygen electrodes have been widely used in research and industry. *Steam-sterilizible* DO probes are used in bioreactors.

22.4.1 Review of DO Electrodes

The original Clarke electrode had a platinum (Pt) cathode and a reference electrode of Ag/AgCl, that is calomel, dipped in 1M KOH solution. The exposed tip of the electrode was covered with a polyethylene membrane which is selectively permeable to oxygen. It was used for blood oxygen measurement. In 1959 Carrit and Kanwisher used a modified Clarke electrode for DO measurement in Chesapeake Bay water. A different type of membrane covered electrode, the *galvanic* electrode, was introduced by Mancey et al in 1962. Unlike the polarographic electrode, the galvanic probe did not require external voltage biasing. Mackereth probe, a variant of Mancey probe, has been widely used for monitoring and controlling oxygen in cultivation media [7].

There was a period of quiescence, during which *polarographic* Clarke and *galvanic* Mancey electrodes became the standards. Of late there has been a renewed interest in DO measurement and the literature is replete with new electrode designs, choice of electrode materials and their structure and use of *gel* or *paste* or *solid* electrolytes. The method of measurement also underwent changes like pulsed excitation and consequent transient response measurement and signal recovery. Further, transient response measurement required accurate modeling of the electrode including effects of surface deposition and got due attention. Bioprocess monitoring of DO using a computerized pulsing membrane electrode has been reported by Chen and Wang (1993 [8]). Lee et al (1991 [9]) reported development of an adaptive DO concentration algorithm taking into account DO electrode dynamics and response time delay.

Wang and Li (1989 [10]) developed a four-layer diffusion model for galvanic electrodes to study transient mode dissolved oxygen measurement. More complicated diffusion models have been discussed by Hale and Hitchmann (1980 [11]). Koehler and Goepel (1991 [12]) used mixed valent tungsten oxides as new electrode material for potentiometric detection of dissolved oxygen at temperatures below 35° C. Mizutani et al (2000, [13]) devised a DO sensor using platinum electrode with polymethylsiloxane coating. A graphite electrode with a thin layer of palladium-gold applied by vacuum sputtering, is used for oxidizing hydrogen peroxide in enzyme electrodes (Yang, 1989 [14]. Prien et al (2001 [15]) have used mesoporous microelectrode DO sensor for marine applications. Silicon thin film sensor for measurement of DO has been described by Wittkampf et al (1997 [16]). Similarly, screen-printed amperometric dissolved oxygen sensor utilizing an immobilized electrolytic *gel* and membrane has been described by Glasspool and Atkinson (1998 [17]).

A new development in DO measurement is the appearance of biosensors. A catalase based biosensor for alcohol determination in beer samples has been devised by Akyilmaz and Dinckaya (2003 [18]). When ethanol is added to the medium, catalase catalyzes the degradation of both ethanol and hydrogen peroxide producing a new steady state DO concentration. A thin film of unicellular algae whose growth and photosynthetic efficiency is related to water composition has been used for water quality monitoring (Ory et al, 1996 [19]). In the same vein oxygen-sensitive fluorescent dyes, monolithically integrated in the sensing substrate, made possible optical measurement of dissolved oxygen (Chang-Yen and Gale, 2002 [20])

22.4.2 Principle of DO Measurement

If a fixed voltage in the plateau region of the voltage-current diagram is applied to the cathode, then the current output of the electrode can be calibrated to the dissolved oxygen (see Fig.22.18). Galvanic electrodes (Fig.22.19) are different from polarographic type in that it does not require an external voltage source for oxygen reduction at the cathode. The voltage generated across the electrode pair in an electrolyte is sufficient for spontaneous reduction of oxygen at the cathode surface. Oxygen is reduced via four electron reaction. The probe life depends on the available surface area of the anode. Provided that the oxygen diffusion is controlled by the membrane covering the cathode, the current in the probe is proportional to the oxygen activity or the partial pressure, in the liquid medium.

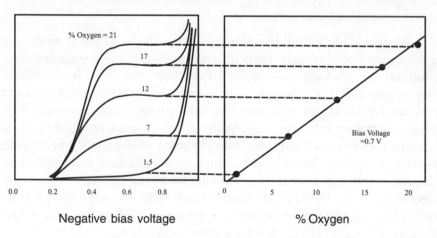

Fig. 22.18 Polarogram and Calibration curve

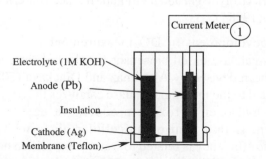

Fig 22.19 Basic arrangement for galvanic electrode

The behaviour of the probe can be explained based on a simplified electrode model [7]. With the cathode well polished and the membrane snugly fitting over the cathode surface and no entrapped electrolyte layer in between, a single layer electrode model will be good enough. Suppose the electrode is immersed in a well agitated liquid and, at time zero, the oxygen partial pressure of the liquid changes from 0 to p_0. At steady state, the pressure profile in the membrane is linear (Fig.22.20a) and the electrode current is proportional to the oxygen partial pressure of the bulk liquid. The pressure profile (p) and the current output (I_s) under steady state condition are given by:

$$p/p_0 = x/d_m, \tag{16}$$

where d_m is the membrane thickness and

$$I_s = NFA \ (P_m /d_m)p_0, \tag{17}$$

where N, F, A and P_m are the number of electrons per mole of oxygen reduced, Faraday's constant, surface area of the cathode, and oxygen permeability of the membrane respectively. Equation (17) forms the basis for DO probe measurements. Another important consideration is the time response of the probe. Probe response depends on the probe constant, k, defined as follows:

$$k = i^2 D_m /d_m^2, \tag{18}$$

where D_m is the oxygen diffusivity of the membrane. A large k, which means a thin membrane and/or a high D_m, results in a fast probe response. These conditions weaken the assumption of membrane-controlled diffusion.

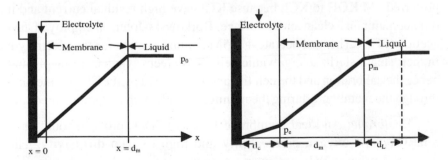

Fig. 22.20 (a) One-layer electrode model (b) three-layer electrode model

Often there is a finite thickness of electrolyte layer between the cathode and the membrane because of the thickness of the cathode surface (Fig.22.20b). A new overall mass transfer coefficient K_0 is defined by

$$1/K_0 = d_L/P_L + d_m/P_m + d_e/P_e, \tag{19}$$

where d_L, d_e, P_L, and P_e are liquid film thickness, the electrolyte thickness, the oxygen solubility of the liquid film and that of the electrolyte layer respectively. The condition for membrane controlled diffusion becomes:

$$d_m/P_m >> d_l/P_L + d_e/P_e \tag{20}$$

This mean that a relatively thick membrane with low oxygen permeability is required which contradicts the requirement for a fast probe response [7].

22.4.3 Design of Electrode

The membrane covered DO electrode basically consists of a cathode, an anode and the electrolyte. The three main constituents of the DO probe are as below:

Electrode metals: Sawyer and Interrante [7] studied the reduction of dissolved oxygen at Pt, Pd, Ag, Ni, Au, Pb and other metal electrodes. For *galvanic probes*, silver as the cathode and lead as the anode are most common. The useful life-time depends on the current drain just as in an electrical battery. In other words, a probe can be used longer when it is used for monitoring low, rather than high oxygen tension. Dead probes can be rejuvenated by dissolving away the oxide layer and re-depositing lead as in a lead-acid battery.

Electrolyte: Since the electrode reaction occurs in the electrolyte solution, the composition and the volume of the electrolyte are directly related with probe stability. The solubility of the electrode metals in the electrolyte solution has to be low for probe stability. For galvanic probes, Mancey preferred 1M KOH to KCl; because KCl gave high residual current and it did not maintain a clean anode surface. Borkowski-Johnson employed 5M acetic acid + 0.1M lead acetate + 0.5M sodium acetate for the electrolyte in their steam-sterilizable galvanic cell. This electrolyte has a low pH and hence the calibration and the cell life are not affected by the CO_2 permeating through the membrane during the monitoring of DO_2 in cultivation medium.

Membrane: An ideal membrane for use in DO_2 probes should have relatively low oxygen permeability and high oxygen diffusivity. The permeability has to be low to ensure membrane control of oxygen diffusion. Since the current output is directly related to the thickness and the oxygen permeability of the membrane, the probe sensitivity is directly affected by change in membrane properties. The water permeability of the membrane has to be low to prevent loss of water from the electrolyte solution, which causes an increase in electrolyte concentration and early failure of the probe.

Teflon, polyethylene, and polypropylene have been most popular as the membrane material. According to data, polypropylene is better than Teflon in several aspects: it has lower oxygen permeability, a lower CO_2 permeability and yet higher oxygen diffusivity. Teflon seems to be more popular in steam-sterilizable probes for its higher heat resistance. An added advantage of the Teflon is its extremely low water permeability.

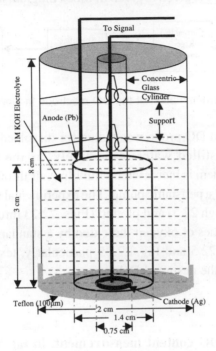

Fig. 22.21 The DO electrode designed in the laboratory

22.4.4 Design, Fabrication and Testing of the Laboratory DO Electrode and its Associated Instrumentation

The laboratory designed DO probe details are depicted in Fig. 22.21 [21]. The designed electrode had a small circular silver cathode of 0.5 mm thickness, a lead anode in the form of a circular cylinder formed from a lead sheet and 1M KOH solution as electrolyte. The lead anode will finally be formed from a gridded plate into which lead will be deposited. After a particular life period, the cylindrical lead anode will be taken out, developed and lead will be re-deposited for further use, the same technique as is being used for rejuvenating car batteries. The membrane used is Teflon as it is required to withstand high temperature during steam-sterilization of the bioreactor. The electrode will be dipping into the bioreactor cultivation medium or substrate fluid, which is under continuous 'see-saw' motion.

The DO probe is basically a current measuring instrument. It uses a current-to-voltage converter followed by an amplifier to give the desired signal level voltage of 1 to 5 volts. However, in the laboratory designed see-saw bioreactor, the voltage signal is converted to standard 4-20 mA current signal for transmission to the computer USP for control purposes. The signal processing circuitry is given is given in block diagram form in Fig. 22.22.

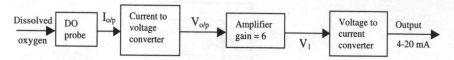

Fig. 22.22 Instrumentation of the DO electrode in block diagram form

The fabricated DO probe was tested for dissolved oxygen content of tap water and distilled water respectively, as the dissolved oxygen concentration in them is increased by bubbling air through a three pronged feed tube driven by a peristaltic pump. The experimental results are tabulated (Tables 22.3 through 22.5) and plotted (Figs. 22.23 through 22.25) side by side with ppm values of oxygen indicated by a standard DO sensor in the laboratory. Table 22.5 and Fig. 22.25 particularly, tests the repeatability characteristics of the designed DO sensor.

Observations

Table 22.3 DO content measurement. in tap water by the electrode in the laboratory

Peristaltic pump speed (rpm)	Output Voltage (V)
0	0.36
10	0.365
20	0.371
30	0.375
40	0.395
50	0.405
60	0.415
70	0.428
80	0.438
90	0.445

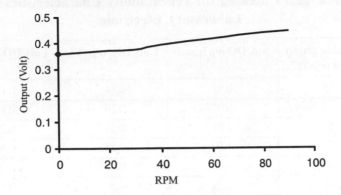

Fig.22.23 Plot of the table 22.3 data

Table 22.4. DO content Measuremement in distilled water by the laboratory electrode vis-a-vis the Standard electrode.

Peristaltic pump speed (rpm)	Lab. DO output (V)	Std. DO output (ppm)
0	0.28	1.02
10	0.37	1.1
20	0.42	1.21
30	0.54	1.28
40	0.68	1.32
50	0.81	1.37
60	0.98	1.43
70	1.12	1.48
80	1.25	1.53
90	1.32	1.64

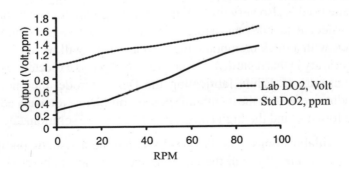

Fig.22.24 Plot of the table 22.4 data

Table 22.5 Checking the repeatability Characteristics of the Laboratory Electrode

Peristaltic pump speed (rpm)	Std. DO o/p (ppm)	Lab. DO$_1$ o/p (V)	Lab. DO$_2$ o/p (V)
0	0.92	0.14	0.15
10	0.97	0.19	0.18
20	1.01	0.28	0.28
30	1.10	0.35	0.36
40	1.19	0.42	0.40
50	1.29	0.50	0.47
60	1.32	0.65	0.68
70	1.37	0.76	0.75
80	1.42	0.84	0.83
90	1.46	0.91	0.89

[Series 1: Std. DO probe output in ppm
Series 2: Lab. DO probe output in volts, first set of readings
Series 3: Lab. DO probe output in volts, second set of readings.]

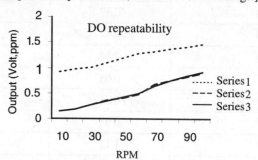

Fig.22.25 Plot of the table 22.5 data

The designed electrode has a glass body and is fragile. The Teflon membrane used is also very thin adding to its fragility. An electrode with a stainless steel metal body and a thicker Teflon membrane is recommended. However, with a thick membrane the response time will be more. A trade off is necessary in this regard. A porous silver cathode is expected to increase the output current. While fabricating the DO electrode, the choice of a proper adhesive to stick the membrane posed a problem. Not many adhesives are able to withstand the high *causticity* of 1M KOH solution.

The initial readings in all the graphs, when the peristaltic pump is not forcing air bubbles through the aqueous medium, are non-zero. This is because of the fact that the sample water had some dissolved oxygen. While in the tap water the DO concentration is fairly high, that in the distilled water sample was lower. No zero suppression circuitry is built in. The repeatability of the instrument is found to be fairly good. During the experiment, special care was taken to keep the water stirred to avoid stagnant liquid layer in the vicinity of the electrode. Sufficient time was given to get the dissolved oxygen to come to a stable value. As the laboratory was air-conditioned, the room temperature was assumed to remain constant.

We checked the performance of the designed electrode with respect to a commercially available standard DO electrode. Inaccuracies of the standard electrode will reflect on the results. There was no *absolute* measurement. The dynamic response of the probe is yet to be checked. The next step will be to go for pulsed response, though a galvanic probe measures the DO content continuously.

22.5 A Sliding Mode Observer for Bioprocess Variables

22.5.1 Review of Nonlinear Observers

Various approaches for the design of the controller are developed based on the assumption that each state of the system is available. Many of such states are not measurable in practice. A dynamical system that performs estimation of non-measurable states is called a state observer or estimator. Alternatively, observer based measurement is also termed 'soft sensing'. The control inputs and the error between measurable states (output) and their estimated values act as input for the state observer.

Convergence and stability of an observer is largely dependent on the model of the plant considered in the design. The observer design for bioprocess which is a nonlinear and time varying parameter system, is very difficult. Various approaches for the design of nonlinear observers like extended Kalman filter, linearized observer, adaptive observer, set theoretic approach, sliding mode observer, etc. are evaluated based on their computational requirement and robustness properties against certain classes of uncertainty in [22, 23]. Performance of extended Kalman filter (EKF) is not guaranteed, as its convergence is questionable with parameter uncertainty. EKF also requires extensive calculation [24]. Linearized observer [1] is sensitive to modeling error. Adaptive observer requires the assumption of parameters for uncertain systems and substantial amount of real-time computations.

In comparison, the sliding mode technique is more attractive for the observer design [26]. A sliding mode observer (SMO) has good robustness against uncertainty in parameters. The technique is also less complex in design. In recent years, several methods for synthesis of observer based on this technique have been addressed.

22.5.2 Introduction to Variable Structure Control

In synthesis of the controller for a given plant, major problems occur because of discrepancies between the actual plant and the mathematical model used for design. There are several approaches developed in the area of *robust control* to handle such situations. One such approach is variable structure control, which was evolved in Russia in early 1960s. Variable structure control systems are a class of systems in which the control law is deliberately changed during the control process according to some definite rules depending upon the state of the system [25].

In the sliding mode technique the decision rule, known as switching function, changes according to system behaviour. The switching function produces an output that is used in the system at that instant of time. This concept has been successfully extended for the design of robust regulators, model reference adaptive systems, state observer design and fault detection and isolation systems. The controller forces the system states to reach and remain on a predefined surface, called the *sliding surface*. It has mainly two advantages: the dynamic behavior of the system can be set as per the requirement by choosing proper switching functions and the response becomes insensitive to particular class of uncertainties.

22.5.3 Walcott and Zak SMO

Design of observers based on the variable structure system theory and sliding mode concept can be classified in two categories:

(i) Equivalent control based method

(ii) Observer design based on the method of Lyapunov

Comparisons of both the approaches are given in [27]. In the equivalent control based method, the estimation error may be controlled within an acceptable level by suitable choice of gain under assumption of uncertainty being small enough. The performance is not guaranteed because it is difficult to ensure that the uncertainty is small always.

Walcott and Zak SMO design is based on the method of Lyapunov. It requires each unknown input element (uncertainty channel) to be compensated by each output directly and uses remaining outputs to design observer for the unknown input free subsystems. If the number of unknown inputs equal the number of outputs, like the problem considered in this work, the unknown input free subsystem itself has to be stable. This means invariant zeros of the system must be stable, if any. As such, Walcott-Zak observer imposes strong structural constraint on the system but its performance is guaranteed against uncertainty. Performance of sliding mode observers based on Lyapunov's method is guaranteed compared to equivalent control based design [27]. Synthesis method of such observer given in [26], is used for design.

22.5.3.1 Basic Theory

Consider a class of nonlinear system with uncertainty described as below:

$$\mathbf{x}(t) = \mathbf{A}\mathbf{x}(t) + \mathbf{B}\mathbf{u}(t) + \mathbf{D}\mathbf{f}(t,\mathbf{x},\mathbf{u})$$
$$\mathbf{y}(t) = \mathbf{C}\mathbf{x}(t), \tag{21}$$

where

$\mathbf{A} \in \Re^{n \times n}, \mathbf{B} \in \Re^{n \times m}, \mathbf{C} \in \Re^{p \times n}, \mathbf{D} \in \Re^{n \times q}$ with $p \geq q$. The matrices B, C and D are assumed to be of full rank. Here, the number of states and outputs are n and p respectively. The uncertainty function $\mathbf{f}$ (t, $\mathbf{x}$, $\mathbf{u}$) is unknown but bounded. If there exists a linear change of coordinates (the detail have been worked out in sub section 22.5.3.2), the system equation (21) can be rewritten as below:

$$\mathbf{x}_1(t) = \overline{\mathbf{A}}_{11}\mathbf{x}_1(t) + \overline{\mathbf{A}}_{12}\mathbf{y}(t) + \overline{\mathbf{B}}_1\mathbf{u}(t)$$
$$\mathbf{y}(t) = \overline{\mathbf{A}}_{21}\mathbf{x}_1(t) + \overline{\mathbf{A}}_{22}\mathbf{y}(t) + \overline{\mathbf{B}}_2\mathbf{u}(t) + \overline{\mathbf{D}}_2\mathbf{f}(t,\mathbf{x},\mathbf{u}) \tag{22}$$

with $\mathbf{x}_1 \in \Re^{n-p}$ and $\mathbf{y} \in \Re^p$.

In the above structure the unknown input, that is, the uncertainty affected the output dynamics only. If the unknown input free subsystem is stable, a stable sliding motion exists on the error subspace. According to Walcott and Zak [26] if there exist a gain matrix G and some matrix F with dimension $\mathbf{G} \in \Re^{n \times p}, \mathbf{F} \in \Re^{m \times p}$ and positive definite matrices P and Q such that

1. $\mathbf{A}_0 = (\mathbf{A} - \mathbf{GC})$ has stable eigenvalues,
2. $\mathbf{A}_0\mathbf{P}^T + \mathbf{P}\mathbf{A}_0^T = -\mathbf{Q}$ and
3. $\mathbf{C}^T\mathbf{F}^T = \mathbf{PB}$,

there exists an asymptotical observer for the system given in eqn (21). The dynamical equation for the observer is as below:

$$\hat{\mathbf{x}}(t) = \mathbf{A}\hat{\mathbf{x}}(t) + \mathbf{B}\mathbf{u}(t) - \mathbf{G}(\mathbf{C}\hat{\mathbf{x}}(t) - \mathbf{y}(t)) + \mathbf{P}^{-1}\mathbf{C}^T\mathbf{F}^T\mathbf{v} \tag{23}$$

with the discontinuous function

$$v = \begin{cases} f(t,\mathbf{u},\mathbf{y})\dfrac{FCe}{\|FCe\|} & ; FCe \neq 0 \\ 0 & ; \text{otherwise} \end{cases},$$

where

$$\mathbf{e} = \hat{\mathbf{x}}(t) - \mathbf{x}(t);$$

$\hat{\mathbf{x}}(t) = $ estimated states and

$\mathbf{x}(t) = $ original states

The above concept is attractive for state observation but have a computational problem on establishing the gain matrix, G, fulfilling the above stated conditions. An explicit solution for the synthesis is given in [27]. The synthesis procedure has been explained in details in the next sub section. The synthesis method has been used to convert the system into a canonical form as well to find out gain matrices and the Lyapunov matrix.

22.5.3.2 Canonical form Algorithm

The primary existence condition for a stable sliding motion on the error subspace is that there must be linear transformation of the co-ordinate axes such that the system could be transformed into the form given in eqn. (22). The algorithm for such a coordinate transformation is presented in the form of a flow chart in Fig. 22.26 [28].

22.5.3.3 Synthesis of Observer

After applying linear change of coordinates as above to the system given in eqn. (22), the system triple with respect to the new coordinates gives the canonical structure. With the earlier assumptions of stable eigenvalues of matrix A_0 and Lyapunov pair (P, Q), an observer can be written in the form,

$$\hat{\mathbf{x}}_1(t) = \overline{A}_{11}\hat{\mathbf{x}}_1(t) + \overline{A}_{12}\hat{\mathbf{y}}(t) + \overline{B}_1\mathbf{u}(t) - \overline{A}_{12}\mathbf{e}_y(t)$$

$$\hat{\mathbf{y}}(t) = \overline{A}_{21}\hat{\mathbf{x}}_1(t) + \overline{A}_{22}\hat{\mathbf{y}}(t) + \overline{B}_2\mathbf{u}(t) - (\overline{A}_{22} - \overline{A}^s_{22})\mathbf{e}_y(t) + \mathbf{v}, \qquad (24)$$

where $\overline{A}^s_{22}$ is a stable design matrix and $\mathbf{e}_y = \hat{\mathbf{y}}(t) - \mathbf{y}(t)$. If $P_2 \in \Re^{p \times p}$ is symmetric positive definite Lyapunov matrix for the design matrix $\overline{A}^s_{22}$ then the discontinuous vector, $\mathbf{v}$, is defined as

$$\mathbf{v} = \begin{cases} -\rho(t,\mathbf{y},\mathbf{u})\|\overline{D}_2\|\dfrac{P_2\mathbf{e}_y}{\|P_2\mathbf{e}_y\|} & ; \mathbf{e}_y \neq 0 \\ 0 & ; \text{otherwise} \end{cases}, \qquad (25)$$

where $\overline{D}_2 \in \mathfrak{R}^{p \times p}$ is defined in the canonical transformation (algorithm sub section 22.5.3.2). The error dynamics can be derived by subtracting state dynamical eqn. (22) from eqn. (24).

$$e_1(t) = \overline{A}_{11}e_1(t)$$
$$e_y(t) = \overline{A}_{21}e_y(t) + \overline{A}^s{}_{22}e_y(t) + v - D_2 f \qquad (26)$$

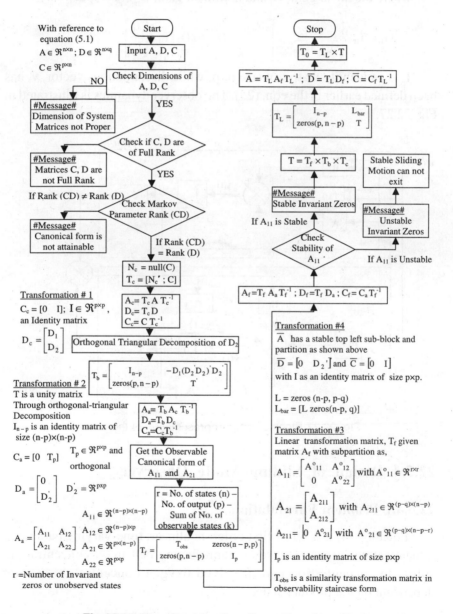

Fig 22.26 Flow Chart for Coordinate Transformation

The dynamical equation for the sliding mode observer is given in eqn. (27). In comparison with observer eqn. (23), this equation gives explicit solution. Linear gain, G_1, and nonlinear gain, G_n, can be found from the transformation matrix, T_0.

$$\hat{\mathbf{x}}(t) = A\hat{\mathbf{x}}(t) + B\mathbf{u}(t) - G_1(C\hat{\mathbf{x}}(t) - \mathbf{y}(t)) + G_n\mathbf{v} \qquad (27)$$

where the linear gain, G_1 and nonlinear gain, G_n are respectively

$$G_1 = T_0^{-1}\begin{bmatrix} \overline{A}_{12} \\ \overline{A}_{22} - \overline{A}^s_{22} \end{bmatrix} \qquad G_n = \|D_2\| T_0^{-1}\begin{bmatrix} 0 \\ I_p \end{bmatrix},$$

I_p is an identity matrix of size p×p, while discontinuous vector, $\mathbf{v}$, has been defined earlier in the eqn. (25). The observer dynamics is illustrated in Fig. 22.27.

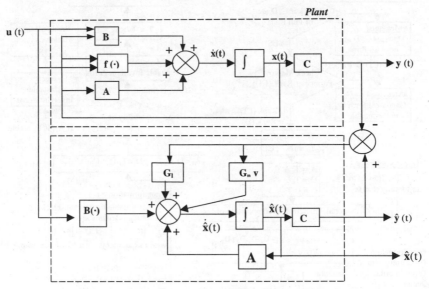

Fig. 22.27 Block diagram representation of the SMO

22.5.4 Design of Sliding Mode Observer

22.5.4.1 Problem Formulation

Equation (4) is a set of first order, nonlinear, time-varying parameter dynamical equations. The states representing cell mass concentration, substrate concentration and dissolved oxygen concentration are not dependent on the product state.

Thus, the product state emerges as a redundant variable. For this situation state $x_4(t)$ has been omitted reducing the order of a problem by one. So this is problem of observer design for three states only, i.e., cell mass concentration, dissolved oxygen concentration and substrate concentration from the single measurement of dissolved oxygen concentration. State $x_5(t)$ was not considered as a variable as it was held constant. The product state $x_4(t)$ could be mathematically constructed from the other observed states. The dynamics of the bioprocess also supports this reduction in order. After removing the recycle flow and product state from eqn. (4), the reduced order model in continuous mode of operation is as given below:

$$
\left.
\begin{aligned}
\frac{dx_1(t)}{dt} &= \frac{\mu_m x_1(t)x_2(t)x_3(t)}{(K_s + x_2(t))(K_c + x_3(t))} - K_d x_1(t) + \frac{F_i(t)}{x_5(t)} X_{in} - \frac{F_0(t)}{x_5(t)} x_1(t) \\
\frac{dx_2(t)}{dt} &= -\frac{\mu_m x_1(t)x_2(t)x_3(t)}{Y(K_s + x_2(t))(K_c + x_3(t))} - m_s x_1(t) - \frac{\alpha x_1(t)}{Y_p} - \\
&\quad \frac{\beta\mu_m x_1(t)x_2(t)x_3(t)}{Y_p(K_s + x_2(t))(K_c + x_3(t))} + \frac{F_i(t)}{x_5(t)} S_{in} - \frac{F_0(t)}{x_5(t)} x_2(t) \\
\frac{dx_3(t)}{dt} &= -\frac{\mu_m x_1(t)x_2(t)x_3(t)}{Y_0(K_s + x_2(t))(K_c + x_3(t))} - m_{s0} x_1(t) - \frac{\alpha x_1(t)}{Y_{p0}} - \\
&\quad \frac{\beta\mu_m x_1(t)x_2(t)x_3(t)}{Y_{p0}(K_s + x_2(t))(K_c + x_3(t))} + \frac{F_i(t)}{x_5(t)} O_{in} - \frac{F_0(t)}{x_5(t)} x_3(t) + K_l a(O_2^* - x_3(t))
\end{aligned}
\right\} \quad (28)
$$

Rearranging the linear and nonlinear parts of equation (5.8) and rewriting in the regular form,

$$
\begin{aligned}
\mathbf{x}(t) &= \mathbf{A}\mathbf{x}(t) + \mathbf{B}(t,\mathbf{x},\mathbf{u})\mathbf{u}(t) + \mathbf{d}f(t,\mathbf{x},\mathbf{u}) \\
y(t) &= \mathbf{c}^T\mathbf{x}(t) ,
\end{aligned}
\quad (29)
$$

where the states $\mathbf{x} \in \mathfrak{R}^3$, the control input $\mathbf{u} \in \mathfrak{R}^3$, the output $y \in \mathfrak{R}$ and function $f(t, \mathbf{x}, \mathbf{u})$ represents the nonlinear term. $\mathbf{d} \in \mathfrak{R}^3$ is the nonlinearity distribution vector and $\mathbf{c}^T \in \mathfrak{R}^3$ is the output vector. With reference to eqn. (28) the matrices are,

$$
\mathbf{A} = \begin{bmatrix} -K_d & 0 & 0 \\ -(m_s + (\alpha/Y_p)) & 0 & 0 \\ -(m_{s0} + (\alpha/Y_{p0})) & 0 & 0 \end{bmatrix}
$$

$$
\mathbf{B} = \begin{bmatrix} X_{in}/x_5(t) & -x_1(t)/x_5(t) & 0 \\ S_{in}/x_5(t) & -x_2(t)/x_5(t) & 0 \\ O_{in}/x_5(t) & -x_3(t)/x_5(t) & (O^* - x_3(t)) \end{bmatrix} \quad \mathbf{c} = \begin{bmatrix} 0 \\ 0 \\ 1 \end{bmatrix},
$$

$$\mathbf{d} = \begin{bmatrix} \mu_m \\ -[(\mu_m / Y) + (\beta\mu_m / Y_p)] \\ -[(\mu_m / Y_0) + (\beta\mu_m / Y_{p0})] \end{bmatrix} \tag{30a}$$

while the nonlinear term as a function of states is

$$f(t, \mathbf{x}, \mathbf{u}) = \frac{x_1(t)x_2(t)x_3(t)}{((x_2(t) + K_s) + (x_3(t) + K_c))} \tag{30b}$$

22.5.4.2 Design Matrices

According to the value of various parameters given in the fermentation process (section 22.2.3.2), the matrix A and $\mathbf{d}$ are as below:

$$A = \begin{bmatrix} -0.001 & 0 & 0 \\ -0.1713 & 0 & 0 \\ -0.1613 & 0 & 0 \end{bmatrix} \qquad \mathbf{d} = \begin{bmatrix} 0.6102 \\ -1.0170 \\ -0.8717 \end{bmatrix}$$

With the given value of saturation constants K_c and K_s the nonlinear term is

$$f(t, \mathbf{x}, \mathbf{u}) = \frac{x_1(t)x_2(t)x_3(t)}{((x_2(t) + 0.31) + (x_3(t) + 0.01))}$$

The linear transformation of coordinates T_0 has been applied to the system as explained. The system triple has been converted to the canonical form. The system shown in eqn. (28) with above values of parameter, has two stable invariant zeros at [-1.14, 0]. If the system has invariant zeros, the poles of the sliding motion are given by the invariant zeros of the system triple (Proposition #3.4 in [27]). The gain vectors, when the error estimation pole is placed at [-1.0], are

$$\mathbf{g}_l = \begin{bmatrix} -0.6993 \\ 0.9949 \\ 1.1670 \end{bmatrix} \qquad \mathbf{g}_n = \begin{bmatrix} -0.6102 \\ 0.7627 \\ 0.8717 \end{bmatrix} \qquad \text{and} \qquad p_2 = [0.500]$$

The observer dynamical Eqn. (27) is solved with the system matrices and gain vectors. The initial conditions for the various states have been considered as per Table (22.6). The observed states are plotted along with the simulation plot of the concentration of cell mass $x_1(t)$, substrate $x_2(t)$ and dissolved oxygen $x_3(t)$ respectively. The product state, $x_4(t)$, is obtained by solving the first order dynamical equation with the values of the states from the observer.

Table 22.6 Initial values given to original and estimated states respectively in simulation experiment.

Concentration of	Value of x(t)	Value of $\hat{x}$ (t)
Cell mass : $x_1(t)$	0.002	0.0005
Substrate : $x_2(t)$	0.43	0.435
Oxygen : $x_3(t)$	0.006	0.0075
Product : $x_4(t)$	0.0001	0.00015

22.5.4.3 Results

Original and estimated states for the fermentation process are shown superimposed in Fig. 22.28a. The dissolved oxygen concentration measurement was considered without noise. The error plot Fig. 22.28b reveals high frequency chattering near the sliding surface in error space. *High frequency chattering is a limitation of sliding mode technique.*

The robustness of the observer is checked by adding random noise to the output, that is, the measured value of the dissolved oxygen concentration. Fig.22.29a shows the original and estimated states while Fig.22.29b corresponds to the estimation error under noisy measurement condition.

The problem of chattering in the observed states is reduced by scaling down the rate of change of the states at switching. Fig. 30 shows the error phase trajectories of the states. A combined error plot for error estimation with pole placed at -1.2 and -0.8 are given in Fig. 22.31.

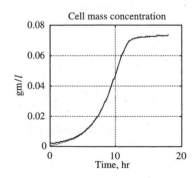

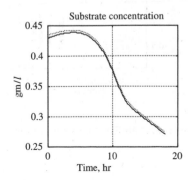

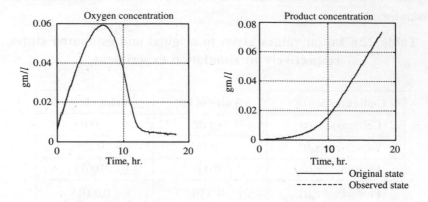

Fig. 22.28a Original and observed states with estimation pole at [-1.0]

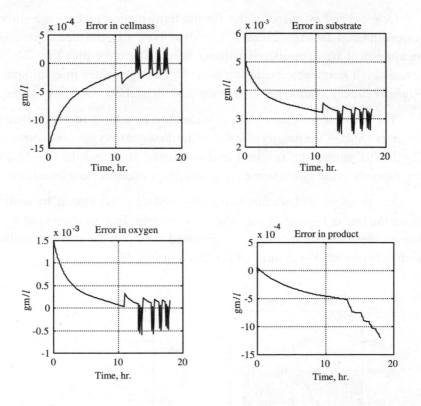

Fig. 22.28b Error between original and observed states with pole at [-1.0]

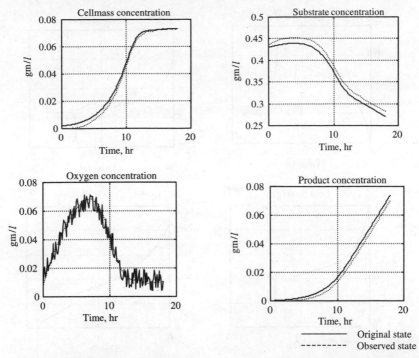

Fig. 22.29a Original and observed states with measurement noise in output

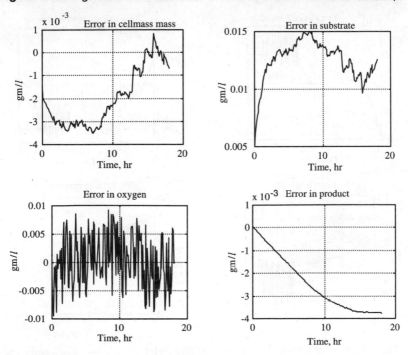

Fig. 22.29b Error between original and observed states with noise in output

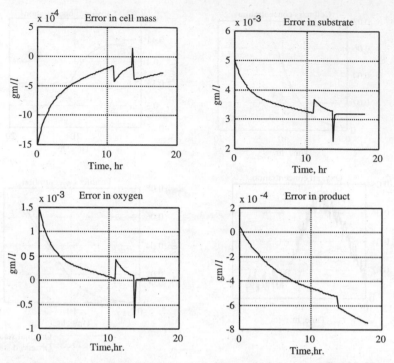

Fig. 22.30 Error between original and observed state with reduced chattering

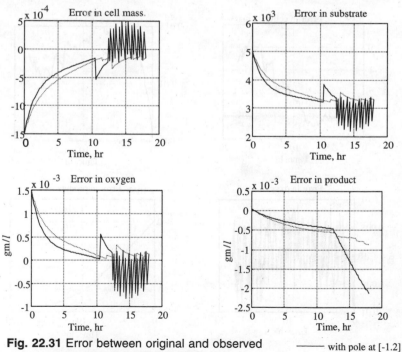

Fig. 22.31 Error between original and observed
states with poles at [-0.8] and [-1.2]

——— with pole at [-1.2]
——— with pole at [-0.8]

The above shows that 'soft sensing' is possible in bioprocess experiments rather than time consuming and discrete-in-time laboratory assaying.

22.6 Conclusions

A few concluding lines on 'control' of bio-reaction processes may not be out of place here. Proportional plus integral plus derivative (P-I-D) controller is the accepted method of control for the bioprocesses the world over. We did try a few alternative control strategies but not with much success.

We started with time delay control (TDC) for bioprocesses [29]. Except when the bioprocess is operated in mixed mode, in all other cases P-I-D controller scored over or matched the TDC performance. Mixed mode means alternating between batch and continuous mode of operation which is not usually the practice. It helps follow the optimized model reference system [1] though. Genetic algorithm with a defined fitness function is used to get the optimized time profiles of the state variables. TDC control law is also not easy to implement.

Alternatively, we could have had the sliding mode controller. The only important process variable which is in the control loop, is the dissolved oxygen content, that is, x_3 (t) and the control is by changing the see-sawing rate of the bioreactor fluid. To define a Lyapunov function for the same alone is not easy. Model following does not work here. We tried applying quantitative feedback theory (QFT) to bioprocess control. The resultant controller comes out to be of very high order, what with a pre-filter and a post-filter and a linearization network for the inherent nonlinearity of the system.

All said and done, the fixed order P-I-D controller with stability boundaries defined for robust control [30] appears to be the best choice. The discretized P-I-D control is also easily implemented for integrated computer control of the bioreactor through the visual basic platform [31].

22.7 Reference

1. Goutam Saha, "Time Delay Control for Bioprocesses and its application to a newly developed See-Saw Bioreactor in the Laboratory", PhD Thesis, Department of Electrical Engineering, Indian Institute of Technology, Kharagpur, India, 1999.

2. Goutam Saha, Alok Barua, Satyabroto Sinha, B C Bhattacharyya and S Ray, "Modeling of the Oxygen Transfer Characteristics of a 'See-Saw' Bioreactor", J. Chem. and Engg. Tech., vol. 24, 2001, pp. 97-101.

3. Ujjal Kr. Bhowmik, Goutam Saha, Alok Barua and Satyabroto Sinha, "On-line Detection of Contamination in Bioprocess using Artificial Neural Network", J. Chem. and Engg. Technol. Vol. 23, 2000, pp. 543-549.

4. G. Bastin and D. Dochain, "On line Estimation and Adaptive Control of Bioreactors", Elsevier Science, Amsterdam, 1990.

5. Biswajit Kar, "Bioprocess Simulation and Control", M. Tech. Thesis in Instrumentation, Electrical Engineering Department, Indian Institute of Technology, Kharagpur, April 2004.

6. Partha Sarathi Pal, Aniruddha Chakrabarty, Alok Barua and Satyabroto Sinha, "Study of See-Saw Bioreactor Fluid Dynamics, Shear Force Constraint and Oxygen Absorption Characteristics" Proc. National Systems Conference NSC-2002, November 18-19, 2002, Hyderabad, India.

7. Lee, Y H and Tsao, G T : Dissolved Oxygen Electrodes, pp35-86 in Advances in Biochemical Engineering, Mass Transfer and Process Control, Vol. 13, Ed. Ghose, T K, Fiechter, A K and Blakebrough, N, Springer-Verlag, Berlin, 1979.

8. Chen, J, and Wang, H Y, : Bioprocess monitoring of dissolved oxygen using a computerized pulsing membrance Electrode, Biotechnology Progress, v 9, pp 75-80, Jan-Feb, 1993.

9. Lee, S C, Hwang, Y B, Chang, H M and Chang Y K: Adaptive control of dissolved oxygen concentration in a bioreactor, Biotechnology and Bioengineering, v 37, pp 597-607, Mar 25, 1991.

10. Wang, H Y and Li, X M : Transient measurement of dissolved oxygen using membrane electrodes, Biosensors, v 4, pp 273-285,1989.

11. Hale J M and Hitchman M L : Some consideration of the steady-state and transient behavior of membrane-covered dissolved oxygen detectors, Electroanalytical Chemistry, Elsevier Science, v 107, pp 281-294, 25 February 1980.

12. Kohler, H and Geopel, W : Mixed valent tungsten oxides: new electrode materials for the potentiometric detection of dissolved oxygen at temperature below 35 degree C, Conference: Eurosensors IV '90, pp 345-54, 1-3 Oct. 1990, Karlsruhe, West Germany.

13. Mizutani, F, Yabuki, S and Lijima, S : Dissolved oxygen sensor using platinum electrode coated with polydimethylsiloxane, Transactions of the Electrical Engineers of Japan, Part E, v 120-E, pp 487-7, Oct. 2000.

14. Yang, X : Measurement of dissolved oxygen in batch solution and with flow injection analysis using an enzyme electrode, Biosensors, v 4, pp 241-149, 1989.

15. Prien, R D, Pascal, R W, Attard, G S, Birkin, P R, Denuault, G, Cook, D and Offin, D: Development and first result of a new mesoporous

microelectrode DO-sensor, Oceans, 2001. MTS/IEEE Conference and Exhibition, v 3, pp 1910-1914, 2001

16. Wittkampf, M, Chemnitius, G C, Cammann, K, Rospert, M and W. Mokwa : Silicon thin film sensor for measurement of dissolved oxygen, Sensors and Actuators B: Chemical, Elsevier Science, v 43, pp 40-44, September 1997

17. Glasspool, W and Atkinson, J : A Screen-printed amperometric dissolved oxygen sensor utilising an immobilised electrolyte gel and membrane, Sensors and Actuators B: Chemical, Elsevier Science, v 48, pp 308-317, 30 May 1998

18. Akyilmaz, E and Dinckaya, E : Development of a catalase based biosensor for alcohol determination in beer samples, Talanta, The International Journal of Pure and Applied Analytical Chemistry, Elsevier Science, 23 May 2003

19. Ory, J M, Jacques, F and Boudey, Y : A Biosensor for water quality monitoring, Instrumentation and Measurement Technology Conference, 1996. IMTC-96. Conference Proceedings. 'Quality Measurements: The Indispensable Bridge Between Theory and Reality', IEEE, pp 1354-1359, 4-6 Jun 1996

20. Chang-Yen, D A and Gale, B K : A Novel integrated optical dissolved oxygen sensor for cell culture and micro total analysis systems, Micro Electro Mechanical Systems, 2002. The Fifteenth IEEE International Conference, pp 574-577, 2002.

21. Partha Sarathi Pal, Srijan Bhatnagar, Alok Barua and Satyabroto Sinha, "Design and Fabrication of Low Cost Dissolved Oxygen (DO) Sensor", Proceedings National System Conference, NSC-2003, Electrical Engineering Department, Indian Institute of Technology, Kharagpur, 17-19 December 2003.

22. E.A.Misawa and J.K.Hedrick: Nonlinear observers - a state -of- the art survey, Transaction ASME : Journal of Dynamic Systems, Measurement, and Control, vol.111, pp. 344-352, 1989

23. B. Walcott, M. Corless, S. Zak : Comparative study of nonlinear state observation techniques, International Journal of Control, vol. 45, pp. 2109-2132, 1987.

24. Manish Agarwal, "Study of Software Sensor or Observer Design for a Bioprocess, " M. Tech. Thesis, Electrical Engineering Department, Indian Institute of Technology, Kharagpur, December 1999.

25. C.Edwards and S.K.Spurgeon: "Sliding Mode Control: Theory and Application", Taylor & Francis, London, 1998.

26. B.L.Walcott and S.H.Zak: State observation of nonlinear uncertain dynamical systems, IEEE Transaction on Automatic Control, vol. AC-32, no-2, pp. 166-170, 1987.

27. Yi Xiong and Mehrdad Saif: Sliding mode observer for uncertain systems,

Part-I, Proc. Conference on Decision and Control, Sydney, Australia, pp. 316-321, December 2000.

28. Jignesh B Patel, "Sliding Mode Observer for Bioreactors", M. Tech Thesis, Electrical Engineering Department, Indian Institute of Technology, Kharagpur, January 2002.

29. Partha Sarathi Pal, G. Saha, A. Barua and S Sinha, "Implementation of a suitable process controller along with contamination detector in a novel see-saw bioreactor for achieving very high yield from animal cell line culture", Proceedings International Conference on Control, Instrumentation and Information Communication, CIIC 2001, Department of Applied Physics, University of Calcutta, 13-15 December 2001.

30. S P Bhattacharyya: New approaches to the design of fixed order controllers, Report Dept. of Elec. Engg., Texas A & M Univ, College Station, Tx 77843-3128, December 2003.

31. Partha Sarathi Pal: Studies on certain aspects of a newly development See-Saw bioreator for animal cell culture, Ph.D thesis, Department of Electrical Engineering, Indian Institute of Technology, Kharagpur, 2005.

| Chapter 23 |

Measurement of Two Phase Flow Parameters
(Void Fraction and Flow Rate)

P. K. Das

Abstract

In this article the techniques for measuring void fraction and flow rate of two phase flow have been discussed. Emphasis has been given on gas liquid two phase flow, though time to time the measurement techniques for the flow of other two phase mixtures have been illustrated. Numerous techniques exist for the measurement of the above parameters. It is not possible to present an elaborate account of all the techniques within the limited scope of this article. Therefore, the commonly adopted techniques have been dealt with a greater detail and only brief mentions have been made about the relatively less common techniques. Relative merits and demerits of the techniques have also been discussed. However, as this is a topic of active research and due to its industrial importance and omission of some of the possible methodologies is not unlikely.

23.1 Introduction

Multiphase flow or simultaneous flow of several phases is commonly encountered in a variety of engineering processes. Simultaneous flow of as many as four phases namely, water, crude oil, gas and sand is not uncommon during oil exploration, though flow of two phase mixtures is the most common occurrence in industry. The following are the different variations of two phase flow:

(a) Gas-liquid flow - involves boiling, condensation as well as adiabatic flow. They are common in power and process industries, refrigeration, air-conditioning and cryogenic applications.

(b) Gas-solid flow - pneumatic conveying, combustion of pulverized fuel, flow in a cyclone separators are examples of this category of two phase flow.

(c) Liquid-solid flow - this type of flow is encountered in slurry transportation, food processing as well as in various processes in biotechnology.

(d) Liquid-liquid flow - This type of flow is also characterized by the presence of a deformable interface (similar to gas-liquid flow) and processes several features similar to other two phase flow phenomena. Liquid-liquid flow is common in petroleum industries and chemical reactors.

In spite of the extensive volume of past research activity, two phase flow is not yet an area in which theoretical prediction of flow parameters is generally possible. Indeed, this situation is likely to persist for the foreseeable future. Thus, the role of experiment and parametric measurement is particularly important.

The techniques of measurement for single phase flow are well established. Based on these techniques, various meters and instruments have been developed which are successfully employed for industrial measurement as well as for R&D activities. Unfortunately, these instruments can not be directly used for multiphase flow measurement. Most of the problems in multiphase flow measurements arises from the fact that the parameters characterizing it are many times larger than those in single phase flows. In single phase flow, the flow regimes encountered are laminar, turbulent and a transition region between them. In multiphase flow, numerous flow regimes are possible. The flow regimes observed under a specific set of flow conditions is dependant on flow geometry (size and shape) and orientation (vertical, horizontal and inclined), flow direction in a vertical or inclined flows (up or down), phase flow rates and properties (density, viscosity, surface tension). Figs.23.1 and 23.2 show some typical flow regimes one encounters in two phase gas liquid flow in circular conduits (Collier et. al 1994). It may be noted that the flow regimes change drastically from the adiabatic case when phase transition (due to heat transfer) is involved. In general, the following features, which complicates the flow situation, may be identified

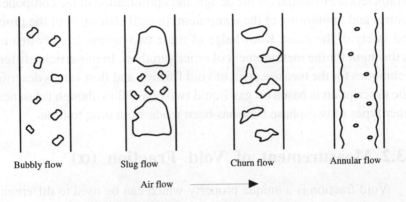

| Bubbly flow | Slug flow | Churn flow | Annular flow |

Air flow ⟶

Fig.23.1 Flow pattern in vertical up flow

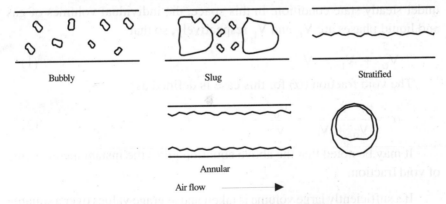

Bubbly   Slug   Stratified

Annular

Air flow ⟶

Fig. 23.2 Flow pattern in horizontal flow

1. Even for the steady flow rates of the phases at the inlet, the local flow phenomenon may become intermittent or random.

2. The distribution and velocity of the phases may change both with space and time.

3. In case of phase change, the flow regimes vary substantially along the flow direction.

From the brief description of gas-liquid flow given above, it is obvious that methods of flow measurement conventionally adopted for single phase flow are grossly inadequate.

This has given rise to the development of a number of techniques especially suited for the measurement of two phase flow parameters. In the limited scope of this discussion, it is not possible to consider the principles of measurements of all the parameters. However, void fraction and flow rate are two parameters of unique importance. Information regarding these parameters are essential for the design and optimization of the components, control and monitoring of the equipment, overall efficiency of the process and safety of the plant. Knowledge of these two parameters is often used as the input for the measurement of other variables. In this article, different techniques for the measurement of void fraction and flow rate is described. The description is based on gas-liquid two phase flow though reference to other types of two-phase flow has been made from time to time.

23.2 Measurement of Void Fraction (α)

Void fraction is a unique property which can be used to differentiate two phase flow from single phase flow phenomena. Let us, assume gas-liquid two phase flow is taking place through a flow channel of volume V under steady state condition. In this space, the individual volumes of gas and liquid phases are V_G and V_L respectively, so that

$$V_G + V_L = V \tag{1}$$

The void fraction (α) for this case is defined as

$$\alpha = \frac{V_G}{V_G + V_L} = \frac{V_G}{V} \tag{2}$$

It may be noted that the above equation gives the instantaneous value of void fraction.

If a sufficiently large volume is taken and average values over a suitable duration is considered, average void fraction can be defined as

$$\langle \alpha \rangle = \frac{V_G}{V} \tag{3}$$

$\langle \alpha \rangle$ can be taken as a characteristic of the two phase flow phenomena. The importance of $\langle \alpha \rangle$ can be realized easily. All average properties of the mixture like density, viscosity, specific heat enthalpy, etc. are direct functions of void fraction. Moreover, the flow regime and the interfacial area are also dependant on this. The knowledge of void fraction is particularly necessary for the design of costly components. In a nuclear reactor, the absorption of neutron depends directly on it. Therefore, it is not surprising that in two phase flow, most of the efforts for instrumentation has been made for the measurement of void fraction. Before describing the method of void fraction measurement, different definitions are explained below.

23.2.1 Definitions of Void Fractions

Though the term void fraction refers to the volume averaged property by definition, it is not always possible to measure $\langle \alpha \rangle$ over a volume element. The measurements then obtain averaged values with respect to different space dimensions and with time which are subsequently processed to obtain the volumetric averaged parameter. This gives rise to the following definitions of void fraction in a flow conduit:

Volume average void fraction

It is obtained as the fraction of the channel volume occupied by the gas phase at any instant of time. In mathematical terms, the definition can be expressed as -

$$\langle \alpha \rangle_V = \frac{\text{Volume of liquid in mixture}}{\text{Total volume of mixture}} = \frac{\int_{V_G} dv}{\int_{V_G + V_L} dv} = \frac{V_G}{V} \qquad (4)$$

This is the most used definition of void fraction in industrial designs and gives the overall composition of the flowing mixture.

Area average void fraction

It is the fraction of the channel cross sectional area occupied by the gas phase at any instant of time. This average property is usually determined by impedance method and optical techniques. This is the volume averaged value for infinitesimal axial length of the test section and is equal to the volumetric average void fraction when that does not vary with the length of the conduit. For this, eqn. (4) can be approximated as

$$\langle \alpha \rangle_A = \frac{\int_{A_G} dA}{\int_{A_L + A_G} dA} = \frac{A_G}{A_L + A_G} \qquad (5)$$

Chordal average void fraction

It is defined as the composition of the mixture along a particular chord of known length. It is obtained when it is difficult to measure the area or volume average values. It is used particularly in connection with the radiation attenuation and scattering techniques. It is converted to the area average values either by mathematical manipulation or by the use of multiple beams and is mathematically expressed as

$$\langle\alpha\rangle_{\text{chordal}} = \frac{\int_{l_G} dl}{\int_{l_L+l_G} dl} \tag{6}$$

Time average void fraction

This is obtained by measuring the void fraction of the mixture at a particular point in the two phase flow field as a function of time. This is usually required to obtain the void fraction profile in any system since a knowledge of it adequately describes the structure of the flow field. Any point in the field can only be occupied by one phase at a time. Therefore, by definition, the void fraction of a phase at this point can only be either zero or unity. For a negligible thickness of the gas-liquid interface, the variation of the void fraction from zero to one and vice versa will be instantaneous and assume a square wave form with respect to time. The point average void fraction or local void fraction with respect to time is, therefore, meaningless. The void fraction under this situation is defined as the average fraction of a particular interval of time during which the point is occupied by the liquid phase. It does not carry the sense of volume or area but is related to time only. It can be mathematically expressed as

$$\langle\alpha\rangle_t = \frac{\int_{t_G} dt}{\int_{t_L+t_G} dt} = \frac{1}{\Delta T}\sum_{i=n}^{n}\Delta t_1 = \frac{\Delta T_G}{\Delta T} \tag{7}$$

where i = 1, 2, ... n and n is the number of periods during which liquid phase exists at a particular point. In Fig. 23.3, different void fractions are illustrated. The measurement techniques for different types of void fraction are described pictorially in Fig. 23.3. Fig 23.3(a) shows the scheme of measuring volume averaged void fraction. The probing is done over a volume of the conduit. If measurement is done for a cross sectional area as shown in (b), one gets area averaged void fraction. In this area, if measurement is done along any chord length (not necessarily along the diameter) chordal average is obtained. If the probing area is much smaller compared to the cross sectional area (Fig. 23.3b) the obtained measurement approaches local (point) value.

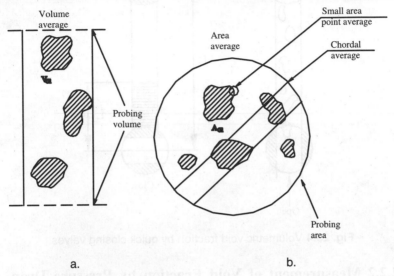

a.   b.

Fig. 23.3 Different schemes of void fraction measurement

23.2.2 Mechanical Techniques

23.2.2.1 Direct Volume Measurement by Quick Closing Valve Technique

This is the most widely used method for measuring holdup for adiabatic gas-liquid as well as vapour-liquid flows. It has been used by a majority of the researchers in the past. In this method, the flowing two phase mixture is instantaneously trapped between a pair of valves placed at the beginning and end of the test section. The relative proportion of the two phases is then obtained by noting the fraction of the channel volume occupied by either of them after gravity separation of the phases. The main drawbacks of this method include a finite time requirement for closure of the valves. The closing down of the system for each measurement and bringing the system back to the steady state which might require considerable time between successive runs. This is also not suitable for transient situations. Fig. 23.4 depicts the arrangement.

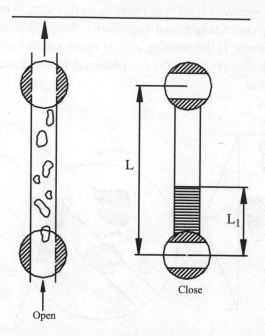

Fig. 23.4 Volumetric void fraction by quick closing valves

23.2.2.2 Measurement of Void Fraction by Pressure Drop

Volume average void fraction or the overall void fraction for a section of conduits can be estimated from the differential pressure of the section. The scheme of the measurement is explained in Fig. 23.5.

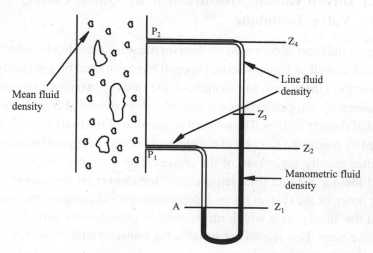

Fig. 23.5 Void fraction estimation by pressure drop measurement

Making a pressure balance at Section A one gets,

$$p_1 + (z_2 - z_1)g\rho_C = p_2 + (z_4 - z_3)g\rho_C + (z_3 - z_1)g\rho_m \qquad (8)$$

Rearranging we have,

$$p_1 - p_2 = (z_3 - z_1)g(\rho_m - \rho_C) + (z_4 - z_2)g\rho_C \qquad (9)$$

From the hydrodynamics of two phase flow in the conduit, considering no accelerational pressure drop

$$p_1 - p_2 = g\rho_t(z_4 - z_2) + g\rho_t h_e \qquad (10)$$

h_e is the head loss due to friction. Where ρ_t is the mixture density and is given in terms of volume average void fraction,

$$\rho_t = (1 - \langle\alpha\rangle_v)\rho_\ell + \langle\alpha\rangle_v \rho_G \qquad (11)$$

Equating equations (9) and (10), one gets,

$$(z_3 - z_1)g(\rho_m - \rho_C) + (z_4 - z_2)\rho_C$$

$$= g\{(1 - \langle\alpha\rangle_v)\rho_\ell + \langle\alpha_v\rangle\rho_g\}\{(z_4 - z_2) + h_e\} \qquad (12)$$

From the above equations, the overall void fraction may be determined if h_e is known. The head loss due to friction may be neglected in a number of cases or it can be determined approximately using a suitable correlation. The above method is simple and is well suited for situations where

(i)   frictional pressure drop is small,

(ii) accelerational pressure drop is small,

(iii) there is no drastic change in void fraction along the channel length.

It is mandatory to ensure that the manometer lines are filled up by a single phase fluid. In case of gas-liquid flow, it is generally filled up with the fluid, which forms the continuous phase. This method is also suitable for transient cases if the manometer is replaced by a pressure transducer. This technique is applied for gas-liquid, solid-liquid and gas-solid flows.

23.2.3 Radiation Absorption and Scattering Methods

Attenuation of a beam of gamma rays is one of the widely used techniques of void fraction measurement. Attenuation of gamma ray passing through the two phase medium occurs by three distinct processes.

1. Photoelectric effect,

2. Production and absorption of positron-electron pair,
3. Compton effect.

Details about these effects are elaborated in Hewitt (1978).

The absorption of a collimated beam of initial intensity I_o (photons / m^2s) is described by a exponential relationship.

$$I = I_o \exp(-\mu z) \tag{13}$$

where μ is the linear absorption coefficient and z is the distance traveled. Absorption coefficient on the other hand depends on the density of the medium.

Estimation of void fraction, can be made purely from a theoretical point of view. However, that needs gamma spectrometry for separating out the secondary photons produced in the process. This is expensive. In most of the cases, in-situ calibration is done by measuring intensity of the absorbed beams for tubes full of liquid (I_L) and full of gas (I_G) respectively. If the intensity for a two phase system is I, then void fraction is given by

$$\alpha = \frac{\ell n(I / I_L)}{\ell n(I_G / I_L)} \tag{14}$$

This method void fraction measurement is very versatile and can be used for a wide range of applications containing two or more phases. However, there are some inherent sources of error and difficulties in this measurement.

1. The system is in general costly and needs specially trained manpower for its operation.

2. There are difficulties of handling high-energy radiation.

3. Normal fluctuations of photons gives rise to an error which can be minimized by using strong sources and using long counting times. This makes the process unsuitable for transient measurement

4. If the phase interface is parallel to the beam, void fraction is given by

$$\alpha = \frac{I - I_L}{I_G - I_L} \tag{15}$$

This induces some error in the measurement.

5. Presence of the metal wall also induces some error when averaging is done.

6. Gamma ray absorption also has some limitations for liquid-liquid systems.

Some of the above shortcomings can be overcome by the use of calibration, multiple sources, etc.

In specific applications, neutron attenuation and the use of x-ray produce good results. However, availability of neutron source in the former case and safety in the later are major concerns.

23.2.4 Optical Techniques

A large number of measurement systems have been developed using optical techniques. Both local and global as well as steady state and transient measurements are possible by suitable design of the instrumentation system. Some of the important optical methods are described below.

23.2.4.1 Photographic Techniques

This is suitable for identifying the flow regimes where the two phases are not thoroughly mixed like stratified flow. By suitable arrangement, one can use the technique for complex gas-liquid regimes like annular-mist flow. Fig.23. 6 describes one such arrangement. In this arrangement a section of the tube is illuminated. Photograph is taken from the top of the conduit, which is closed by a transparent cover. The camera is focused on the illuminated portion of the conduit. Photographic technique provides a good visual appraisal at the flow phenomena.

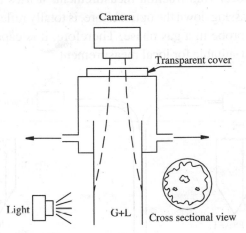

Fig. 23.6 Photographic technique for void fraction measurement

The photographic technique got a boost up in recent times due to the availability of laser sources and digital camera at lower prices. Using simple optical arrangement thin light sheet can be generated from a laser beam.

By this illumination of a particular cross section is possible. This gives better information regarding phase distribution. The arrangement is shown in Fig.23.7. Digital photography has also made the post processing of photographed information easier. With the videographic recording, the transient phenomena can be analysed while applying digital image analysis a better quantification of the flow regime is possible.

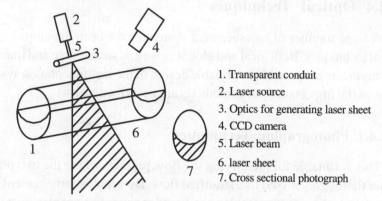

1. Transparent conduit
2. Laser source
3. Optics for generating laser sheet
4. CCD camera
5. Laser beam
6. laser sheet
7. Cross sectional photograph

Fig. 23.7 Lighting by laser sheet

23.2.4.2 Local Measurements by Intrusive Probes

Optical probes of small dimension, as shown in Fig. 23.8, have been developed for local void fraction measurement. It uses the principle that incident light, passing down the optical fibre, is totally reflected and returned back when the probe in a gas phase. Therefore, it is capable of detecting minute void and suitable for local measurement.

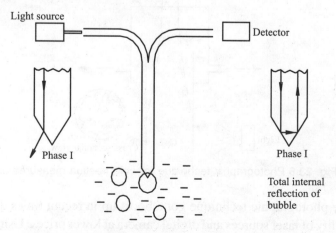

Fig. 23.8 Optical probe for phase detection

An innovative design based on the combination of optical and isokinetic sampling technique is illustrated in Figure 23.9. In this probe, a small volume of the two phase mixture is isokinetically sucked and possed through a capillary tube. Inside the tube, the size of the bubble is determined with the help of two light sources. This gives an estimation of the size distribution and flow rate of the gas phase (Steevers et. al. 1995).

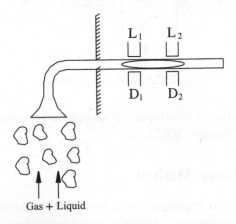

Fig. 23.9 Operation of photo suction probe

23.2.4.3 Fluorescence Technique

Instantaneous and highly localized, measurements of film thickness can be obtained using the fluorescence technique. Blue light from a mercury vapour lamp is passed through a microscope illuminator and focused in a conical beam into the liquid film. The circulating water in the apparatus contains fluorescein dye-stuff and the incident beam excites a green fluorescence in the film. The fluorescent illumination is separated in a spectrometer and its intensity is a direct measure of the liquid film thickness.

23.2.4.4 Scattering Technique

Scattering technique forms the basis of different instruments for continuous drop size distribution. A light beam passing through a dispersion will have an extinction proportional to the effective superficial area. This is true for both transparent dispersed phases and opaque ones. Provided the received beam is exactly in line with the transmitted one, the transparent dispersed elements will scatter the incident beam and behave as though they are opaque with regard to the received beam. This method of light scattering is suitable for both bubbly and drop flow. Fig. 23.10 illustrate the use of light scattering technique for mist annular flow.

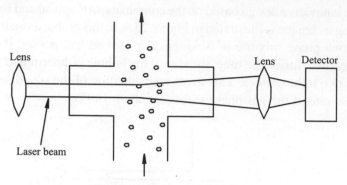

Fig. 23.10 Scattering technique

Some of the recent developments in optical technique are reported in Mayinger and Feldmann (2001).

23.2.5 Impedance Method

As the electrical impedance of a two phase mixture is a function of concentration, measurement of impedance can form a basis for the estimation of void fraction. Several instruments for the measurement of void fraction and associated parameters have been developed based on impedance technique. Impedance technique has the following advantages:

1. It is a low cost technique.
2. It is suitable for transient measurement.
3. Large variations of electrode design are possible making the method appropriate for different flow situations and geometry.
4. Both intrusive and non-intrusive measurements are possible.
5. Point measurement as well as global measurement can be made by suitable design of the probe.
6. Same principle (sometimes the same probe) may be used for the measurement of associated parameters like,
   (a) flow regime identification,
   (b) bubble size and frequency,
   (c) bubble velocity.

An impedance probe can operate either in resistance (conductance) or in capacitance mode. If the liquid phase is the continuous one and electrically conducting, then the probe is used in the resistive mode. If the gas phase is continuous or the liquid phase is non-conducting, then the probe is used in capacitance mode. A large variety of probe design is possible some of them are described below.

23.2.5.1 Separated Flow of Gas and Liquid

In annular flow, stratified flow, film flow, the gas and liquid phases are separated by a well defined interface and generally the liquid phase does not contain any gas bubble in dispersed condition. In such situation, impedance probe gives a good estimate of the liquid film thickness. The film thickness may be obtained either from direct calibration or from a theoretical analysis. The theoretical analysis will be described shortly. Some of the probe geometry suitable for the measurement of film thickness are shown in Fig. 23.11.

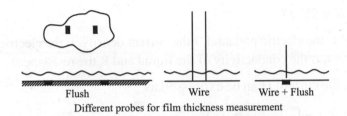

Flush   Wire   Wire + Flush

Different probes for film thickness measurement

Fig. 23.11 Probes for local film thickness

The same principle can be used for two phase flow measurement through a conduit. For conduits of circular section arc electrode probes, parallel wire probe or ring electrode probe may be used. The methodology of measurement can be explained based on the work of Gupta et. al. (1994) in which a pair of arc electrode probe were used to measure the liquid height in stratified gas-liquid system. In Fig. 23.12 stratified flow in a circular tube is shown. Flush mounted arc electrode probes may be used for the estimation of liquid fraction.

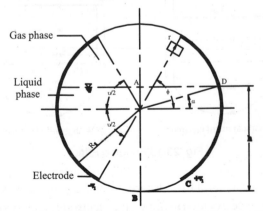

Schematic representation of the stratified gas-liquid

Fig. 23.12 Arc electrode probe

From electrostatics one gets,

$$\nabla^2 V = 0 \tag{16}$$

$$E = -\nabla V \tag{17}$$

$$J = \in E \tag{18}$$

From Ohm's Law,

$$I = \iint_S Jds$$

and $R = 2V_1 / I$

where V is the electric potential, J the current density, E the electric field, I the current, σ the conductivity of the liquid and R the resistance.

Finally, the current I can be expressed as

$$I = L\sigma \int_{-\theta/2}^{\alpha} E_\gamma \left(R_1 d\phi \right) \tag{19}$$

From the above relationships, one can calculate the resistance for a given liquid height and liquid conductivity. Fig. 23.14 is the circuit for the probe. It may be noted that the probes induces 3-D effect due to fringing from the end of the probes as shown in Figure 23.13.

Fringing

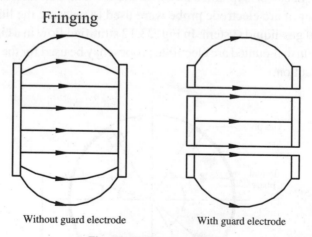

Without guard electrode   With guard electrode

Fig. 23.13 Fringing Effect

However, it is difficult to take care of the fringing effects by the theory. This difficulty can be avoided using guard electrodes as shown in the figure. A probable measurement scheme is shown in Fig. 23.14. This arrangement renders the electrical field two dimensional along a small axial length.

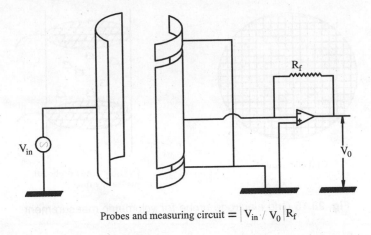

Probes and measuring circuit $= \left| V_{in} / V_0 \right| R_f$

Fig. 23.14 Measuring circuit with guard electrodes

The resistance can be determined from the following formula:

$$R = \left| V_{in} / V_O \right| R_f \tag{20}$$

The resistance on the other hand can be related to the void fraction. Impedance probes can also be used when the phases are well mixed. When the phases are well mixed from Maxwell's theory, one gets

$$\alpha = \frac{A - A_c}{A + 2A_c} \cdot \frac{\in_G + 2 \in_L}{\in_G + \in_L} \tag{21}$$

where A_c is the admittance of the gauge when immersed in the liquid phase alone. ε_G and ε_L are the conductivities of the gas and liquid phases if conductivity is dominating. On the other hand, one should use the dielectric constants if capacity is important. The above equation is suitable for bubbly flow. Accordingly arrangement for volumetric measurement of void fraction can be made as shown in Fig. 23.15.

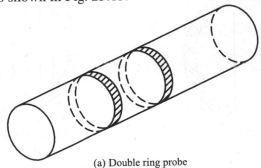

(a) Double ring probe

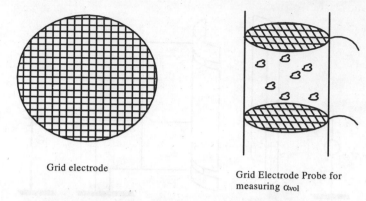

Grid electrode

Grid Electrode Probe for
measuring α_{vol}

Fig. 23.15 Grid electrode probe for volumetric measurement

The arrangement in Fig 23.15(a) depicts ring type of electrodes. The void fraction between two rings can be estimated. The arrangement is suitable for bubbly, slug, annular and stratified flow. A pair of grid electrodes is shown in Fig 23.15(b). It is well suited for measuring the volumetric void fraction in bubbly flow. People have also used this probe for slurry flow.

For liquid droplet flow through a gas, one gets

$$\alpha = 1 - \frac{(A \in_L - A_c \in_G)}{(A \in_L + 2A_c \in_G)} \cdot \frac{(\in_L + 2 \in_G)}{(\in_L - \in_G)} \tag{22}$$

Another probe configuration widely used in gas-liquid flow is shown in Fig. 23.16 (a). It uses two thin parallel conducting wires spanning the cross section. This arrangement is suitable for the measurement in annular, stratified and slug flow. In case of curved interface in stratified flow multiple wires may be used as shown in Fig. 23.16 (b).

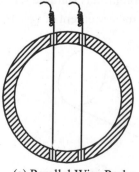

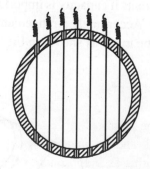

(a) Parallel Wire Probe   (b) Multiple Parallel Wires

Fig. 23.16 Wire probes

Conductivity principle can be used for local measurement. For this purpose, endoscopic probes with single or multiple needle electrodes as shown in Fig. 23.17 are used.

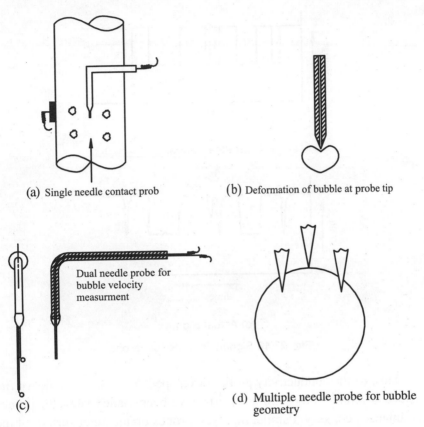

(a) Single needle contact prob

(b) Deformation of bubble at probe tip

Dual needle probe for bubble velocity measurment

(c)

(d) Multiple needle probe for bubble geometry

Fig. 23.17 Single and multiple needle contact probes

The needle electrode which is a fine needle with only its uninsulated tip exposed to the two-phase mixture forms the heart of the measurement scheme. The other electrode is so large that always some part of it is in contact with the conducting phase of the mixture. If the small tip of the probe comes in contact with the non-conducting phase (say gas bubble) the circuit is broken and there is a change in signal level. Theoretically, the probe should produce a square wave signal, if the probe tip is infinitesimally small and the interface is infinitesimally thin. But in practice the bubble deforms as it approaches the probe tip as shown in Fig. 23.17 (b). Also finite time is needed for wetting the probe tip. Hence the signal is often distorted and careful signal processing is needed for deriving qualitative information. Fig. 23.17 (c) shows double needle probe, which may be used for determining the bubble velocity. Using multiple probe, as shown in Fig. 23.17 (d) bubble geometry can be determined. Fig. 23.18 shows the typical nature of the signal from a needle probe.

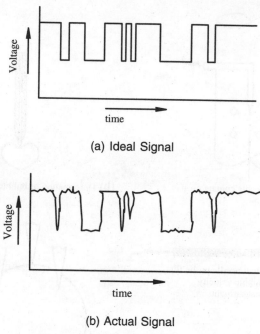

(a) Ideal Signal

(b) Actual Signal

Fig. 23.18 Signals from needle probe

Most of the conductivity probes developed so far are suitable for flat surface or circular tubes. Some efforts have been made to develop probes for annular geometry. Parallel ring type probes on the outer surface of the inner tube and inner surface of outer tube have been used. Das et.al. (2000) have used a unique parallel plate type probe. The probe is made from fine strips cut from double sided printed circuit board (PCB) and can be supported only from the outer tube due to its inherent rigidity. Das et.al. (1998,2002) have also used it for the estimation of bubble velocity and bubble shape.

One of the major drawbacks of the conductivity probes is the change of conductivity with temperature and impurity. As the probe is exposed to a conducting medium there is electrochemical interaction (McNaughtan et.al. 1999) between the two. This is commonly known as double layer formation. As it is difficult to model the double layer effect it makes the quantitative prediction difficult. However, conductivity probes are extensively used for qualitative prediction like flow regime identification (Das et.al. 1999a, 1999b).

Some of the difficulties of conductivity probes may be avoided using the probe in the capacitance mode. This is suitable for non conducting fluids and the probes can also be mounted outside the conduit wall (it need not contact the two phase mixture). Some of the popular probe geometries (Sami et. al. 1980) are shown in Fig. 23.19. However, the dielectric constants of the two phases do not vary widely and there are stray capacitance effects. These make the measurement challenging.

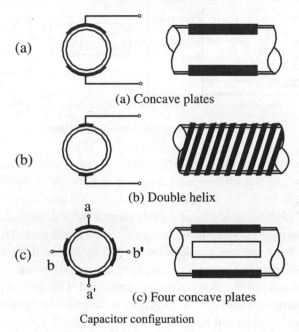

(a) Concave plates

(b) Double helix

(c) Four concave plates

Capacitor configuration

Fig. 23.19 Different types of capacitance probe

23.2.6 Other Methods

Apart from the above techniques, hot wire principle, micro thermocouple, microwave and radiowave attenuation as well as NMR technique and ultrasonic probes are also used for void fraction measurement.

23.2.7 Tomographic Imaging

As local and spatial fluctuations are inherent to any two phase flow phenomenon, none of the methods described above can give the instantaneous value of void fraction over a volume correctly. Moreover, they can supply only a partial or distorted picture regarding the flow regime as some short of averaging is done in all the above measurements. To overcome this limitation, scientists have adopted tomographic imaging technique for scanning the entire flow passage.

Tomographic imaging of a flow passage may be obtained by a variety of basic void measurement techniques like impedance, optical, radiation attenuation, etc. The basic principle of all these systems are the same though the detail arrangement may vary. In Fig. 23.20, the typical arrangement for a tomographic measurement with conductivity or capacitance probe is shown.

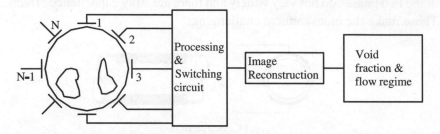

Fig. 23.20 Tomographic imaging system

In any of these methods, a large number of miniature sensors / probes are used. Two of the probes are energized in turn. A probable sequence could be 1-2, 1-3, ...1-N; 2-3, 2-4,...........2-1 and so on. This allows collection of high accuracy local signals from a three dimensional probing volume. By using tomographic techniques, the physical property of interest is recovered from the observations integrated along the different paths of probe measurement for each plane (slice) of the measurement volume. The three dimensional information is then reconstructed. Theoretical background as well as hardware details of different tomographic systems may be obtained from the classical book by Williams and Beck (1995).

23.3 Flow Rate Measurement

Measurement of volume flow rate or mass flow rate is a basic requirement in two phase flow as it is in single phase flow. For single phase flow through a conduit one uses the following relationships:

$$Q = \overline{V}A \tag{23}$$

and, $M = \rho\overline{V}A$ \hfill (24)

As the density is generally known a-priori,(in case of an incompressible flow) the measurement of flow rate finally boils down to the measurement of the average velocity.

In case of two phase flow

$$Q = A(1-\alpha)\overline{V}_L + A\alpha\overline{V}_G \tag{25}$$

and, $M = \rho_L A(1-\alpha)\overline{V}_L + \rho_g A\alpha\overline{V}_G$ (26)

As in general, the velocity of the two phases may vary considerably, one needs to measure the average velocity of the phases along with the void fraction or concentration. Moreover, the inherent difficulties of two phase flow measurement enhance the complexity.

In general, the techniques of two phase flow measurement can be divided into these broad categories:

(i) Separation of the phases and measurement of the phase flow rates using the single phase measuring techniques.
(ii) Flow stabilization and homogenization and measurement of the composition and velocity over a cross section.
(iii) Measurement of mass flow rate in the natural state using innovative techniques.

Some methods pertaining to each of the above category are described below.

23.3.1 Separate Estimation of the Phase Flow Rates

In a number of situations this can be done either before the equipment or after the equipment by separating the phases in a phase separator. Fig. 23.21 shows as air lift loop where the liquid flow is generated by the motion of the gas stream. In this case, the air flow rate can be measured before it is introduced to the loop. After air is separated, the water flow rate can be measured in the down comer.

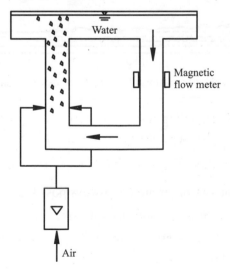

Fig. 23.21 Separate measurement of the flow rates of air and water in an air lift loop

In case of film flow or annular flow, the flow rate of the liquid film can be measured by sucking it through a porous section as shown in Fig. 23.22.

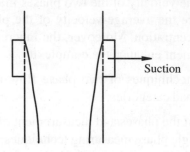

Fig. 23.22 Porous wall film suction device

Alternatively, in-situ measurement is possible. Film thickness can be measured by any conventional techniques while the film velocity can be measured by tracer injection. Fig. 23.23 explains the principle.

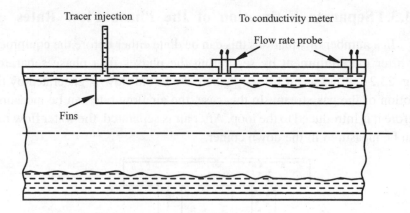

Fig. 23.23 Salt dilution method for measurement of local film flow rate

By iso-kinetic sampling two phase samples may be collected from a small probing area. After separation their volume flow rate can be determined.

The present methodology has the following limitations;

(i) Separation of phases are not always possible.

(ii) In general, the techniques are intrusive.

(iii) It does not necessarily provide the in-situ information like flow fluctuations.

23.3.2 Measurement after Flow Stabilization or Homogenization

The difficulty in the measurement of two phase flow arises from the facts that the average phase velocities and the local composition may vary widely even over a cross-section. This is so as one phase tries to slip past another phase and moreover gravity tries to segrigate the lighter and the heavier phase. If both the phases are "well mixed", then the flow of the mixture can be treated as a "single phase flow". Then the techniques well known for single phase measurements can be adopted. However, two simultaneous measurements are necessary for estimating the average composition and avarage velocity.

For this purpose, several homogenizers are designed. Some of them are static while others require auxiliary power. If the measurement is done within a short distance of the homogenizer, a good accuracy can be expected. Rajan et. al. (1993) provides adequate references for the design of homogenizers. In the present method, as composition and velocity are measured separately and the flow rate is inferred from these two measurements, it is also referred as Inference Technique. In the inference technique, the void fraction or composition may be measured by any of the techniques described in the previous sections. Methods for velocity measurement are described below.

23.3.2.1 Velocity Measurement

Excellent reviews on the different methods for velocity measurement of two phase flow are available in the literature (Rajan et. al., 1993; Beck, 1981; Thron et. al., 1982). The measurement can be done using both intrusive and non-intrusive techniques. The different methods available can be listed as-

Intrusive methods:

(a) Tracers

(b) Heat probes

(c) Pitot tubes

(d) Orifices, venturies, nozzles

(e) Momentum bar and turbines.

Non-intrusive methods:

(a) Electromagnetic meter
(b) Ultrasonic method
(c) Laser method
(d) Cross-correlation technique.

Non-intrusive methods of measurement are particularly desirable in a pneumatic conveyor since any protrusion into the pipeline can act as a catalyst to the formation of a blockage. Intrusive instrumentation also results in an increased pressure drop in the pipeline which has to be compensated for by an increase in compressor power. Moreover, since most solid-gas flows are highly abrasive, any inserted probes will have to be replaced frequently because of wear. The different available non-intrusive methods for two phase flow measurements are discussed below.

23.3.2.2 Electromagnetic Method

Faraday's law of induction is used here as the working principle. A strong magnetic field is applied across the electrically conducting fluid. The voltage induced in the electrodes in contact with the fluid is proportional to the mean velocity of the fluid. The electrical power requirement can be expensive in this method. Also, special design and installation are required. The main disadvantage of this method is that the fluid should be electrically conducting. Accuracy upto ±2.0% for measurement for gas-liquid two phase flow has been reported.

23.3.2.3 Ultrasonic Method

Ultrasonic flowmeters can be Doppler shift type or time-of-flight type. But this method has been successfully used in two phase flow involving liquids only. In gas-solid flow, this method has met only limited success. The major difficulty encountered is that of obtaining efficient energy coupling into the gas flow from the ultrasonic transmitter. Error is introduced by change in the temperature of the gas conveying the solids.

23.3.2.4 Laser Method

In laser Doppler velocity meters, the frequency shift of the scattered light from flowing particles is measured to obtain the velocity of the flow. This technique is well suited for measurement of fluid velocity at a point. This is also suitable for gas solid systems.

Accuracy : ±0.5%

Range : Reports are available that it can measures the velocity of the particles in the size range of 100 mm to 500 mm diameter being conveyed over a velocity range of 2-100 meters/sec in 42 mm diameter pipe.

Advantages : Capable of achieving non-intrusive point velocity measurements of extremely high accuracy.

Does not require any calibration.

Measurement is unaffected by variation of temperature of air conveying solids.

Disadvantages : Too expensive.

Difficulties may arise with the flows with a high solids loading.

With the above discussions on the different techniques of velocity measurement in two phase flow, it is apparent that most of them are not suitable for measurements with gas-solid flow. The cross-correlation technique belongs to an altogether different category where a statistical method is used for determining the velocity of the two phase flow. This technique with its twin advantages of low cost and reliable operation can offer a useful method for two phase velocity measurement.

23.3.3 Cross-correlation Technique

Cross-correlation technique inolves the correlation of the fluctuation of any property in the flowing fluid between two points. The main advantage of this technique is its invariance to temporal fluid property changes as long as the property remains the same between the two measurement points. The rejection of spurious interference is the most important advantage of correlation. Cross-correlation flowmeter are based on the principle of measurement of the transit time of a tagging signal between two known points. The transit time is measured by the correlator. The instrument consists of two transducers, spaced in the direction of the flow, to detect fluid disturbances which are assumed to retain their identity between the sensors and induce similar signals at each transducer. If x(t) and y(t) are two signals derived from upstream and downstream transducers respectively, the cross correlation function between the two signals is given by (Beck, 1981; Coulthard, 1983; Chen, et. al., 1994; Thorn, 1992).

$$R_{xy}(\tau) = T \xrightarrow{\lim} \infty \frac{1}{T} \int_O^T x(t-\tau)y(t)dt \qquad (27)$$

where $R_{xy}(\tau)$ is the value of the cross-correlation function when the upstream signal is delayed by a time τ. The basic principle of cross-correlation method of velocity measurement is explained.

It can be shown (Beck, 1981) that the function $R_{xy}(\tau)$ attains the maximum value when the variable time lag τ is equal to the transit time τ^* of the tagging signals; hence the flow velocity is given by

$$\bar{V} = L/\tau^* \qquad (28)$$

where L is the spacing between the two sensors.

The computation of the cross-correlation to obtain the transit time can be done in different ways. In fact, on-line computation of the cross-correlation function in the analog domain as in equation (26) is difficult. So, some of the alternatives are using

(i) Digital cross-correlation technique.
(ii) Polarity cross-correlation technique.
(iii) Henry's algorithm.

23.3.3.1 Digital Cross-correlation

The main advantage of this method is that the computation with the digitized signal by using microprocessor or microcomputer is easier. Thus, it is more suitable for on-line computations. The digital expression for cross-correlation function (26) is

$$R_{xy}(j) = \frac{1}{N} \sum_{n-O}^{N} x(n-j)y(n) \qquad (29)$$

where   N : number of sampled data for cross-correlation
         j : shift in number of samples.

The digital cross-correlation technique is powerful but it has some inherent problems. The main problem is with the sampling where the loss of information due to reduction of bandwidth causes a loss of accuracy of transit time measurement.

23.3.3.2 Polarity Cross-correlation

In the polarity cross-correlation technique (Beck, 1981; Thorne, 1992), the signals are both quantized in either the binary value or the sign of the deviation from the mean value. Polarity cross-correlation technique is computationally much faster compared to the digital cross-correlation one. The use of complex hardware circuit like analog to digital converter can be avoided in polarity cross-correlator. The polarity cross-correlation function $R'_{xy}(j)$ is defined as

$$R'_{xy}(j) = \frac{1}{N} \sum_{n-O}^{N} \text{sgn } x(n-j) \text{sgn } y(n) \qquad (30)$$

where sgn indicates the sign of the function over the mean value. It is clear that the absolute value of the polarity cross-correlation function will not be the same as the digital cross-correlation function. However, the important issue is that the peak position of the functions coincide. From flow measurement point of view, it is the peak position of the correlation function that is of importance. By reading the binary values, the amplitude information of the data is lost and the statistical accuracy of the cross-correlation function will suffer. This can be compensated by increasing the data observation time by a factor of about 2.2 over the time required for the digital cross-correlation.

23.3.3.3 Henry's Algorithm

Henry's method (Beck, 1981) is an alternative to the polarity cross-correlation technique to improve its speed and resolution. Henry's arrangement is based on the principle that the only significant information used by a polarity cross-correlator is the actual zero crossing times of the signals. This information can be stored in the computer by using an internal clock generator as time reference. This avoids the need for sampling the data at close intervals that were previously thought to be necessary to give adequate resolution.

From the above discussion, it can be concluded that the cross-correlation technique is the most suitable method for velocity measurement for pneumatic conveyors. There are of course few other statistical methods (Barschdorff and Wetzlar, 1983) like complex coherence technique, cepstrum analysis, etc. for finding out the transit time in two phase flow, but they are basically the frequency domain techniques and not suitable for on-line measurements. One advantage of these methods is that they take care of the sensor dynamics and necessary corrections can be made. On the other hand, the cross-correlation technique is a time-domain approach and is more suitable for on-line measurements. But the accuracy of the cross-correlation flowmeter depends upon several factors like

     (i)   flow profile,

     (ii) velocity / time trade-off,

   (iii) mismatching of two sensor characteristics,

   (iv) bandwidth of the sensor circuits,

    (v) spacing between two transducers, etc.

Different methods for overcoming these difficulties have been discussed in (Beck, 1981). It has been suggested that the optimum spacing between the two sensors should be 3-5 times the tube diameter. The length of the transducer should be so chosen so that the spatial averaging effect is minimum. The bandwidth of the sensor electronics should be larger than the spatial averaging bandwidth. Due to the factors mentioned above, a discrepancy arises between the correlation velocity and the mean velocity. As a result, a calibration factor is normally used to obtain the mean velocity.

The greatest advantage of the cross-correlation instrument is that, unlike many other types of instruments the gain stability of the sensors is not important because the cross-correlator simply measures the time delay between the two sensor signals and this is not dependent on the amplitude of the signals.

23.3.3.4 Selection of Cross-correlation Flowmeter Sensors

It has already been pointed out that two sensors detect the tagging signal at two different points which are located along the flow axis. The sensors may be of three broad major categories (Beck, 1981), namely,

    (i) Modulation of radiation by the flowing fluid,

   (ii) Emission of radiation by the flowing fluid,

  (iii) Modulation of electrical and thermal properties of the fluid.

Modulation of Radiation by the Flowing Fluid

Radiation techniques are attractive, because they can interrogate the flow across a large part of the pipe section, radiation sensors can frequently be mounted externally to the pipe.

  (a) Modulation of the acoustic radiation: In this method, ultrasound is passed through the flowing fluid, the received ultrasound is modulated by the turbulent eddies in the fluid. The transmitting and the receiving sensors can be mounted on the outside wall and the time shift between these two signals give the velocity of the fluid.

But the major problem of this method arises when gas is used as the medium, since due to the acoustic mismatch between the gaseous medium and the pipe wall, the sensors should be placed in contact with the gas.

(b) **Modulation of light:** When steam or suspended solid particles are present in the flow, turbulent clouds are generated and the velocity of the flow can be measured from a remote point by cross-correlating the output signals from two photodetectors which view through telescopes at two positions along the flow axis. Here, the main problem is the soiling of the optical surfaces.

(c) **Modulation of gamma rays:** When solid particles are present in a fluid flow, they can be used to modulate a beam of radiation which is directed perpendicular to the flow and signals from two sensors spaced axially along the flow can be cross-correlated to get the velocity. But in this method we need strong radiation sources which limits its applicability.

Emission of Radiation by the Flowing Fluid

(a) **Thermal radiation:** The turbulence in the thermal radiation emitted by hot gases can be detected by infrared sensors. This technique has limited applications.

(b) **Ionization radiation:** This method can be used in highly radioactive liquid in nuclear fuel separations plants where variations in the emitted radiation are caused by either solid particles or gas bubbles present in the flowing fluid and this variation can be detected by scintillation detectors.

Modulation of Electrical and Thermal Properties of the Fluid

(a) **Electrical conductivity sensor:**In two phase gas-liquid or liquid-solid systems discontinuous phases in the fluid causes instantaneous variations in the electrical conductivity across the electrodes exposed to the flowing fluid and changes can be measured and cross-correlated to give the flow velocity. The electronic circuitry is simple but sometime the method could be intrusive.

(b) **Thermal sensor to detect turbulent eddies:**Turbulent eddies in fluids cause corresponding random variations in the rate of heat exchange with a hot body. If a thermistor is placed into the flow, then its temperature change can be correlated. It is very cheap and simple device to use but suffers from limited frequency response.

(c) **Thermal sensors to cross-correlate injected heat pulse:** By injecting a tagging signal into a flow in form of heating pulses in a pseudo-random manner and detecting the temperature pattern by thermocouple or thermistor at the two points along the flow axis; the detectors output can be cross-correlated to get the velocity.

(d) **Capacitance sensors:** Capacitance sensors are particularly used to measure the flow of gasses with entrained solids. The solids from clouds due to turbulence and these clouds modulate the capacitance of a plate which is exposed to the flowing fluid. By cross-correlating the capacitance modulation of two plates, the flow velocity can be measured. The advantage is that capacitance electrodes do not protrude into the flow and the necessary insulating materials are available to withstand most of the conditions due to temperature, corrosion and abrasions. The electrodes can be mounted either inside or outside the pipe wall. The externally mounted plates provide a fully non-intrusive method of measurement, but the sensitivity becomes less.

(e) **Electrodynamic sensor:** An electrodynamic sensor (Xie et. al., 1989; Yan et. al., 1995) detects the natural electrostatic charge on the moving particles. Whenever solid particles are entrained in a flowing stream of gas / air, a considerable amount of electrostatic charge accumulates on the particles. This phenomenon of charge generation of particles can be attributed mainly to the collisions between the flowing particles and the pipe wall and to the friction between the particles and air stream. The magnitude of the charge depends on the physical properties of the particles, including their shape, size, density, conductivity, permittivity, humidity and composition as well as on the pipe wall roughness, pipe diameter and the pipe length traversed by the particles. The solids velocity and concentration are also major factors contributing to the magnitude of the charge. The charge carried on the particles can be detected by the passage of the charged flowing particles. Recently, use of electrodynamic sensors for the measurement of velocity of solids (sands and glass) in gravity falling system has been reported in (Yan et. al., 1995). The detailed theory of the electrodynamic sensor, the sensing mechanism and the modeling of sensor has also been discussed in this paper. It has been reported that an electrodynamic sensor can achieve higher degree of sensitivity required in the mass flow rate measurement of dilute phase solids and is also adaptable to large diameter pipes used in plants (Yan et. al., 1995). Another advantage of the electrodynamic sensor is that it sees only the moving particles and is little affected by the stationary solids accumulated on the pipe wall. The electrodynamic sensors reported so far (Xie et. al., 1989; Yan et. al., 1995) are intrusive since the electrodes are fixed inside a metallic wall. In this case, however, the stringent requirement of a totally non-intrusive scheme of measurement on a non-metallic pipe wall makes the problem more challenging. Mass flow measurement in a pneumatic conveying system has been made by Sen et. al. (2001) using inference technique adopting electrodynamic sensors. The scheme is illustrated in Fig. 23.24. The details of the technique is availaible in final project report (1997).

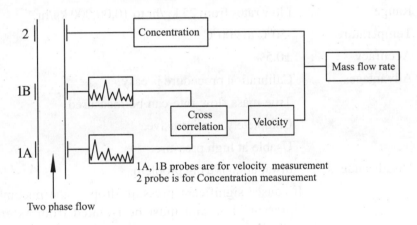

Fig. 23.24 Mass flow measurements in a pneumatic conveying system

23.3.4 True Mass Flowmeters

The total mass flow rate of a two phase mixture can be measured by using some innovative techniques. These methods sense the mass flow rate by measuring the forces exerted on fluid particles within the sensing volume. It uses fundamental physical principles to relate these induced forces to measurable forces or displacement.

23.3.4.1 Coriolis Technique

In the commercially available Coriolis force flowmeter there is a horizontal U-shaped sensor tube through which the fluid flows. The curved part of this tube is oscillated in a vertical direction at its natural frequency by an external energy source (Fig.23.24). As the fluid particles move down the inlet leg of the pipe, they experience increasing vertical velocity and acceleration due to induced oscillation in the tube. A deceleration of the particle is produced as the fluid moves through the outlet leg of the pipe. The acceleration and deceleration of the particles are equal in magnitude and opposite in direction. The fluid resists this acceleration and exerts forces F equal in magnitude but opposite in direction to these accelerations. These forces are linearly proportional to the mass flow rate of the fluid. The forces produce a vertical couple that twists the U-tube as it oscillates. The degree of twist is a measure of mass flow rate.

The basic specifications of commercially available Coriolis flowmeter, their advantages and disadvantages are described below.

Range	:	Flow rates from 25 kg/hr to 10,00,000 kg/hr.
Temperature	:	-50^0C to 200^0C
Accuracy	:	$\pm 0.5\%$
Advantages	:	Calibration procedure is easy.
		True mass flow rate can be measured.
		Completely non-intrusive.
		Usable at high pressure.
Disadvantages	:	Installation is a big problem.
		It causes significant pressure drop in the pneumatic pressure line and must be isolated from external vibration.

23.3.4.2 Gyroscopic Technique

The gyroscopic mass flowmeter has only reached to the experimental stage for pneumatic conveying. The measurement section of the flow conveyor is formed into a circle of radius r and rotated at a constant angular velocity ω about an axis through the inlet and outlet pipes. The fluid flowing in the circular loop has an instantaneous angular momentum about the z-axis that reacts with the input angular velocity, ω by the usual gyroscopic equations to produce an instantaneous momentum M with the y-axis. This momentum produces a couple with resultant force F acting on the input bearings. Then the mass flow rate m is proportional to F. The accuracy claimed for this device is ±0.25% of full scale. Some unique techniques for the online mass flow rate measurement of two phase flow may be obtained in OFMPS' 98 (1998).

23.4 Concluding Remarks

Some basic techniques for the measurement of mixture composition and flow rate during two phase flow have been discussed in the previous sections. Some of these techniques are to be improved substantially so that they can give reliable prediction in an industrial atmosphere. The status of two phase flow measurement is still in its early development stage. The tomographic measurement shows enough promise as one can measure velocity, temperature and composition using the same principle. However, more research is needed to produce rugged instrument within affordable price.

Measurement of other parameters like temperature, pressure, heat flux, heat and mass transfer coefficient and wall shear stress in two phase flow are equally challenging. Some techniques suitable for the above measurement can be found in Hewitt (1978).

23.5 Nomencletues

V	Volume
V	Velocity
A	Area
t,T	Time
p	Pressure
g	Gravitational acceleration
z	Elevation, distance
I	Intensity
Q	Volume flow rate
M	Mass flow rate
α	Void fraction
ρ	Density
ε	Conductivity
μ	Absorption coefficient

Sub-script

L, l	Liquid
G, g	Gas

Acknowledgement

Some of the investigations reported in this article were conducted in collaboration with my colleagues Prof. Gargi Das, Department of chemical engineering, Prof. Biswajit Maiti, Department of Mechanical Engineering and Prof. Siddhartha Sen, Department of Electrical Engineering. Their contribution is gratefully acknowledged. The financial help provided by RDCIS, Ranchi, Steel Authority of India for the project "Design and Development of a Two Phase Mass Flowmeter (gas-solid) for pneumatic conveying system" is gratefully acknowledged. Mr. Arup Kr. Das an MS student of the Department of Mechanical Engineering needs special mention for his meticulous help in preparing the manuscript.

23.6 References

1. Barschdorff, D. and Wetzlar, D. (1983) "Statistical methods for single component velocity determination on two-phase flow" Proc, of Flomeko, Budapest, Hungary, pp. 203-210.

2. Beck, M. S. (1981) "Correlation in instruments: cross-correlation flow meters" Journal of Physics (pt. E) vol. 14, pp. 7-19.

3. Chen, J., Tornberg, J., Kivimma, J, and Karras, M. (1994) "Automatic Industrial Cross-correlator for flow measurement" ISA Transactions, vol. 16 pp. 214-218.

4. Collier, John. G.and Thome, John. R., (1994) "Convective Boiling Condensation", Third Edition, Claredon Press, Oxford.

5. Coulthard, J. (1983) "Cross-correlation Flow Measurement- a history and the state of art" Measurement and Control, vol. 2, pp. 103-106.

6. Das, G., Das, P. K., Purohit, N. K. and Mitra, A. K. (1998), "Rise of a Taylor bubble through concentric annulus", Chem. Engg. Sci., Vol. 53, No. 5,

7. Das, G., Das, P. K., Purohit, N. K. and Mitra, A. K. (Dec. 1999a) "Development of flow pattern during cocurrent gas liquid flow through a vertical concentric annulus Part I Experimental Investigations", Trans. ASME, J. Fluids Engg. , vol.121, pp 895-901.

8. Das, G., Das, P. K., Purohit, N. K. and Mitra, A. K. (Dec. 199b9) "Development of flow pattern during cocurrent gas liquid flow through vertical concentric annulus Part II Mathematical Models", Trans. ASME, J. Fluids Engg. , vol.121, pp 902-907.

9. Das, G., Das, P. K., Purohit, N. K. and Mitra, A. K. (2000) "Distribution of gas-liquid mixture in a concentric annulus - inception and termination of asymmetry",- Int. J. of Multi-phase Flow , Vol.26, pp 857-876.

10. Das, G. Das, P. K., Purohit, N. K. and Mitra, A. K. (Feb 2002) "The geometry of Taylor bubbles rising through liquid filled concentric annuli", - AIChE Journal. Vol. 48 , pp411-420.

11. Final Project Report, (1997) "Design and Development of Two-phase

Mass Flow Meter (gas-solid) for Pneumatic Conveying System" Electrical and Mechanical Engg, Indian Institute of Technology, Kharagpur.

12. Gupta, D., Sen, S. and Das, P.K., (1994) "Finite-differences resistance modeling for liquid level measurement in stratified gas-liquid systems" Measurement Sc. & Tech. 5 pp 574-579.

13. Hewitt, G. F. (1978) "Measurement of Two Phase Flow Parameters", Academic Press Inc., New York.

14. Mayinger,F. and Feldmann, O. (2001) " Optical Measurements-Techniques and Applications" Springer, New York.

15. McNaughtan, Meney, K., Grieve, B., (1999) "Electrochemical Issue in Impedance Tomography, 1st World Congress on Industrial Process Tomography", Buxton, April 14-17, pp.344-347.

16. OFMPS'98 (1998) , Proceedings of 1st International Symposium on " On line Flow Measurement of Particulate Solids", The University of Greenwich, UK.

17. Rajan, V. S. V., Ridley, R. K., and Rafa, K. G. (1993) " Multiphase flow measurement techniques- a review" Journal of Energy Resources technology, Vol. 115, pp. 151-161.

18. Sami., Abouelwafa, A., John, E., Kendall, M., (1980) "The use of capacitance sensors for phase percentage determination in multiphase pipelines" IEEE Transactions on Instrumentation and Measurement, vol. IM-29, no. 1, March.

19. Sen, S., Das, P. K., Dutta, P. K., Maity, B., Chaudhury, S., Mandal, C. and Roy, S. K. (2000) " PC-based gas-solids two-phase mass flowmeter for pneumatically conveying systems" Flow Meaurement and Instrumentation, 11, pp 205-212.

20. Steevers, M., Gaddis, E. S. and Vogelpohl, A. (1995) " Fluid Dynamics in an impinging-stream reactor" Chemical Engineering and Processing, 34, pp. 115-119.

21. Thorn, R., Beck, M. S. and Green, R. G. (1982) "Non-intrusive methods of velocity measurement in pneumatic conveying" Journal of Physics (pt. E), pp. 1131-1139.

22. Thorn, R. (1992) "The influence of micro-electronics on the development of multi-component flowmeters" Journal of Institute of Engineers (I), (Pt-E), vol..72, pp. 130-136.

23. Williams, R. A. and Beck, M. S. (1995) "Process Tomography: Principles, Techniques and applications" Butterworth-Heinemann Ltd.

24. Xie, C. G., Stott, A. L., Hung, S. M., Plaskowski, A. and Beck, M. S. (1989) " Mass-flow measurement of solids using elctro-dynamic and capacitance transducers" Journal of Physics (pt. E) vol. 22, pp. 712-719.

25. Yan, Y., Byne, B., Wodland, S. and Coulthard, J. (1995) "Velocity measurement of pneumatically conveyed solids using electrodynamic sensors" Measurement Science and Technology, vol. 6, pp. 515-537.

Index